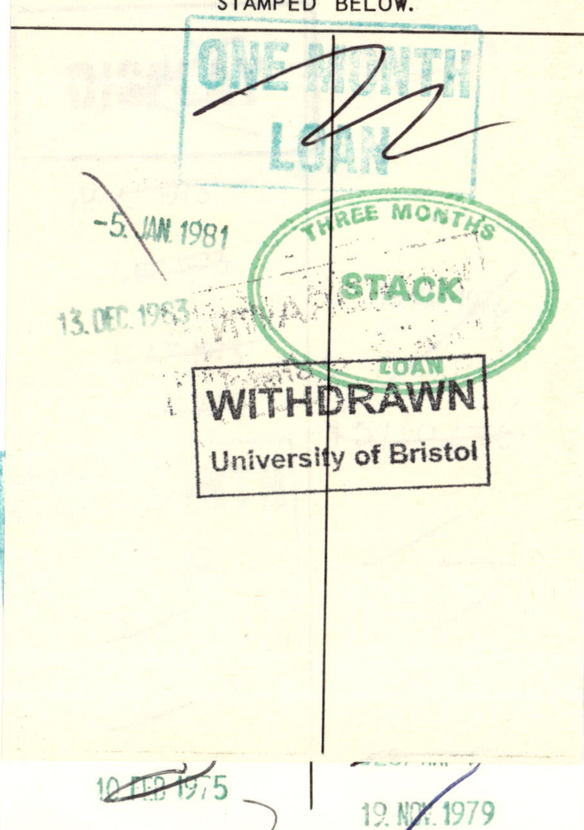
ONE MONTH LOAN
-5 JAN 1981
THREE MONTHS STACK LOAN
13. DEC. 1983
WITHDRAWN
University of Bristol
10 FEB 1975
19. NOV. 1979

ROLE OF MEMBRANES IN SECRETORY PROCESSES

ROLE OF MEMBRANES IN SECRETORY PROCESSES

Proceedings of the 1971 meeting of the International Conference on Biological Membranes

Editors

Liana Bolis

Institute of General Physiology
University of Rome, Italy

R.D. Keynes, W. Wilbrandt

1972

NORTH-HOLLAND PUBLISHING COMPANY – AMSTERDAM · LONDON
AMERICAN ELSEVIER PUBLISHING COMPANY, INC. – NEW YORK

Library of Congress Catalog Card Number 72-79722

ISBN North-Holland 7204 4116 1
ISBN American Elsevier 0444 104 038

Publishers:
NORTH-HOLLAND PUBLISHING COMPANY – AMSTERDAM
NORTH-HOLLAND PUBLISHING COMPANY, Ltd. – LONDON

Sole distributors for the U.S.A. and Canada:
AMERICAN ELSEVIER PUBLISHING COMPANY, INC.
52 Vanderbilt Avenue
New York, N.Y. 10017

PRINTED IN THE NETHERLANDS

EDITORIAL PREFACE

This volume is based on the Fourth Meeting of the International Conference on Biological Membranes in Gargnano, Italy, June 20–27, 1971.

The meeting was devoted to the role of membranes in secretory processes. The term "secretory" here is not meant to be restrictive in the sense of a relation to the activity of glands. It is used to describe various aspects of the participation of cell organisation in transport. This viewpoint led naturally to the inclusion of membrane synthesis and turnover. The discussion of the central problem of "secretory membrane transport" was subdivided into "transmembrane" transport (into cells and out of cells) and "transepithelial" transport (across cells). The powerful position that genetics have attained as a tool in transport research, particularly in micro-organisms, stimulated the organisation of one session on genetics.

Finally the increasing body of evidence for the dependence of various transport systems on gradients of sodium and other ions motivated a separate discussion of the general problems of coupling in biological transport.

Again, as in previous meetings, the exchange of ideas was catalyzed by the beautiful surroundings, which helped to create a friendly and relaxed atmosphere. The thanks of all participants are particularly due to Professor Liana Bolis, our indefatigable general secretary, for her excellent, always cheerful, and highly efficient organisation.

R.D. Keynes
W. Wilbrandt

LIST OF CONTRIBUTORS

CONTENTS

Part I

MEMBRANE BIOSYNTHESIS AND TURNOVER

INTRODUCTION

P. SIEKEVITZ
The Rockefeller University, New York, U.S.A.

The theme of this first session is the dynamic organization of membranes. Heretofore we have viewed the architecture of a membrane as being a relatively fixed structure, obeying certain chemical and physical laws of interactions between protein and protein, between protein and lipids, and between lipids and lipids. We have thus tended to look upon a membrane as having a certain invariant organization of its constituent molecules, with these molecules being in stoichiometric proportions to each other. However, in the last few years many papers have appeared, too numerous to list, whose conclusions, either implicitly or explicitly, would argue with this hypothesis. For example, once the time element is introduced into the problem, it becomes clear that singular cellular membranes go through regular developmental, or if you will, differentiating patterns, that their constituent molecules are not in fixed proportions to each other, that at any one time in the life of the cell, the architecture of a particular membrane is probably not the same as it is at other times.

The talks in the session will deal directly with this aspect of membrane constituency and architecture. Dallner will discuss the changes with time in the endoplasmic reticulum membranes of the liver, the possible arrangement of the proteins there, the possible mosaicism in membrane organization (whether there are membrane "patches" which are different from other "patches"), and the methods whereby we can come to grips with this problem. Ohad will talk on the development of chloroplast membranes, of how their constituent proteins are synthesized and are fitted into an increasing membrane mass, and the possible regulation of membrane biogenesis. Wirtz will discuss the recent literature on the exchangeability of membrane lipids, of how different membranes within the cell are in a constant lipid exchange with other membranes in the cell, and of what this means for our concept of membrane architecture.

I would just like to mention one topic which will probably not be brought up by the speakers, and that is the turnover of membrane constituents. The proteins and lipids of membranes are in a constant state of turnover, as has been found in the case of endoplasmic reticulum membranes, mitochondrial membranes, and plasma membranes; the lists of papers are too numerous to mention [cf. reviews 1,2]. More to the point, in the case of the endoplasmic reticulum membranes, three enzymes have been purified from the membranes and their turnover rates determined under the same experimental conditions. What was found is that the rates are different for each of these three membrane proteins [3–7]; futhermore, if all the membrane proteins were solubilized and seperated on gels, it was found [5,8] that even these proteins, some of which could be called "structural" in a sense, are turning over at rates not in consonance with each other. Going further, it was found that in the case of one of these proteins, cytochrome b_5, the heme and apoprotein moieties were turning over at different rates [9]. All this turnover data, combined with the lipid exchange data mentioned by Wirtz, and combined with the data on differentiating membrane systems mentioned by Dallner and Ohad, can only lead, in my mind, to the view of a membrane as a structure having a plasticity of architecture, a structure that paradoxically combines specificity and stability on the one hand with lability and dynamism on the other. Any formulation which tries to relate function with structure of membrane will have to incorporate within its model this dynamic organization as exemplified by the recent literature and by the remarks of the following speakers.

References

[1] R.T. Schimke, in: Mammalian Protein Metabolism, Vol. 4, ed. H.N. Munro (Academic Press, New York, 1970) p. 177.
[2] P. Siekevitz, in: Ann. Rev. Physiol. 34 (1972) 117.
[3] H. Jick and L. Shuster, J. Biol. Chem. 241 (1966) 5366.
[4] T. Omura, P. Siekevitz and G.E. Palade, J. Biol. Chem. 242 (1967) 2389.
[5] I.M. Arias, D. Doyle and R.T. Schimke, J. Biol. Chem. 244 (1969) 3303.
[6] Y. Kuriyama, T. Omura, P. Siekevitz and G.E. Palade, J. Biol. Chem. 244 (1969) 2017.
[7] K.W. Bock, P. Siekevitz and G.E. Palade, J. Biol. Chem. 246 (1971) 188.
[8] E.D. Kiehn and J.J. Holland, Biochem. 9 (1970) 1716.
[9] K.W. Bock and P. Siekevitz, Biochem. Biophys. Res. Comm. 41 (1970) 374.

PHYSICOCHEMICAL AND ENZYMIC PROPERTIES OF THE ENDOPLASMIC RETICULUM IN RELATION TO MEMBRANE BIOGENESIS

Lennart ERIKSSON, Hans SVENSSON, Anders BERGSTRAND and Gustav DALLNER

Department of Pathology, Sabbatsberg Hospital, Karolinska Institutet, and Department of Biochemistry, University of Stockholm, Stockholm, Sweden

The dynamic nature, richness and keyfunctional role of the endoplasmic reticulum in the liver cell makes the system very suitable for study of a number of biological problems, among them membrane biogenesis [1]. On the other hand, a special problem, and one which is unfavorable from several points of view, arises during the isolation procedure. Unlike the majority of cellular organelles, this membrane system cannot be separated in the intact form [2]. The elongated tubular and cisternal structure of the intact hepatocyte breaks up into small pieces, and the resulting structures, the microsomal vesicles, represent a new morphological picture. Theoretically, this sequence of events should prevent the study of a number of functional phenomena since the intactness of the limiting membrane may be damaged. Interestingly, this is not the case. The breakage occurring during homogenization is not a simple mechanical process. The closure occurs in such a way and is of such a speed that the intraluminal content (plasma proteins in the case of liver, and digestive enzymes in the case of exocrine pancreas) is not released. This phenomenon, which still remains to be explained, is called by Palade and Siekevitz in their classic paper (1956) an "active pinching off process" [3].

Abbreviations: ER, endoplasmic reticulum; cyt., cytochrome; PLP, phospholipid, G6P, glucose-6-phosphate; ATP, adenosine triphosphate; NADH, reduced nicotinamide adenine dinucleotide; NADPH, reduced nicotinamide adenine dinucleotide phosphate.

1. The intactness of microsomal membranes

One of the best indicators of the intact character of the microsomal membranes is the fact that various water compartments with different permeability properties can be distinguished in the isolated microsomal pellet [4].

Liver microsomes isolated in 0.25 M sucrose have a relatively small intramicrosomal water compartment, 1.4 μl H_2O/mg dry weight (table 1). This space includes the sucrose accessible space, which is up to 76%. Since the hydration water is estimated to be about 0.2 μl/mg dry weight, practically no water is left to be involved in osmotic shifts.

Table 1
Water compartments of liver microsomal pellet

Water compartment	μl/mg dry weight	% of total	% of intra-microsomal water
Total microsomal water	3.30 ± 0.08	100	
Extramicrosomal water	1.86 ± 0.05	56	
Intramicrosomal water	1.44 ± 0.06	44	100
Sucrose space	1.10 ± 0.05		76
'Osmotic space' and hydration water	0.34 ± 0.03		24

Data taken from ref. [4].

The passive permeability of isolated microsome membranes is both selective and limited. Non-charged substances, with a molecular weight up to at least 600, penetrate the intravesicular compartment freely (table 2). On the other hand, charged substances display a completely different behavior. Acetate, for instance, which has a molecular weight of less than 100, is completely impermeable to ER fragments.

As with other biological membranes, the limited permeability of ER fragment can be easily abolished by influencing the structural arrangement of lipids [5,6]. Incubation in warm hypotonic media followed by cooling has been found effective for the removal of vesicle content, without visible rupture of the membrane. In fact, mere water treatment abolishes the permeability barrier for charged substances, such as acetate, mevalonate, and G6P (table 3). This modification, however, does not imply complete removal of the permeability barrier, since detergents are still required for the full activity of a number of membrane-bound enzymes.

The intactness of microsomal membranes can be retained only by appropriate handling of the fraction, as shown in table 4. Smooth microsomes

Table 2
Apparent penetration of various substances through the microsomal membranes

Substance	Intramicrosomal space (µl/mg dry weight)	% of total intramicrosomal water
Intramicrosomal water	1.44	100
Glycerol[a]	1.08	75
Glucose[b]	1.18	82
Sucrose[b]	1.09	76
Maltotriose[b]	0.95	66
Oubain[a]	1.19	83
Inulin[b]	0	0
Acetate[a]	−0.04	−3
Mevalonate[b]	0.09	6
Glucose-6-phosphate[b]	0–0.43	0–30

Every centrifuge tube contained in 11 ml microsomes from 2 g liver, carrier solutions in final concentration of 0.05 M, and the radioactive substance (5 µCi of ^3H and 0.5 µCi of ^{14}C labeled compounds). Data taken from ref. [4].

a ^3H label; b ^{14}C label.

Table 3
Permeability of H_2O-treated microsomal membranes

Substance	Intramicrosomal space			
	Control		H_2O-treated	
	µl/mg dry weight	% of intra-micr. H_2O	µl/mg dry weight	% of intra-micr. H_2O
None	1.42	100	0.92	100
Acetate	0.06	4	0.65	71
Mevalonate	0.06	4	0.79	86
Glucose-6-phosphate	0–0.39	0–30	0.46–0.69	50–75

Data taken from ref. [7].

become aggregated if they are suspended at a low protein concentration or if the sucrose concentration is below the critical threshold. This aggregation can be counteracted to a certain extent by the presence of protective colloids such as albumin. Smooth microsomes do not tolerate pelleting by ultra-centrifugation.

It is an absolute requirement to exercise care when microsomes are sub-

Table 4
Stability of rat liver smooth microsomes

Experiment	Microsomal concentration (g/ml)	Sucrose concentration (M)	% aggregated
1. Smooth microsomes	0.05	0.15	81
	0.1	0.15	65
	0.2	0.15	36
	0.4	0.15	22
2. Smooth microsomes	0.05	0.88	36
	0.1	0.88	30
	0.2	0.88	25
	0.4	0.88	19
3. Smooth microsomes + 0.3% albumin	0.05	0.15	40
4. Smooth microsomes, pelleted + resuspended	0.4	0.88	80

The degree of aggregation was determined by filtration of microsomes through 0.45 μ millipore filter. Data taken from ref. [8] and [9].

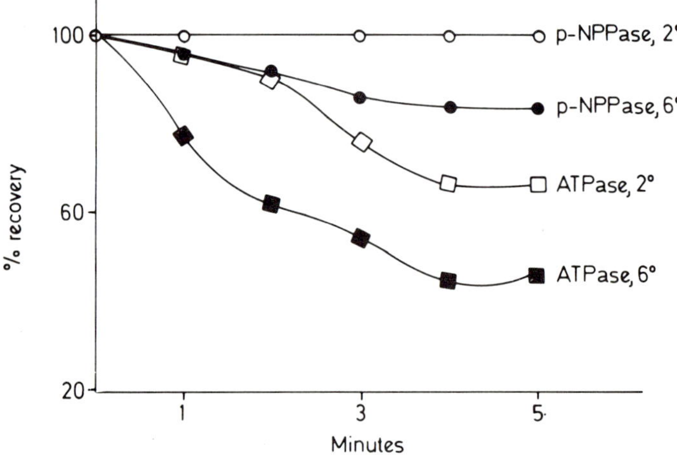

Fig.1. Effects of sonication on *p*-NPPase and ATPase activities of rough microsomes. Batches of rough microsomes (3.5 ml, from 3.5 g liver were sonicated at 1.5 amp. in the chamber described, keeping the temperature at 2° or 6°C. Sonication was interrupted every minute for mixing, and 0.1 ml was removed for enzyme activity measurements. Data taken from ref. [10].

jected to various treatments. Heat generation during sonication, for example, may easily inactivate membrane bound enzymes, thereby causing artifacts. ATPase activity displays a 20% decrease after only a few minutes of sonication if the temperature is elevated as moderately as 4 ° (fig. 1).

It is apparent that the isolated ER represents a labile structure which, depending on the environmental conditions, can be modified and acquire new properties that are non-existent in the *in vivo* state. However, taking these possibilities into account by performing appropriate control experiments, the information given by the vaious analyses are highly reliable.

2. Relationship of rough and smooth membranes

Ultrastructurally, the isolated microsomes contain two main components: rough and smooth microsomes, the former having attached ribosomes on the

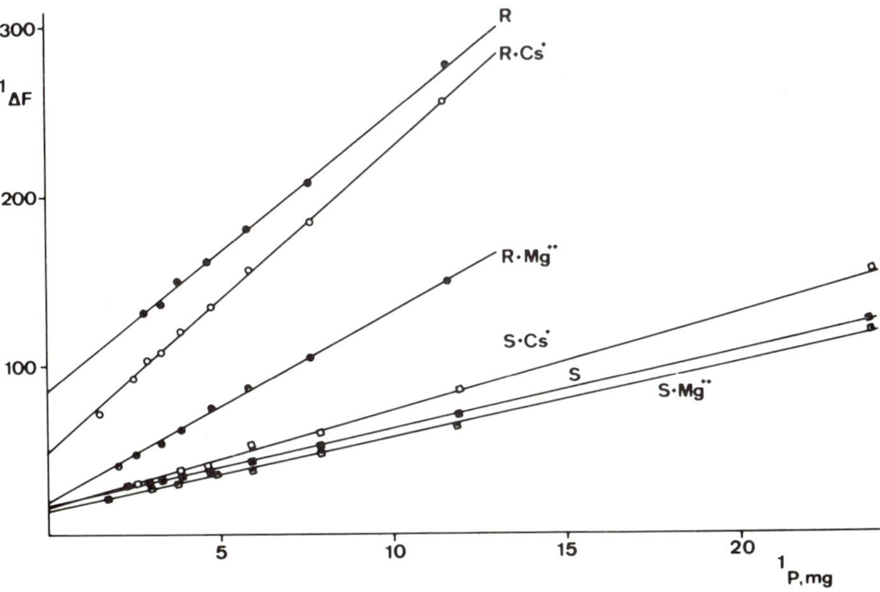

Fig. 2. Relative quantum yields of ANS fluorescence bound to microsomes. 5 μl of microsomal suspensions containing 17 mg protein/ml were added several times to a solution of 20 μM ANS in 0.25 M sucrose and the resulting fluorescence increase was plotted in reciprocal values against the reciprocal of the final concentration of protein added. Corrections were made for the light scattering of protein in the absence of ANS. CsCl and $MgCl_2$ concentrations were 10 mM. Data taken from ref. [11].

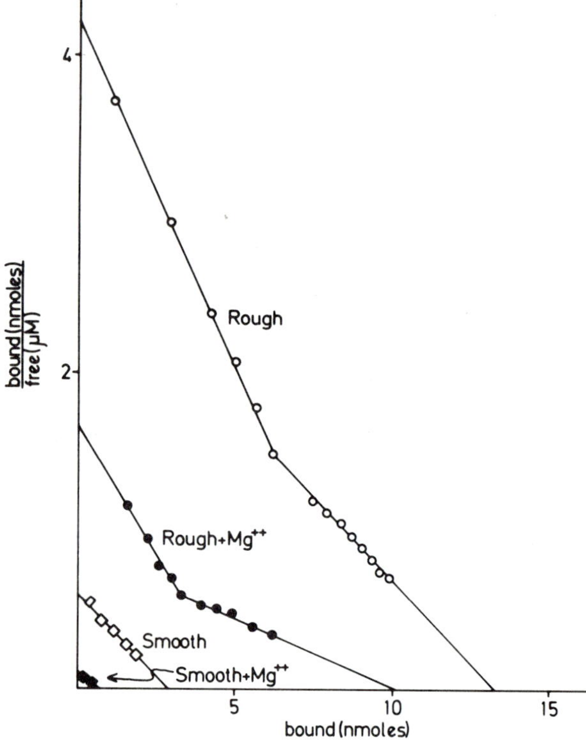

Fig. 3. Scatchard plots of ethydium bromide binding to rough and smooth microsomes. Data taken from ref. [11].

outer surface. During the last ten years, a great number of investigations have been performed in order to determine the properties of these two fractions [6]. As regards enzymic pattern, the two subfractions are very similar. Physicochemically, on the other hand, they represent two different entities.

When the interaction of the fluorescent probe ANS with microsomes is studied, the results indicate a larger negative surface charge density among rough than among smooth microsomes [11]. Plotting the reciprocal of fluorescence against the reciprocal of microsomal protein concentration, the value extrapolated at zero abscissa represents the fluorescence of the amount of ANS added, all of which is bound to the membrane (fig. 2). It is apparent that the relative quantum yield of rough microsomes is five times less than that of the smooth variety, but in the presence of cations the two membranes are indistinguishable. These findings can be explained on the basis of different

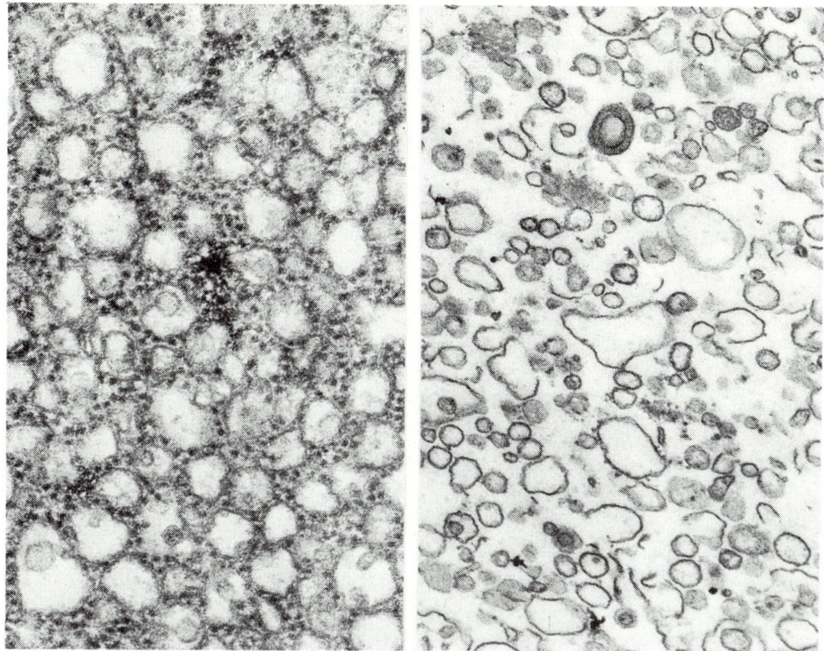

Fig. 4. a) Isolated rough microsomes before EDTA treatment. × 100 000. b) Rough microsomes after treatment with EDTA. × 80 000.

surface charges, that is, ANS would penetrate deeper in the smooth microsomes and thus react with more hydrophobic regions. When the interaction of the positive probe ethydium bromide (fig. 3) with microsomes is studied, the Scatchard plot demonstrates a significantly larger number of binding sites in rough microsomes compared to smooth microsomes, regardless of whether high or low affinity binding sites are studied. This again indicates that the negative charge density on the surface of the latter is less.

The findings, like those with fluorescence probes, do not mean that the membrane itself in the two microsomal subfractions is different. The properties of ribosomes in the case of rough microsomes could be simply additive, or, alternatively, the presence of bound ribosomes might change the structural state of the membrane. A complete removal of bound ribosomes can be obtained by EDTA treatment of rough microsomes prepared in the absence of Mg^{2+} [12]. Electromicroscopically, a small proportion of the vesicles are found to be opened, but the majority is intact, both as regards vesicle form

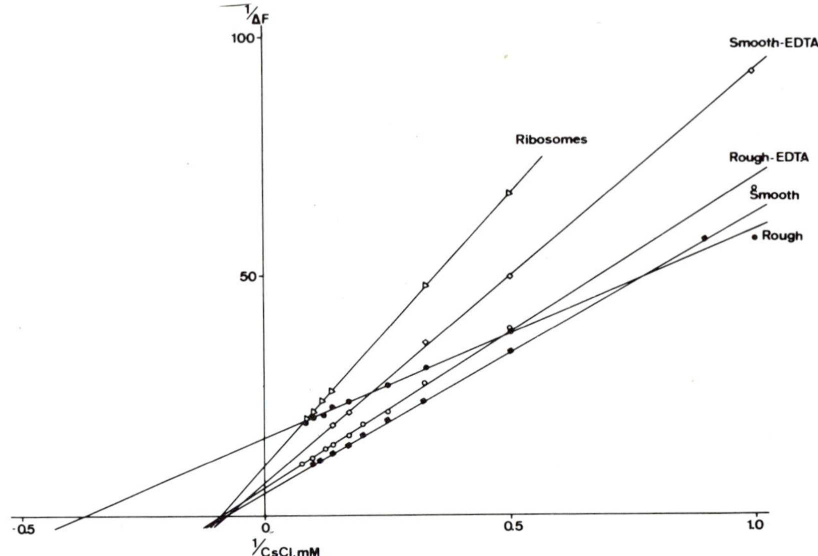

Fig. 5. Affinity of microsomal membranes for CsCl. CsCl was added in aliquots of 5 μl, and the resulting fluorescence increase was plotted in reciprocal values against the reciprocal of the final concentration of CsCl added. ANS was present in a concentration of 1.5 μM. Data taken from ref. [12].

and structure of the membranes, at least at the degree of resolution employed (fig. 4). These rough membranes are not aggregated and do not seem to have lost PLP or membranous protein, and a comparison of these properties with smooth microsomal membranes is therefore considered justifiable.

The EDTA-treated rough membranes can be used to study the type of interaction occurring between Cs^+ and membranes, utilizing the fluorescence probe ANS. When the values of l/fluorescence are plotted as a function of the reciprocal of Cs^+-concentration, the dissociation constant is 12 nmoles for EDTA-treated and non-treated smooth microsomes, EDTA-treated rough microsomes, and ribosomes (fig. 5). On the other hand, rough microsomes have four times smaller dissociation constant, that is, a much higher affinity for Cs^+ than the membranes avoiding ribosomes.

The findings with EDTA-treated rough membranes show that a number of pronounced physicochemical properties of rough microsomes are entirely connected to the presence of attached ribosomes. Similarly to the chemical and enzymic findings, the physicochemical properties also point to the close relationship which exists between the two types of membranes. Therefore,

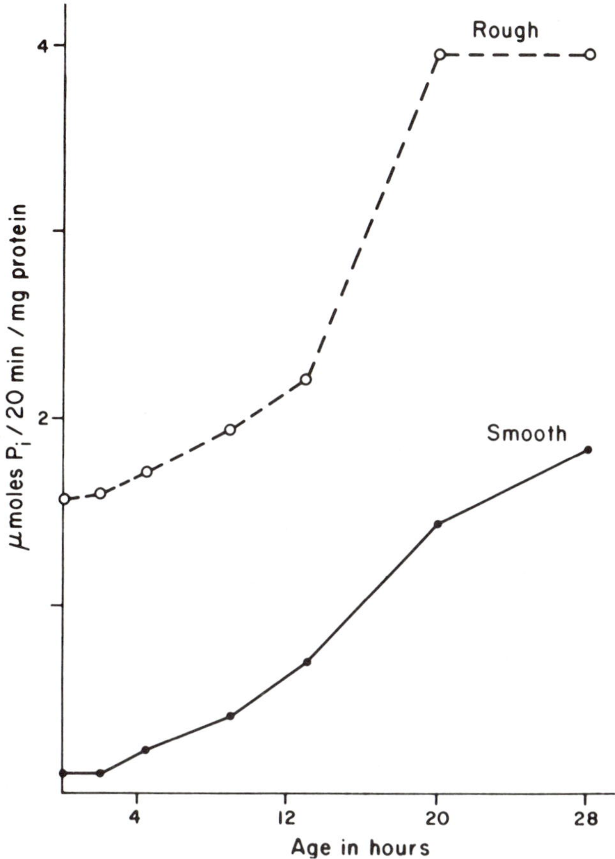

Fig. 6. Glucose-6-phosphatase activity of rough and smooth microsomes during the 1st day after birth. Data taken from ref. [14].

they should also be related to each other in the biosynthesis of their protein and lipid components.

The similarity between rough and smooth membranes, however, does not mean that the idea of two main microsomal compartments should be rejected. A number of specific properties of the rough microsomes cannot be explained by the simple addition of the properties of rough membranes and ribosomes. The type of combination which occurs *in vivo* appears to modify the characteristics of the membrane, giving them qualitatively new properties. Since the kinetic and thermodynamic properties of the membrane-bound

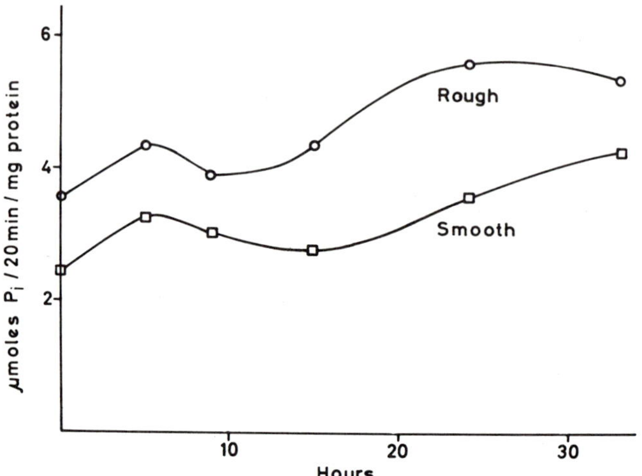

Fig. 7. Distribution of G6Pase activity in rough and smooth microsomes after alloxan treatment. Alloxan was given intraperitoneally (40 mg/100 g). Data taken from ref. [18].

enzymes are partly determined by the environment, one and the same enzyme may exhibit different properties in different locations. In fact, Stetten and Ghosh found that the pH-optimum of G6Pase in rough and smooth microsomes is not the same [13].

The close relationship between rough and smooth microsomal membranes is apparent from study of their biogenesis. The newly appearing glucose-6-phosphatase activity of liver ER in the newborn rat, which is the result of *de novo* synthesis of the enzyme, is localized completely in the rough microsomal membranes (fig. 6). During the rapid increase of activity which is already apparent some few hours after birth, hydrolysis also occurs in the smooth fraction and it takes days before the two fractions display equal activities. More convincing are the experiments of Kuriyama and Omura, who measured the appearance of the protein label in purified NADPH-cytochrome *c*-reductase and cytochrome b_5' of rough and smooth microsomes [15]. The label appeared first in the rough microsomes and rapid equilibration with the smooth ER occurred after no more than two hours. The similar turnover of total protein and phospholipids of the two subfractions also indicates the close biogenetic relation among these membranes [16].

The mechanism of transfer from the site of synthesis to another membrane structure is not known. Membrane-bound enzymes do not seem to follow the path described, for example, for catalase which moves freely from the lumen of rough microsomes to the peroxysomes [17].

The differential appearance of an enzyme activity may also be utilized to distinguish between *de novo* synthesis and enzyme activity enhancement for secondary reasons [18]. In the first phase of alloxan diabetes, when events occur mainly because of the direct effect of alloxan on the microsomal membranes, G6Pase activity increases simultaneously in both rough and smooth fractions (fig. 7). In the second phase (after 15 hr), when *de novo* synthesis dominates, activity enhancement is restricted primarily to rough microsomes.

3. Heterogeneity of rough and smooth microsomes

The similar ultrastructural appearance of rough and smooth microsomes does not mean that the various enzymes are necessarily distributed in the vesicles in random fashion. The idea that metabolically interrelated enzymes are concentrated in specific parts is a tempting one. The main problem is to isolate these theoretical patches, since enzymically heterogeneous pieces of the membrane do not necessarily differ physicochemically. Density, size, permeability, and surface charge are the properties utilized mainly in ultracentrifugation procedures, but these characteristics are not always related to enzyme composition [19,20]. On the other hand, the rapidly developing field of gradient centrifugation provides such a variety of methods that even membranes with only slight physical differences can be separated from each other.

Rough microsomes placed on a stabilizing sucrose gradient may be subjected to zone centrifugation, which results in distribution of the various vesicles on the gradient according to their sedimentation velocity [21]. The sedimentation coefficient (S_0) of the subfractions in 0.25 M sucrose lies between 0.4×10^3s and 1.2×10^3s. The morphological appearance of the top and bottom parts differ also (fig. 8). In the upper part of the gradient, there are smaller vesicles varying between 50 and 150 mμ, and the number of ribosomes per unit area is relatively low. On the bottom of the gradient, the vesicles are larger (ranging between 100 and 350 mμ) and they are more heavily coated with ribosomes. The electronmicroscopic heterogeneity is also followed by a differentiated distribution of electron transport enzymes. These are more concentrated in the slowly sedimenting vesicles, in comparison with vesicles in the lower part of the gradient or in the pellet (fig. 9). The distribution pattern is almost identical in the newborn rat, in spite of the fact that the various electron transport enzymes are synthesized at differing rates. After phenobarbital injection, NADPH-cyt. *c* reductase activity is first somewhat more enriched in the upper fractions, but very soon an increase is noticed in all parts of the gradient without preferential location.

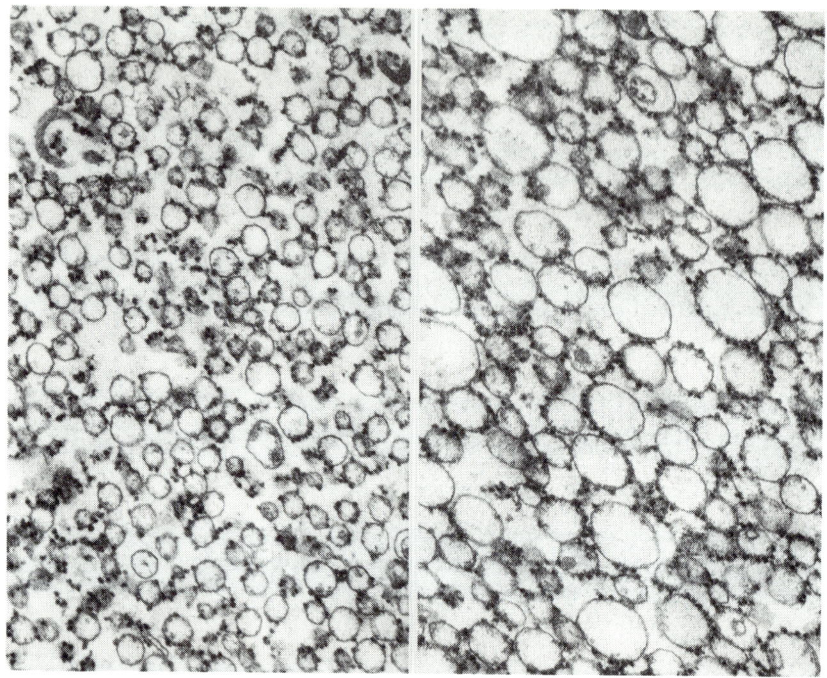

Fig. 8. Rough microsomal subfractions after zone centrifugation. a) upper part; b) lower part of the gradient. × 60 000.

Phenobarbital administration leads to an extensive increase in the ER membranes [22,23], and one would expect there to be preferential increase of the incorporation of protein and PLP precursors in those parts where new segments are accumulated. The complex phenomenon of membrane biosynthesis (or perhaps the mere appearance of an increased number of membranes) clearly cannot be interpreted in such simple terms. The *in vivo* incorporation of leucine-^{14}C after pulse labeling mainly occurs into the slowly sedimenting rough microsomes after phenobarbital injection (fig. 10). Injecting a precursor of PLP synthesis, one would expect a similar pattern of change as after the injection of protein precursor since the increased amount of protein is exactly parallelled by an increase in the amount of PLP. This, however, is not the case (fig. 11). In the control, there is a relatively even incorporation rate of glycerol-^{3}H into PLP, while after phenobarbital treatment the incorpotation is decreased in the same part of the gradient where the leucine-^{14}C incorporation into protein is increased. In the other part, the rate is unchanged.

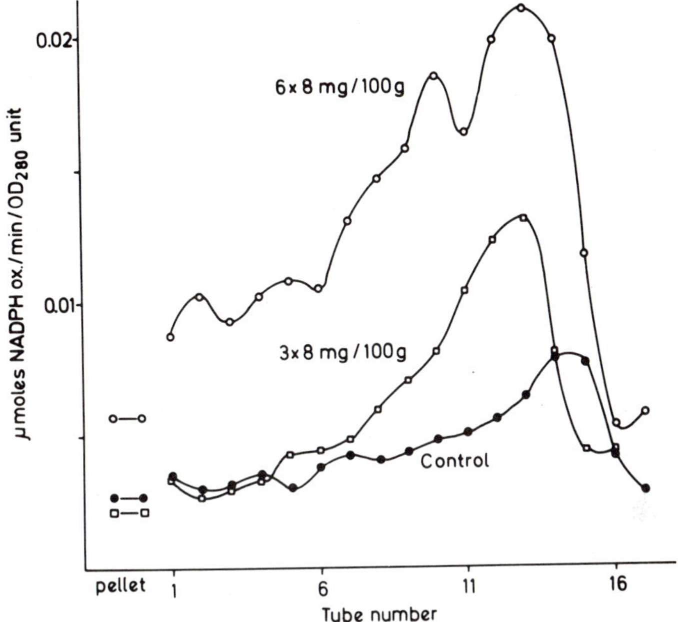

Fig.9. NADPH-cytochrome *c* reductase activity in rough microsomal subfractions from control and phenobarbital-treated rats. Phenobarbital was given three times (8 mg/100 g/day) and six times (8 mg/100 g/day). Data taken from ref. [21].

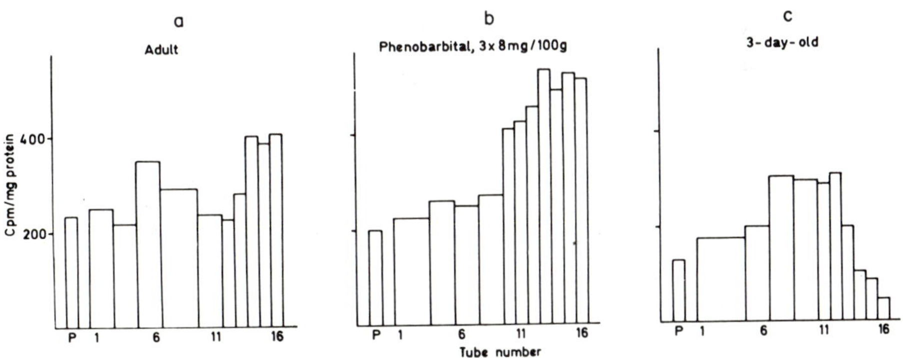

Fig. 10. Incorporation of leucine-^{14}C into proteins of rough microsomes from a) adult, b) phenobarbital-treated (3 × 8 mg/100 g/day), and c) 3-day-old rats. The *in vivo* incorporation time was 8 min. As indicated by the width of the columns, two or more fractions were combined at the lower part of the gradient. Data taken from ref. [21].

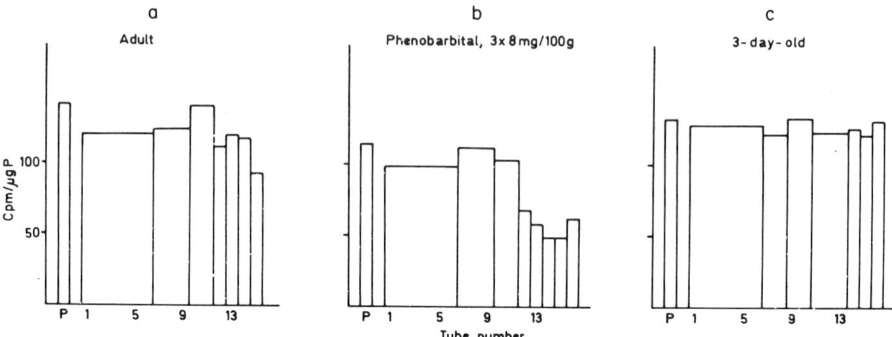

Fig. 11. Incorporation of glycerol-^{3}H into total lipids of rough microsomes from a) adult, b) phenobarbital-treated (3 × 8 mg/100 g/day), and c) 3-day-old rats. The *in vivo* incorporation time was 10 min. The width of the column indicates the number of fractions used for assay. Data taken from ref. [21].

The rough microsomal subfractions are clearly heterogeneous and also seem to be influenced differentially under various experimental conditions. On the other hand, there are no parts isolated so far which have a specific role in enzyme or membrane synthesis.

Smooth microsomes exhibit properties regarding heterogeneity and phenobarbital induction which resemble those of rough microsomes from many points of view. Zone centrifugation separates peak activities of ATPase, G6Pase, and electron transport enzymes at different locations on the gradient [9]. Glycosyl transferases are highly enriched in the low density fractions after isopycnic equilibrium centrifugation [24]. After phenobarbital treatment a highly increased NADPH-cyt. *c* reductase activity is apparent in all subfractions, and no preferential sites for incorporation of protein and lipid precursors can be separated.

4. Submicrosomal particles

The enzymic pattern of rough and smooth microsomes displays a certain degree of heterogeneity, which immediately raises the question as to what the basic unit for a single function is. It is quite possible that the microsomal vesicle, after homogenization is too large to house only one enzymic system, and the study of membrane organization may therefore require the production of smaller units. An approach to this problem is the preparation of

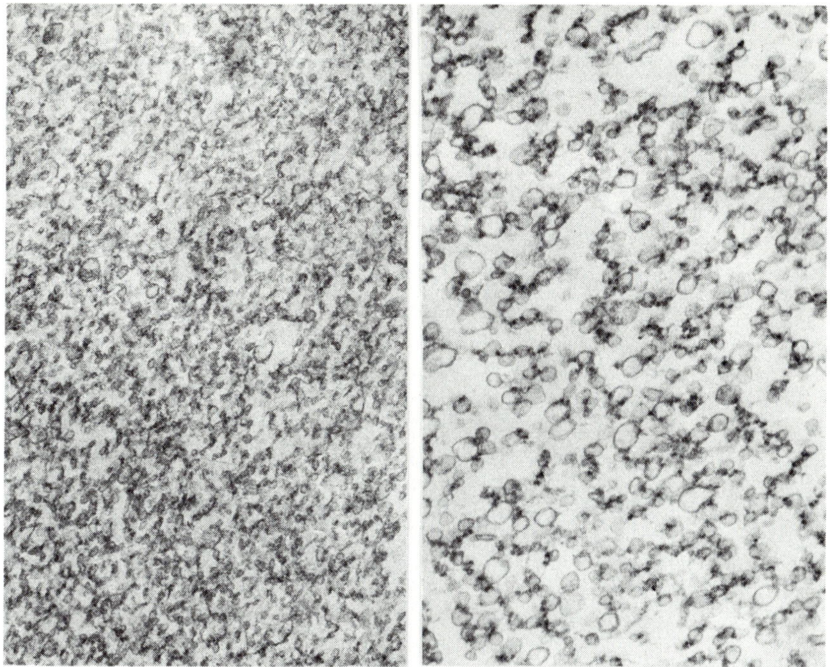

Fig. 12. Appearance of subfractions of sonicated rough microsomes. a) Upper third; b) lower third of the gradient. × 90 000.

submicrosomal particles by sonication under appropriate conditions [10,25]. If sonication causes breakage at specific points and the liberated membrane fragments with different enzyme patches exhibit different physicochemical properties, specific membrane segments with highly enriched enzyme activities might be obtained. Such an approach may also produce membrane pieces which represent newly synthesized units of the constant renewal cycle.

Sonicated rough microsomes exhibit a heterogeneous distribution after both density gradient and zone centrifugation. The smallest particles in the upper part of the density gradient are small vesicles with a diamater of 30 mμ, but no free nonvesicular membrane fragments are apparent (fig. 12). In the bottom part of the gradient, vesicle structures with a diameter of up to 110 mμ are visible. Sonication is therefore effective for the production of submicrosomal particles since the average particle size is estimated to be reduced more than 100-fold.

Analysis of the subfractions reveals a type of heterogeneity which is absent

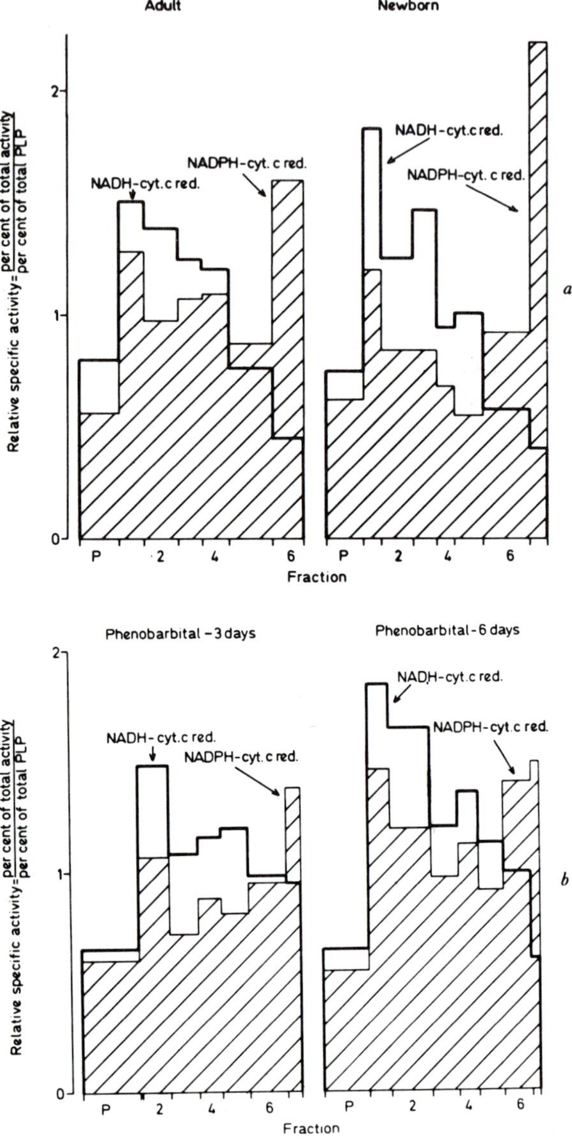

Fig. 13. Distribution of NADH- and NADPH-cytochrome *c* reductase activity in subfractions of sonicated rough microsomes. Fig. 13a: Distribution of enzyme activities in the adult (left) and newborn, 3-day-old (right). Fig. 13b: Distribution of enzyme activities in adult, treated with phenobarbital (8 mg/100 g/day) for 3 days (left) and 6 days (right). Data taken from ref. [25].

in the non-treated rough microsomes. A certain degree of seperation appears between the NADH- and NADPH-linked electron transport chains, the former being concentrated in the larger, denser and also in the more rapidly sedimenting particles, while the small vesicles on the top are enriched in enzymes of the NADPH-linked sequence (fig. 13). Conditions under which certain electron transport enzymes are synthesized at a rapid rate, such as directly

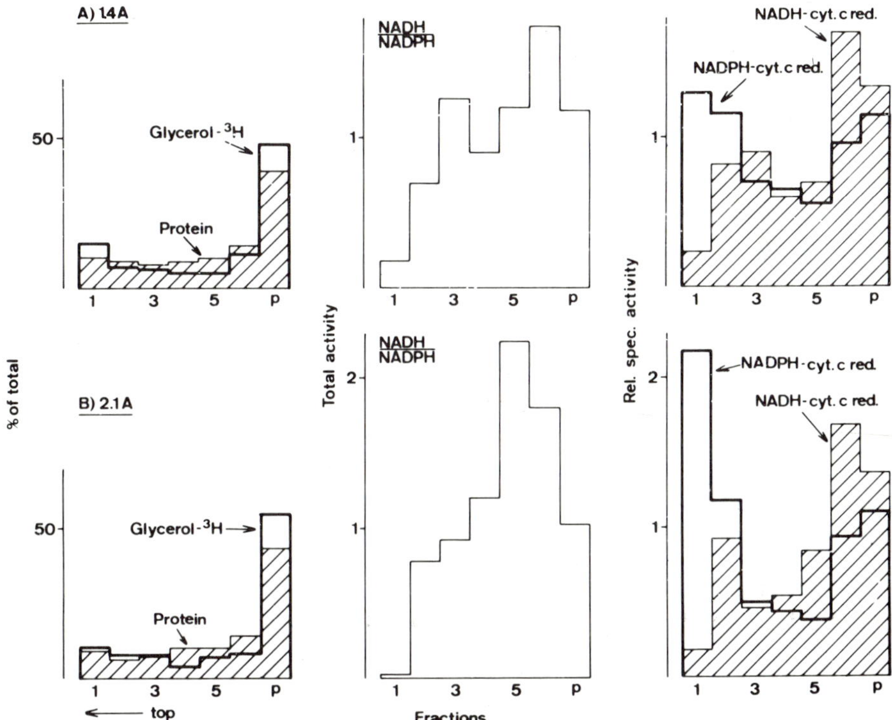

Fig. 14. Cytochrome *c* reductases of the high sedimentation velocity submicrosomal particles after resonication and recentrifugation on a density gradient. Pellets after zone centrifugation (glycerol-^{3}H labeled rat, intraperitoneal injection of 300 μCi/100 g, 20 min before decapitation) were suspended in i ml H_2O per pellet and resonicated at 1.4 amp. (exp. a) and 2.1 amp. (exp. b) for 5 min. The first column shows the distribution of the protein and radioactivity. The second column shows NADH/NADPH cyt. *c* red. activity ratios. NADH and NADPH stand for the appropriate pyridine nucleotide reductases. The third column demonstrates the relative specific activity of the two cytochrome *c* reductases on a protein basis in the individual subfractions. Data taken from ref. [10].

after birth and during phenobarbital treatment, do not give evidence for the presence of specific, newly synthesized membrane segments after similar analyses. It will appear from fig. 13 that the degree and type of heterogeneity among the two cytochrome *c* reductase activities are about the same as in the control.

Thus, similarly to non-sonicated subfractions of rough and smooth microsomes, the various types of submicrosomal particles also failed to show the presence of "new" membrane units differing from the "old" ones.

The failure to obtain a higher degree of enrichment of some enzymes may be interpreted as still not sufficient breakdown after sonication or poor resolution during ultracentrifugation. When the NADH-cyt. *c* reductase enriched pellet after zone certrifugation is resonicated and recentrifuged, one is able to approach the questions as to whether it is possible to make a further subdivision, to separate these units, and to release preferentially one specific electron transport enzyme.

Zone centrifugation of the resonicated pellet releases 50–60% of the protein and lipid content into the gradient, but the patterns of ratios and specific activities do not exhibit signigicant differences. There are several alternative explanations for this. The vesicles may be homogeneous, they may be heterogeneous but are of uniform size or they may be heterogeneous both in size and in enzyme composition but the procedure used for separation does not resolve them. A choice among these alternatives can be made after isopycnic centrifugation of the resonicated pellet (fig. 14). Analysis of the ratios and specific activities of the cytochrome *c* reductases reveals a high degree of specialization among the particles. The low density particles in the upper part of the gradient after sonication at high energy level exhibit a 10-fold enrichment in NADPH-cyt. *c* reductase activity compared to the NADH counterpart, while the high density fractions are enriched in NADH-cyt. *c* reductase activity.

5. Conclusions

The endoplasmic reticulum, particularly in the liver, is characterized by a number of specific properties such as large surface [26], very complex enzymic equipment [2], limited permeability [4], specific synthetic and transport functions for macromolecules [27], and sensitivity to external agents [28]. There are many indications, but few proofs, that the structure, organization, and actual state of membrane components influence functions just as much as a direct enzyme effect. This is the basic explanation for our interest in the structure of endoplasmic membranes, including their physical, chemical and enzymical aspects.

The most unique property of the ER in relation to all the other cytoplasmic organelles is the non-random distribution of most of its enzymes and enzymic systems [6]. This heterogeneous distribution of functions turns out to have important implications for the understanding of the behavior of the ER in biogenetic processes and its response to internal and external agents. We can only guess, but not yet decide, what the smallest units are. Are they devoted to one single function and to what extent may these units morphologically or functionally cooperate and thereby build up a larger unit?

Taking into consideration that after sonication a part of the smallest vesicles, which are 300 A large and have a surface area of 2.5×10^5 A^2, are enzymically homogeneous, an approximate calculation of the number of enzyme molecules present can be made [10,25]. Knowing the amount of lipid (6×10^{-18} g) and protein (9×10^{-18} g) in these vesicles, the physical properties of the individual enzymes, the stoichiometric relationship among individual components of electron transport chains, as well as making a number of assumptions, it is possible to estimate how many electron transport chains may possible be present in one 300 A large vesicle. This homogeneous unit is certainly not large: in one case the vesicle contains 15 complete NADPH and 2 NADH electron transport chains, and in the other case 25 complete NADH and 3 NADPH chains. In other words, if the basis for heterogeneity in the isolated microsomal vesicle is only the unit isolated after sonication, the heterogeneity would appear to be of only moderate extent. On the other hand, sizeable enrichment of electron transport enzymes or phosphatases can be obtained in isolated microsomal subfractions. If we accept that the membrane obtained after sonication is the unit, one has to assume an accumulation of units in localized parts of the ER system.

From the point of view of biogenesis, the presence of small units may also have implications. If membranes or constitutive enzymes are synthesized or removed only in the form which correspond to the unit isolated by sonication, the changes in a selected part of endoplasmatic membrane surface will be masked. In this sense, changes in turnover of chemical components or in composition of membrane enzymes will appear all along the ER system and consequently in all isolated microsomal subfractions. This would be one explanation for the failure in isolating newly synthesized membrane pieces.

What we finally have to decide is what we mean by heterogeneity. At the individual molecular level, naturally, everything is heterogeneous. A high degree of enrichment in one enzyme system can be obtained only at the level when isolated patches do not contain more than 100 to 300 enzyme molecules. If this is sufficient for definition of the non-random distribution, the microsomal membranes can be defined as heterogeneous. Any further enrichment in a vesicle necessitates the assumption of non-random distribution of these basic units. However, still very little is known about the enzyme topography of membrane architecture, since we know too little about how the isolated microsomal membrane pieces can be identified as parts of the lamellar and cisternal system of the endoplasmic reticulum.

Acknowledgement

The work reported from the authors laboratory is supported by grants from the Swedish Medical Research Council and the Magn. Bergvall's Foundation.

References

[1] P. Siekevitz, G.E. Palade, G. Dallner, I. Ohad and T. Omura, in: The Biogenesis of Intracellular Membranes, eds. H.J. Vogel, J.O. Lampen and V. Bryson (Academic Press, New York, 1967) p. 331.
[2] P. Siekevitz, Ann. Rev. Physiol. 25 (1963) 15.
[3] G.E. Palade and P. Siekevitz, J. Biophys. Biochem. Cytol. 2 (1956) 171.
[4] R. Nilsson, E. Pettersson and G. Dallner, FEBS Letters 15 (1971) 85.
[5] M. Schramm, B. Eisenkraft and E. Barkai, Biochim. Biophys. Acta 135 (1967) 44.
[6] G. Dallner and L. Ernster, J. Histochem. Cytochem. 16 (1968) 611.
[7] R. Nilsson, E. Pettersson and G. Dallner, In press.
[8] G. Dallner and R. Nilsson, J. Cell. Biol. 31 (1966) 181.
[9] H. Glaumann and G. Dallner, J. Cell. Biol. 47 (1970) 34.
[10] H. Svensson, G. Dallner and L. Ernster, In press.
[11] G. Dallner and A. Azzi, Biochim. Biophys. Acta 255 (1972) 589.
[12] G. Dallner, A. Bergstrand, L. Ernster and A. Azzi, In press.
[13] M.R. Stetten and S.B. Ghosh, Biochim. Biophys. Acta 233 (1971) 163.
[14] G. Dallner, P. Siekevitz and G.E. Palade, J. Cell Biol. 30 (1966) 97.
[15] Y. Kuriyama and T. Omura, J. Biochem. 69 (1971) 659.
[16] T. Omura, P. Siekevitz and G.E. Palade, J. Biol. Chem. 242 (1967) 2389.
[17] T. Higashi and T. Peters, J. Biol. Chem. 238 (1963) 3952.
[18] S. Jakobsson and G. Dallner, Biochim. Biophys. Acta 165 (1968) 380.
[19] C. de Duve, J. Berthet and H. Beaufay, Progr. Biophys. Biophys. Chem. 9 (1959) 325.
[20] C. de Duve, Harvey Lect. 59 (1965) 49.
[21] G. Dallner, A. Bergstrand and R. Nilsson, J. Cell. Biol. 38 (1968) 257.
[22] H. Remmer and H.J. Merker, Science 142 (1963) 1657.
[23] S. Orrenius, J.L.E. Ericsson and L. Ernster, J. Cell Biol. 25 (1965) 627.

[24] R.R. Wagner, E. Petterson and G. Dallner, In press.
[25] P.R. Dallman, G. Dallner, A. Bergstrand and L. Ernster, J. Cell Biol. 41 (1969) 357.
[26] E.R. Weibel, W. Staubli, H.R. Gnagi and F.A. Hess, J. Cell Biol. 42 (1969) 68.
[27] G.E. Palade, P. Siekevitz and L.G. Caro, in: Ciba Foundation Symposium on the Exocrine Pancreas, eds. A.V.S. de Reuck and M.P. Cameron (Churchill, London, 1962) p. 23.
[28] E.A. Smuckler and M. Arcasoy, in: Intern. Rev. Exper. Path., eds. G.W. Richter and M.A. Epstein (Academic Press, New York, 1969) vol. 7, p. 305.

BIOGENESIS AND MODULATION IN CHLOROPLAST MEMBRANES

Itzhak OHAD*
Department of Biological Chemistry, the Hebrew University of Jerusalem, Jerusalem, Israel

1. Introduction

During the past decade the efforts of many investigators have been focused on the problem of biological membrane biosynthesis. The challenge offered by the complexity of the problem has been met not only because of its theoretical interest but also because it has become evident that the process of membrane biosynthesis underlies and plays a major role in a great variety of cellular activities including energy transduction and bulk transport, genome replication, cell division, growth and differentiation.

Different experimental systems have been developed and analyzed including most of the cellular membrane systems such as the plasma membrane of the mycoplasma [1, 2], bacteria [3–5] and chromatophores of photosynthetic bacteria [6–9] the plasma membrane of eucariotic cells [10] the endoplasmic reticulum of liver cells [11–14], the membranes of the secretory granules of the exocrine pancreas [15] and parotid gland (16) and the membrane system of the mitochondria [17, 18] and chloroplasts [19–27]. The last two systems have some common features. They both belong to semi-autonomous organelles, the formation of which can be induced experi-

* This lecture is based on results obtained through the cooperation and efforts of a group of students during the past 5 years, including, in alphabetical order: Miss S. Bar-Nun, Mrs. R. Broza, Mr. G. Eytan, Dr. I. Goldberg, Dr. R. Jennings, Miss B. Kill, Mr. A. Lalazar, Mr. S. Schuldiner and Mr. D. Wallach.
I wish to express my thanks to them for their patience, skill and critical consideration of this work at all its stages.

mentally in absence of cell division by use of natural triggers such as oxygen pressure [28] or availability of light [20]. In addition, both organelles contain a membrane system specialized in one major function which, because of its dependence on the structural integrity of the membrane can be used as a very sensitive indicator for minor changes in the organisation of the membrane, that is electron flow coupled with ion translocation and synthesis of ATP. The membranes forming the crystae of the mitochondria and the thylakoids and grana of the chloroplast might be regarded as a homogenous system in terms of function and structure in the sense that the specific arrangement of the components within the unit area of the membrane capable to carry all its functions is bound to repeat itself in a homogenous way throughout the membrane system. This will permit an easier correlation to be made between changes induced in the gross composition of the membranes with changes in its function and structure. On the other hand, membrane systems such as the plasma membrane or endoplasmic reticulum designed to fulfill different non-linked functions are basically heterogenous in nature. Modulation of their gross compostion during growth and differentiation will be the result of changes in the specific composition of the separate units superimposed on changes in the relative ratio of the different types of functional units within the membrane.

The development of the photosynthetic membranes in the *y*–1 mutant of *Chlamydomonas reinhardi* can be used as a model system for the biogenesis of a homogenous membrane. This system has been analyzed in detail in an attempt to gain information on the changes in the lipid, pigment and protein composition as well as ultrastructure and function of the growing membranes [19–25, 29–32]. From these studies, a picture emerges indicating that the photosynthetic membranes grow by random intussusception of new material, within pre-existing membranes. Although the process of growth occurs normally by simultaneous addition of all building blocks to the growing membranes one can interfere with this normal pattern by use of specific protein synthesis inhibitors or alteration of light and dark cycles [24,30,37]. In this case, a stepwise growth process can be induced. However, a certain sequence in the addition of the missing components appears to be obligatory. Due to this feature of the system, membranes can be formed having an altered ratio of pigments and specific types of proteins as compared with the normal membranes. This will result in the formation of membranes altered functionally and structurally which can be, however, repaired by the secondary addition of the missing components. Some of the data on which these conclusions are based will be considered.

2. Description of the basic experimental system

The *y*–1 mutant of *Chlamydomonas reinhardi* differs from the wild type cells by the loss of ability to convert protochlorophyll to chlorophyll en-

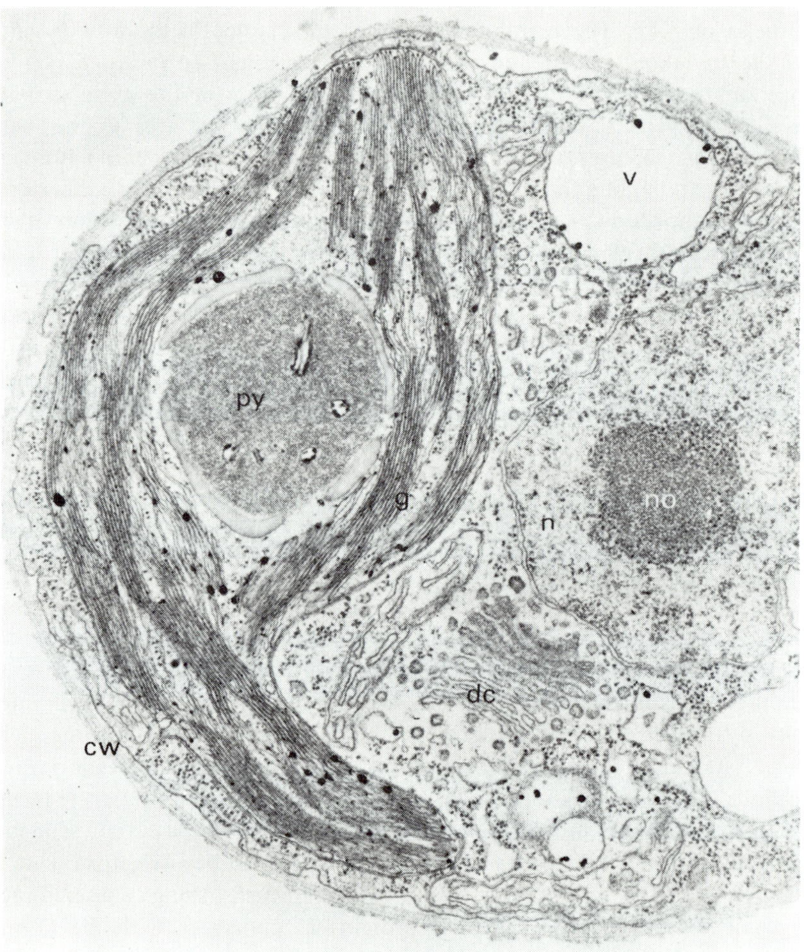

Fig. 1. An electron micrograph of a section through the basal part of a light grown *Chlamydomonas reinhardi y*-1 cell showing the chloroplast which contains numerous grana (g) and a pyrenoid (py); starch granules (sg); no, nucleoulus; n, nucleus; dc, dictiosomes; v, vacuole; m, mitochondria, cw, cell wall. For experimental details see ref. [19].

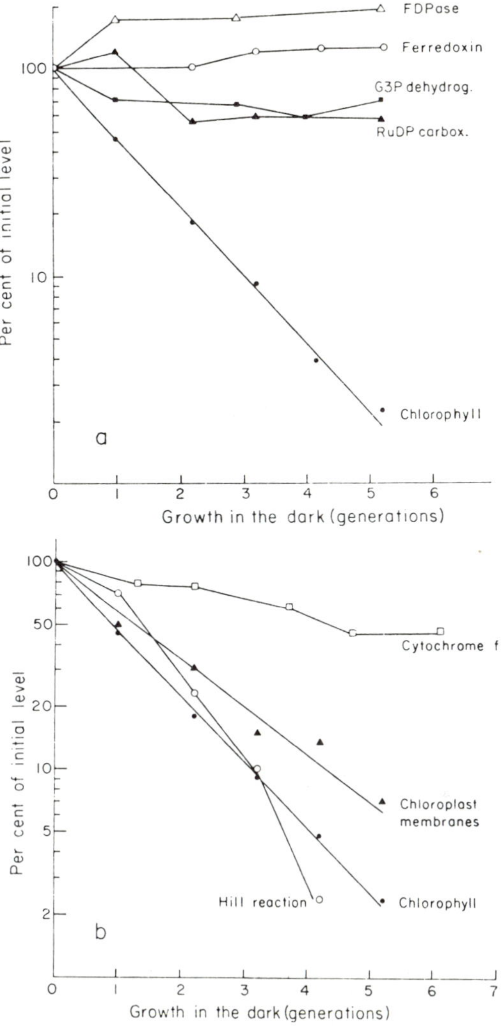

Fig. 2. Changes in chlorophyll, chloroplast membranes and various enzymes in *Chlamydomonas reinhardi y*-1 cell as a function of number of generations of growth in the dark. The initial levels (100%) were as follows: (a) chlorophyll, 34 μg/10^7 cells; FDPase, *p*H 8, 0.33 μmoles Pi/30 min/10^7 cells. RuDPcarboxylase, 50.3 μmoles CO_2-fixed/10 min/10^7 cells; ferredoxin, 42 mμmoles NADP-reduced/min/10^9 cells; G-3P dehydrogenase, 0.072 μmoles NADP reduced/min/10^7 cells. (b) Hill reaction, 0.088 μmoles DCI reduced/min/10^7 cells; cytochrome *f*, 2×10^{-3} OD/10^7 cells; chloroplast membranes index, 1.9 intersections/cm^2 for a total of 2217 membrane profiles counted. For experimental details see ref. [19].

zymatically in the absence of light and thus cannot form photosynthetic membranes when grown in the dark [19]. However, chlorophyll can be formed by a non-enzymatic photochemical reduction of protochlorophyll [33]. Thus, when grown in the light the mutant cells are undistinguishable from the wild type cells (fig. 1). Since chlorophyll cannot be synthesized in the dark the transfer of light grown cultures to the dark will cause a dilution

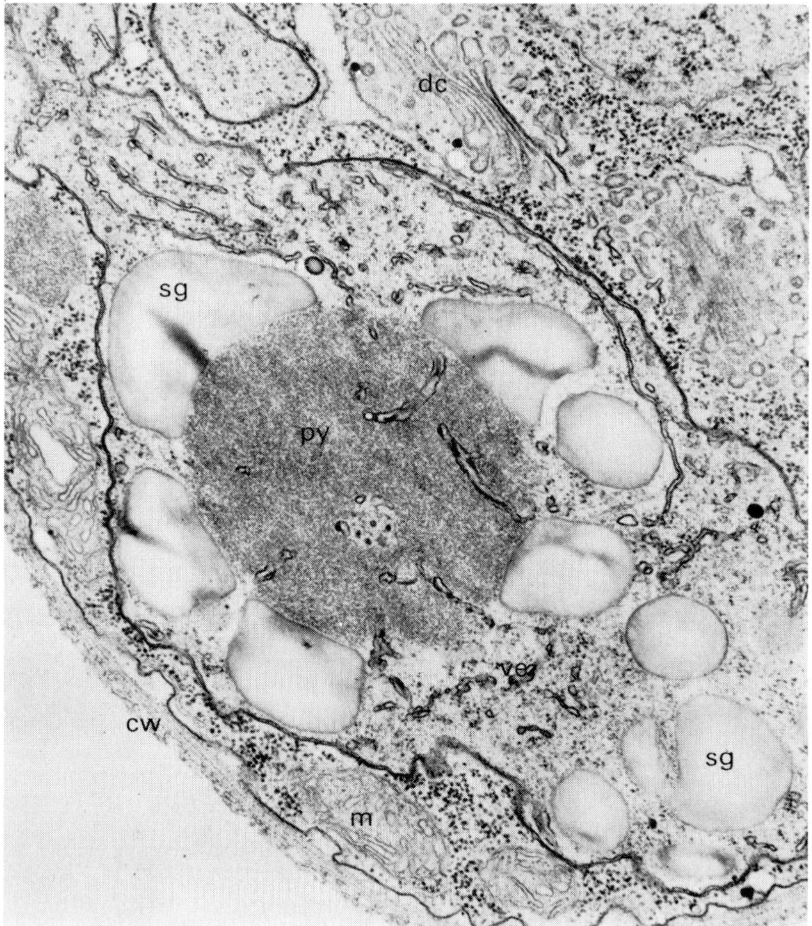

Fig. 3. Electron micrograph of a section through the basal part of a *y*-1 cell from a culture grown in the dark for 5 generations. Only few vesicular elements (ve) are left from the photosynthetic lamellar system of the chloroplast. Same experimental conditions and abbreviations as in fig. 1.

of the photosynthetic membranes through division (fig. 2). However, certain membrane proteins such as cytochrome *f* and soluble chloroplast proteins such as ferredoxin, RuDPcarboxylase, alkaline fructose 1.6-diphosphate phosphatase and NADP-dependent glyceraldehyde-3-phosphate dehydrogenase can be formed and accumulated relative to the remnants of the photosynthetic lamellae (fig. 2, [19]). After 5–6 generations of growth in the dark on a medium containing acetate as a carbon source only a few vesicular elements

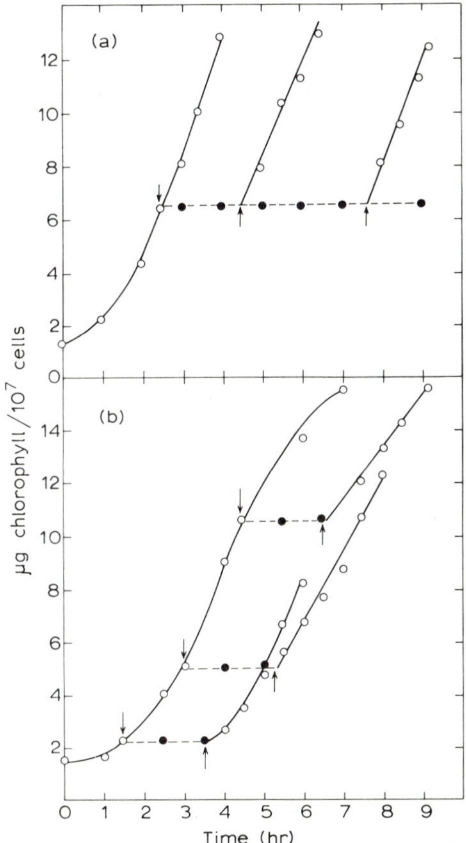

Fig. 4. Chlorophyll synthesis during alternative exposure of dark grown *y*-1 cells to the light or dark. The rate of synthesis achieved at the time of transfer of the greening cells to the dark is resumed upon reexposure to the light (fig. 4b) and preserved even after 5 hr of incubation in the dark (fig. 4a). For experimental details see ref. [34]. ↓, light off; ↑ light on.

can be found in the plastid which has accumulated large amounts of starch. However, all the other cell organelles are not affected and appear morphologically identical to those found in light grown cells (fig. 3). When dark grown cells are transferred to the light, chlorophyll synthesis is resumed following a relatively short lag period of about 1.5 to 3.5 hr. The synthesis will stop immediately upon cessation of illumination and resume upon reexposure to the light [20, 34]. In this case, the rate of synthesis will be equal to that attained at the time light was shut of (fig. 4). This suggests that the enzymatic apparatus responsible for chlorophyll synthesis is still present after the interspaced dark period and can be utilised immediately by onset of illumination.

3. Synthesis of lipids and pigments during the greening process

Light grown photosynthetic cells contain large amounts of glycolipids and poly-unsaturated fatty acids as compared with the dark grown cells. When the lipid and pigment content of the cells was analyzed at different times of the greening process it was found that during the lag phase the lipid and pigment composition of the cells changed rapidly. During the linear phase of the greening glycolipids, poly-unsaturated fatty acids, carotenoids and chlorophylls are synthesized proportionally to the synthesis of chlorophyll a [22]

Table 1
Molar ratio of various lipids and pigments to total chlorophyll during light induced synthesis of membranes in *Chlamydomonas reinhardi y*–1

Compound	Light (hr)												Data from literature [35]
	1½	2	2½	3	3½	4½	5	6	6½	7	7½	9	
Chlorophyll a *	0.74			0.75		0.70		0.74			0.72		0.70
Chlorophyll b *	0.26			0.25		0.30		0.26			0.28		0.30
Lutein *	0.60			0.32		0.26					0.11		0.09
MG **						1.35			1.25			1.15	1.50
DG **			1.87			0.71			0.55			0.50	0.62
SL **			0.45			0.33			0.24			0.27	0.20
Total glycolipids			2.24			2.26			1.90			1.82	2.30
FA C18:2 ***		6.70			5.75		4.05			3.50		2.28	–
FA C18:3 ***		4.15			4.65		3.45			2.70		1.82	–

The data are computed from three different experiments. The initial and final chlorophyll content ($\mu g/10^7$ cells) was: * initial, 1.0, final, 18.0; ** initial, 0.6, final, 11.1; *** initial, 1.9, final, 27.5.

(table 1). This indicates that the lipid composition of the initially present membranes remnants is rapidly altered. As they accept new membrane components and grow, their lipid and pigment composition become typical of that of normal photosynthetic membranes as found in light grown cells [22, 32].

4. Synthesis of membrane proteins during the greening process

Gel electrophoretic analysis of the protein composition of a chloroplast membrane fraction shows that during the greening several new proteins are formed. A major lamellar protein (L protein [24] seems to be formed by cytoplasmic ribosomes since its synthesis is inhibited by cycloheximide but not by chlormaphenicol [24, 29, 30, 36] (fig. 5). Several other proteins peaks: 3–5, 10 and 11–16, (fig. 5) appear to be formed by chloroplast ribosomes since their synthesis is inhibited by chloramphenicol [24, 36]. During the greening process the synthesis of L protein follows very closely the synthesis of chlorophyll (fig. 6). The question arises whether cessation of chlorophyll synthesis due to transfer of cells to the dark will cause an immediate cessation of L protein synthesis. The results of an experiment in which the rate of incorporation of labeled precursors into L protein was measured during the greening before and after transfer of the greening cells to the dark indicate that L protein can be formed in the dark for a short period of time and accumulate in the membranes relative to the chlorophyll [37] (fig. 7). Thus, one can induce the formation of membranes enriched in L protein in two ways: a) when greening is carried out in the presence of chloramphenicol which blocks synthesis of chloroplast-made proteins, inhibits only partially synthesis of chlorophyll and has no effect on the synthesis of L protein; b) following transfer of greening cells to the dark when only chlorophyll synthesis is completely blocked.

Addition of cycloheximide to cells greening under continuous illumination at any time of the greening process immediately blocks the synthesis of chlorophyll and both types of membrane proteins, those of cytoplasmic origin as well as chloroplast-made proteins [24]. However, addition of cycloheximide to cells which have been previously exposed to the light in the presence of chloramphenicol and thus formed membranes enriched in cytoplasmic-made L protein does not block either synthesis of chloroplast-made proteins (fig. 5) nor formation of chlorophyll as it is shown in fig. 8.

These results indicate that concomitant synthesis or presence of excess L protein within the growing membranes is a prerequisite for the addition of

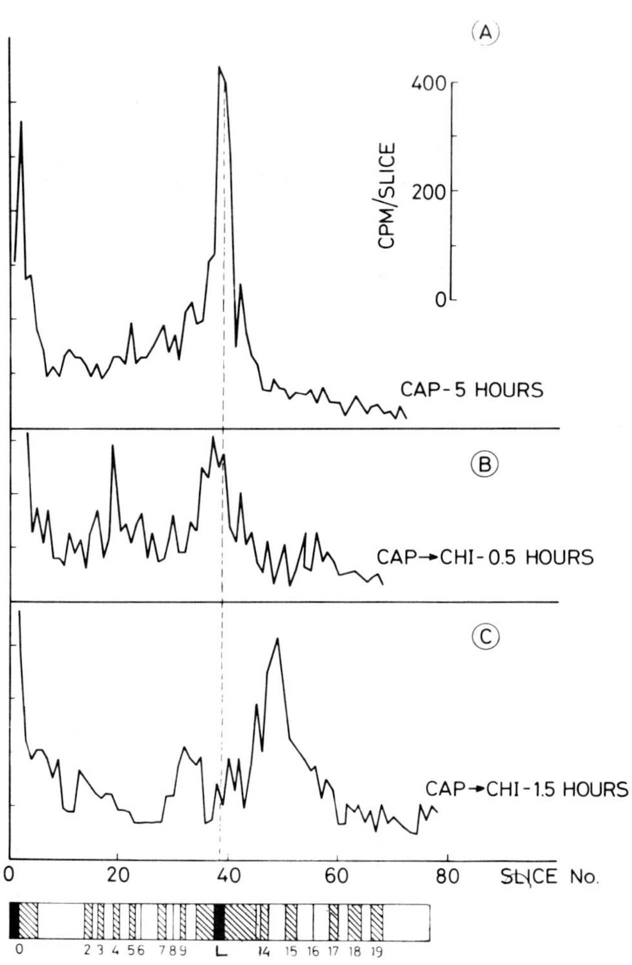

Fig. 5. Pulse labeling of chloroplast membrane proteins with ^{3}H acetate in greening *y*-1 cells incubated first in the presence of chloramphenicol (CAP) and then transferred to cycloheximide (CHI). A, B and C, distribution of radioactivity in acrylamide gel electrophoretograms of membrane proteins labeled before the transfer and after ½ and 1½ hr after transfer of the cells from CAP to CHI respectively. The heavy black band marked L in the schematic representation of the stained gel indicates the location of the L protein. For experimental details see ref. [24].

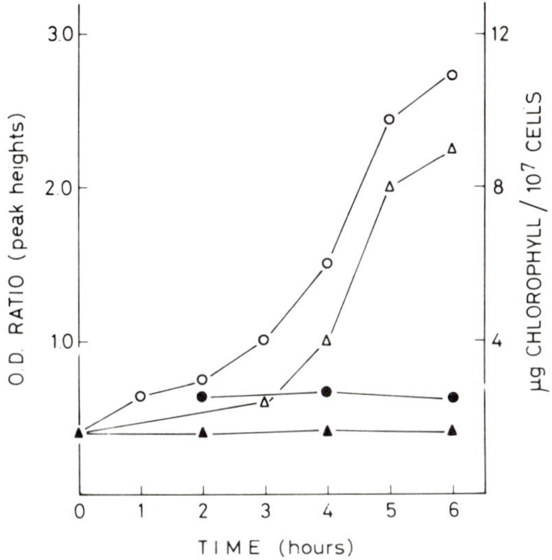

Fig. 6. Parallel increase in the relative amounts of L protein and chlorophyll synthesis during greening of *y*-1 cells. The amount of L protein is represented as the ratio of optical density of the L peak to peak 7 which does not change during the greening, as calculated from densitometer tracings of electrophoretograms of membrane proteins. For experimental details see ref. [24]. Circles, L protein; triangles, chlorophyll; open and dark figures, cells incubated in the light and dark respectively.

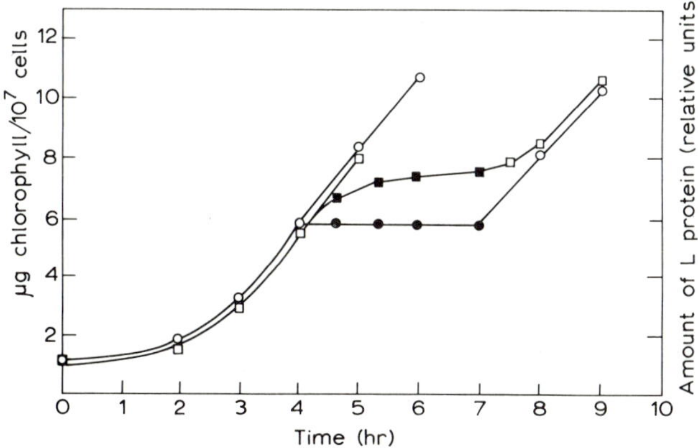

Fig. 7. Synthesis of chlorophyll (○) and L protein (□) during alternate exposure of greening *y*-1 cells to the light and dark. Black figures, cells incubated in the dark. For experimental details see ref. [37].

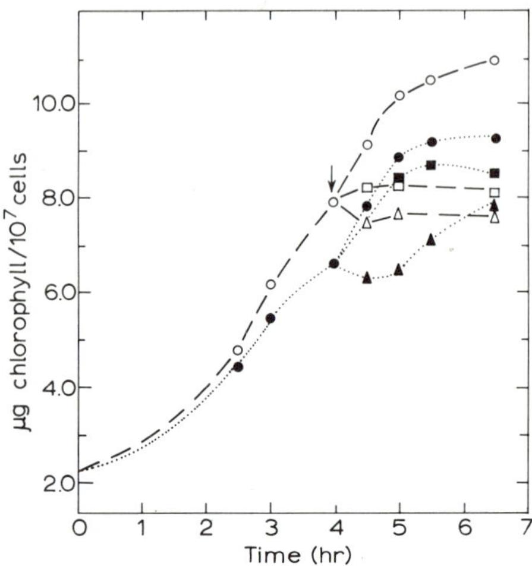

Fig. 8. Effect of different concentrations of cycloheximide on chlorophyll synthesis in cells greening in the absence or presence of chloramphenicol. Control cells greening in the light (○) were transferred at the time marked by the arrow to fresh medium containing 0.2 μg/ml (□) and 2.0 μg/ml cycloheximide (△). Cells greening in the presence of chloramphenicol (●) (100 μg/ml) were transferred at the time indicated by the arrow to fresh medium containing 0.2 μg/ml (■) and 2.0 μg/ml (▲) cycloheximide. Notice the complete inhibition of chlorophyll synthesis by cycloheximide in the control cells as compared with the reduced effect at both concentrations used on cells previously incubated with chloramphenicol. For experimental details see ref. [24].

chlorophyll and proteins of chloroplast origin. The question arises whether excess of L protein is the only requirement for the synthsis and integration of the chloroplast-made membrane proteins or in addition light (that is – chlorophyll synthesis) is also necessary. In order to answer this question advantage was taken of the fact that membranes formed in the presence of chloramphenicol and thus containing L protein lipids and chlorophyll but lacking the chloroplast-made proteins are photosynthetically inactive [24, 30, 38, 39]. However, the inactive membranes can regain their photosynthetic activity through addition of chloroplast-made proteins if the protein synthesis by the chloroplast ribosomes is released from chloramphenicol inhibition [30, 38, 39, also fig. 5]. This experimental system was used in order to establish whether the synthesis of the chloroplast-made proteins required for the activation of the photosynthetically inactive membranes depends on concomitant synthesis of chlorophyll or L protein. The results of an experiment devised to answer these questions are shown in fig. 9. It appears that the synthesis of the

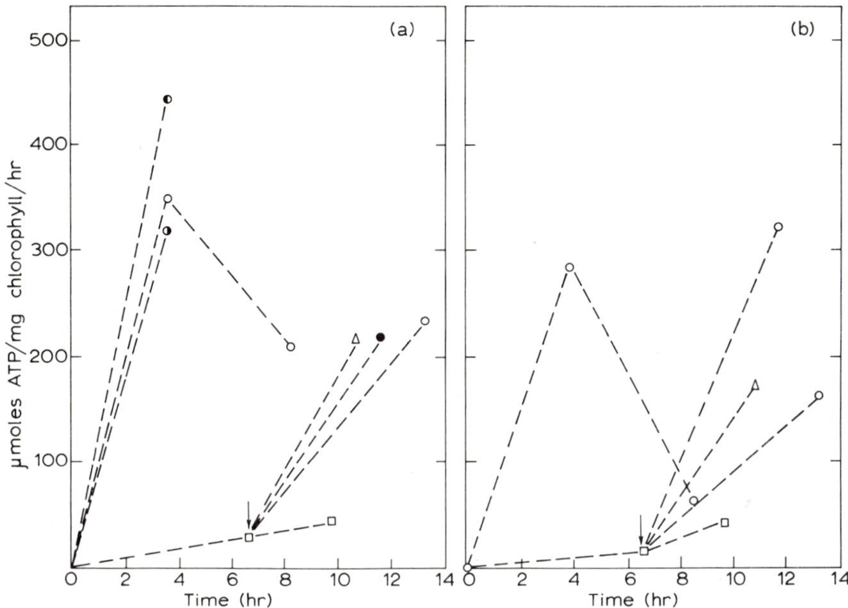

Fig. 9. Changes in photophosphorylation activity of photosystem I and II following transfer of *y*-1 cells greening in the presence of chloramphenicol to different incubation conditions: Cells incubated in absence of inhibitors; activity measured with diquat (◑– – –◑), pyocyanine (◐– – –◐), PMS-ascorbate ((○– – –○), fig. 9a) and ferricyanide ((○– – –○), fig. 9b); cells incubated in the dark in fresh growth medium; activity measured with PMS-ascorbate ((●– – –●), fig. 9a) and ferricyanide ((●– – –●), fig. 9b); cells incubated in medium containing chloramphenicol; activity measured with PMS-ascorbate ((□–□), fig. 9a) and ferricyanide ((□–□), fig. 9b; cells incubated in fresh growth medium containing cycloheximide; activity measured with PMS-ascorbate (△–△), fig. 9a); activity measured with ferricyanide ((△–△), fig. 9b). The time of transfer is indicated by the arrow. Photophosphorylation was measured using an "open cell preparation". For experimental details see ref. [38].

activation proteins can occur in the dark in absence of chlorophyll synthesis. The activation proteins can be made also in absence of concomitant L protein synthesis if the repair process is carried out in the light in the presence of cycloheximide.

Additional support for the conclusion that synthesis of activation proteins does not require illumination comes from the results of an experiment similar to that described in fig. 7 in which L protein was allowed to accumulate in the dark. If during the dark incubation, activation proteins will be also

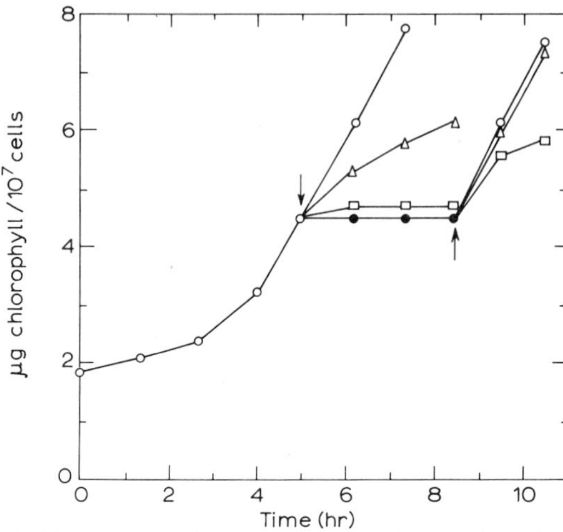

Fig. 10. Effect of chloramphenicol and cycloheximide on chlorophyll synthesis before and after incubation of greening cells in the dark. ○ control, light; ● control, dark; △, cells incubated in the presence of 100 μg/ml chloramphenicol; □, cells incubated in the presence of 0.2 μg/ml cycloheximide. Arrows, time of addition of the inhibitors and transfer of the cells to the dark (↓) or reexposure to the light (↑). For experimental details see ref. [37].

synthesized and integrated within the membranes in absence of chlorophyll synthesis one should expect that following a second illumination period chlorophyll will be synthesized and photosynthetic activity will increase even if L protein or chloroplast protein synthesis will be now blocked by addition of protein synthesis inhibitors. Indeed, chlorophyll synthesis and photophosphorylation activity were less inhibited if the drugs were added after a dark interspaced incubation period than if the addition occurred before transfer of the greening cells to the dark (fig. 10, table 2). These results demonstrate that during the dark incubation both L protein and activation proteins were synthesized and accumulated in excess to chlorophyll which cannot be made in the dark. However, it was shown before that the dark incubation alone without an additional light incubation does not increase the photosynthetic activity [39] indicating that the addition of chlorophyll was necessary. The above experiments show that each one of the three classes of components, L protein, chloroplast proteins and chlorophyll can become limiting factors either for the formation of the other components and membrane growth as well as for the expression of the potential photosynthetic activity. Based on the results presented above one can conclude that the membrane proteins of cytoplasmic origin (L protein) the activation proteins of chloroplast origin

Table 2
Effect of protein synthesis inhibitors on development of photophosphorylation activity during a discontinuous greening

Treatment		Photosystem II			Photosystem II + I		
First incubation	Second incubation	Total activity μmoles ATP/ 10^8 cells/hr	Inhibition (%)	Sp. activity μmoles ATP/ mg chlorophyll/hr	Total activity μmoles ATP/ 10^8 cells/hr	Inhibition (%)	Sp. activity μmoles ATP/ mg/chlorophyll/hr
5 hr of greening	Light control	19.0	–	383	18.3	–	390
	Light + cycloheximide	13.0	29	540	12.3	33	513
	Light + chloramphenicol	7.5	59	182	9.6	48	234
5 hr of greening and then 3 hr of dark incubation	Light control	18.3	–	390	18.7	–	393
	Light + cycloheximide	18.3	0	520	16.3	12	465
	Light + chloramphenicol	16.8	3	370	13.3	29	296

Dark grown cells were first incubated for 5 hr in the light or 5 hr light and 3 hr dark and then washed by centrifugation and resuspended in fresh medium without or with addition of chloramphenicol (100 μg/ml) or cycloheximide (0.5 μg/ml) and incubated again in the light for 2 hr. At the end ot the second incubation period the photophosphorylation activity was measured with ferricyanide (photosystem II) or diquat (photosystem II + I) using an "open cell preparation" [38]. For experimental details see ref. [37].

and chlorophyll can be added stepwise while the presence of L protein in excess serves as an obligatory acceptor frame.

5. Homogeneity of the membranes during growth

It was demonstrated before that newly synthesized lipids are incorporated into the growing membrane by a process of random intussusception [23]. Statistical analysis of autoradiographic grain distribution over membrane profiles following short pulse-chase labeling by ^{3}H acetate during the linear phase of the greening process showed that single thylakoids as well as fused thylakoids forming grana can accept newly synthesized lipids. The incorporation was in a random fashion over the membranes surface and it appeared that the rate of growth of non-fused thylakoids was somewhat higher (table 3). Random distribution of newly incorporated molecules can be the results

Table 3
Autoradiographic grain distribution in different types of membranes in the chloroplast of cells labeled with ^{3}H-acetate during the greening process

Treatment	Chloroplast inner membranes		
	Total length (μ/cell)	Specific activity * of the thylakoids	
		Single	Paired
3.5 hr light 10′ pulse	43.50 ± 6.10	1.19 ± 0.15	0.71 ± 0.25
6.0 hr light 10′ pulse	59.00 ± 16.50	1.30 ± 0.22	0.72 ± 0.35
6 hr light 10′ pulse 1.0 hr chase	75.00 ± 16.80	1.25 ± 0.25	0.78 ± 0.30

Dark grown cells were exposed to the light for periods as indicated in the table and then pulse labeled with ^{3}H-acetate. Samples were fixed immediately or after further incubation for one hour in the presence of unlabeled acetate (chase). The fixed material was processed for electron microscopy autoradiography and the distribution of autoradiographic grains over the chloroplast and chloroplast membrane profiles analyzed. The specific activity of the different growing membranes is not influenced by the time of the pulse nor by the chase period indicating that the distribution of the grains over the membrane profiles is random. However, the rate of labeling of unpaired thylakoids is significantly higher than that of paired thylakoids indicating that the former grow faster. For experimental details see ref. [23].

* Specific activity was calculated for each type of membrane as % of grains on the membrane profiles from total grains in the chloroplast divided by the length of profiles (μ) from total length of all chloroplast membrane profiles.

of two different processes: growth by addition of new molecules within randomly distributed growth centers resulting in the formation of a mosaic of new and old membrane regions; b) random incorporation at molecular level resulting in a homogenous distribution of the new membrane components within the growing membrane. Electron microscopy autoradiography cannot distinguish between these two possibilities. However, it appears that establishing which of the two alternatives is operative might serve as an indicator for the distribution of recognition and acceptor sites within the membrane for its specific components and give some hint to the mechanism whereby their selective incorporation is achieved. The concomitant synthesis of L protein with other membrane components such as chlorophyll and activation proteins or its preaccumulation appear to be a requirement for membrane growth. This might indicate that L protein serves as an acceptor for the other components and that its acceptor capacity can be and actually is saturated under normal growth conditions.

The distribution of such acceptor sites in a distinct as opposed to a homogenous way has different implications on the mechanism of membrane growth. Thus, one can visualize the membrane as a continuum of complexes of rigid composition, structural organisation and function consisting of cytoplasmic (L) proteins, lipids pigments and activation proteins. Addition of L protein and lipid will form new and distinct acceptor sites which might become saturated by addition of the other components in exact proportions. Another possibility is that an L protein-lipid complex can within certain limits accept the other components in varying proportions forming a homogenous membrane exhibiting flexibility of composition, structure and function.

The observation that membranes formed under different conditions can be separated by density gradient centrifugation [29] offered an oportunity to attempt a differentiation between the mosaic or homogenous way of membrane growth. It was found that membranes remaining from the lamellar system present in light grown cells after dilution through growth in the dark have an altered structure and composition and are unable to exhibit photooxidation of cytochrome *f*, photophosphorylation or light dependent *p*H rise activity [21]. An indication for the impaired photosynthetic activity can be found also in the relative high fluorescence of the dark grown cells [37]. When centrifuged in a linear density gradient membranes obtained from dark grown cells band at a density of 1.120 g/cm^3. Following onset of illumination, the relative fluorescence of the cells decreases rapidly while the specific photosynthetic activity develops and reaches a peak after about 3 hr of illumination (fig. 11). At the same time, the banding density of the membranes increases gradually to 1.140 g/cm^3 (fig. 12). When the greening is

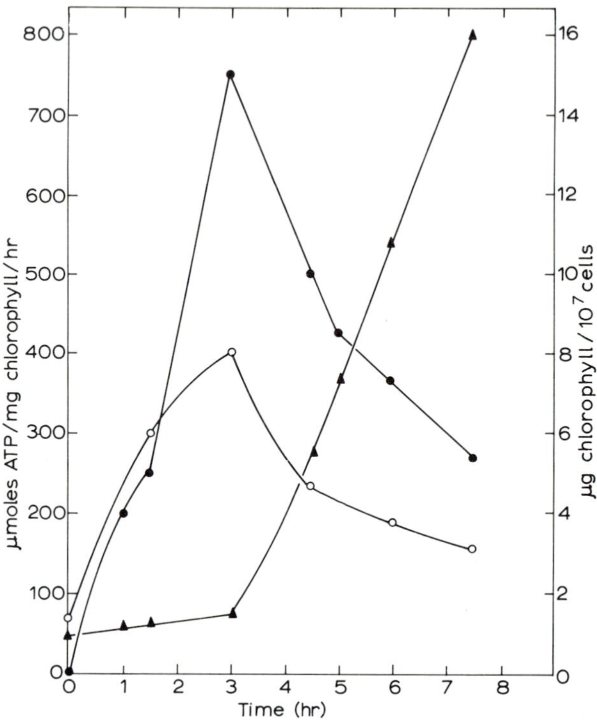

Fig. 11. Development of photophosphorylation activity during greening of *y*-1 cells. ○, activity measured with ferricyanide (photosystem II); ●, activity measured with PMS and ascorbate in the presence of DCMU (photosystem I); ▲, chlorophyll. For experimental details see ref. [38].

carried out in the presence of chloramphenicol no photosynthetic activity develops as already mentioned. At the same time, the relative fluorescence does not decrease and the initial density of 1.120 g/cm^3 does not increase (*cf.* Fig. 14). During the repair of photosynthetic activity after chloramphenicol removal from the incubation medium again fluorescence decreases while the banding density rises gradually to 1.140 g/cm^3 [37]. Thus the changes in the banding density of the membranes reflect changes in their composition and photosynthetic activity. If addition of the missing components during repair of inactive membranes (either at the beginning of the greening process or during the repair of membranes formed in the presence of chloramphenicol) will result in the formation of a mosaic of repaired membrane regions having an increased density one should be able to separate them from the rest of the

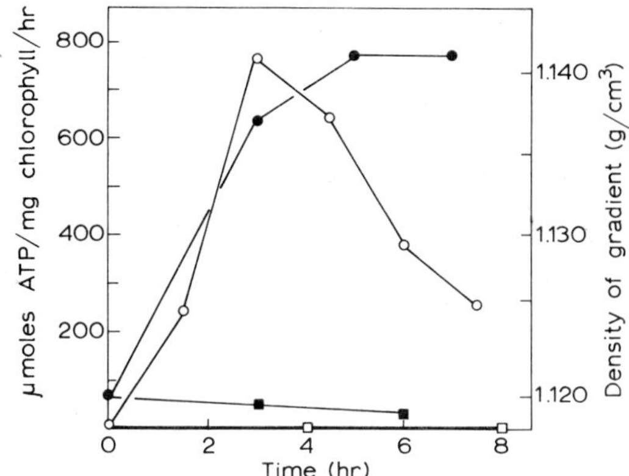

Fig. 12. Changes in the photophosphorylation activity and apparent buoyant density of chloroplast membranes during greening of *y*-1 cells. Circles, cells incubated in the light; squares, cells incubated in the dark; open figures, photophosphorylation; dark figures, apparent buoyant density. For experimental details see ref. [37].

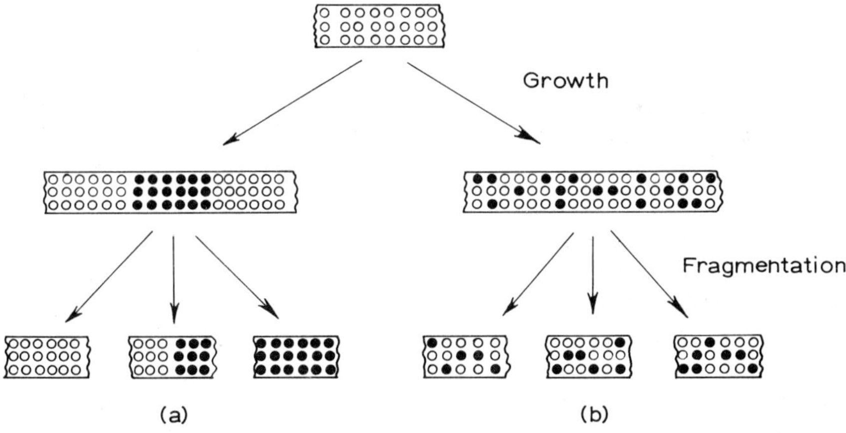

Fig. 13. Schematic representation of the expected distribution of newly added components within fragments of growing membranes. a, growth occurs by addition of new material to distinct acceptor sites resulting in its nonhomogenous distribution within the fragments. b, growth occurs by random homogenous addition of new material resulting in its homogenous distribution within the fragments.

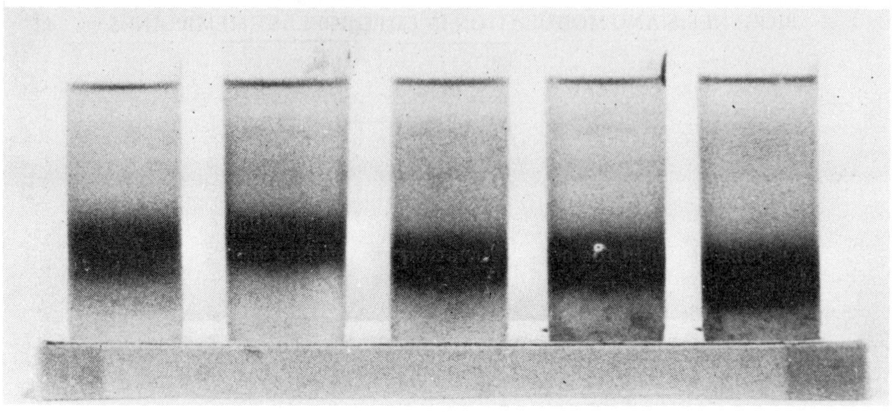

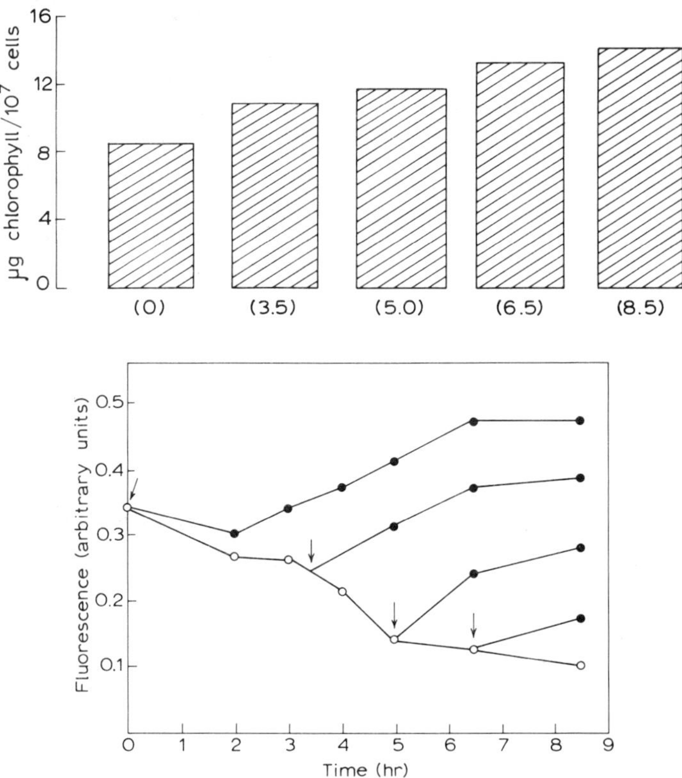

Fig. 14. Effect of chloramphenicol on the changes of fluorescence and apparent buoyant density of chloroplast membranes during greening of *y*-1 cells. Dark grown cells were exposed to light for 8.5 hr. At different times as indicated by the arrows (lower figure) chloramphenicol was added to portions of the cells suspension and incubation continued as shown in the graph. ○, fluorescence of cells incubated in absence of chloramphenicol; ●, fluorescence of cells incubated with cloramphenicol; bars, chlorophyll content of the cells at the end of the experiment (8.5 hr). The numbers in brackets indicate the time of addition of the inhibitor and the corresponding photographs of the gradient tubes (upper figure) show the location of the membranes in the linear density gradient. For experimental details see ref. [37].

membrane after mechanical fragmentation of the membrane and centrifugation on a linear density gradient. On the other hand, if the addition occurs in a homogenous way no matter how small the fragments will be they should have all the same banding density throughout the repair process and move down the gradient as a single band (fig. 13). The results of an experiment devised to distinguish between these two possibilities indicate that vesicular fragments obtained by sonication from membranes formed in the presence of chloramphenicol added at different stages of the greening process do not separate into bands of different density and exhibit a gradual decrease in the apparent buoyant density as the formation of altered membranes continues (fig. 14). A control experiment in which inactive membranes were admixed with repaired membranes before sonication and the mixed fragments centrifuged on the linear density gradient showed that the two types of membranes did separate in two bands of the corresponding densities 1.120 and 1.140 g/cm^3 indicating that if fragments of different density were present one should be able to separate them under the experimental conditions used. After sonication the average diameter of the vesicles was about 700 Å. While significant amounts of the vesicular fragments did have diameters smaller than 700 Å, many of them were as small as 300 Å diameter [37]. Yet they did not separate from the rest of bigger membrane fragments. Since the membrane of a vesicle of a diameter of 300 A might comprise about 10^3 globular molecules of 25 Å diameter one can interpret the results of the above experiments to mean that if inhomogeneity exists during addition of new components to the growing membrane, it should be restricted to membrane areas comprising less than about 10^3 molecules.

6. Relationship between the light harvesting system and the photosynthetic electron transfer chain and energy coupled reactions during membrane growth

The composition and structure of functional photosynthetic membranes permit efficient utilization of light energy harvested by the pigment system for electron flow through an electron carrier chain and the reduction of acceptors and trapping of the energy released by the electron flow for the formation of ATP or ion translocation. The question arises whether the modulation of membrane composition will be reflected in the relative activities of the above partial reactions and whether one could gain information on the flexibility of membrane molecular organization from the modulation of these functions.

Measurements of maximal photophosphorylation and reduction of acceptors rates will give information on the total amount of active membranes and serve as an indicator for the increase in the content of activation proteins during repair of inactive membrane in which these proteins are present in limiting amounts. Measurements of relative fluorescence will serve as an indication for the ability of the system to channel the absorbed light energy into the electron carrier chain. Finally, the intensity of illumination required for the saturation of photosynthetic activity can be used as an indicator for the relationships between the light harvesting system and the electron carrier chain and for the efficiency of the whole system.

It was shown above that the specific photosynthetic activity per chlorophyll content of the inactive membranes remnant after cell division in the dark increases rapidly during the greening process reaching a peak several times higher than that of light grown cells after about 3 hr of illumination. Following additional chlorophyll synthesis as the greening process cotinues, the specific activity of both photosystems decreases to values usually found in membranes obtained from light grown cells (fig. 11).

Measurements of the incident illumination intensity required in order to achieve half of the photosynthetic activity obtained at saturation light intensity ($I_{s½}$) at different times of the greening process shows that the system requires higher illumination intensities before or at the time of peak activity than at the end of the greening process (fig. 15). These results can be interpreted as a change in the relative content of the active light harvesting units as compared with active electron carrier chains. Thus, one can assume that during the initial phase of the greening, missing protein components are added and activate the membrane remnants which are rich in components of the electron transfer chain [19] but poor in chlorophyll. Through this activation process, membranes can be obtained in which the pigment system is a limiting factor. On the other hand, when chlorophyll and L protein continue to be accumulated toward the end of the greening process, some protein components related to the electron transfer and energy coupling activities eventually become limiting and thus saturation by light can be obtained at lower illumination intensities. The question arises what will be the fate of chlorophyll accumulated in membranes formed in the presence of chloramphenicol which have been shown to be relatively rich in L protein but photosynthetically inactive because of lack of the activation proteins. If this chlorophyll is organized as in the normal membranes one would expect during the repair process, as activation proteins are synthesized and integrated in the membrane, a situation in which the light harvesting potential is higher than the electron transfer potential. This will allow saturation to be obtained

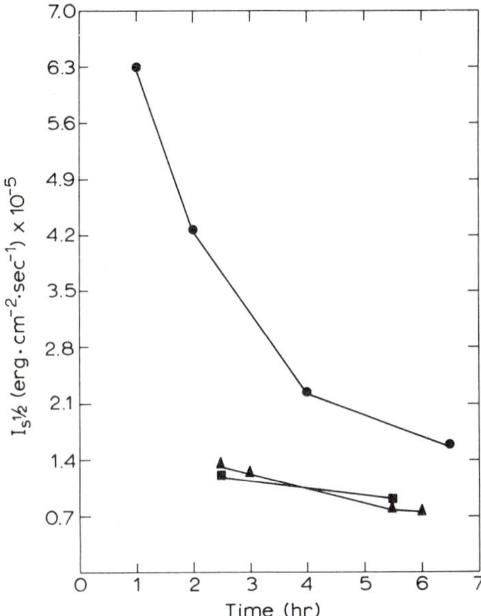

Fig. 15. Changes in the illumination intensity required for the saturation of photophosphorylation activity during greening of *y*-1 cells. ●, activity measured with PMS and ascorbate in the presence of DCMU (photosystem I); ■, ▲, activity measured with ferricyanide or diquat (photosystem II or II + I) respectively. For experimental details see ref. [39].

initially at illumination intensities even lower or at least equal to those necessary for saturation of membranes at the end of the greening process. On the other hand, if this chlorophyll is "misplaced" and cannot be utilized immediately, one might expect to obtain active membranes requiring higher illumination intensity, the requirement being higher as the amount of active chlorophyll is smaller.

Measurements of the illumination intensity required for the saturation of photosynthetic activity of inactive membranes during the repair process disclosed two different situations. If additional chlorophyll was not synthesized during the repair period saturation could not be obtained at all by the illumination intensity available under our experimental conditions throughout the entire length of the experiment (6 hr) (fig. 16). This will indicate that most of the chlorophyll and related proteins incorporated in these membranes are organized in such a way as to be inefficient for channeling the light energy into the electron carrier system. However, if small amounts of chlorophyll

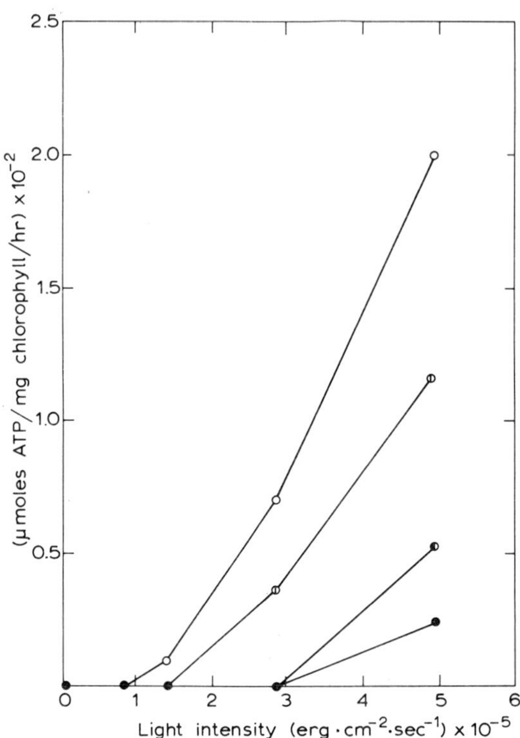

Fig. 16. Dependence of photophosphorylation activity on the illumination intensity during repair of inactive membranes formed in the presence of chloramphenicol. Dark grown cells were incubated in the light with chloramphenicol for 5 hr and then transferred to fresh medium containing 0.5 μg/ml cycloheximide and further incubated in the light. Photophosphorylation activity was measured with PMS, ascorbate and DCMU 1½ hr after transfer, ●, 3 hr ◐, 4½ hr ⦶ and 6 hr ○, after transfer of the cells to the chloramphenicol free medium. During the incubation with cycloheximide no additional chlorophyll was synthesized. For experimental details see ref. [39].

were synthesized during the repair process saturation could be obtained and the $I_{s½}$ value decreased during the repair process (see fig. 17). This would indicate that the inactive membranes can accept and allow chlorophyll molecules to be organized together with the activation protein in an efficient system.

The reactivation of photosynthetic membranes involves not only addition of missing components but also a reorganization of the membranes which might be at least partially reversible. This is suggested by results of an experi-

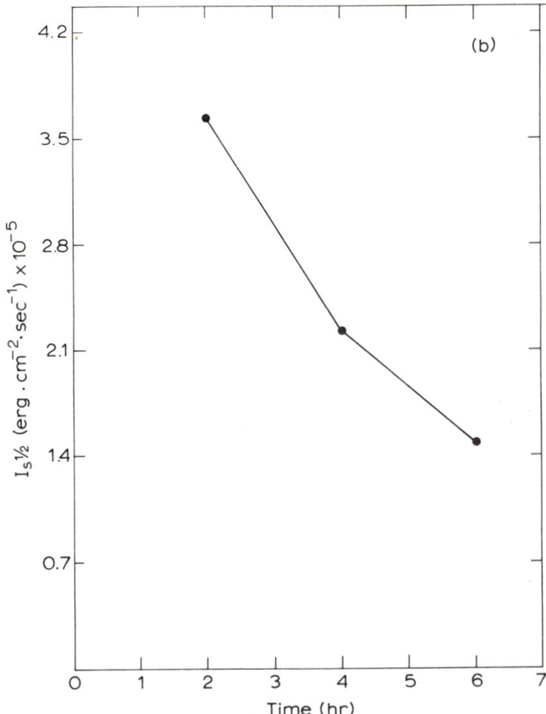

Fig. 17. Decrease in the $I_{s}½$ during repair of inactive membranes formed in the presence of chloramphenicol. Same experimental conditions as in fig. 16. The cells were transferred to a medium containing 0.5 μg cycloheximide/ml after greening in the presence of chloramphenicol (150 μg/ml). During the repair process the chlorophyll content of the cells increased by 3 μg.

ment in which the persistance of membrane activity developed during the greening was assessed by transferring the cells to the dark at different times from onset of illumination and their activity measured again after several hours of dark incubation.

It was found that in the cells transferred to the dark during the early phase of the greening the activity decayed rapidly while it remained rather stable if the cells were allowed to green for longer periods of time (fig. 18). Since chlorophyll and inactive proteins are not degraded in the dark one can explain these findings as an indication for a reversible change in the structural organisation of the membranes. These results indicate that during the development of membranes the light harvesting system as well as the electron

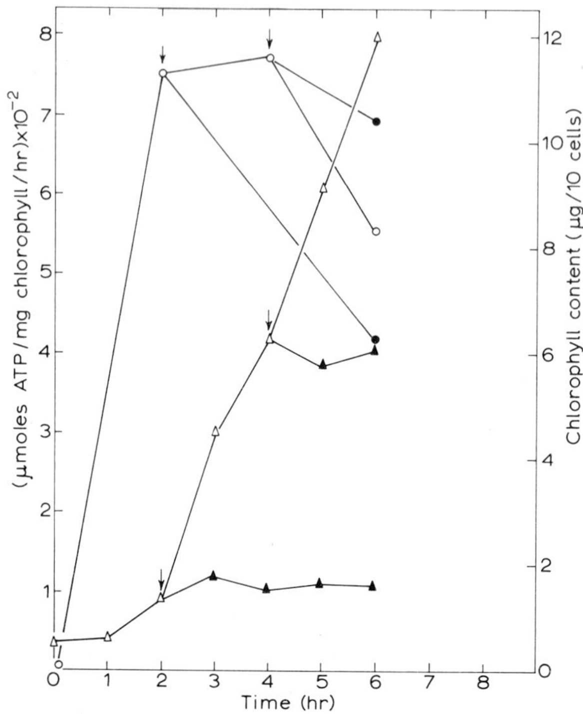

Fig. 18. Decay of photophosphorylation activity following transfer of greening cells to the dark. Photophosphorylation was measured with PMS and ascorbate in the presence of DCMU. Circles, photophosphorylation activity; triangles, chlorophyll content; open figures, before transfer to the dark; dark figures, after transfer to the dark.

carrier and energy trapping system can become limiting factors of the photosynthetic activity. The relative ratio of these systems can be changed during the repair of inactive membranes obtained by prolonged growth of the *y*-1 cells in the dark or during greening of dark grown cells in the presence of chloramphenicol. The repair or activation of the inactive membranes include addition of the missing components and reorganisation of the membrane components which in itself appear to be a slow and partially reversible process. The different possible stages in the formation of the photosynthetic membranes in *Chlamydomonas reinhardi y*-1 are shown schematically in fig. 19.

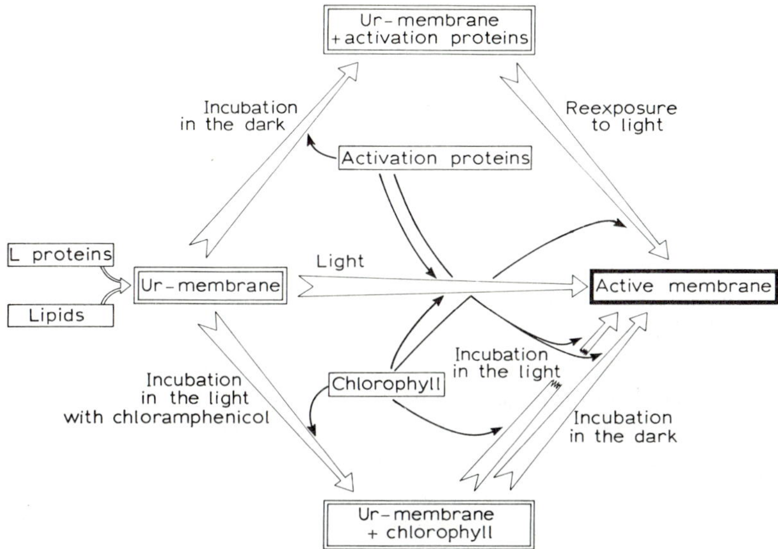

Fig. 19. Schematic representation of the modulation of membranes composition and activity during their formation under different experimental conditions. The term "Urmembrane" [11] designates the basic membrane structure composed of L protein of cytoplasmic origin and lipids which might serve as an obligatory acceptor frame for the addition of chlorophyll and activation proteins of chloroplast origin.

7. Concluding remarks

Based on all the above data, it appears that modulation of membrane composition results also in a modulation of their activity. Modulation of membranes composition and function as shown in this work is not an isolated case specific for the *y*-1 mutant of *Chlamydomonas reinhardi.* Recently Schor et al. [31] have reported modulation of the photosynthetic electron carrier activity by alternative exposure to the light or dark also in synchronized cultures of the wild type cells. The fact that during formation of membranes several transient states can be obtained some of which seem to be reversible, reflect a certain degree of flexibility built in the intimate structure of the membrane. Such flexibility will be compatible with more than one of the models proposed for the structure of biological membranes. It appears that accommodation with the subunit model will encounter the most difficulties while a model comprising distribution of protein or lipoprotein complexes embedded within a frame of lipid bilayer will encounter the least. In addition, the absolute dependence of the membrane growth on the presence of L protein suggest that a lipoprotein complex might serve as an organizing frame which will be recognized by the components to be added to it thus forming what Siekevitz et al. have called an "Urmembrane" [11]. The existence of such an acceptor surface might also provide an answer to the problem of target-recognition by the molecular species synthesized in different cellular compartments so as to enable them to be incorporated in the membrane for which they have been specifically synthesized. The results presented here raise more questions than they have actually attempted to answer. One of the basic questions is what is the mechanism by which the synthesis of the different membrane components is regulated and balanced and what is the role of light in this process. Since a detailed discussion of this topic is beyond the scope and available time of this lecture, I will mention only briefly the main points which seem to emerge from results of experiments devised to answer these questions [34, 40]. The synthesis of L protein appears to be coded by a rapidly turning over RNA whose synthesis is coupled to a continuous conversion of protochlorophyll to chlorophyll and hence the requirement for light. On the other hand, the presence of L protein seems to exert a direct control on the enzymatic system responsible for the synthesis of protochlorophyll which in itself is also turning over but can be formed in the dark [34]. However, much more work will be required in order to elucidate the exact mechanism of these control pathways.

References

[1] S. Razin, Z. Neeman and I. Ohad, Biochim. Biophys. Acta 193 (1969) 277.
[2] I. Kahane and S. Razin, Biochim. Biophys. Acta 183 (1969) 79.
[3] D. Morrison and H.J. Morowitz, J. Mol. Biol. 49 (1970) 441.
[4] L. Mindich, J. Mol. Biol. 49 (1970) 415.
[5] L. Mindich, J. Mol. Biol. 49 (1970) 433.
[6] A. Gorchein, A. Neuberger and G.H. Tait, Proc. Roy. Soc. B 170 (1968) 311.
[7] A. Gorghein, A. Neuberger and G.H. Tait, Proc. Roy. Soc. B. 171 (1968) 111.
[8] J. Oelze and G. Drews, Biochim. Biophys. Acta 203 (1970) 189.
[9] J. Oelze, J. Schroeder and G. Drews, J. Bacteriol. 101 (1970) 669.
[10] L. Warren and M.C. Glick, J. Cell. Biol. 37 (1968) 729.
[11] P. Siekevitz, G.E. Palade, G. Dallner, I. Ohad and T. Omura, in: Organizational Biosynthesis, eds. H.J. Vogel, J.O. Lampen and V. Bryson (Academic Press, New York, 1967) p. 331.
[12] T. Omura, P. Siekevitz and G.E. Palade, J. Biol. Chem. 242 (1967) 2389.
[13] G. Dallner, A. Bergstrand and R. Nilsson, J. Cell Biol. 38 (1968) 257.

[14] H. Glauman and G. Dallner, J. Cell Biol. 47 (1970) 34.
[15] J. Jamieson and G.E. Palade, J. Cell Biol. 48 (1971) 503.
[16] A. Amsterdam, I. Ohad, Z. Selinger and M. Schramm, J. Cell Biol. 50 (1971) 187.
[17] G. Schatz, in: Membrane of Mitochondria and Chloroplasts, ed. E. Racker (Van Nostrand Co., 1970) p. 251.
[18] D.O. Woodward, D.L. Edwards and R.B. Flavell, in: Symposia of the Society for Experimental Biology. Control of Organelle Development, ed. P.L. Miller (Cambridge University Press, 1969) p. 55.
[19] I. Ohad, P. Siekevitz and G.E. Palade, J. Cell Biol. 35 (1967) 522.
[20] I. Ohad, P. Siekevitz and G.E. Palade, J. Cell Biol. 35 (1967) 553.
[21] S. Schuldiner and I. Ohad, Biochim. Biophys. Acta 180 (1969) 165.
[22] I. Goldberg and I. Ohad, J. Cell Biol. 44 (1970) 563.
[23] I. Goldberg and I. Ohad, J. Cell Biol. 44 (1970) 572.
[24] G. Eytan and I. Ohad, J. Biol. Chem. 245 (1970) 4297.
[25] S.J. Surzycki, V.W. Goodenough, R.P. Levine and J.J. Armstrong, in: Symposia of the Society for Experimental Biology Control of Organelle Development, ed. P.L. Miller (Cambridge University Press, 1969) p. 13.
[26] J. Schiff, in: Symposia of the Society for Experimental Biology. Control of Organelle Development, ed. P.L. Miller (Cambridge University Press, 1969) p. 277.
[27] R.M. Smillie and S. Scott, in: Progress in Molecular and Subcellular Biology, ed. F.E. Hahn (Springer-Verlag, Berlin, 1969) Vol. 1 p. 136.
[28] G. Lenaz, G.P. Littarru and A. Casteli, FEBS Letters 2 (1969) 198.
[29] J.K. Hoober, J. Biol. Chem. 245 (1970) 4327.
[30] J.K. Hoober, P. Siekevitz and G.E. Palade, J. Biol. Chem. 244 (1969) 2621.
[31] S. Schor and P. Siekevitz, Proc. Nat. Acad. Sci. (U.S.A.) 66 (1970) 74.
[32] B. de Petrocellis, P. Siekevitz and G.E. Palade, J. Cell Biol. 44 (1970) 618.
[33] N.K. Boardman, in: The Chlorophylls, eds. L.P. Vernon and G.R. Seely (Academic Press, London, 1966) p. 437.
[34] I. Goldberg, Ph.D. Thesis, Hebrew University, Jerusalem, 1971.
[35] H.K. Lichtenthaler and R.B. Park, Nature (London) 198 (1963) 1070.
[36] J.K. Hoober and G. Blobel, J. Mol. Biol. 41 (1969) 121.
[37] G. Eytan and I. Ohad, J. Biol. Chem. 247 (1972) 112.
[38] D. Wallach, S. Bar-Nun and I. Ohad, Biochim. Biophys. Acta 267 (1972) 125.
[39] S. Bar-Nun, D. Wallach and I. Ohad, Biochim. Biophys. Acta 267 (1972) 138.
[40] G. Eytan and I. Ohad, J. Biol. Chem. 247 (1972) 122.

EXCHANGE OF PHOSPHATIDYLCHOLINE BETWEEN MITOCHONDRIA AND MICROSOMES: STIMULATION BY A SPECIFIC PROTEIN

K.W.A. WIRTZ and H.H. KAMP
Laboratory of Biochemistry, State University of Utrecht
Utrecht, The Netherlands

It is generally accepted that the endoplasmic reticulum is the principle site of phosphatidylcholine biosynthesis in the rat liver [1–3]. One of the key enzymes in its biosynthesis, CDP-choline: 1,2-diglyceride cholinephosphotransferase (E.C. 2.7.8.2) has been shown to be located in this subcellular organelle [4,5]. This implies that other subcellular organelles, such as the mitochondria, depend on the endoplasmic reticulum for their supply of phosphatidylcholine. Rather surprisingly, it is observed, however, that this phospholipid among other phospholipids becomes very rapidly labeled in liver mitochondria after injection of ^{32}P-phosphate [6, 7]. The investigations of Jungalwala and Dawson [8] as well as our own studies [9] have provided evidence that this rapid labeling *in vivo* is due to an exchange of phosphatidylcholine between the endoplasmic reticulum and the mitochondria.

Exchange of phospholipids between mitochondria and microsomes has been observed also *in vitro* [1, 10–17]. Incubating either ^{32}P-labeled mitochondria with unlabeled microsomes (table 1) or ^{32}-P-labeled microsomes with unlabeled mitochondria (table 2) resulted in a redistribution of phospholipid-^{32}P radioactivity among the subcellular particles. The drop in the total phospholipid specific activity of the originally labeled particles is a measure for the exchange of phospholipid. For example, in table 1 a drop in specific activty of (923 – 549 =) 374 means that 374/923 × 100% = 40% of the mitochondrial phospholipid has exchanged in 90 min at 37°. Examination of the exchange of the individual phospholipid classes showed that the exchange of phosphatidylcholine contributed to this drop in specific activity particularly.

Table 1
^{32}P-Labeled mitochondria incubated with unlabeled microsomes
A mixture containing 168 μg of mitochondrial PLP (= phospholipid phosphorus), 389 μg of microsomal PLP, and 78 μg of 105 000 × g supernatant PLP was incubated for 90 min in a total volume of 10 ml 0.25 M sucrose-1 mM EDTA (*p*H 7.4). For experimental details, see ref. [10].

Incubation temperature	Specific activity (cpm/μg PLP)		Distribution of phospholipid-^{32}P radioactivity (% nonincubated mitochondria)	
	Mitochondria	Microsomes	Mitochondria	Microsomes
	923*			
0°	842	45	91	11
37°	549	139	60	35

* Specific activity of mitochondrial PLP before incubation.

To make sure that the entire phosphatidylcholine molecule was involved in the exchange, microsomes containing doubly labeled phosphatidylcholine (^{3}H-label in the choline moiety, ^{14}C-label in the fatty acid moiety at the 2-position) were incubated with unlabeled mitochondria (table 3). The identical specific activity ratios of phosphatidylcholine from microsomes before and after incubation (0.42 and 0.40, respectively) and from mitochondria after incubation (0.44) indicated that the entire molecule was transferred in the course of the exchange process. The extent of phosphatidylcholine ex-

Table 2
^{32}P-Labeled microsomes incubated with unlabeled mitochondria
A mixture containing 205 μg of mitochondrial PLP, 300 μg of microsomal PLP, and 69 μg of 105 000 × g supernatant PLP was incubated for 90 min in a total volume of 10 ml 0.25 M sucrose-1 mM EDTA (*p*H 7.4). For experimental details, see ref. [10].

Incubation temperature	Specific activity (cpm/μg PLP)		Distribution of phospholipid-^{32}P radioactivity (% nonincubated microsomes)	
	Mitochondria	Microsomes	Mitochondria	Microsomes
		1030*		
0°	81	863	5	84
37°	304	737	20	72

* Specific activity of microsomal PLP before incubation.

Table 3
Exchange of phosphatidylcholine between ^{14}C, ^{3}H-phosphatidylcholine-labeled microsomes and unlabeled mitochondria

The doubly labeled microsomes were obtained by injecting intraperitoneally 80 μC ^{3}H-choline into a starved male rat followed after 30 min by an intraportal injection of 20 μC ^{14}C-linoleate complexed to albumin. The liver was excised 3 min after the last injection and the microsomes were isolated. A mixture containing 18.9 mg mitochondrial protein, 4.7 mg microsomal protein and 15.6 mg *p*H 5.1 supernatant protein was incubated for 30 min at 37° in a total volume of 6 ml 0.25 M sucrose – 1 mM EDTA (*p*H 7.8). At the end of incubation mitochondria were sedimented at 15 000 × g for 10 min and washed twice. The phospholipids were extracted from the mitochondria and the mitochondrial supernatant (microsomes) according to Bligh and Dyer [18]. Phosphatidylcholine was isolated and the specific radioactivity determined as described previously [11].

Subcellular fraction	Specific activity (cpm/μmol PC*)		Ratio of specific activity
	^{14}C	^{3}H	
Microsomes before incubation	1685	4045	0.42
Microsomes after incubation	1035	2580	0.40
Mitochondria after incubation	590	1325	0.44

* PC = phosphatidylcholine

change is given by the decrease in microsomal phosphatidylcholine specific activities.

Under the conditions, where we measured exchange of phospholipids, a 105 000 × *g* supernatant from rat liver was present in the incubation medium. Omitting this supernatant from the incubation medium resulted in a greatly reduced exchange. In fact, we observed that the exchange of phospholipids between ^{32}P-labeled mitochondria and unlabeled microsomes increased nearly proportionally with the amount of supernatant (fig. 1). This exchange-stimulating activity was also apparent in experiments where we measured exchange between mitochondria and lipoproteins [11] and between mitochondria and liposomes prepared from egg-yolk phosphatidylcholine (unpublished observations). Apart from this, Zilversmit [19] has recently shown that the *p*H 5.1 fraction of the 105 000 × *g* supernatant (*p*H 5.1 supernatant) stimulated exchange of phospholipids between chylomicrons and liposomes and between phosphatidylcholine-stabilized oil emulsions and liposomes. Whatever the mode of action of the active factor in the 105 000 × *g* supernatant is, natural membranes do not appear to be a pre-

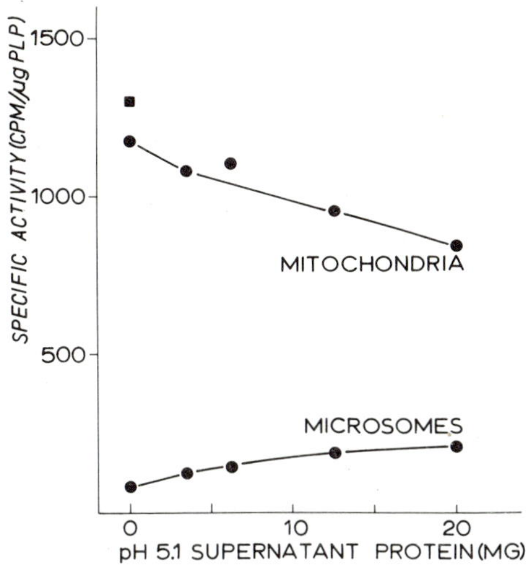

Fig. 1. The effect of various amounts of *p*H 5.1 supernatant protein on the exchange of phospholipids between ^{32}P-labeled mitochondria and unlabeled microsomes as reflected by the specific activities of mitochondrial and microsomal PLP. A mixture containing 52 µg of mitochondrial PLP, 96 µg of microsomal PLP, and *p*H 5.1 supernatant protein was incubated for 40 min at 37° in a total volume of 4 ml 0.25 M sucrose - 1 mM EDTA *p*H 7.8). ■, specific activity of mitochondrial PLP before incubation. For experimental details see ref. [11].

requisite for exchange. A direct interaction of the factor with the exchangeable phospholipid in the interface seems likely.

Evidence for the active factor being a protein came from the following observations. If one pre-incubated the *p*H 5.1 supernatant at too high a temperature the stimulating activity diminished drastically. Heating for 10 min. at 60° decreased its effect on phospholipid exchange by 50%. Furthermore, digestion of the *p*H 5.1 supernatant protein with trypsin markedly reduced the ability of the supernatant fraction to promote exchange. We felt that these observations warranted the attempt to isolate the exchange-stimulating factor denoted as phospholipid-exchange protein (PLEP) from the cytosol. As a source for PLEP we used beef liver. A 30% homogenate was made in 0.25 M sucrose with a Waring Blendor. The homogenate was centrifuged at 600 X *g* and 15 000 X *g* to remove nuclei and mitochondria, respectively. The *p*H of the mitochondrial supernatant was adjusted to 5.1 with 3N HCl and the ensuing precipitate was sedimented by centrifugation at 15 000 X *g*. The virtually phospholipid-free supernatant (the *p*H 5.1 supernatant) was used as the starting material in the purification of PLEP.

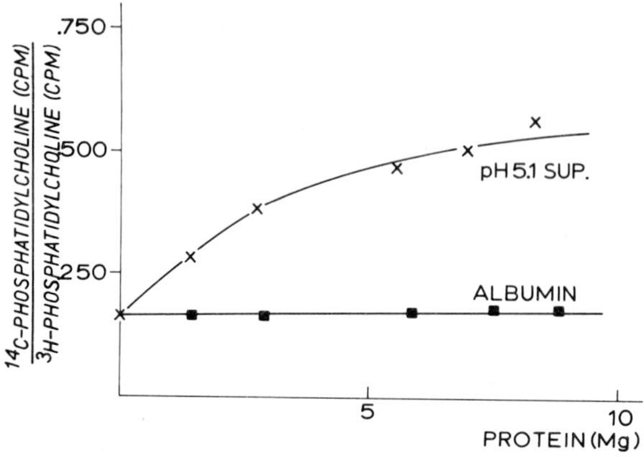

Fig. 2. The effect of various amounts of either *p*H 5.1 supernatant protein or albumin on the exchange of phosphatidylcholine between ^{14}C-phosphatidylcholine-labeled mitochondria and unlabeled microsomes. The exchange is reflected by the ^{14}C-phosphatidylcholine radioactivity transferred from mitochondria to microsomes i.e., mitochondrial supernatant. ^{14}C-labeled mitochondria were isolated from a rat liver 40 min after an intraperitoneal injection of 20μC ^{14}C-choline into a rat. A mixture containing 11 mg mitochondrial protein, 2.5 mg microsomal protein and *p*H 5.1 supernatant protein (albumin) was incubated for 20 min at 37° in a total volume of 3 ml 0.25 M sucrose - 1 mM EDTA (*p*H 7.8). At the end of incubation the mitochndria were sedimented at 10 000 × *g* for 10 min. The ^{14}C-phosphatidylcholine-labeled mitochondrial supernatant was mixed with a fixed volume of a ^{3}H-phosphatidylcholine-labeled microsomal suspension. The phospholipids of the mixture were extacted according to Bligh and Dyer [18]. The lipid extract was dissolved in 0.3 ml methanol and mixed with 16 ml of toluene containing 0.5% PPO and 0.03% POPOP. Radioactivity was measured with a Packard-Tricarb liquid scintillation spectrometer and the ^{14}C/^{3}H ratio of phosphatidylcholine was calculated. The ratio reflects the tranfer of ^{14}C-phosphatidylcholine from mitochondria to microsomes. This transfer is normalized by adding ^{3}H-phosphatidylcholine as an internal standard at the end of incubation. The assay used in the purification of phosphatidylcholine-exchange protein (PLEP) as summarized in table 4 was based on the increase of the ^{14}C/^{3}H ratio with increasing amounts of phospholipid-exchange activity. By reading the ratio from the ^{14}C/^{3}H ratio – *p*H 5.1 supernatant standard curve, the purification of PLEP in the various steps (table 4) could be calculated.

To determine the exchange activity of the protein fractions in the various steps of the purification, we measured the transfer of ^{14}C-phosphatidylcholine from the ^{14}C-phosphatidylcholine-labeled mitochondria to the microsomes as a function of the amount of protein in the incubation medium. For details on the assay, see the legend to fig. 2. The ^{14}C/^{3}H-ratio, an index for the exchange activity, increases with increasing amounts of *p*H 5.1 supernatant protein. It follows from fig. 2 that an unrelated protein such as bovine serum albumin has absolutely no effect on the transfer of ^{14}C-phosphatidylcholine.

The actual purification of PLEP was started by making use of an anion-exchanger, DEAE-Sephadex A-50. The *p*H 5.1 supernatant in 0.25 M sucrose – 0.05 M Tris (*p*H 8.4) was pumped directly on the anion exchange column which was equilibrated with the eluting buffer (0.1 M Tris-0.01 M β-mercaptoethanol, *p*H 8.4). The column was rinsed with this buffer until no more protein appeared in the eluate. PLEP was eluted from the column by increasing the ionic strength of the buffer (0.05 M sodiumchloride). (step 2).

PLEP was concentrated by saturating the eluate with solid ammonium sulfate (60 g/100 ml). The precipitated protein was dissolved in a buffer containing 0.01 M sodium phosphate - 0.01 M β-mercaptoethanol (*p*H 7.8) and dialysed overnight against this buffer (Step 3). Next, aliquots of the dialysate were applied to a Sephadex-G75 column. The protein fractions were eluted from this column with the phosphate-mercaptoethanol buffer and tested for phosphatidylcholine-exchange activity (fig. 3). The eluates in tubes No 44 – 50 were pooled (Step 4). The pooled eluate was again saturated with solid ammoniumsulfate (60 g/100ml). The precipitated protein was dissolved in the phosphate-mercaptoethanol buffer (Step 5). After dialysis against this

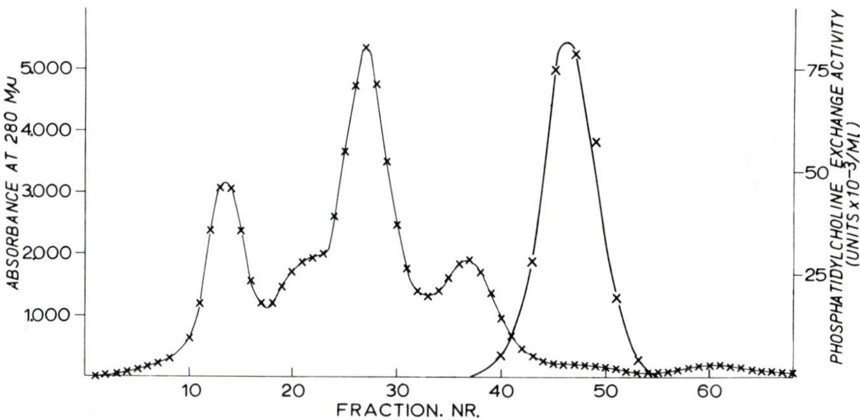

Fig. 3. Fractionation of beef liver PLEP on Sephadex-G75. 30 ml (27 mg protein/ml) of the active material of Step 3 were applied to a Sephadex-G75 column (72×5 cm). Protein was eluted with 0.01 M phosphate – 0.01M β-mercaptoethanol (*p*H 7.8) at a flow rate of 24 ml/hr. 12 ml fractions were collected. Absorbance was measured at 280 mμ. ×-×-×, units × 10^{-3} of phosphatidylchline-exchange activity per ml of eluent. A unit of activity is defined as a mμmole of phosphatidylcholine transferred from mitochondria to microsomes per minute.

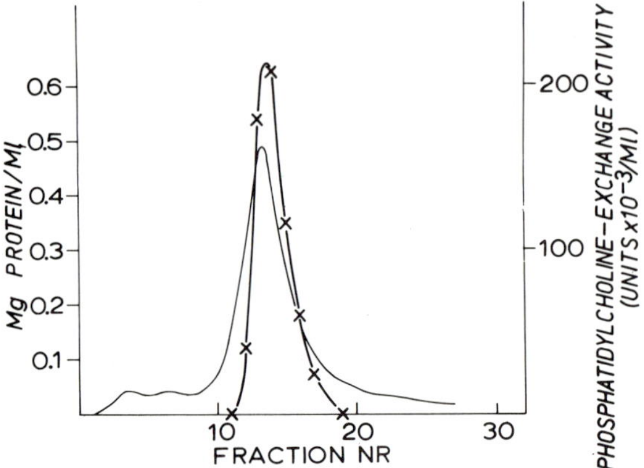

Fig. 4. Fracionation of beef liver PLEP on Sephadex-G50. 5.2 mg protein of Step 5 in 1 ml were applied to a Sephadex-G50 column (35 × 2.3 cm). Protein was eluted with the phosphate-mercaptoethanol buffer at a flowrate of 3.6 ml/hr 1.8 ml fractions were collected. Protein was determined according to Lowry [20]. X-X-X, units × 10^{-3} of phosphatidylcholine-exchange activity per ml of eluate.

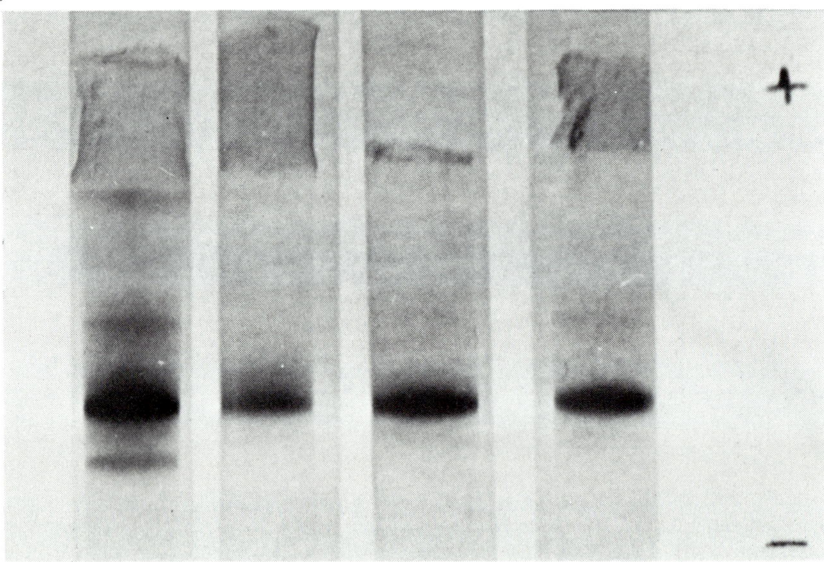

Fig. 5. Disc-gel electrophoresis of purified PLEP after fractionation on a Sephadex-G50 column. Protein of Step 5 (left hand gel) and of Step 6 (tubes 14,15 and 16, see fig. 4) were subjected to polyacrylamide electrophoresis as described by Ornstein and Davis [21].

Table 4
Summary of purification procedures for beef liver phosphatidylcholine-exchange protein. For assay system, see legend to fig. 2.

Purification step	Volume (ml)	Protein (mg)	Specific activity	Recovery (%)	Purification factor
1. *p*H adjustment	1000	27,750	2	100	–
2. DEAE-Sephadex-A50	1220	4,188	9	70	4.5
3. $(NH_4)_2SO_4$ fractionation	125	3,313	13	76	6.5
4. Sephadex-G75	375	56	286	29	143
5. $(NH_4)_2SO_4$ fractionation	11	37	364	25	182
6. Sephadex-G50	54	14	547	16	273

buffer aliquots were applied to a Sephadex-G50 column and PLEP was eluted (fig. 4). Disc gel electrophoresis according to Ornstein and Davis [21] showed that one component was present in tubes 14 – 16 which contained most of the PLEP activity (fig. 5).

Table 4, which summarizes the purification of PLEP indicates that the protein was 273–fold purified and that 16% of the exchange activity was recovered. In addition, PLEP had a final specific activity of 547, i.e. 547 mμmoles of phosphatidylcholine will be transferred from the mitochondria to the microsomes per minute per mg of protein at 37°. Preliminary to this purification a partial purification of PLEP from beef heart had been achieved [22].

In the exchange of phospholipids between mitochondria and microsomes it was observed that the *p*H 5.1 supernatant stimulated both the exchange of phosphatidylcholine and the exchange of phosphatidylinositol and of phosphatidylethanolamine. Since in the assay of phospholipid-exchange activity only the transfer of ^{14}C-phosphatidylcholine was determined, it was of interest to know if purified PLEP could stimulate the exchange of other phospholipids than phosphatidylcholine. Therefore, PLEP was incubated with ^{32}P-labeled microsomes and unlabeled mitochondria and its stimulating activity on the exchange of the different phospholipid classes was compared with that of the *p*H 5.1 supernatant (table 5). With the *p*H 5.1. supernatant 3% of the ^{32}P-phosphatidylethanolamine-, 10% of the ^{32}-phosphatidylinositol-, and 13% of the ^{32}P-phosphatidylcholine radioactivity originally pres-

Table 5

Specificity of phosphatidylcholine-exchange protein

A mixture containing mitochondria (12 mg protein), ^{32}P-labeled microsomes (2.8 mg protein) and either *p*H 5.1 supernatant (3.7 mg protein) or PLEP (0.028 mg protein) was incubated for 40 min at 37° in a total volume of 3 ml 0.25 M sucrose – 1 mM EDTA (*p*H 7.8). At the end of incubation mitochondria were sedimented at 10 000 × g for 10 min and washed twice. The phospholipids were extracted from the mitochondria and the unincubated ^{32}P-labeled microsomes according to Folch [23]. Lipid phosphorus and ^{32}P-radioactivity in the individual phospholipids were determined as described previously [11].

Subcellular fraction	Exchange protein in incubation medium	Distribution of ^{32}P-radioactivity (cpm)		
		Phosphatidyl-ethanolamine	Phosphatidyl-inositol	Phosphatidyl-choline
Microsomes (2.8 mg prot.)	–	15,940 (22.6%)	1,690 (2.4%)	52,870 (75.0%)
Mitochondria (12 mg prot.)	*p*H 5.1 supernatant	454 (5.9%)	176 (2.3%)	7,078 (91.8%)
Mitochondria (12 mg prot.)	PLEP	14 (0.2%)	18 (0.2%)	7,236 (99.6%)

ent in the microsomes were transferred to the mitochondria. 91.8% of the total transferred radioactivity was presented in the phosphatidylcholine. PLEP, however, stimulated exclusively the transfer of phosphatidylcholine (99.6% of the total transferred radioactivity). Further work is now in progress to answer the question if the specificity of PLEP apparent in the mitochondrion-microsome incubation mixture is valid also for model membrane systems. In addition, the mechanism by which PLEP stimulates exchange, is under investigation.

Acknowledgements

The author (K.W.A. Wirtz) is grateful to Dr. D.B. Zilversmit (Cornell University, Ithaca, New York) under whose supervision the exchange experiments have been started. The authors wish to thank Professor Dr. L.L.M. van Deenen for his interest and advice in the purification of PLEP.

References

[1] W.C. McMurray and R.M.C. Dawson, Biochem. J. 112 (1969) 91.
[2] F.B. Jungalwala and R.M.C. Dawson, European J. Biochem. 12 (1970) 399
[3] M.L. Williams and F.L. Bygrave, European J. Biochem. 17 (1970) 32.
[4] G.F. Wilgram and E.P. Kennedy, J. Biol. Chem. 238 (1963) 2615.
[5] M.G. Sarzala, L.G.M. van Golde, B. de Kruyff and L.L.M. van Deenen, Biochim. Biophys. Acta 202 (1970) 106.
[6] G.L. Ada, Biochem. J. 45 (1949) 422.
[7] M.I. Gurr, C. Prottey and J.N. Hawthorne, Biochim. Biophys. Acta 106 (1965) 357.
[8] F.B. Jungalwala and R.M.C. Dawson, Biochem. J. 117 (1970) 481.
[9] K.W.A. Wirtz and D.B. Zilversmit, Biochim. Biophys. Acta 187 (1969) 468.
[10] K.W.A. Wirtz and D.B. Zilversmit, J. Biol. Chem. 243 (1968) 3596.
[11] K.W.A. Wirtz and D.B. Zilversmit, Biochim. Biophys. Acta 193 (1969) 105.
[12] M. Akiyama and T. Sakagami, Biochim. Biophys. Acta 187 (1969) 105..
[13] D.S. Beattie, J. Membrane Biol. 1 (1969) 383.
[14] A.B. Abdelkader and P. Mazliak, European J. Biochem. 15 (1970) 250.
[15] K.W.A. Wirtz, L.M.G. van Golde and L.L.M. van Deenen, Biochim. Biophys. Acta 218 (1970) 176.
[16] M.C. Blok, K.W.A. Wirtz and G.L. Scherphof, Biochim. Biophys. Acta 233 (1971) 61.
[17] M.T. Sauner and M. Levy, J. Lipid Res. 12 (1971) 71.
[18] E.G. Bligh and W.J. Dyer, Can. J. Biochem. Physiol. 37 (1959) 911.
[19] D.B. Zilversmit, J. Biol. Chem. 246 (1971) 2645.
[20] O.H. Lowry, N.J. Rosebrough, A.L. Farr and R.J. Randall, J. Biol. Chem. 193 (1951) 265.
[21] L. Ornstein and B.J. Davis, Ann. N.Y. Acad Sci. 121 (1964) 305.
[22] K.W.A. Wirtz and D.B. Zilversmit, FEBS Letters 7 (1970) 44.
[23] J. Folch, M. Lees and G.H. Sloane Stanley, J. Biol. Chem. 226 (1957) 497.

SYNTHESIS AND INTERACTIONS OF CYTOPLASMIC MEMBRANES IN THE ACINAR CELLS OF THE GUINEA PIG PANCREAS

Jacopo MELDOLESI and Dario COVA
Istituto di Farmacologia, Università di Milano
e Centro di Citofarmacologia del C.N.R., Milano, Italy

Biochemical and morphological evidence clearly indicates that in pancreatic acinar cells [1–4], as well as in other protein secreting cells [5–7], secretory proteins are always segregated by membranes. Thus, following their synthesis in the attached ribosomes, such proteins are released within the cisternae of the rough endoplasmic reticulum (RER) and then transported through a series of functionally interconnected membrane-bound compartments which involve, in sequence, the RER, the Golgi complex and finally the secretory granules. The content of the latter is ultimately discharged in the extracellular space by exocytosis.

While it is clear that the different membranes specifically involved play a central role in these processes, unequivocal information concerning their synthesis and interactions is still lacking. Based on indirect evidence it is generally assumed that these membranes are synthetized in the RER. Some of them would then separate from RER and transform into Golgi membranes, with a balance maintained by a simultaneous transformation of other Golgi membranes into secretory granule membranes [8–11].

However, other mechanisms have also been envisaged. According to Jamieson and Palade [3] in pancreatic acinar cells the transport of secretory proteins from the RER to the Golgi complex and immature zymogen granules (ZG) could be effected through intermittent tubular connections or, alternatively, by discrete vesicles that shuttle between the two compartments. In this case no membrane transformation would be necessary. We decided to look into this problem by investigating whether the membranes bounding the different cell compartments involved in secretion are synthetized independently from one another or rather derive from pre-existing membranes. In our work we followed basically the approach used previously by Jamieson and Palade for studying the kinetics of intracellular transport of secretory proteins [3,4]. Thus the proteins of tissue slices prepared from the pancreas of starved guinea pigs were pulse labeled *in vitro* for a short time (5 min) with ^{14}C-L-leucine and then incubated in chase medium for various times (20 to 150 min). It is known that in such a system a true pulse labeling of proteins can be obtained. Therefore changes in protein specific radioactivity of cell fractions during post-pulse incubation are not due continued synthesis but rather to intracellular transport [3]. *

From the pioneer work of Palade and his associates [3,12] it is known that most of the radioactive leucine incorporated into microsome and ZG proteins (*in vivo* as well as in tissue pancreas slices incubated *in vitro*) is in fact accounted for by non-membrane proteins, particularly by secretory enzymes. Since in our work we were interested in the synthesis of microsome and ZG membranes as well as in the kinetics of their interactions, it was necessary to develop procedures for the isolation of membrane fractions from rough microsomes (derived from fragmented and resealed cisternae of the RER), smooth microsomes (primarily contributed by elements of the Golgi complex) and from ZG. The procedures that we developed involve only mild treatments. Rough and smooth microsomes were resuspended in 0.5 mM puromycin at high ionic strength, *p*H 6, and incubated at 4°C for 4 hr. Partially purified membranes were recovered by density gradient centrifugation and then washed with 0.2 M $NaHCO_3$, *p*H 7.8. For ZG the resuspension in 0.2 M $NaHCO_3$, *p*H 7.8 sufficient for extracting most of the secretory proteins. Membrane ghosts were then purified by density gradient centrifugation. Details on the procedures are given in [13] and [14].

The purified membrane fractions either are completely free or contain only traces of non-membrane proteins present in microsomes and ZG: secretory proteins, ribosomal proteins (table 1 and 2) and absorbed soluble proteins (not shown) [14]. On the other hand these membranes do not seem to have been severely damaged during the purification, particularly by large-scale removal of protein. Morphologically they are composed of vesicles, bounded by a unit membrane, they contain the bulk of the PLP originally present in microsomes or ZG and have PLP: protein ratios comparable with those found in membrane fractions isolated from other sources. Moreover, most of the membrane-bound enzymes known to be present in microsomes and ZG are

* The leucine concentration of membrane proteins is about the same in all 3 fractions and accounts for 9.8 ± 1%.

Table 1
Recovery of protein, phospholipid, RNA and α-amylase in purified rough and smooth microsome membranes

	Protein	PLP		RNA		α-amylase	
	%	%	μmoles/ mg protein	%	μg/mg protein	%	μg/mg protein
Rough microsomes	100	100	132.0	100	270.2	100	11.8
Rough microsome membranes	29.8	70.8	324.0	4.6	41.0	1.3	0.5
Smooth microsomes	100	100	277.8	100	43.5	100	12.2
Smooth microsome membranes	34.6	71.0	572.5	3.4	4.3	0.6	0.2

Table 2
Recovery of protein, phospholipid and chymotrypsinogen in purified zymogen granule membranes

	Protein	PLP		Chymotrypsinogen	
	%	%	µg/mg prot.	%	U/mg protein
ZG	100	100	13.5	100	20.1
ZG membranes	1.1	52.4	460.0	0	0

not solubilized during the purification of the membranes but rather recovered quantitatively in the membrane fractions [13, 14].

The kinetics of pulse labeling of microsome and ZG membrane proteins is summarized in fig. 1. It is evident that at the end of the 5 min pulse incorporation a high protein specific radioactivity is associated with both the rough

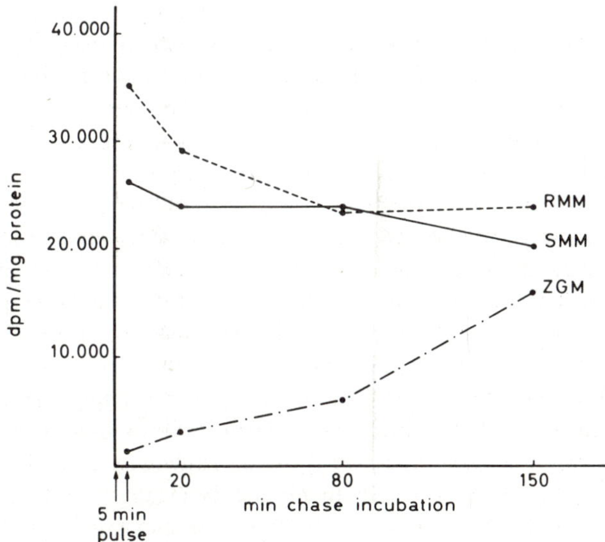

Fig. 1. Specific radioactivity of proteins in membrane fractions isolated from pancreatic slices pulse labeled for 5 min and incubated up to 150 min in chase. Values given are averages of either 5 (0 min in chase), 3 (20 and 80 min in chase), or 2 (150 min in chase) experiments. Variations from one experiment to another were less than 15%. RMM = rough microsome membranes; SMM= smooth microsome membranes; ZGM = zymogen granule membranes.

and smooth microsome membranes, the ratio of rough to smooth being ~1.3. In rough membranes a consistent decrease of protein specific radioactivity was found after 20 min of chase incubation; between 20 and 80 min there was an additional small decrease, with no further change at the last time point (150 min). In smooth microsome membranes the early decrease of protein labeling was less marked. No change could be detected at 80 min whereas between 80 and 150 min there was a further moderate decrease.

A completely different curve was found with ZG membrane proteins. These were practically unlabeled at the end of the pulse; during chase incubation a virtually linear increase of labeling was observed, so that after 150 min the protein specific radioactivity of ZG membranes approached that of smooth microsome membranes.

Taken as a whole these data seem to indicate that the mechanisms of synthesis of the different membranes involved in the secretory processes could be at least two fold. As expected, rough microsome membranes are highly labeled after a short pulse incorporation. RER appears therefore to be the site of an active membrane synthesis. The same conclusion also applies to the Golgi complex, since smooth microsomes are also highly labeled at the end of the pulse. In contrast, ZG membranes, which are unlabeled at the end of the pulse and pick up their radioactivity during chase incubation seem to be not synthetized as such but rather derive from pre-existing structures.

These conclusions, however, appear open to a number of questions. In particular we were concerned about the possibility that the traces of highly labeled secretory proteins, which are still present on our membrane fractions, particularly in rough microsomes (table 1), could influence substantially our results. Another question which had to be answered is related to the interactions between RER and Golgi membranes. In fact our data do not rule out the possibility that some RER membranes might be transformed into Golgi membranes. The way this transformation could occur is not clear, but it seems possible that, besides other processes, it could imply the insertion of "new" proteins into "old" membranes. Therefore, the high protein specific radioactivity of Golgi membranes at the end of the pulse could be due not to the independent synthesis of these membranes but rather to the insertion of a few highly labeled "transforming" proteins on pre-existing, non-labeled ER membranes. In order to shed some light on these problems, we decided to separate the proteins of the different membrane fractions by polyacrylamide gel electrophoresis. This work [15] enabled us to establish that the pattern of membrane proteins isolated from rough and smooth microsomes is rather similar. In the electrophoretic system that we used [16] (which is a modification of the method originally described by Takayama et al. [17]) most of the

Rough microsome membranes

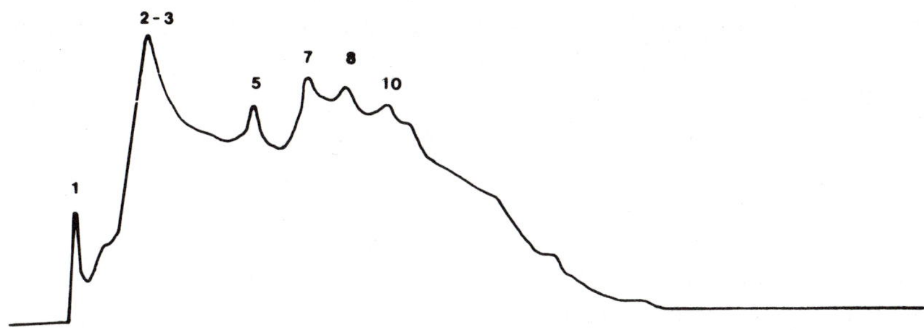

Smooth microsome membranes

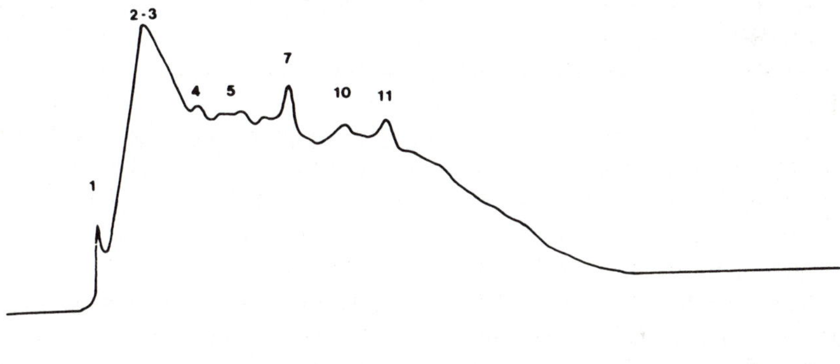

ZG membranes

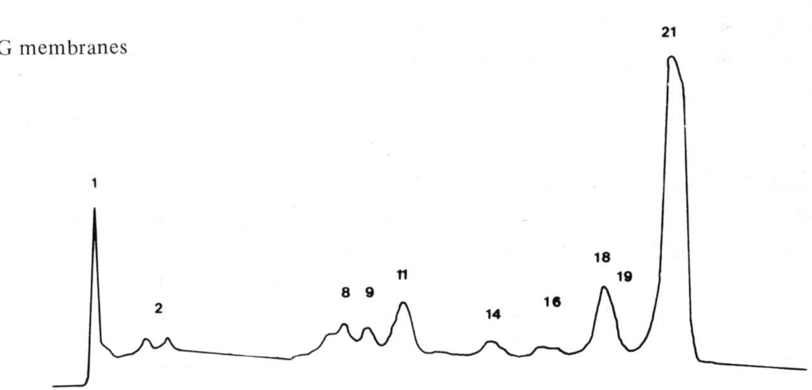

Fig. 2. Typical densitometric tracings of polyacrylamide gels of membrane fractions isolated from guinea pig pancreas. The major peaks are indicated by numbers. A detailed description of the experiments is given in 15. Staining: 1% amido black.

proteins migrate slowly and the numerous band do not appear well separated (fig. 2). In contrast, the electrophoretic pattern of ZG membranes appears very simple; most of the protein separates in a narrow, fast moving band (fig. 2). It is worth noting that in microsomal membranes, rough as well as smooth, we could never detect the presence of a band having a similar mobility. *

In a further series of experiments, radioactive membrane proteins obtained from microsomes and ZG isolated from pulse labeled slices either at the end of the pulse or after 150 min of chase were separated by acrylamyde gel electrophoresis. 35 to 40 slices, 1.5 mm thick, were cut from each gel, and their radioactivity was determined [18].

The results of these experiments, which will be reported in detail elsewhere [19], can be summarized as follows:

a) In all membrane fractions isolated at the end of the pulse there was a good correlation between absorbancy determined by densitometry of stained gels and distribution of radioactivity.
b) In both microsome membranes the distribution of radioactivity found at the end of the pulse was analogous to that found after 150 min of chase, except for a moderate decrease in that portion of the gels where secretory proteins are known to move.
c) In ZG membranes the increase of radioactivity taking place during chase in the specific, fast moving band is analogous to that observed with total membrane proteins.

These results obtained with radioactive gels strengthen the conclusions of the experiments carried out with total membrane proteins. Thus, were the high labeling of the Golgi membranes derived from the insertion of a few, highly labeled "transforming" proteins into non-radioactive RER membranes one would have expected in Golgi membranes a poor correlation between absorbancy and radioactivity at the end of the pulse as well as a marked change of the radioactivity pattern after 150 min of chase, when labeled RER membranes could have begun to be transformed into Golgi membranes. Furthermore, since during 150 min of chase the bulk of secretory proteins are known to be transported from microsomes to ZG the observation that the

* Since the isolation of microsomal membranes includes a long incubation of microsomes in puromycin at high ionic strenght (1 M KCl) the question was raised wether such a treatment realeased the fast moving protein(s). However, this does not seem to be the case since such a fast moving protein(s) was not detected even in microsomal membranes partially purified by simple washing the microsomes with 0.2 M $NaHCO_3$, without incubation at high ionic strength.

distribution of radioactivity changes only slightly during chase in all fractions clearly indicates that the traces of secretory proteins have only a minor influence on our results.

In conclusion, our data indicate that in guinea pig pancreas slices incubated *in vitro* the cytoplasmic membranes involved in secretion are actively synthetized. The membranes of the Golgi complex appear to be synthetized independently from the bulk of RER membranes. Since the Golgi complex itself is not endowed with protein synthetizing capacity the membrane proteins could be synthetized elsewhere and immediately assembled in the Golgi together with phospholipids to yield complete membranes. Possible sites of membrane protein synthesis could be either the free ribosomes (as already suggested in a different system [20]) or specialized areas of the RER from which the membranes would be immediately transferred to the Golgi complex. Morphological data which seem consistent with the latter hypothesis have been reported recently in other cell types [21, 22].

The result with ZG membranes are more puzzling. Strong morphological evidence suggests that secretory granules could derive their membranes from the Golgi complex [22–24]. However, it is difficult to reconcile this hypothesis with the greatly different electrophoretic patterns found in ZG membranes and smooth microsome membranes. Our data on the kinetics of labeling of the two fractions are also unconclusive. Were the ZG membranes derived from Golgi membranes one would have expected the increase of radioactivity of the former to be paralleled by an analogous decrease in the latter membrane fraction while only a small decrease was found in Golgi membranes. The latter discrepancy could be only apparent and simply depend on the fact that in pancreatic acinar cells of fasted guinea pig there are many more Golgi than ZG membranes. Therefore the origin of ZG membranes from Golgi membranes cannot be excluded. This however would imply either a profound transformation of the protein composition of the membranes of the existence in the Golgi complex of specialized portion(s) specifically devoted to the assembly of ZG membranes. These areas could be too small for permitting the detection of the protein band specific for ZG membranes in smooth microsomes membranes.

In any event our results seem to rule out the possibility that, within the acinar cell, cytoplasmic membranes involved in secretion flow along with secretion products, from the site of synthesis in the RER to ZG and the plasma membrane. Rather, they seem much more consistent with the possibility that membranes involved are synthetized independently, and are able to interact with one another by means of repeated non random fusions and fissions, thus permitting the intracellular transport of secretory proteins [3, 23–25].

The skilfull technical assistance of Mr. Guido Ramellini is gratefully acknowledged.

Note added in proof

Recently we have found that the proteins bound to the membranes of rough and smooth microsomes and zymogen granules can be separated very well by SDS polyacrylamide gel electrophoresis. The two microsomal membranes are resolved in similar, but not identical, patterns composed of ~40 separate bands, while the pattern of zymogen granule membranes (23 bands) is clearly different (J. Meldolesi and D. Cova, J. Cell Biol., in press).

Using this method with membranes pulse-labeled with ^{3}H-L-leucine we have now observed that, contrarywise to our previous conclusions, in both microsomal membrane fractions (but not in zymogen granule membranes) a sizable amount of TCA-insoluble radioactivity is contributed by secretory proteins. However, when the results on microsomal membranes were corrected by subtracting the contribution by secretory proteins they turned out to be still consistent with the hypothesis put forth in this paper, i.e., that the membranes of the RER and of the Golgi complex of pancreatic acinar cells are synthesized simultaneously and independently from each other.

References

[1] L.G. Caro and G.E. Palade, J. Cell Biol. 20 (1964) 473.
[2] G.E. Palade, P. Siekevitz and L.G. Caro, in: Ciba Found. Symp. on the Exocrine Pancreas, eds. A.V.S. de Reuck and M.P. Cameron (J. and A. Churchill Ltd. London, 1962) p. 23.
[3] J.D. Jamieson and G.E. Palade, J. Cell Biol. 34 (1967) 577.
[4] J.D. Jamieson and G.E. Palade, J. Cell Biol. 34 (1967) 597.
[5] S.R. Wellings and J.R. Philip, Z. Zellforsch. 61 (1964) 871.
[6] M. Schramm, Ann. Rev. Biochem. 36 (1967) 307.
[7] J. Racadot, L. Olivier, E. Porcile and B. Droz, C.R. Acad. Sci. 261 (1965) 2972.
[8] A.B. Novikoff and W.Y. Shin, J. Micr. 3 (1964) 178.
[9] S.N. Grove, C.E. Bracker and D.J. Morré, Science, 161 (1968) 171.
[10] D.J. Morré, H.H. Mollenhauer and C.E. Bracker, in: Results and Problems in Cell Differentiation, eds. J. Reinert and H. Ursprung (Springer Verlag, 1971) vol. III, p. 82.
[11] A. Claude, J. Cell Biol. 47 (1970) 745.
[12] P. Siekevitz and G.E. Palade, J. Biophys. Biochem. Cytol. 7 (1960) 619.
[13] J. Meldolesi, J.D. Jamieson and G.E. Palade, J. Cell Biol. 49 (1971) 109.

[14] J. Meldolesi and D. Cova, J. Cell Biol. 51 (1971) 396.
[15] J. Meldolesi and D. Cova, J. Cell Biol., in press.
[16] G. Eytan and I. Ohad, J. Biol. Chem. 245 (1970) 4297.
[17] K. Takayama, D.H. Mclennan, A. Tzagoloff and C.D. Stoner, Arch. Biochem. Biophys. 114 (1966) 223.
[18] P.V. Tischler and C.J. Epstein, Anal. Biochem. 22 (1968) 89.
[19] J. Meldolesi and D. Cova, in preparation.
[20] G. Ragnotti, G.R. Lawford and P.N. Campbell, Biochem. J. 112 (1969) 139.
[21] R.G. Kessel, J. Ultr. Res. 34 (1971) 260.
[22] C.G. Flickinger, J. Ultr. Res. 34 (1971) 260.
[23] G.E. Palade, in: Subcellular Particles, ed. T. Hayashi (Ronald Press, New York, 1959) p. 64.
[24] F.S. Sjöstrand, in: The Membranes, eds. A.J. Dalton and F. Haguenau (Academic Press, New York and London, 1968) p. 151.
[25] J. Meldolesi, J.D. Jamieson and G.E. Palade, J. Cell Biol. 49 (1971) 150.

SPECIFICITY IN BASE-EXCHANGE REACTIONS OF MEMBRANE PHOSPHOLIPID METABOLISM

G. PORCELLATI

Istituto di Chimica Biologica, Università di Perugia, Italy

1. Introduction

A calcium-dependent base-exchange reaction has been detected in brain subcellular membranes [1,2], which produces new phospholipid moieties at the expenses of endogenous phosphoglycerides and free nitrogenous bases. Membranes from the endoplasmic reticulum were found to be noticeably active, whereas those from mitochondria or other subcellular structures were rather poor in the exchanging activity [2]. Properties of the enzymic system have been described [2,3], as well as solubilization procedures have been discussed in a recent Symposium [4]. Thus far, little is known about the structural requirement of both the free base and the exchanging phospholipid molecules within the enzymic system, apart from some indications which pointed to the greater exchanging capability of L-serine and ethanolamine, when compared with other free bases [2,3]. The present work extends these findings and gives some insight as to whether the fatty acyl residues and the bound-bases of phospholipids may affect the rate of the exchange reaction.

2. Experiments and results

It was shown already [2] that free serine and ethanolamine exchange relatively rapidly with the bases of the endogenous phospholipids in purified

* This work was aided by grants from the Consiglio Nazionale delle Ricerche, Rome (contract No. 70/01810/04/115.3381).

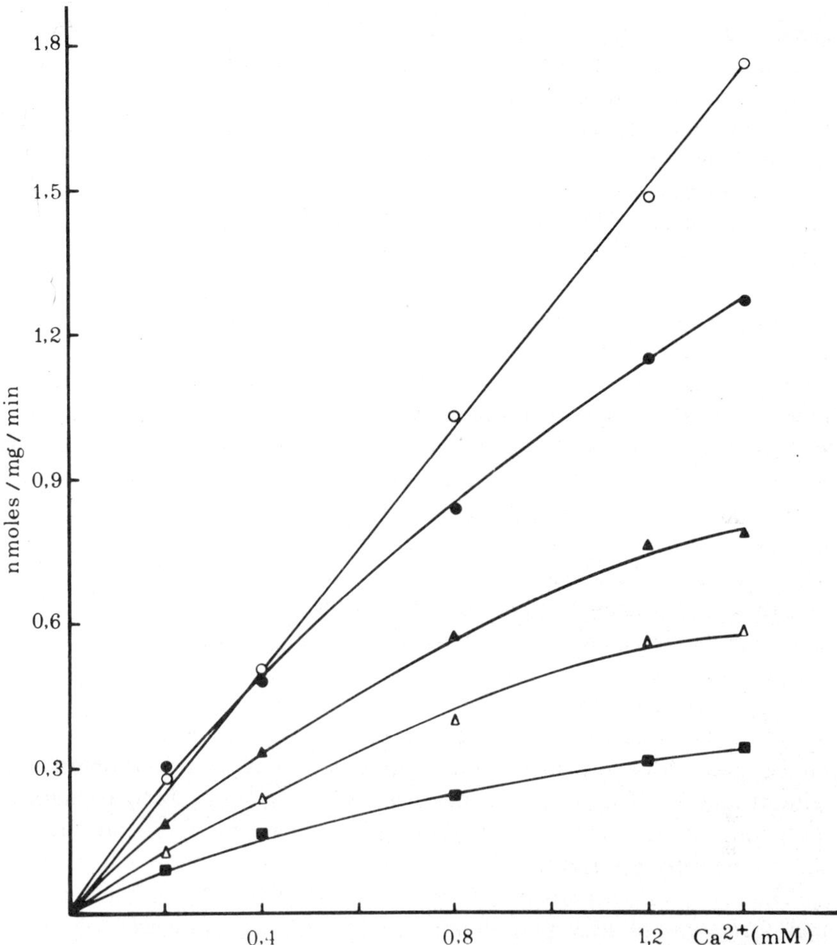

Fig. 1. Exchange of nitrogenous bases with diacyl-GPS in a lipid-depleted enzymic system of chick brain microsomal membranes. The enzymic system, corresponding to about 100 mg fresh wt of original material, prepared according to previously described procedures [4], was incubated with the ^{14}C-labelled nitrogenous bases (1.5 mM), 20 mM-tris-HCl buffer (*p*H 8.5), 2 mM-beta-mercaptoethanol, 0.4 mM of a previously sonicated [4] phosphatidylserine from bovine brain, and $CaCl_2$, as indicated. Enzyme and other components were added in the indicated order, and incubation was carried out for 30 min. Final volume: 0.5 ml. Activity was determined as described elsewhere [2,4], and is expressed as nmoles labelled phospholipid/mg protein/min.

–○–○–○–○–, L-serine; ●–●–●–●–●–, ethanolamine; –▲–▲–▲–▲–, monomethyl-ethanolamine; △–△–△–△–△–, dimethylethanolamine; ■–■–■–■–■–, choline.

chick brain microsomal membranes. Furthermore, it was shown also that the two amino-compounds compete probably for the same enzymic site [2,4].

Experiments have now indicated that the incorporation of L-serine into chick brain microsomal phospholipids by exchange reaction is competitively inhibited by ethanolamine and slightly affected or unaffected by D-serine, choline, inositol and carnitine. Similarly, the incorporation of ethanolamine into diacyl-glycerophosphorylethanolamine (GPE) is inhibited by L-serine, slightly affected by choline, and unaffected by D-serine, inositol and carnitine.

With the use of a solubilized and partially delipidized enzyme obtained from the brain microsomes [4], experiments have been performed on the ability of the enzymic system to exchange different labelled nitrogenous bases with diacyl-glycerophosphorylserine (GPS) added to the incubation system. Fig. 1 shows that different degrees of incorporation are obtained when various nitrogenous bases are incubated, at increasing calcium ion concentrations. More precisely, decreasing levels of incorporation are obtained on changing the structure of the non-methylated free base (serine or ethanolamine) towards the *pluri*-methylated free base, *i.e.* choline. However, similar activity-concentration relationships are obtained for all the bases (fig. 1), on changing the levels of calcium ions, thus indicating that calcium probably forms complexes of a similar kind with the nitrogenous bases, before the enzymic exchange takes place.

It is interesting to mention, in this connection, that higher values of exchange have been obtained during our experiments, on preincubating for various lengths of time the various bases with Ca^{2+}-ions, before incubating with the enzymic system, thus again signifying that the Ca^{2+}-ions probably link to the bases prior to the enzymic exchange.

The action of purified lipid classes upon the yield of the exchange reaction and formation of new phospholipid molecules was successively examined. The lipid classes were synthesized by chemical procedures [5], and purified by column chromatography on silicic acid columns. They were successively sonicated under appropriate conditions at a 5 mM initial concentration in a volume of 10 ml for 30 min in a 100 Watt MSE ultrasonic disintegrator (20 Kcycles, 8 microns peak-to-peak on amplitude meter) in a 0.32 M-sucrose-40 mM-Veronal buffer (*p*H 8.5)-4 mM-dithiothreitol solution at +2 °C, before adding to the incubation system. The results of table 1 indicate that, when diacyl-GPS and diacyl-GPE are used as the exchanging phospholipids, a noticeable exchange with L-serine occurs, whereas less evident incorporation of the amino-compound is obtained with other lipid classes. These findings bear

Table 1
The effect of adding phospholipid classes upon the serine exchange in a lipid-depleted enzymic system of chick brain microsomes *

Addition	Activity (nmoles/mg protein/min)
None	0.11
	0.18
	0.08
1,2-dipalmitoyl-GPE	0.31
	0.35
1,2-dipalmitoyl-GPMME	0.28
	0.35
1,2-dipalmitoyl-GPDME	0.14
	0.16
1,2-dipalmitoyl-GPE	0.12
	0.18
1,2-dipalmitoyl-GPS	0.44
	0.48
1,2-dipalmitoyl-GPT	0.37

* Phospholipids were sonicated as explained in the text, and added to the incubation system (0.4 mM final concentration). Incubation was carried out as reported in fig. 1 with 0.5 mM $CaCl_2$ for 20 min at 37°C with the solubilized and partially lipid-depleted microsomal system corresponding to 50 mg fresh wt original material.
Abbreviations: GPE, glycerophosphorylethanolamine; GPMME, glycerophosphorylmonomethylethanolamine; GPDME, glycerophosphoryldimethylethanolamine; GPC, glycerophosphorylcholine; GPS, glycerophosphorylserine; GPT, glycerophosphorylthreonine.

evident similarities with those reported in fig. 1, indicating a certain degree of specificity of the lipid-bound base for the enzymic system.

Further experiments have been carried out on the effect that different purified molecular species of diacyl-GPE would have exerted on the exchange reaction. L-serine was used as the free base, whereas the phospholipid species, synthesized by chemical means [5] and listed in table 2, were sonicated as described above, and incubated in the same conditions described in table 1. The results of table 2 show that the diacyl-GPE containing two identical saturated fatty acid substituents exchanges with L-serine, as do the molecular species which contain two identical unsaturated fatty acid residues. However, the fully saturated species exchange, on the whole, less readily than the unsaturated ones. This difference, which is not very large, does not depend, in our opinion, of different response to the sonication procedure given by the

Table 2
The effect of adding phospholipid molecular species upon the serine exchange in a lipid-depleted enzyme system of chick brain microsomes *

Addition	Activity (nmoles/mg protein/min)
None	0.11
	0.18
	0.08
1,2-dipalmitoyl-GPE	0.31
1,2-distearoyl-GPE	0.26
1,2-dioleoyl-GPE	0.41
1,2-dilinoleoyl-GPE	0.37
α'-stearoyl, β-oleoyl-GPE	0.48
α'-oleoyl, β-stearoyl-GPE	0.51
α'-stearoyl, β-linoleoyl-GPE	0.46
α'-palmitoyl, β-oleoyl-GPE	0.54
α'-palmitoyl, β-linoleoyl-GPE	0.52

*For abbreviations and technical informations, see table 1.

unsaturated lipids as compared to the saturated species, which are known to be less susceptible to sonication under comparable conditions. The mixed species of diacyl-GPE are even better utilized in fact than the fully unsaturated forms (table 2). The data of table 2 rather indicate that structural differences in the fatty acyl residues of phospholipids might affect the exchange property of the enzymic system. Table 2 illustrates also that positional isomers of diacyl-GPE react equally well with L-serine in the presence of the enzymic system, and that therefore differences in fatty acid positioning of the intact phospholipid molecule do not seem to affect the exchange of lipid with the externally added serine. *

3. General considerations

The incorporation of serine as well as that of other nitrogenous bases into phospholipid by the chick brain system is probably catalyzed by a single

* It must be added, in connection with the data of tables 1 and 2, that adding external phospholipid classes to the lipid-depleted incubation system has often not always resulted in exchange activity with L-serine. Probably physical factors may be at play in explaining divergent results.

enzyme. The mechanism of action is based on a calcium-dependent exchange reaction of free serine and of other bases with endogenous phospholipids (or with externally added phospholipids, when a lipid-depleted enzymic system is used, in place of the whole microsomal membranes, but see footnote). This single enzyme system does not distinguish in exchanging the phospholipids with the free bases ethanolamine, serine and others, but shows different degrees of specificity with regard to the structure of the lipid-bound and free bases, and of the fatty acyl residues of the phospholipid. The degree of the specificity of the hydrocarbon chains of the reacting phospholipids as well as of the bound nitrogenous bases may have some importance in regulating the rate of the enzymic exchange, and therefore in operating an adequate and controlled displacement of the phospholipid bases at the membrane level.

References

[1] G. Arienti, M. Pirotta, D. Giorgini and G. Porcellati, Biochem. J. 118 (1970) 3P.
[2] G. Porcellati, G. Arienti, M. Pirotta and D. Giorgini, J. Neurochem. 18 (1971) 576.
[3] M. Pirotta, D. Giorgini and G. Porcellati, Boll. Soc. It. Biol. Sper. 45, issue n. 20 bis, communic. n. 208 (1969).
[4] G. Porcellati and F. di Jeso, in: Membrane-bound Enzymes, eds. G. Porcellati and F. di Jeso (Plenum Press Corp., New York, 1971) p. 208.
[5] E. Bear, J. Amer. Oil Chem. Soc. 42 (1965) 257.

TOPOLOGY OF BACTERIAL MEMBRANE SYNTHESIS

F. AUTISSIER and A. KEPES
Institut de Biologie Moléculaire, Laboratoire des Biomembranes, Paris, France

Bacteria unlike higher organisms exhibit very little turnover of their macromolecular cell constituents, except for messenger RNA which has a very short half life even relative to the generation time. Proteins in exponentially growing cell populations are renewed to the extent of a fraction of one per cent to a few per cent per generation time. This holds also for membrane proteins. The stability of membrane phospholipids is somewhat less. In *E. coli* the major phospholipid, phosphatidyl ethanolamine is completely stable, while phosphatidyl glycerol, the second major component exhibits a turnover which cannot be described by a single exponential. The fast component of this turnover would give a complete renewal in about half a generation time.

These facts together with the rigid geometry of the bacterial cell make the bacterial membrane particulary suitable for a topological study of membrane growth. An additional motive for such a study stems from the hypothesis of Jacob et al. [1], according to which, in the absence of a mitotic apparatus, in bacteria, the membrane could play the main role in distributing invariably the nuclear material in daughter cells. A necessary condition for fulfilling this function is that DNA be attached to the membrane, and that a localized growing zone of the membrane put an adequate distance between the attachment points of two nuclei intended for different daughter cells. The attachment of DNA to membrane has since been amply and unambiguously documented [2], while the demonstration of a pattern of localized membrane growth, attempted by many workers, remained controversial due to contradictory results [3].

Our approach to this problem [4] was to observe the distribution among the progeny of the parental membrane with the help of inducible membrane markers of functional value, such as inducible transport systems. If membrane growth occurs at a small number of *loci*, parental membrane will be frag-

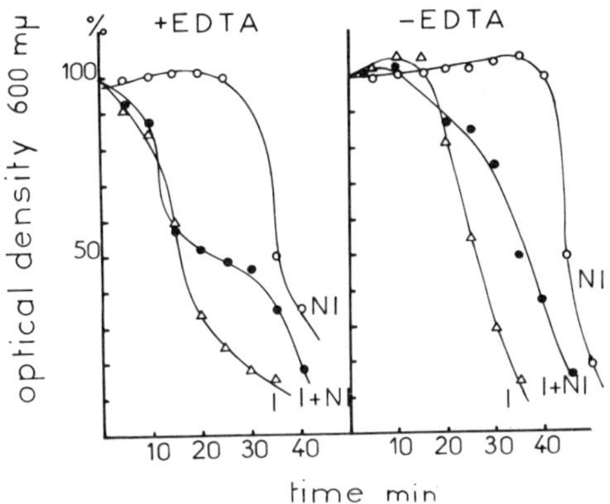

Fig. 1. comparison of the time course of lysis of bacteria, induced or non induced for Lac-permease and a 1:1 mixture of the two. *E. coli* K 12 strain 3 000 was grown for 6 generation times in medium 63 glycerol B1, with or without 0.2 mM IPTG. When optical density at 600 nm reached 1.0, bacteria were filtered on a millipore filter 0.45 μ pore size, washed with Tris HC1 buffer 0.12 M *p*H 7.6 and resuspended with the same buffer in one tenth of the original volume. EDTA 10^{-3} M was added to one part of each suspension and after 1 min of vigourous shaking at 37 ° the suspension was diluted in 9 volumes of medium 63 [13] containing penicillin (1 000 units per ml) and lactose 5 mM, and placed on a reciprocating shaking bath. At intervals O.D. readings at 600 nm were made in a Zeiss PM QII spectrophotometer in cuvettes of 1 cm light path. Left: bacteria treated with EDTA, right: without EDTA. ○—○ non induced; △—△ induced; ●—● mixture.

mented in a small number of coherent zones and only a limited number of the descendents will inherit parental membrane. The membrane markers used provide to the cells a selective advantage in the presence of the transport substrate which is a carbon source. The detection of individuals within a population possessing this selective advantage and their separation from those devoid of the marker have been achieved by a penicillin treatment which causes the lysis of growing bacteria, while non growing individuals survive.

If bacteria induced for lactose or non induced or a mixture of these populations are submitted to the penicillin treatment with lactose as the sole source of carbone, the results depicted in fig. 1 are observed. Induced bacteria lyse fast, non induced bacteria lyse after a lag period necessary for the induction of new permease by lactose, and the mixed population lyses in two distinct steps. The distinction between the two parts of the population is due to the presence of Lac-permease since a permease deficient mutant, induced for β-galactosidase behaves like a non induced wild type population.

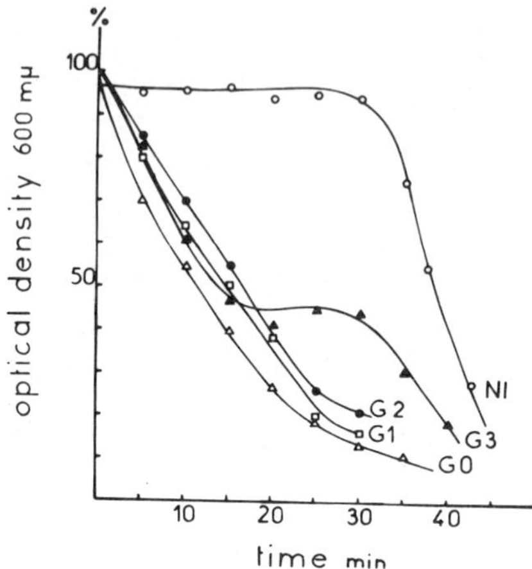

Fig. 2. Time course of lysis of populations of *E. coli* 3 000 after various times of deinduction. Fully induced bacteria were centrifuged and washed with medium 63, then transferred into medium 63 glucose B1 at 37°. After 0, 1, 2, 3 generation times (G0, G1, G2, G3) EDTA treatment and lysis in medium 63 penicillin lactose were performed as described in the legend of fig. 1. Non induced bacteria were used as control (NI).

If fully induced *E. coli* is grown in non inducing conditions, during two generation times the population behaves as a homogeneous induced population. After three generation times a striking heterogeneity is observed (fig. 2).

A similar experiment with melibiose permease gives similar results, namely a heterogeneity after three generation times, but with this marker it is possible to avoid *de novo* induction of the permease during the penicillin treatment by carrying out this step at 40°, a non permissive temperature for Mel-permease synthesis [5]. In these conditions the lysis of the second half of the population is delayed, and actually it does not occur within the duration of the experiment. This again ascertains that permease is the factor which determines penicillin lysis, since α-galactosidase is normally synthesized at 40°. Moreover this experiment provides a better facility to isolate by centrifugation the surviving half of the population, and its permease content can be measured along the penicillin treatment. It is shown in fig. 3 that TMG uptake completely disappears when the first half of the population is lysed.

Similar heterogeneity was detected after three generations of deinduction with the following carbon sources; succinate, glucose-6-phosphate, D-glucuronate, D-gluconate, D-mannitol, trehalose, maltose. Some of the transport systems implicated have never been described (gluconate, trehalose...). One of

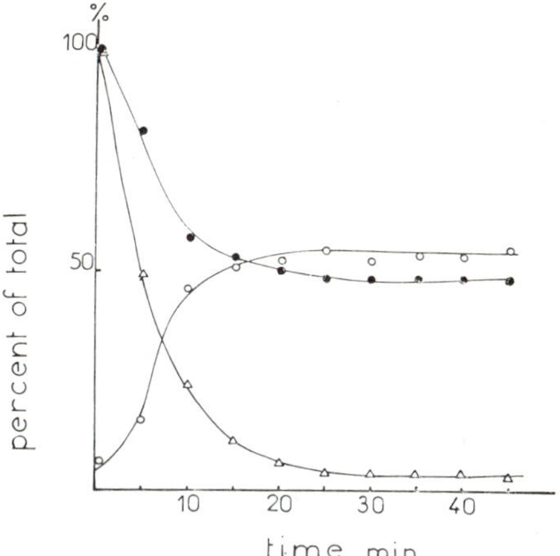

Fig. 3. Distribution of the Mel-permease in the survivors of a population deinduced three generation times (G3). Treatment with penicillin in the presence of melibiose was carried out at 39°. Content of Mel-permease is measured by the initial velocity of uptake of Me ^{14}C thiomethylgalactoside (TMG) according to Kepes (1960). β-galactosidase content of the lysed bacteria was measured in the filtrate by the rate of ONPG hydrolysis and expressed as per cent of β-galactosidase activity of the original bacteria after toluene treatment. •—• absorbance; △—△ – permease activity; ○—○ β-galactosidase in lysate.

them (D-mannitol), is of the class of the phosphotrasferase systems [6].

No heterogeneity was observed with glycerol, L-arabinose, glucose and fructose. The last two are known to have constitutive transport systems [7]. Glycerol is generally believed to penetrate by diffusion [8]. The arabinose uptake involves a periplasmic binding protein instead of a firmly membrane bound component [9], and this might be the reason of its different distribution.

The possibility to physically separate from a mixed population the part which does not carry parental membrane can be used to follow the fate of isotopically labeled membrane material. Phospholipid label was measured in permease negative segregants using Lac-permease or Mel-permease as the selective markers. Label in the backbone of the phospholipid: ^{3}H glycerol and chloroform-methanol extractable ^{32}P, disappears from the survivors after the lysis of the permease rich half of the population (figs. 4 and 5) while ^{14}C oleic acid and chloroform-methanol extractable ^{14}C derived from ^{14}C acetate appear to be evenly distributed between permease rich and permease poor descendents (fig. 5). A rather conspicuous turnover of fatty acids on stable glycerophosphate backbones can be inferred from this experiment

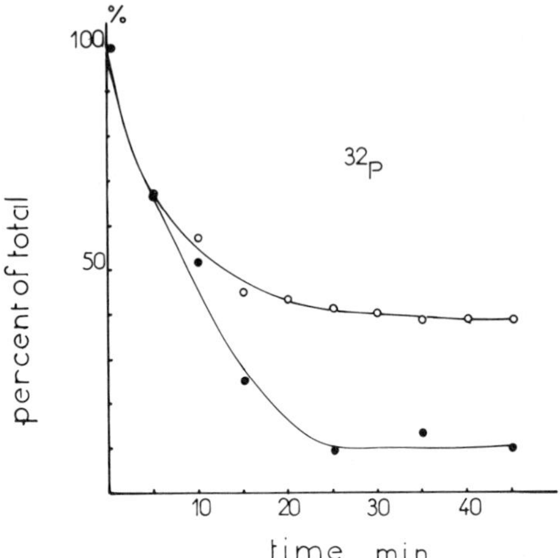

Fig. 4. A culture of *E. coli* K 12 300 P (Lac-permease negative) was grown for 6 generation times in medium 63 glycerol plus melibiose at 30°. ^{32}P orthophosphate was added one hour before the end of this induction period. After washing, the bacteria were grown for an additional three generation times in 63 glycerol, then submitted to the EDTA treatment described in legend of fig. 1 and resuspended in 63 melibiose plus penicilline. Absorbance readings ○–○ and chloroform – methanol extractable ^{32}P in centrifugal pellets ●–● were measured at close intervals and expressed as per cent of the values obtained at the time zero.

(presumably by reacylation of lysophosphatides). This might be an explanation of the failure of some labeling experiments to demonstrate a localized membrane growth [10].

Fig. 6 shows the time course of growth of permease containing versus total population during the deinduction of Lac-permease. The two increase in parallel for two generation times, whereafter the permease rich part of the population stops further increase, while the total population continues exponential growth.

An indication that growing zones in the membrane are correlated with the nucleus can be derived from the pattern of segregation in a rich medium (fig. 7). It is known that in rich medium the number of nuclei per cell is higher than in minimal medium [11] and this can be expected to be accompanied by a higher number of growing zones in the membrane. In the experiment of fig. 7, the population behaves as homogeneous with respect to Lac-permease during three generation times, and only after four doubling times, the heterogeneity is observed which appears usually after three generation times on minimal medium.

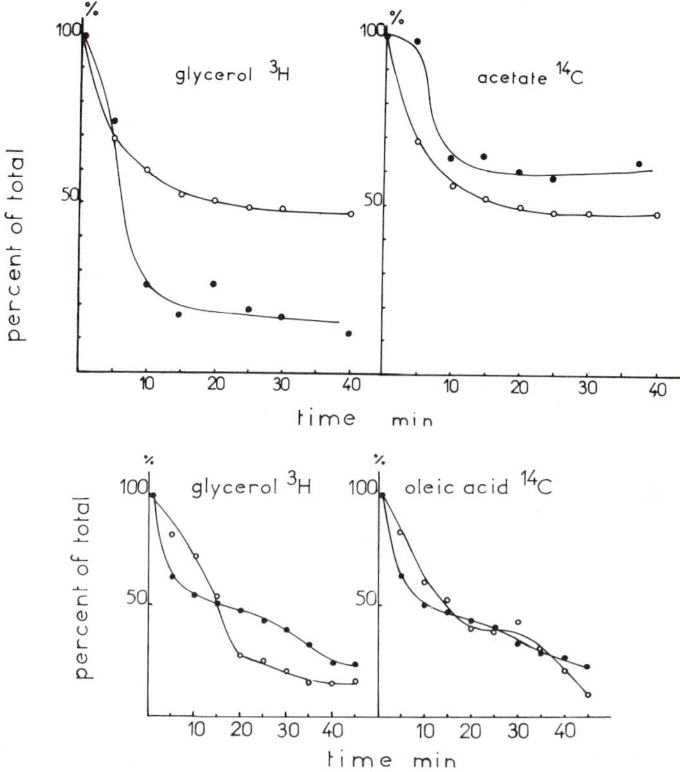

Fig. 5a. *E. coli* K 1059 (φ 80 d lac) induced with IPTG and labeled during the last generation time of the induction period with ^{3}H glycerol or ^{14}C oleic acid was deinduced three generation times and submitted to EDTA treatment followed by incubation with lactose and penicillin. ●–● absorbancy reading; ○–○ radioactivity in pellet. 5 b. *E. coli* 300 P grown in 63 glycerol and melibiose and labeled during the last generation time of the induction period with ^{3}H glycerol or ^{14}C acetic acid. Deinduction and lysis were carried out as described in legend of fig. 3. Absorbancy measurement ○–○; and counting of radioactivity in the pellet (^{3}H glycerol) of chloroform methanol extract (^{14}C acetate) ●–●.

Three possible models of membrane growth can account for the observations reported (fig. 8). The first is directly inspired by the postulated role of membrane growth in the separation of nuclei [1] after a round of DNA duplication. It includes a median growig zone active during one generation time and finally becoming the initiator of a new septum, while two growing zones appear *de novo* in the median plans of the daughter cells.

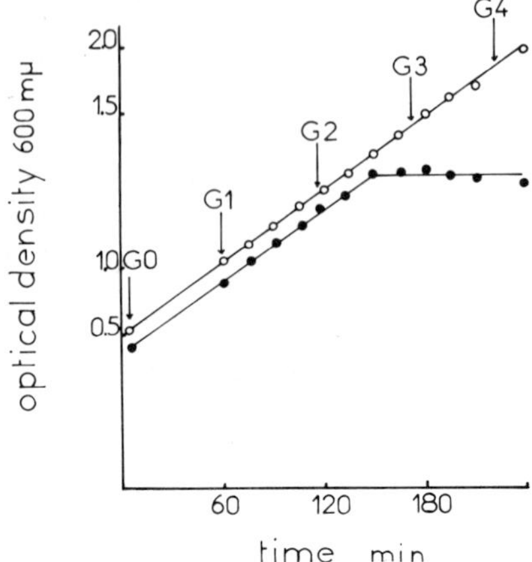

Fig. 6. Growth curve during deinduction of *E. coli* 3 000 previously induced with IPTG ○–○. At various times samples were submitted to EDTA treatment followed by lysis in 63 lactose penicillin. The permease containing part of the population ●–● was calculated from a calibration curve made with artificial mixtures of induced and non induced populations.

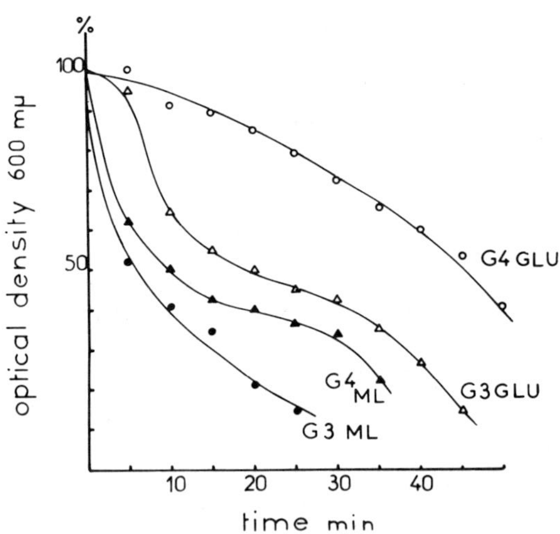

Fig. 7. Comparison between the time course of lysis of populations of *E. coli* 3 000 fully induced by IPTG deinduced 3 and 4 generation times for Lac-permease in medium 63 glucose or in rich medium (ML).

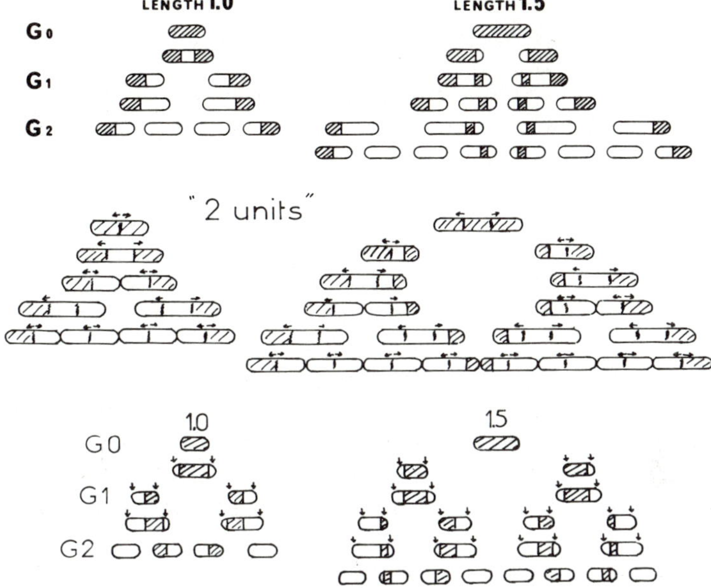

Fig. 8. Three models of membrane growth compatible with the observed segregation of membrane markers. Parental membrane is the hatched area.
a) One median growing zone arising *de novo* in daughter cells and ceasing activity after transformation into a septum forming zone.
b) Two unidirectional growing zones arising from the duplication of one of the growing zones of the mother cells. Septum forming zone arises *de novo*.
c) Two polar growing zones, one arising *de novo* after septum formation.

The second model is the adaptation of a model proposed by Donachie and Begg [12] for a "two unit" cell (presumably containing two nuclear bodies) and based on the microscopic observation of the direction of cell elongation and the visualization of growing zones of the murein layer (surrounding the cell membrane) by penicillin. In this model each growing zone derives from the doubling of a growing zone of the mother cell, while septa are formed by *de novo* initiation zones. The peculiarity of this model is that a growing zone is unidirectional, it deposits new membrane toward the equator of the cell but remains at constant distance from the adjacent pole.

The third model includes two polar growing zones and it is quoted for completeness because, like the other two, it predicts segregation after two generation times of individuals devoid of parental membrane. In contrast with the first two models, this one is difficult to correlate with nuclear division and to extrapolate to situations with more than two nuclei entailing an increase in the number of growing zones.

On the basis of the observations here reported, it can be conjectured that at least three factors have contributed to the difficulty of the demonstration of well defined growing zones in the membrane of *E. coli*. First a cell population is seldom homogeneous with respect to the number of nuclei and consequently with respect to the number of growing zones. In our experiments bacteria growing on minimal medium with glycerol give four descendents carrying parental membrane, when the carbon source is glucose, this number reaches 5–6 indicating that a fraction of the population has a more complexe growing pattern (e.g. tetranucleated) and growing on nutrient broth gives 8 descendents carrying the parental membrane.

The second difficulty is that the growing zone might not be an ideal line in a plan exactly perpendicular to the long axis of the cell, but a zone of some width with diffuse edges, so that, in the first model its appearance in F2 cells might not occur exactly at the limit of parental and F1 membrane so that a marginal fraction of parental markers contaminates the "wrong" descendent. Indeed a minor amount of parental markers is generally found associated with the part of the population poor in these markers. The model of Donachie due to the continuity of growing zones in time and their generation by duplication of a preexisting growing zone would provide a more efficient guarantee against such contamination.

The third difficulty is the apparent turnover of fatty acids of otherwise stable phospholipids. The demonstration of this turnover provided by the comparison of fig. 4 and 5 is only preliminary, it should be confirmed by double labeling in a single experiment. The rate of such turnover will be exceedingly difficult to measure, since our technique based on physical separation of the cells is only possible after more than two generation times. The exact location of the growing zone remains to be found by detailed ultrastructural investigations. But more than the exact location, the intriguing question is the molecular basis of the regulation which controls either the *de novo* initiation of a growing zone or the duplication of a preexisting one, as well as the control of possible extinction of an active growing zone and/or possible initiation of septum formation which might be a different kind of growing zone obeying a different control. Finally these control mechanisms are correlated with the control of termination and/or reinitiation of a round of DNA replication, but this correlation is not rigid, providing for a number of different situations. E. g. when the cells have a larger number of nuclei, it means that septum formation is delayed comparatively to DNA duplication by one, two or three generation times.

References

[1] F. Jacob, A. Ryter and F. Cuzin, Proc. Roy. Soc. (London) 164 (1966) 267.
[2] A.J. Ganesan and J. Lederberg, Biochem. Biophys. Res. Commun. 18 (1965) 824.
C. Lark and K.G. Lark, J. Mol. Biol. 10 (1964) 120.
A. Ryter and F. Jacob, C. R. Acad. Sci. (Paris) 257 (1963) 3060.
[3] M.E. Beachey and R.M. Cole, J. Bact. 92 (1966) 1245.
K.L. Chung, R.Z. Hawirko and P.K. Isaac, Can. J. Microbiol. 10 (1964) 43 and 473.
R.M. Cole, Sciences 43 (1964) 820.
R.M. Cole and J.J. Hahn, Sciences 143 (1962) 820.
J.W. May, Exptl. Cell. Res. 31 (1963) 217.
[4] F. Autissier, A. Jaffe and A. Kepes, Molec. Gen. Genetics 112 (1971) 275.
F. Autissier and A. Kepes, Biochim. Biophys. Acta (1971).
[5] C. Burstein, Science Thesis, Paris (1967).
A. Pardee, J. Bact. 73 (1957) 376.
[6] S. Tanaka and E.C.C. Lin, Proc. Natl. Acad. Sci. (U.S.) 57 (1967) 913.
[7] W. Kundig and S. Roseman, Proc. Natl. Acad. Sci. (U.S.) 52 (1964) 1067.
[8] J.P. Koch, S.J. Hayashi and E.C.C. Lin, J. Biol. Chem. 239 (1964) 3106.
[9] D.E. Sheppard and E. Englesberg, J. Mol. Biol. 25 (1967) 443.
[10] N. Tsukagoshi, P. Fielding and C.R. Fox, Biochem. Biophys. Res. Commun. 44 (1971) 497.
[11] K.G. Lark and C.C. Lark J. Mol. Biol. 13 (1965) 105.
[12] W.D. Donachie and K.J. Begg, Nature 227 (1970) 1220.
[13] L. Leive, Biochem. Biophys. Res. Commun. 18 (1965) 13.
M. Jacquet and A. Kepes, Biochem. Biophys. Res. Commun. 36 (1969) 84.

FUNCTIONAL REQUIREMENT FOR PHOSPHOLIPIDS OF HORMONE-SENSITIVE ADENYLATE CYCLASE

V. TOMASI, A. RÉTHY *, A. TREVISANI and O. BARNABEI
Istituti di Fisiologia Generale, Università di Ferrara e Roma, Italy

Lipids are known to play a major role in the molecular organization of membranes and in the activity of some membrane-bound enzymes. Previous work of Sutherland's group [1] and the results of Pohl et al. [2] as well as Marinetti et al. [3], has shown that in liver cell adenylate cyclase is localized in the plasma membrane. Studies with other tissues suggest that lipids are important in all hormone-sensitive adenylate cyclase systems. Solubilization of the membrane enzyme with detergents, for example, is always accompanied by a loss of the hormonal sensitivity of the enzyme [4,5].

Elucidation of the relationship between membrane lipids and adenylate cyclase will undoubtedly be required in order to obtain a better understanding of the initial effects of hormones. As a matter of fact lipids may be involved in the binding of hormones to biological membranes, and in the conversion of this binding reaction into an effect such as stimulation of adenylate cyclase activity.

For the study of the lipid requirement of adenylate cyclase we have employed the following procedures: removal of lipids with organic solvents under mild conditions and treatment of membranes with phospholipase A and C. There is evidence that phospholipases are able to hydrolyze specifically phospholipid substrates which are attached to biological membrane like erythrocytes ghosts [6] or endoplasmic reticulum membranes [7].

The studies described in the present report show that the hormone response of the adenylate cyclase system as well as the ability of membranes to

* Present adress: Medical University of Debrecen, Microbiological Institute, Debrecen 12, Hungary.

bind labeled epinephrine are drastically reduced, following mild extraction and phospholipase A and C treatment, and are restored at least to a large extent by exposing the treated membranes to aqueous dispersions of phosphatidylserine.

The effects of mild extraction, phospholipase A and C on the phospholipid composition of plasma membrane and enzyme activities

The phospholipid composition of control and treated plasma membrane is reported in fig. 1. Mild extraction of membranes caused a selective loss of

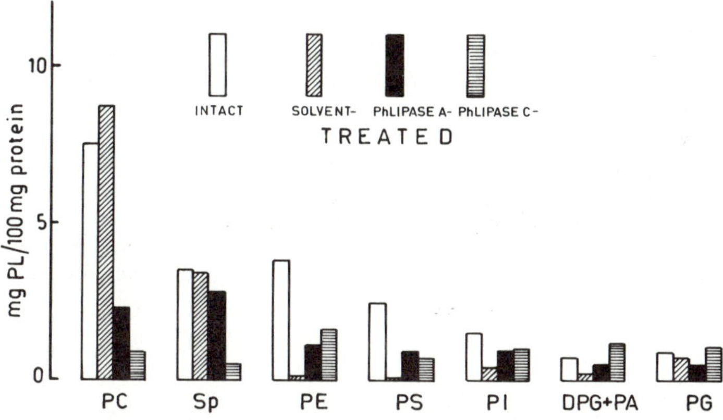

Fig. 1. Phospolipid composition of intact, solvent- and phospholipases-treated plasma membranes. Phospholipid extraction, chromatographic separation and assays were performed as described previously [7a]. Mild extraction was carried out with petrol ether (bp 30–40°)-n-butanol (7:3) for 2 min at 4° [7a]; phospholipase treatment was according to a modified method of Emmelot and Bos [8,9]. Phospholipase A (purified from pancreas) was a gift of Prof. L.L.M. van Deenen, Utrecht, The Netherlands; phospholipase C (from Cl. Welchii) was purchased from Sigma. Phospholipase treatment was performed in the presence of albumin (0.25 mg).

PE *, PS and PI. The first two formers were virtually absent from extracted membranes, while PI was reduced to 1/3 of the initial value. The concentra-

* Abbreviations used: PE, phosphatidylethanolamine; PS, phosphatidylserine; PI, phosphatidylinositol; PC, phosphatidylcholine; Sp, sphingomyelin; PA, phosphatidic acid; DPG, diphosphatidylglycerol (cardiolipin); cyclic AMP, cyclic adenosine 3′,5′-monophosphate; ATP, adenosine triphosphate; ATPase, adenosintriphosphatase; TRIS-HCl, tris(hydroxymethyl)aminomethanechloride.

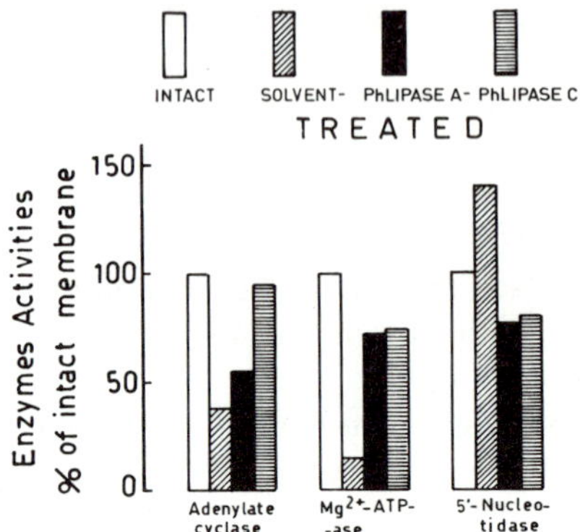

Fig. 2. Effect of mild extraction and phospholipases on membrane-bound enzymes. Adenylate cyclase was assayed as previously described [3] the modified incubation medium containing in a final volume of 0.5 ml: TRIS-HCl 0.05 M, *p*H 8.0, ATP 0.6 mM; Mg^{2+} 4 mM, 0.5 μC of ^{14}C-8-ATP, 80–100 μg of membrane protein and 50 μg of albumin; incubation was carried out at 37° for 15 min. The activity is expressed as: nmoles cyclic AMP/mg membrane protein/hr. Mg^{2+}-ATPase and 5′-nucleotidase were assayed according to Emmelot and Bos [10]. The activity of adenylate cyclase of intact membranes was 6.49 ± 0.3 nmoles/mg protein/hr (mean ± S.E. of 5 experiments). The activity of Mg^{2+}-ATPase and 5′-nucleotidase were respectively 24.1 ± 2.4 and 24.1 ± 2.5 μmoles Pi/mg protein/hr (means ± S.E. of 7 experiments).

tion of PC, Sp and PG was unaffected by the solvent treatment. The comparison of the effect of mild extraction with that of phospholipases shows that the former treatment is more selective than the latter one. In fact phospholipase A hydrolyzed 72% of PC, 18% of Sp, 71% of PE, 64% of PS, 41% of PI, 44% of PG and 20% of DPG + PA. Phospholipase C hydrolyzed 89% of PC, 85% of Sp, 63% of PE, 71% of PS, and 30% of PI. Therefore PC, PE and PS are good substrates for both enzymes; Sp is attacked by phospholipase C and only partially by phospholipase A; PI is hydrolyzed more extensively by the latter (41%) than by the former (30%). As far as PI is concerned it appears that about 1/3 of this phospholipid is very firmly bound to membranes.

As shown in fig. 2, mild extraction or phospholipase A treatment of isolated liver plasma membranes produce a decrease of the baseline adenylate cyclase activity. As far as the activity of two other membrane-bound enzymes is concerned, Mg^{2+} ATPase and 5′-nucleotidase were little affected by both phospholipases, while mild extraction strongly reduced the former and stimulated the latter.

Role of phospholipids in the restoration of adenylate cyclase activity, and in the binding of labeled epinephrine

It can be observed from fig. 3 that there is a correlation between the amount of PI removed from membranes by solvents and phospholipases and the reduction of adenylate cyclase activity. Moreover addition of PI to the treated membranes restored the enzyme activity to 85–87% of control. PS was less effective and PE had no activity.

As previously shown [1–3] plasma membranes of rat liver contain a hormone and fluoride sensitive adenylate cyclase system. This was studied also by measuring the binding of labeled hormones to specific sites of the membrane which very probably represent the regulatory subunits (receptors) [12,13].

Removal of phospholipids from membranes produced a loss of the responsiveness of the enzyme to the epinephrine (fig. 4), glucagon and fluoride (table 1). In fig. 4, the effect of phospholipids in the restoration of epinephrine-sensitive adenylate cyclase and in that of the binding of the hormone to

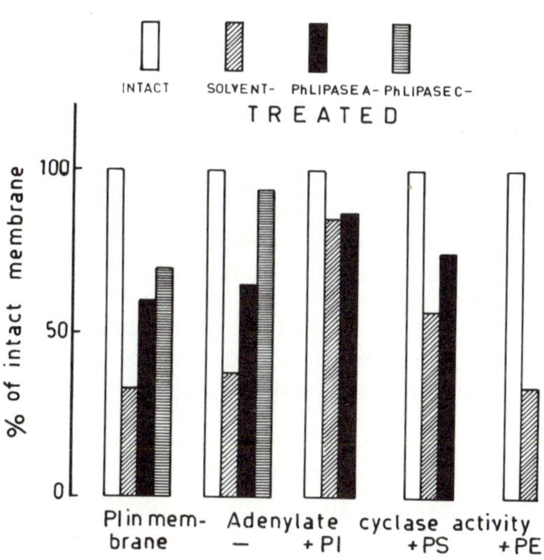

Fig. 3. Role of PI in the restoration of adenylate cyclase activity. For experimental details see fig. 2, and [7a]. Phospholipids were added to the enzyme preparations as aqueous emulsion in buffer according to Lester and Smith [11]. In this and in other experiments the concentration of PI was 8 μg P/mg membrane protein.

Table 1
Role of PS in the restoration of glucagon and fluoride sensitive adenylate cyclase activity [a)]

Additions	Increase of adenylate cyclase activity due to glucagon or fluoride [b)]			
	Control	Solvent treated membranes		
		–	+ PS	+ PI
Glucagon	9.34	0.1	2.3	−0.75
NaF	10.0	0.4	2.7	−0.26
	Control	Phospholipase A-treated membranes [c)]		
		–	+ PS	+ PI
Glucagon	14.1	0.44	4.15	−0.26
NaF	13.3	0.52	2.3	0.2
	Control	Phospholipase C-treated membranes [c)]		
		–	+ PS	+ PI
Glucagon	14.13	0.17	3.0	–
NaF	13.3	0	1.8	–

a) For experimental conditions see fig. 2 and [7a]. Basal enzyme activity was 6.49 ± 0.3 nmoles/mg membrane protein/hr (mean ± S.E. of 5 experiments). The values are differences between the activity of enzyme in presence and absence of glucagon or fluoride.
b) Glucagon (crystalline, insulin-free, gift of dr. R. Chance, E. Lilly, Indianapolis, USA) was 2×10^{-6} M and NaF 1×10^{-3} M.
c) For experimental conditions see fig. 1. The higher stimulatory effect is due to the fact that membranes were preincubated for 45 min at 37° before enzyme assay. The amount of phospholipases was 15 μg/mg membrane protein.

membranes is reported. It can be observed that only PS, but not PI and PE, was able to restore the sensitivity to epinephrine of adenylate cyclase; the restoration was complete (117%) for solvent treated membranes and partial for phospholipase A (58%) and phospholipase C (44%), treated ones. In order to gain a better understanding of the first steps of the action of hormones, the binding of labeled epinephrine to plasma membrane was studied. While the epinephrine sensitivity was completely lost, the binding was reduced only partially (57%, 41% and 46% respectively for solvent, phospholipase A and C

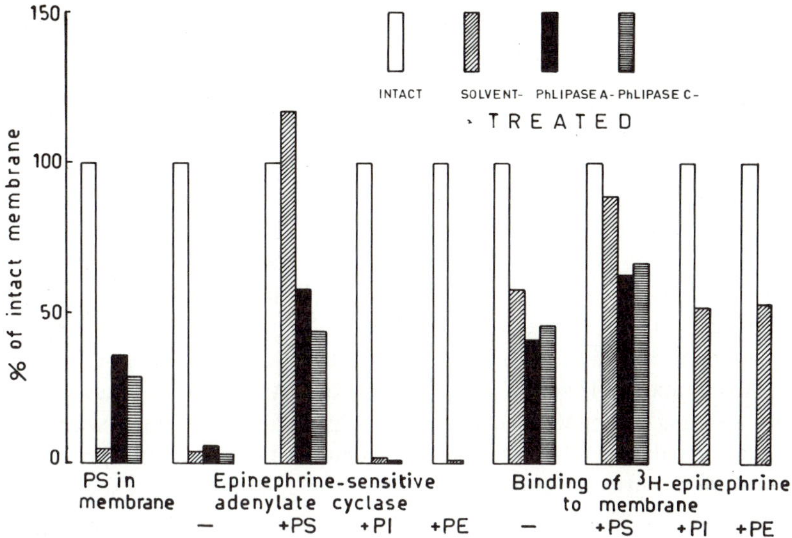

Fig. 4. Role of PS in the restoration of epinephrine-sensitive adenylate cyclase and of the binding of labeled epinephrine to membranes. The amount of PS added in this and in others experiments was 12 μg P/mg protein. Binding of ^{3}H-epinephrine to membranes was evaluated, according to a recently described method [13]. The activity of adenylate cyclase in the presence of 1×10^{-5} M epinephrine was 10.99 ± 0.7 nmoles/mg protein/hr (mean ± S.E. of 3 experiments). The amount of ^{3}H-epinephrine bound to control membranes was 33132 ± 1001 for extracted membranes and 30108 ± 869 (cpm/mg protein/hr; mean ± S.E. of 3 experiments) for phospholipases treated.

treated membranes). As expected only PS had the capacity to restore the binding till to 89% (extracted membranes) and 63% (phospholipase treated membranes) of control. With a different procedure [12] similarly based on the removal of free epinephrine by washing following centrifugation, the extent of restoration was respectively 97% and 85%.

Glucagon and fluoride sensitivity of adenylate cyclase could also be restored by PS (table 1) however the phospholipid was less effective than in the case of epinephrine-sensitive cyclase. Again PI (table 1) and PE (not shown) had no effect on the restoration.

Discussion

The lipoproteic nature of adenylate cyclase, first suggested by Sutherland et al. [4] is clearly confirmed by the present work. In addition the data shown here suggest that PI is essential for the baseline adenylate cyclase activity of plasma membrane, while PS is involved in both the binding of epinephrine and in its stimulatory action on the enzyme activity.

Fisher and Jost [15] have studied the binding of epinephrine to liver cell suspensions using NMR. Their results suggest a strong interaction between plasma membrane components and the phenyl-ring of epinephrine as well as the nitrogen of side chain. Whether PS is the actual "receptor" for epinephrine cannot be stated safely at present, although this idea comes out also from the NMR studies of Hammes and Tallman [16]. As a matter of fact they have found a strong interaction between epinephrine and PS, but not with PE and PC. However the possibility cannot be excluded that PS may represent only one site of the interaction hormone-"receptor", being possibly a second site of polypeptidic nature.

Robison et al. [17] have depicted the adenylate cyclase molecule as composed of at least two distinct subunits: a regulatory and a catalytic subunit. The hormones acting on this enzyme are supposed to interact with the regulatory subunit. The data of the present paper suggest a role of PI in the function of the catalytic subunit and a role of PS in the response to hormones or fluoride. At least in the case of epinephrine, the stimulatory action and the binding capacity were influenced in a parallel way, namely they were reduced in solvent extracted or phospholipase-treated membranes, and were restored by re-addition of PS.

While this manuscript was in preparation, Pohl et al. [18] presented evidence for a role of membrane lipids in the control of adenylate cyclase activity and in the binding of glucagon to plasma membranes of rat liver.

Acknowledgment

The present work was aided by a grant from The Consiglio Nazionale delle Ricerche to Prof. O. Barnabei (Contract 70.0091904).

References

[1] E.W. Sutherland and P.R. Davoren, in: Biochemical Aspects of Hormone Action (Little Brown, Boston, 1964) p. 149.

[2] S.L. Pohl, L. Birnbaumer and M. Rodbell, Science 164 (1969) 566.

[3] G.V. Marinetti, T.K. Ray and V. Tomasi, Biochem. Biophys. Res. Commun. 36 (1969) 185.
[4] E.W. Sutherland, T.W. Rall and T. Menon, J. Biol. Chem. 237 (1962) 1220.
[5] G.S. Levey, Biochem. Biophys. Res. Commun. 38 (1970) 86.
[6] R. Coleman, J.B. Finean, S. Knutton and A.R. Limbrick, Biochim. Biophys. Acta 219 (1970) 81.
[7] A. Martonosi, J. Donley and R.A. Halpin, J. Biol. Chem. 243 (1968) 61.
[7a] A. Rethy, V. Tomasi and A. Trevisani, Arch. Biochem. Biophys. 147 (1971) 36.
[8] P. Emmelot and C.J. Bos, Biochim. Biophys. Acta 150 (1968) 341.
[9] A. Rethy, V. Tomasi, A Trevisani and O. Barnabei, manuscript submitted for publication.
[10] P. Emmelot and C.J. Bos, Biochim. Biophys. Acta 120 (1966) 369.
[11] R.L. Lester and A.L. Smith, Biochim. Biophys. Acta 47 (1961) 475.
[12] V. Tomasi, S. Koretz, T.K. Ray, J. Dunnick and G.V. Marinetti, Biochim. Biophys. Acta 211 (1970) 31.
[13] M. Rodbell, H.M.J. Krans, S.L. Pohl and L. Birnbaumer, J. Biol. Chem. 246 (1971) 1861.
[14] O.H. Lowry, N.J. Rosebrough, A.L. Farr and P.J. Randall, J. Biol. Chem. 193 (1951) 265.
[15] J.J. Fisher and M.C. Jost, Mol. Pharmacol. 5 (1969) 420.
[16] G.G. Hammes and D.E. Tallman, Biochim. Biophys. Acta 233 (1971) 17.
[17] G.A. Robinson, R.W. Butcher and E.W. Sutherland, Ann. N.Y. Acad. Sci. (U.S.A.) 139 (1967) 703.
[18] S.L. Pohl, H.M.J. Krans, V. Kozyreff, L. Birnbaumer and M. Rodbell, J. Biol. Chem. 246 (1971) 4447.

PART II

GENETIC ASPECTS OF TRANSPORT

INTRODUCTION

A. KEPES
Institut de Biologie Moléculaire, Laboratoire de Biomembranes, Paris, France

Genetics have taught us that transmembrane transport is due primarily to specific macromolecules. Errors introduced by mutation into the genetic information may result in structural alterations of a protein involved in a transport function and such alteration can entail the abolition of the function. The genetic deficiencies served and still serve as one of the sharpest tool to distinguish one transport system from another and also to tell it apart from auxiliarly systems.

Other mutations may not cause a complete loss of a transport but introduce a qualitative change such as altered stability, altered specificity or a deficiency in the energy coupling which transforms an equilibrating transport into an active transport. In this respect genetics can help to go much farther than the simple identification and classification of transport systems and can furnish a clue in the understanding of transport mechanisms.

Some transport systems include more than one protein responsible for the function and in these cases again the genetic tool is invaluable. Moreover transport systems are located in a membrane, and alterations of the membrane structure as a result of a genetic deficiency can have a profound influence on transport, ending occasionally in its complete abolition.

Much of the use of the genetic method for the better knowledge of transport has been made in the field of micro-organisms. But even in micro-organisms it would be dangerous to entrust genetics with magic virtues. Mutations can occur not only in the structural genes of transport proteins but also in regulatory genes. Regulatory genes often control the synthesis of more than one gene product, and the deficiencies observed can create confusion. Another kind of confusion may result when more than one system is capable to transport the same substrate.

The situation is even more complexe in higher organisms where the exchange of materials between cells and the *milieu intérieur* is tightly coordinated, a deficiency in transport cannot be easily compensated by an adequate change in the milieu, and therefore a major loss of function has good chances to become lethal and to never be observed. Moreover, transport in higher organisms is controlled not only by the control of the biosynthesis of transport proteins but also by the control of activity through hormones, several of which can have synergistic or antagonistic effects. Such a complexe setup doubtlessly means that transport proteins, besides their substrate binding site, possess recognition sites for regulatory ligands and possibly have one or several regulatory subunits. Therefore the symptomatology of genetic alterations of transport becomes more complexe and its interpretation less unambiguous.

In spite of these weaknesses, genetics have already given valuable contributions to the knowledge of transport systems in higher organisms and the time is ripe to compare and to confront acquisitions in this subject in the field of bacteria, unicellular eukaryotic organisms and multicellular higher organisms. This was the aim pursued in organizing the present session on genetic aspects of transport.

GENETICS OF THE PHOSPHOTRANSFERASE SYSTEM*

Wolfgang EPSTEIN and Susan J. CURTIS
Department of Biochemistry, University of Chicago, Chicago, Illinois, U.S.A.

1. Introduction

In 1964 Kundig, Ghosh and Roseman described a new bacterial phosphorylating system [1]. This phosphotransferase system, here abbreviated PTS, catalyzes the following reactions:

$$\text{Phosphoenolpyruvate} + \text{Hpr} \xleftrightarrow{\text{enzyme I, Mg}^{2+}} \text{P-Hpr} + \text{Pyruvate} \qquad (1)$$

$$\text{P-Hpr} + \text{Sugar}_i \xrightarrow{\text{enzyme II}_i\text{, Mg}^{2+}} \text{P-Sugar} + \text{Hpr} \qquad (2)$$

In the first step a soluble protein, enzyme I, catalyzes the reversible transfer of phosphate from phosphoenolpyruvate to Hpr. Hpr is a heat-stable protein with a molecular weight of 9 500 [2]. Reaction 1 is common to all phosphorylations by this system. In the second step a membrane-bound component called an enzyme II catalyzes the transfer of phosphate from P-Hpr to the sugar. This second step is specific for the sugar, there being a number of different species of enzyme II each with a different range of substrate specificity.

Additional protein components are needed in the phosphorylation of some sugars. The most carefully studied example is the phosphorylation of lactose

* Abbreviations: PTS, the phosphoenolpyruvate-dependent phosphotransferase system; αMG, methyl-α-D-glucopyranoside; cAMP, cyclic 3′,5′-adenosine monophosphate; TMG, methyl-β-D-1-thiogalactopyranoside. All sugars are of the D configuration unless otherwise stated.

in *Staphylococcus aureus*. Here reaction 2 above is replaced by the following two steps:

$$2\ \text{P-Hpr} + \text{III}_{lac} \longleftrightarrow \text{P}_2\text{-III}_{lac} + 2\ \text{Hpr} \tag{3}$$

$$\text{P}_2\text{III}_{lac} + 2\ \text{lactose} \xrightarrow{\text{enzyme II}_{lac},\ \text{Mg}^{2+}} 2\ \text{P-lactose} + \text{III}_{lac} \tag{4}$$

Factor III_{lac} is a soluble protein induced by lactose and specific for lactose phosphorylation. Reaction 3 requires no catalysts besides P-Hpr and factor III. Reaction 4 is analogous to reaction 2 except that the phosphate donor is now sugar-specific just as is enzyme II [3]. Requirements for additional soluble protein components have been reported for the phosphorylation of mannose and mannitol in *S. aureus* [2, 4], and for fructose in *Aerobacter aerogenes* [5].

Not long after the biochemical discovery of PTS it was found that a number of laboratories had been studying bacterial mutants defective in either enzyme I or Hpr without knowing it. These mutant strains were unable to grow on a large number of carbon sources. The mutant strains were defective in the uptake of the carbon sources that they could not utilize [6–8]. With the subsequent demonstration that the defects in the mutants were in the PTS it was natural to conclude that the PTS is needed for the transport as well as the phosphorylation of many substaces [9–11].

Models of transport in which movement of a substance across a membrane barrier is coupled to a covalent chemical change in the substrate are not new, coupled phosphorylation and transport having been proposed as the mechanism of intestinal sugar transport many years ago [12]. The PTS is the first, and so far the only system for which there is good evidence to indicate that transport and covalent chemical modification of the substrate can occur as a single step. Intestinal transport of sugars is not coupled to a chemical change in the transported sugar indicating that this model does not apply to the system for which it was first proposed [13].

In this presentation we will examine the PTS and its transport role, mainly using information obtained from genetic studies of the PTS in *Escherichia coli.* There is now a considerable amount of data about the PTS, and as will be seen below the interpretation of some of it is not so simple as was originally proposed.

2. Mutants in enzyme I and Hpr

Mutants defective in these two common components of the PTS have been obtained in a variety of bacterial species [9–11, 14]. The locus for mutations in *E. coli* affecting enzyme I is called *pts I* by some authors [15, 16] and *ctr* by others [11]. This locus is near minute 46 on the current map of *E. coli* [17] and is the structural gene for enzyme I as shown by the fact that a mutational site in this locus which results in temperature-sensitive growth for PTS-dependent sugars also results in an unusually thermolabile enzyme I [16]. In *Salmonella typhimurium*, whose genetic map is very similar to that of *E. coli*, these mutants are referred to as *carA* and map at a location analogous to the *ptsI* locus in *E. coli* [18]. Mutations in *E. coli* leading to loss of Hpr activity define the *ptsH* locus which is very closely linked to *ptsI* [16]. The Hpr locus in *S. typhimurium*, *carB*, is at the corresponding position on the map of that organism [19]. Fig. 1 shows the locations of the *ptsH* and *ptsI* genes on the *E. coli* chromosome, as well as the positions of loci for several of the enzymes II.

The property of the enzyme I and Hpr mutants that attracted so much initial attention, and that is still not completely understood, is their inability to grow on a large number of carbon sources. Some representative data on growth properties of enzyme I mutants are listed in table 1. This growth

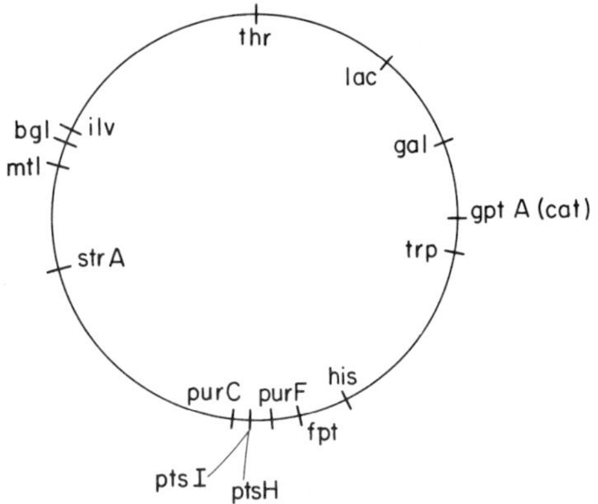

Fig. 1. Loci affecting the PTS in *E. coli*. Inside the circle reading clockwise are the following reference loci [17]: *thr*, threonine; *lac*, lactose; *gal*, galactose; *trp*, tryptophan; *his*, histidine; *purF* and *purC*, the purine F and C loci; *strA*, streptomycin; and *ilv*, isoleucine-valine. Outside the circle are symbols for PTS loci: *gptA*, also called *cat*, glucose-phosphotransferase A; *fpt*, fructosephosphotransferase, or the fructose enzyme II [20]; *ptsH*, Hpr; *ptsI*, enzyme I; *mtl*, mannitol enzyme II; and *bgl*, aryl-β-D-glucoside enzyme II [21].

Table 1
Growth properties of enzyme I mutants *

Species and strain	References	Substrates **									
		Arbutin	Mannitol Sorbitol	Glucose	Mannose Fructose	Lactose	Glycerol Maltose	Melibiose Succinate	Arabinose	Galactose	Glucose-6-P Gluconate Pyruvate
E. coli	this work	–	–	–†	–	–	–	–†	+	+	++
E. coli MM-6	6, 9		–	+	–	+		–		++	++
E. coli 903C	8	–	–	–	–	–	–	–	++	–	++
S. typhimurium SB 703	22		–	–	–		–	–	+	++	++
A. aerogenes ‡ 2050	14		–	+	‡	++	+		++	++	
S. aureus	7		–	++	–	–	#			–	++

* Growth is rated as follows: –, no growth; +, slow growth or growth after a significant lag; ++ growth like that of wild-type strains. No entry means that substrate was not tested, or that it is not utilized by the wild-type strain.

** Not every member of the grouped substrates was tested in all species. For the *E. coli* strains, MM-6 was not tested on melibiose, gluconate or pyruvate, and 903C was not tested on sorbitol or gluconate. *S. typhimurium* SB 703 was not tested on succinate, and the *S. aureus* mutant was not tested on sorbitol, mannose, gluconate, or glucose-6-P.

† Slow growth on low concentrations of glucose and succinate occurs in some *E. coli* strains but not in others.

‡ This strain appears to have a leaky enzyme I mutation accounting for slow growth on several PTS substrates. The strain grows on several PTS substrates. The strain grows slowly on mannose but not on fructose.

The *S. aureus* mutant grows on glycerol but not on maltose.

defect in the mutants is caused solely by the single mutation in enzyme I, since revertants or recombinants that have regained enzyme I activity grow normally. If PTS function is absolutely necessary for the metabolism of a carbon source, then all revertants selected to grow on that compound will be wild-type revertants that have regained enzyme I activity. If, on the other hand, the block to utilization of a substance is an indirect effect of the loss of PTS function, or if an alternate pathway can be created by a single mutation, then some of the revertants will remain defective in enzyme I although able to grow on the selecting compound as well as others for whom a similar situation holds [23].

From genetic and biochemical studies it is possible to classify the compounds not utilized by enzyme I mutants into three classes. In class I we place mannitol, sorbitol, and aryl-β-D-glucosides such as arbutin and salicin. For these compounds the only known catabolic pathway in most enteric bacteria is a PTS-mediated phosphorylation, so that failure to utilize these compounds is a direct consequence of loss of PTS function [15, 21]. Revertants selected to grow on any of these carbon sources are always revertant for enzyme I as expected [23, and our unpublished observations].

Into class II we gather glycerol, lactose, maltose, and melibiose. Failure to grow on these compounds is paradoxical, since it is known from biochemical and genetic studies that these substances are catabolized by steps which do not require the PTS [24–27]. This paradox has been resolved in part by the observation that a primary difficulty is in the induction of the specific enzymes and transport activities needed for the catabolism of these compounds. When the presence of these specific enzymes and activities is assured by the addition of a potent inducer [28], by introduction of a mutation resulting in constitutive synthesis [29], or by the addition of adenosine-3′5′-monophosphate (cAMP) [28, 30], then the enzymes grow on class II substances. Failure of induction in enzyme I mutants is due to catabolite repression, the decreased synthesis of certain catabolic enzymes when the bacteria are growing on a good carbon source such as glucose. This phenomenon can be elicited with virtually any carbon source whenever growth is limited less by the availability of an energy source than by some other factor such as the nitrogen source or an amino acid [31]. This effect is presumably mediated by a lowering of cell cAMP levels, since the addition of cAMP reverses catabolite repression [32].

The effect of cAMP in permitting enzyme I mutants to grow on class II compounts is shown in fig. 1. Removal of cAMP from the culture results in continued exponential growth for a time, followed by transition to linear growth. This is expected if synthesis of the specific enzymes and transport

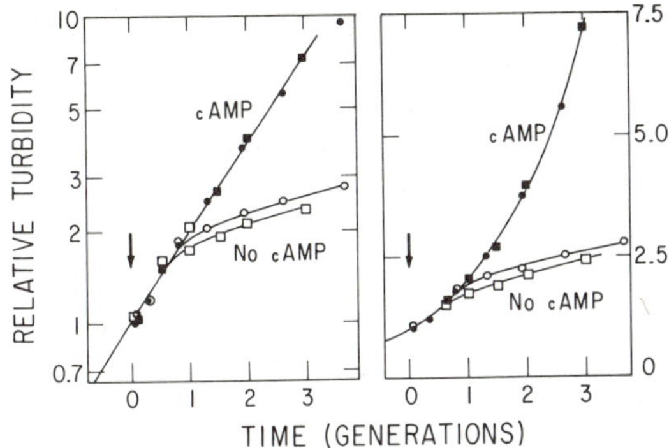

Fig. 2. Growth dependence on cAMP. At time 0 cultures of an enzyme I mutant growing at 37° in medium containing 2g/liter carbon source and 2 mM cAMP were washed and suspended at approximately 10^8 cells/ml in the same medium with or without cAMP. Growth was measured as the increase in turbidity at 610 mμ, plotted relative to the turbidity at zero time. Squares are cultures growing on maltose, and circles cultures growing on lactose. Filled symbols are cultures with cAMP, open circles without it. The data are replotted on a linear turbidity scale at the right to emphasize the linear growth at late times in the absence of cAMP. The time scale is in terms of generation time, which was 78 min for maltose and 68 min for lactose.

activities needed for the catabolism of the substance ceases after cAMP removal, so that the rate at which the substance can be utilized no longer increased with time. If cAMP were required for some biochemical reaction, one would expect growth to cease soon after cAMP is removed (fig. 2).

The failure of induction of enzymes in enzyme I mutants indicates that the mutants have an enhanced sensitivity to catabolite repression. Pyruvate, a substance which produces very little catabolite repression in wild-type *E. coli* produces rather marked catabolite repression in enzyme I mutants (R. Bloom, unpublished observations). The mutants do grow on some sugars the enzymes for which are sensitive to catabolite repression. The best example is L-arabinose. The reason arabinose can be utilized but the compounds of class II cannot probably reflects the mechanism of induction. The true inducer of the arabinose enzymes is arabinose itself [33], while the true inducer of the glycerol and lactose enzymes is not the substrate, but a product formed from the substrate by the action of the first enzyme of the pathway [24, 34]. Thus induction for glycerol and lactose is autocatalytic, and any interference with induction by catabolite repression would be unusually effective in preventing good induction and growth.

Enhanced sensitivity to catabolite repression is not the only factor interfering with the utilization of class II compounds. When enzyme I mutant cells growing in minimal medium are trasferred to minimal medium containing cAMP and one of the compounts of class II as carbon source, growth begins only after a long lag of up to 15 hr. The resulting growth is due to reversion when some strains are tested, but for enzyme I mutants with low reversion rates the cells that finally grow are still mutant. Since the concentration of cAMP used should allow efficient induction of the necessary enzymes, the occurrence of an unusually long lag shows that some other factor is interfering with utilization of these compounds. A growth lag is also seen when enzyme I mutants are transferred to media without cAMP and containing arabinose or galactose, carbon sources which the mutants can utilize [22]. The addition of cAMP shortens but does not eliminate the growth lag indicating that catabolite repression contributes to the lag but that other factors are also operating. This other unknown factor appears to be predominant in *S. typhimurium* since in that species cAMP does not allow enzyme I mutants to grow on compounds of class II [35] although cAMP is effective in abolishing catabolite repression in *S. typhimurium* [32].

Three sugars which are catabolized by the PTS, and for which other major catabolic pathways may exist are in class III: fructose, glucose and mannose. For glucose there is a soluble glucokinase which is synthesized constitutively and has very narrow substrate specificity [36]. However mutants lacking glucokinase grow on glucose, and conversely mutants with normal levels of glucokinase but lacking enzyme I do not grow or grow slowly on glucose [37]. These facts indicate that glucokinase probably plays no important role in glucose catabolism, and that the glucokinase in enzyme I mutants cannot function perhaps because glucose cannot enter the cell (see below in section 5). Pregrowth of the cells on galactose [6], or the addition of fucose [22], a non-metabolizable galactose analog, allows enzyme I mutants to utilize glucose. It has been suggested that galactose and fucose induce a transport system which allows glucose to enter and thus be utilized [22].

Mannose and fructose are sugars for which no pathway other than the PTS has been reported in *E. coli*. There is a low level soluble kinase for these two sugars, but this activity is much too low to allow for growth [37, 38]. Therefore one would expect fructose and mannose to belong to class I and be absolutely dependent on PTS function for their catabolism. However it has been reported that partial revertants that grow on fructose or mannose but not on mannitol can be obtained from enzyme I mutants [23]. These revertants suggest the presence of another pathway for these sugars. Until these revertants are characterized this question cannot be considered resolved.

The inability of many enzyme I mutants to use certain tricarboxylid acid cycle substrates such as succinate has not been explained. Since these substances are not phosphorylated prior to entering common metabolic paths, it may be presumed that indirect effects of the PTS defect are resposible for failure to grow in these compounds.

Mutants defective in Hpr are similar in properties to enzyme I mutants, but all Hpr mutants isolated to date are distinctly leaky (the mutational block is not complete) and grow slowly on the compounds not utilized by enzyme I mutants [16, 35]. This leakiness complicates genetic studies. In the study of mutants it is desirable to use ones in which the mutated function is totally absent; otherwise one may be studying the effects of the marked reduction in the rate of a process which may be different from the results of a total defect. Mutations most likely to be totally lacking a function are deletions, yet no deletions in either enzyme I or Hpr have yet been reported. Among over 100 mutants screened in this laboratory a number have very low reversion rates but none are deletions. A non-reverting mutant defective in both enzyme I and Hpr has been reported [39], but our studies of this strain indicate that it probably carries a mutation in the *ptsI* gene which is definitely not an extended deletion since it recombines with three ptsI and one ptsH point mutations. Whether the mutant has a small deletion, or is a double mutant is not known. The fact that all Hpr mutants are leaky and that no enzyme I deletions have been reported raises the interesting possibility that total loss of function of the PTS is disadvantageous to the cell.

3. The enzymes II

The membrane bound enzymes II are specific for a given sugar or group of sugars. Many questions remain to be answered regarding the composition and functioning of these membrane-bound activities. We will discuss chiefly the *E. coli* enzyme II for glucose, referred to here as glucosephosphotransferase (GPT).

From genetic and biochemical analyses we have identified two different GPT enzymes [40]. The rates of phosphorylation of several sugars by wild-type and some mutant strains of *E. coli* K-12 are shown in table 2. Wild-type strains like W1895 phosphorylate both α-methylglucoside (αMG) and glucose, the latter at a higher rate. In mutant strain W1895D1, derived from W1895 and previously shown to be defective in the phosphorylation of αMG [21], there is also a reduction of glucose phosphorylation approximately equal to the reduction seen with αMG. This suggests that this mutant strain lacks a particular GPT which phosphorylates both glucose and αMG and which we call GPT-A, while retaining a second type, called GPT-B with poor affinity

Table 2
Glucosephosphotransferase activities in *E. coli* *

Strain	Defects **	Glucose added	Substrate			
			αMG	glucose	mannose	*N*-acetyl-mannosamine
			(μmoles/min-g protein)			
W1895	none	–	1.8	4.5	5.7	1.2
		+	3.3	5.9		
W1895D1	*gptA*	–	0.15	2.5	3.4	1.2
		+	0.1	2.3		
ZSC103	*gptA, glk*	–	0.1	2.0	2.3	1.1
ZSC103a	*gptA, gptB, glk*	–		1.1	1.0	0.4

* Cells grown at 37° in phosphate-buffered minimal medium containing 10 g/liter casamino acids with or without glucose 10 g/liter as indicated were harvested in the exponential phase of growth and washed. Sonic extracts were prepared and assayed essentially as described by Fox and Wilson [2]. No entry means assay was not performed.

** *gptA*, loss of glucosephosphotransferase A; *gptB*, loss of glucosephosphotransferase B; *glk*, loss of glucokinase.

Table 3
Inhibition of glucosephosphotransferase by hexoses

Inhibitor	Inhibition of glucose phosphorylation * (%)	
	GPT-A	GPT-B
αMG	> 90	< 10
6-deoxyglucose	> 90	< 10
2-deoxyglucose	70	90
Mannose	> 90	> 90
Galactose	55	74

* The rate of phosphorylation of 1.25×10^{-4} M glucose was measured in the absence and the presence of 10^{-2} M inhibitor in strains W1895, W1895D1, and ZSC103a (table 2). Using the results for strain ZSC103a as a measure of the inhibition of the activity due to neither GPT-A nor GPT-B, percent inhibition, expressed as the reduction in activity in the presence of the inhibitor, was calculated.

for αMG. The data of table 2 show that both GPT-A and GPT-B phosphorylate mannose at rates proportional to those for glucose indicating rather broad substrate specificity. Both types of GPT also have affinity for galactose and 2-deoxyglucose as demonstrated by their inhibition of glucose phosphorylation (table 3). 6-Deoxyglucose is another sugar which has affinity for GPT-A but not for GPT-B (table 3). Strain ZSC103a, selected as a mutant unable to grow on 5 mM glucose from a parent lacking both GPT-A and glucokinase, has much reduced rates of phosphorylation of glucose and mannose. We infer from the data of table 2 that this mutant has lost GPT-B and that the residual activity is of little physiological significance since the mutant does not grow on modest concentrations of glucose or mannose.

The rather broad substrate specificity of these enzymes II contrasts with the stringent substrate requirement of the soluble bacterial kinases [36, 41]. Even the difference in the GPT's with regard to αMG is not absolute, since mutants lacking GPT-A do phosphorylate αMG to some extent, and high concentrations of αMG inhibit glucose phosphorylation by GPT-B. Our data suggest that whereas the K_m of GPT-A for αMG is 1×10^{-4} M, the K_m of GPT-B for this analog is at least 100 times higher. The fructose enzymes II in *A. aerogenes* also seem to have broad specificity. Growth of this species on fructose results in an increase in the V_{max} of fructose phosphorylation and a striking reduction in the K_m from 65 mM to less than 1 mM. Growth on fructose changes a membrane-bound enzyme II and a soluble factor which may function as a factor III. This result has been interpreted to indicate that the soluble factor lowers the K_m of the fructose-specific enzyme II, but we feel that a more likely possibility is that two different types of enzyme II are being assayed, a low K_m fructose enzyme and a high K_m enzyme whose primary specificity is for another sugar. This interpretation is supported by the mannitol inhibition of fructose phosphorylation in mannitol-grown cells [5].

Kundig and Roseman have reported the solubilization and fractionation of enzyme II activity into three components: a IIa protein specific for the substrate, a IIb protein solubilized with difficulty and common to all three substrates studied, and phosphatidylglycerol [42]. Three IIa proteins were purified, one specific for fructose, another specific for N-acetylmannosamine, and a third specific for αMG but with slight activity with N-acetylmannosamine. The data of Kundig and Roseman suggest that their glucose enzyme is GPT-A, while their mannose enzyme is GPT-B. To call these glucose-specific and mannose-specific is misleading, since each of the two has good activity on both mannose and glucose (table 2). It is only when the non-metabolized analogs are used that differences are readily demonstrated. The identification of our two GPT's with their activities is reasonable but not established at present. In particular, we have not excluded the possibility that our mutations affect some soluble component such as a factor III.

There is a striking discrepancy between the *in vivo* and the *in vitro* functioning of the enzymes II that has not yet attracted much attention [9, 43]. In growing cells of *E. coli* the K_m for glucose utilization is approximately 10^{-6} M and the rate of utilization at 30° is 200 μmoles/min-g protein in a typical wild-type strain. In the sonic extracts used by most investigators the K_m is approximately 1.2×10^{-4} M and the V_{max} is in the range of 4 to 10 μmoles/min-gm protein at 30°. A perusal of the literature reveals that the rates of various enzymes II are in the range of 2 to 20 μmoles/min-g protein, always well below the rate at which the sugar is utilized by the intact cell. This difference could be used to argue that the enzymes II are not involved in the catabolism of these sugars much as we dismissed the mannofructokinase above (see section 2), but the genetic evidence established that the enzymes II are needed for catabolism of the sugars. Mutants defective in the enzymes II for a sugar do not grow on that sugar [20, 44], and in the case of the GPT's there is good quantitative correlation in that mutants lacking GPT-A have only 50% of wild-type GPT activity and grow at 60% of the wild-type rate on glucose and mannose. We are thus lead to conclude that *in vivo* these enzymes function at a rate approximately two orders of magnitude higher than that measured in sonic extracts. Gachelin observed that gentle toluene treatment of intact cells resulted in activities up to 10 times higher than those in sonic extracts, and the K_m was generally somewhat lower as well [43]. Although the rates in toluenized cells are still lower than *in vivo* rates, they show that the enzymes II are capable of much higher rates of phosphorylation under suitable conditions. So far the factor or factors which are responsible for the very different kinetic properties of the *in vitro* enzyme II as compared to its *in vivo* functioning have not been identified.

4. Genetic regulation of the PTS

Both of the two common soluble components of the PTS, enzyme I and Hpr, are present at high levels no matter what carbon source is used for growth, but when a substrate of the PTS is present during growth there is approximately a 3-fold increase in each in *S. typhimurium* [35], *A. aerogenes* [14], and *E. coli* [C.F. Fox, personal communication]. The genes for Hpr and enzyme I probably constitute an operon in *S. typhimurium* and *E. coli,* this conclusion being based on partial evidence from one or the other of these two closely related species under the assumption that what is true for one is true for the other. In *S. typhimurium* enzyme I and Hpr vary coordinately as expected for products from genes in an operon, and a UGA nonsense mutation in Hpr results in a 75% reduction in enzyme I. This is probably an example of genetic polarity, the reduced expression of one gene due to a chain-terminating mutation in another gene in the same operon closer to the operator. Very close apposition of the genes as required for an operon has been demonstrated for *E. coli* [16]. The non-coordinate regulation of enzyme I and Hpr in *A. aerogenes* indicates that these two are not the products of a single operon in that species [14].

In enterobacteria the enzymes II for mannitol [15], aryl-β-glucosides [21], and one of these for glucose, namely GPT-A (table 2), are all inducible although the increase in GPT-A is only 3-fold while for the others the increase is much larger. The fructose enzyme II is constitutive in the *E. coli* strain used by Kundig, Ghosh and Roseman [1] but is apparently inducible in other *E. coli* strains since the initial rate of fructose uptake, an indirect measure of enzyme II activity, is inducible [8, 20]. Both the enzyme II and the soluble additional factor necessary for fructose phosphorylation with a low K_m in *A. aerogenes* are inducible [5]. In *S. aureus* both the enzymes II and the factors III for lactose and for mannitol are inducible [4]. However, even GPT-B which is not induced by glucose is subject to regulatory influences; in strains mutant in enzyme I the level of both forms of GPT is elevated approximately 3-fold over the levels in wild-type strains grown on casamino acids (unpublished observations) just as found for the αMG-specific enzyme II in *S. typhimurium* [22].

5. Transport function of the PTS

The PTS appears to mediate coupled transport and phosphorylation but only for substrates of the PTS. The transport of substances that are not PTS substrates does not depend on the PTS. The lactose transport system in an *E. coli* enzyme I mutant functions normally if induced [28], and the normal growth rates of an enzyme I mutant on lactose and maltose in the presence of cAMP (fig. 2) shows that transport of these substances is independent of the PTS. Therefore the earlier proposal that the PTS mediates transport of almost all sugars in enteric bacteria is no longer tenable [2, 22].

The best evidence for coupled transport and phosphorylation is kinetic. When the glucose analog αMG is taken up by intact cells [45] or by membrane vesicles [46] the first form of the analog to appear inside is phosphorylated, and only later do significant amounts of unphosphorylated analog appear. These results are difficult to reconcile with a model in which entry of the substance without chemical modification precedes phosphorylation. There is no evidence for other transport systems for PTS substrates, since enzyme I mutants take up only small amounts of these substances [21, 22]. The residual uptake in the mutants is not so low as to exclude the possibility that PTS substrates enter to reach the same concentration in the cell as outside [21].

Transport and phosphorylation, normally coupled, can be separated in the case of methyl-β-D-thiogalactoside (TMG) phosphorylation in *E. coli*. This lactose analog is accumulated by the β-galactoside permease and small amounts are phosphorylated. Kashket and Wilson isolated mutants in which much larger amounts of TMG are phosphorylated, and showed that phosphorylation was probably performed by the PTS. Kinetic analysis showed that free TMG appeared first in the cell, followed by phosphorylated TMG. Phosphorylation of TMG was dependent on the presence of a functional β-galactoside permease [47]. If this is a PTS-mediated phosphorylation, then the enzymes II of the PTS can phosphorylate without transport of the substrate.

Can the enzymes II mediate transport without phosphorylation? These membrane-bound activities ought to be good carriers, situated in the membrane and with substrate binding site that can fact inside or outside. However, the observation (section 2 above) that enzyme I mutants do not grow on glucose, but will grow on glucose if pre-grown with galactose of fucose suggests that glucose entry is markedly defective in enzyme I mutants even though the enzymes II for glucose are present at high levels [22]. That some other factor may be responsible for the inability of enzyme I mutants to use glucose is implied by the observation that when a lactose-constitutive enzyme I mutant of *E. coli* is exposed to lactose, the glucose resulting from lactose hydrolysis rapidly appears in the medium but is utilized very slowly even after all the galactose has been used [6]. Here there must be a block to glucose utilization independent of transport for if glucose can get out it should be able to get in. At the present time insufficient information is available to decide whether or not the enzymes II can mediate equilibration in the absence of phosphorylation.

6. Conclusions

In the seven years since the first description of the PTS the importance of this system in bacterial metabolism has been established. This system appears to perform coupled transport and phosphorylation for its substrates but does not mediate the transport of other substances. At present our understanding of this system is still far from complete. The reason for the partial or complete block to utilization of non-PTS substrates in enzyme I mutants is not well understood. Neither the mechanism of enhanced catabolite repression nor the nature of the other factors responsible for the properties of the enzyme I mutants is known. The membrane-bound enzymes II, like so many membrane enzymes, remain rather mysterious. In spite of impressive progress in the solubilization and fractionation of several enzymes II, many details of their function, such as how translocation is coupled with phosphorylation, are not known. Especially distressing is the low *in vitro* activity of these enzymes as compared with their *in vivo* rates. A combination of genetic and biochemical techniques should be as fruitful in clarifying these questions as it has been earlier work on the PTS.

Acknowledgment

The work from the authors' laboratory reported above was supported by a predoctoral traineeship (Training Grant No. GM 424) to S. Curtis, a research grant (GM 15766) and a research career development award (GM 10725) to W. Epstein, all from the National Institute of General Medical Sciences of the U.S. Public Health Service.

References

[1] W. Kundig, S. Ghosh and S. Roseman, Proc. Nat. Acad. Sci. (U.S.A.) 52 (1964) 1067.
[2] S.J. Roseman, Gen. Physiol. 54 (1969) 138s.
[3] J.B. Hays and R.D. Simoni, Fed. Proc. 30 (1971) 1062A.
[4] R.D. Simoni, M.F. Smith and S. Roseman, Biochem. Biophys. Res. Commun. 31 (1968) 804.
[5] T.E. Hanson and R.L. Anderson, Proc. Nat. Acad. Sci. (U.S.A.) 61 (1968) 269.
[6] C. Asensio, G. Avigad and B.L. Horecker, Arch. Biochem. Biophys. 103 (1963) 299.
[7] J.B. Egan and M.L. Morse, Biochim. Biophys. Acta 97 (1965) 310.
[8] R.J. Wang and M.L. Morse, J. Mol. Biol. 32 (1968) 59.
[9] S. Tanaka, D.G. Fraenkel and E.C.C. Lin, Biochem. Biophys. Res. Commun. 27 (1967) 63.
[10] W. Hengstenberg, J.B. Egan and M.L. Morse, Proc. Nat. Acad. Sci. (U.S.A.) 58 (1967) 274.

[11] R.J. Wang, H.G. Morse and M.L. Morse, J. Bacteriol. 98 (1969) 605.
[12] W. Wilbrandt and L. Laszt, Biochem. Z. 259 (1933) 398.
[13] R.K. Crane, Physiol. Rev. 40 (1960) 789.
[14] S. Tanaka and E.C.C. Lin, Proc. Nat. Acad. Sci. (U.S.A.) 57 (1967) 913.
[15] E.C.C. Lin, Ann. Rev. Genet. 4 (1970) 225.
[16] W. Epstein, S. Jewett and C.F. Fox, J. Bacteriol. 104 (1970) 793.
[17] A.L. Taylor, Bacteriol. Rev. 34 (1970) 155.
[18] D. Berkowitz, J. Bacteriol. 105 (1971) 232.
[19] M. Levinthal and R.D. Simoni, J. Bacteriol. 97 (1969) 250.
[20] T. Ferenci and H.L. Kornberg, FEBS Letters 13 (1971) 127.
[21] C.F. Fox and G. Wilson, Proc. Nat. Acad. Sci. (U.S.A.) 59 (1968) 988.
[22] R.D. Simoni, M. Levinthal, F.D. Kundig, W. Kundig, B. Anderson, P.E. Hartman and S. Roseman, Proc. Nat. Acad. Sci. (U.S.A.) 58 (1967) 1963.
[23] R.J. Wang, H.G. Morse and M.L. Morse, J. Bacteriol. 104 (1970) 1318.
[24] N.R. Cozzarelli, W.B. Freedberg and E.C.C. Lin, J. Mol. Biol. 31 (1968) 371.
[25] F. Jacob and J. Monod, J. Mol. Biol. 3 (1961) 318.
[26] M. Schwartz, Ann. Inst. Pasteur 112 (1967) 673.
[27] R. Schmitt, J. Bacteriol. 96 (1968) 462.
[28] I. Pastan and R.L. Perlman, J. Biol. Chem. 244 (1969) 5836.
[29] M. Berman, N. Zwaig and E.C.C. Lin, Biochem. Biophys. Res. Commun. 38 (1970) 272.
[30] W. Epstein, S. Jewett and R.H. Winter, Fed. Proc. 29 (1970) 601a.
[31] B. Magasanik, in: The Lactose Operon, eds. J.R. Beckwith and D. Zipser (Cold Spring Harbor Laboratory, 1970) p. 189.
[32] I. Pastan and R. Perlman, Science 169 (1970) 339.
[33] R. Schleif, J. Mol. Biol. 46 (1969) 197.
[34] C. Burstein, M. Cohn, A. Kepes and J. Monod, Biochim. Biophys. Acta 95 (1965) 634.
[35] M.H. Saier Jr., R.D. Simoni and S. Roseman, J. Biol. Chem. 245 (1970) 5870.
[36] C. Asensio, Rev. espan. Fisiol. 16 (1960) Suppl. 2, p. 121.
[37] D.G. Fraenkel, F. Falcoz-Kelly and B.L. Horecker, Proc. Nat. Acad. Sci. (U.S.A.) 52 (1964) 1207.
[38] J. Sebastian and C. Asensio, Biochem. Biophys. Res. Commun. 28 (1967) 197.
[39] G.I. Burd, I.V. Andreeva, V.P. Shabolenko and V.N. Gershanovich, Mol. Biol. 2 (1968) 89.
[40] S.J. Curtis and W. Epstein, Fed. Proc. 30 (1971) 1123a.
[41] M.Y. Kamel, D.P. Allison and R.L. Anderson, J. Biol. Chem. 241 (1966) 690.
[42] W. Kundig and S. Roseman, J. Biol. Chem. 246 (1971) 1407.
[43] G. Gachelin, Biochem. Biophys. Res. Commun. 34 (1969) 382.
[44] S. Tanaka, S.A. Lerner and E.C.C. Lin, J. Bacteriol. 93 (1967) 642.
[45] G. Gachelin, Eur. J. Biochem. 16 (1970) 342.
[46] H.R. Kaback, J. Biol. Chem. 243 (1968) 3711.
[47] E.R. Kashket and T.H. Wilson, Biochim. Biophys. Acta 193 (1969) 294.

INFLUENCE OF LIPID CHANGES ON CITRATE TRANSPORT IN *BACILLUS SUBTILIS*

Klaus WILLECKE *

Department of Biochemical Sciences, Moffet Laboratory, Princeton University, Princeton, New Yersey, U.S.A.

In order to tackle the problem of membrane biosynthesis one can make use of an inducible transport activity. Such an experimental system could serve as a functional probe in the membrane to indicate the influence of the membrane composition or a disturbed membrane biosynthesis. We decided to study the influence of lipid alterations on an inducible transport system in *Bacillus subtilis*. This bacterium is genetically best explored among all gram positive organisms. It has the advantage over *Escherichia coli* that virtually all lipids are localized in a single membrane. Protoplast membranes of *B. subtilis* are easy to purify and probably free of adherent cell wall material.

We discovered a transport system for citrate in *B. subtilis* [1]. In order to obtain accurate transport kinetics one has to prevent the internal catabolism of citrate once it has entered the bacterial cell. We found that this condition is fulfilled in an aconitase-less mutant (*B. subtilis* 60871). These cells can take up citrate but they are not able to degrade it because the citrate synthase reaction is irreversible *in vivo* [2] and the enzyme activities of the citrate fermentation pathway are insignificantly low [1]. Citrate is transported in *B. subtilis* 60871 with a v_{max} of 145 ± 25 μmoles per min, per g, dry weight, of cells and an apparent K_M of 2.3 ± 0.4 mM 37°. These bacteria can accumulate citrate more than 100 fold over the citrate level in the external medium. Induction of citrate transport in *B. subtilis* cannot be followed in the aconitase-less mutant since these cells are always internally induced due to accumulated citrate as a consequence of the aconitase block. When wild type cells of *B. subtilis* were incubated with 1,5-^{14}C-citrate under our standard assay conditions (30 sec uptake at 37°) they took up almost as much radioactivity as the aconitase-less mutant. We could show that the activity from 1,4-^{14}C-citrate

* Present address: Institut für Genetik, 5 Köln 41, Weyertal 121, Germany.

is mainly trapped in the glutamate pool of the bacteria. This observation enabled us to measure citrate transport in wild type cells of *B. subtilis.*

Can citrate transport still be induced when *de novo* phospholipid synthesis is stopped? To answer this question we worked with a double mutant, *B. subtilis* B-42, which is unable to degrade or synthesize glycerol. It can, however, use glycerol for *de novo* lipid synthesis and presumably for teichoic acid synthesis. Mindich, who isolated this strain, found that *de novo* lipid synthesis can be selectively stopped when the cells are deprived of glycerol [3]. Whilst our work was in progress Hsu and Fox reported in a glycerol requiring mutant of *E. coli*, after deprival of glycerol, a sharp decrease in inducibility of the lactose transport system [4].

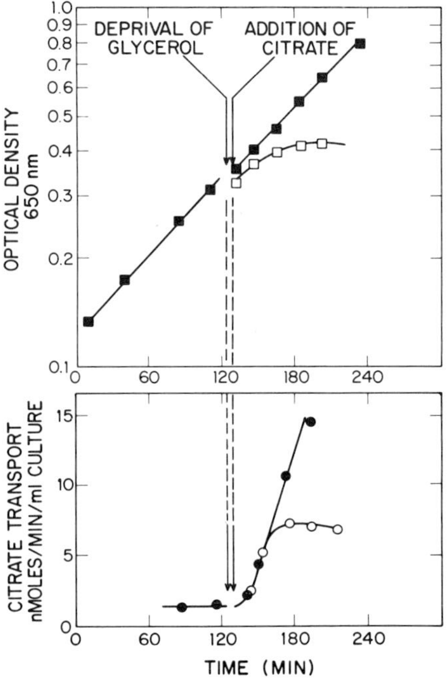

Fig. 1. Induction of citrate transport in the presence and absence of glycerol in the glycerol auxotroph *B. subtilis* B 42. Exponentially growing cells were deprived of glycerol by filtration and incubated again in the presence (closed symbols) and absence of glycerol (open symbols). Growth was estimated by measuring the optical density of the cultures (upper part of this figure). Citrate transport was induced by addition of 5×10^{-3} M citrate. Aliquots were withdrawn at indicated times and the activity of citrate transport was determined (lower part of this figure) by incubation with 1,5-^{14}C-citrate and subsequent Millipore filtration [5].

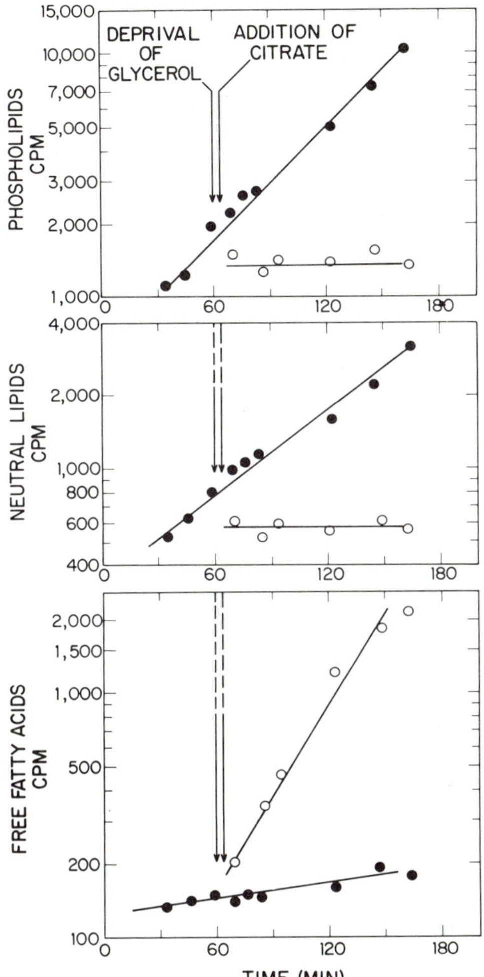

Fig. 2. Syntheses of phospholipids, neutral lipids and free fatty acids in the presence and absence of glycerol in *B. subtilis* B 42. Growth conditions were similar to those shown in the upper part of fig. 1, expect that the cultures were grown, washed and resuspended in media containing ^{14}C-acetate. Aliquots of the cells were withdrawn at indicated times, and the lipids were extracted and determined after thin layer chromatography and autoradiography [5]. Closed symbols: cells grown in the presence of glycerol. Open symbols: cells grown in the absence of glycerol.

Fig. 1 shows the influence of glycerol deprivation on induction of citrate uptake in *B. subtilis* B-42. To our surprise citrate transport could be induced for at least half a doubling time after deprivation of glycerol [5]. On the other hand the *de novo* synthesis of phospholipids and neutral lipids ceased immediately in the glycerol deprived culture as shown in fig. 2. The deprived cells started to accumulate free fatty acids, since their glycerol acceptor molecules were missing. In comparing our results with those of Hsu and Fox [4] it may be significant that the latter authors observed the drastic decrease of lactose transport in *E. coli* only about half a generation after deprivation of glycerol. In their report there is little difference in lactose transport between the deprived culture and the supplemented cells for at least one third of the doubling time after deprival of glycerol [4]. Recently Mindich investigated the induction of lactose transport in a glycerol requiring strain of *Staphylococcous aureus* [6]. He found in the deprived culture that the induced lactose permease activity relative to 6-phosphogalactosidase activity was about 30% to 50% that of the glycerol supplemented culture. Thus we must conclude that bacteria can induce functional transport systems even in the absence of *de novo* glycerol lipid synthesis [5]. In the case of citrate transport we cannot exclude that turnover or storage of appropriate lipid (possibly supplied by a high diffusion rate in the lipid part of the membrane) may provide enough phospholipid to allow full induction for at least half a doubling time. Subsequently citrate transport activity decreased in the culture deprived of glycerol. At the same time, however, all macromolecular syntheses started to slow down (fig. 3). At this late stage after stopping *de novo* glycerol lipid synthesis in *B. subtilis* we did not attempt to distinguish between direct and indirect effects of glycerol deprivation on citrate transport activity.

From recent work of several groups [7–10] it is now well established that the acyl residues in lipids of biomembranes remain in a liquid-like state by the controlled incorporation of unsaturated acyl chains. Mutants of *Saccharomyces cerevisiae* [11] and *E. coli* [12] are known which need unsaturated fatty acids as growth factors. On the other hand no long unsaturated acyl residues are found in *B. subtilis* membranes after growth at 37° [13]. Under these conditions the bacteria contain up to 90% methyl branched acyl residues, the remaining part being straight acyl chains. If these methyl branched fatty acids fulfill the same general role as unsaturated fatty acids for maintenance of membrane function then it might be possible to isolate a mutant of *B. subtilis* which is dependent on the exogenous supply of branched chain fatty acids. Long chain methyl branched acyl residues are synthesized in *B. subtilis* from short chain primer molecules derived from the three branched chain amino acids l-valine, l-isoleucine, and l-leucine. Fig. 4 depicts this path-

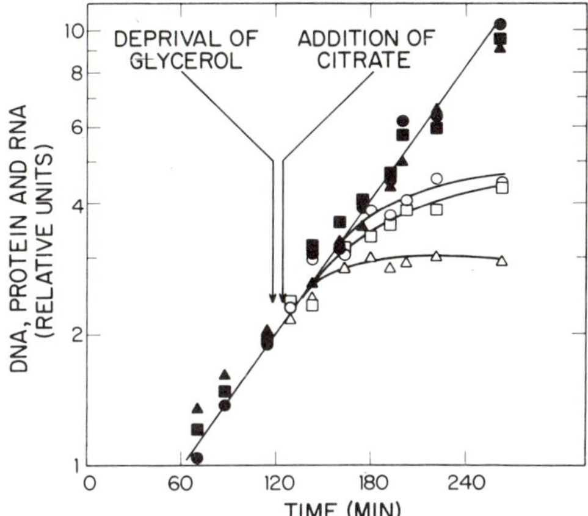

Fig. 3. Syntheses of macromolecules in the presence (closed symbols) and absence (open symbols) of glycerol in *B. subtilis* B 42. Growth conditions were similar to those shown in the upper part of fig. 1. The amounts of DNA (circles), protein (squares), and RNA (triangles) were chemically determined in aliquots of the cultures.

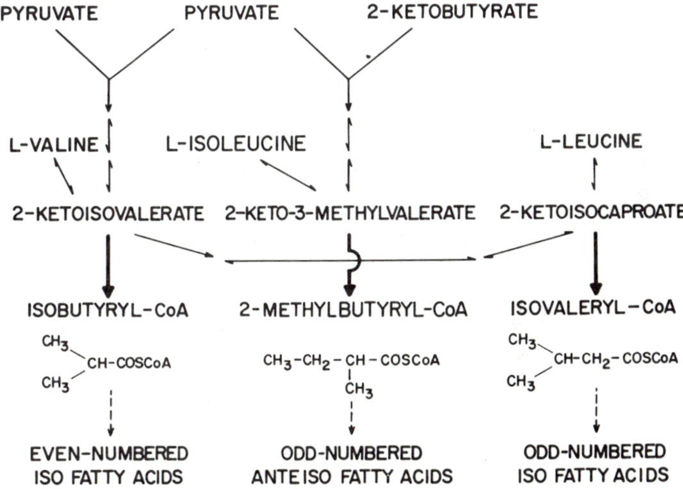

Fig. 4. Metabolic origin of branched chain "primer" molecules for fatty acid biosynthesis in *B. subtilis*. The heavy arrows symbolize the branched chain α-keto acid dehydrogenase activity which is missing in the new mutant. The dotted arrows stand for the chain-lengthening process of fatty acid biosynthesis.

way. We succeeded in isolating a new mutant *B. subtilis* 626 which is dependent for growth on the short primer fatty acids isobutyrate, 2-methylbutyrate, and isovalerate [14]. Crude extracts of the mutant were assayed for branched chain α-keto acid dehydrogenase. As expected no activity could be

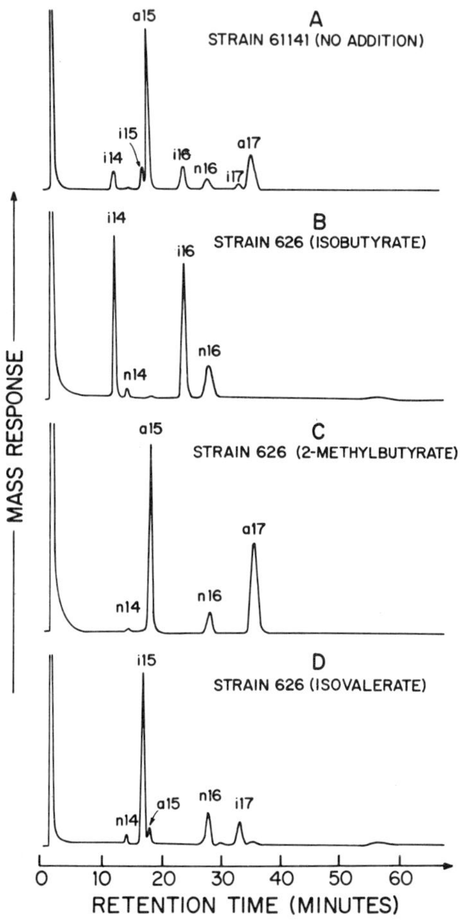

Fig. 5. Gaschromatograms of methylated fatty acid extracts obtained from cells of *B. subtilis* grown in a glucose-acetate medium supplemented with amino acids [14]. *B. subtilis* 626 is the branched chain fatty acid-requiring mutant, *B. subtilis* 61141 is the parent strain. Additions of fatty acids (0.1 mM) to the medium are listed in parentheses after the strain number. All cultures were grown for at least 10 generations under these conditions. Notations of peaks on the gaschromatograms, for example, i 15 means iso 15:0, a 15 means anteiso 15:0, and n 16 means 16:0.

detected whereas the parent strain, *B. subtilis* 61141, contained the normal level of activity.

One of the short primer compounds is sufficient to support growth of the mutant. Fig. 5 shows gas chromatograms of methylated fatty acid extracts obtained from these bacteria after growth on different primer substrates. The corresponding long chain methyl branched fatty acids can also fulfill the growth requirement. It is obvious that this new mutant allows wide alterations of the fatty acid composition in *B. subtilis* membranes. When grown on long branched chain fatty acids *B. subtilis* 626 contained only 20% to 40% branched chain acyl residues as opposed to more than 90% in the wild type. Thus *B. subtilis* cells synthesize normally much more branched chain fatty acids than necessary for growth at 37°C.

Out of a series of tested analog primer compounds only three could serve as growth factors of the mutant: 2-methylvalerate, 2-ethylbutyrate, and trimethylacetate. The chain lengthened products derived from these analog primer substrates were characterized by gas chromatography and mass spectrometry (table 1) (in collaboration with Dr. Iain Campbell, University of Pittsburgh, Pa.). Straight chain fatty acids do not support growth of the mutant. From studies of monolayer films it is known that methyl branched long acyl residues occupy a larger surface area than their straight chain isomers [15]. Moreover a methyl branch in a long straight chain fatty acid causes a decrease in the melting point of this compound.

Table 1

Fatty acid compositon of *B. subtilis* 626 after growth in the presence of 2-ethylbutyrate, 2-methylvalerate or trimethylacetate. Growth conditions were the same as given in fig. 5. The fatty acid composition was calculated from gaschromatograms of the methylated fatty acid extracts. For each major component the corresponding contribution to total fatty acids is given in %

Fatty acid addition (0.1 mM) to medium	Fatty acid composition of whole cells (% of total fatty acids)			
	14:0	16:0	18:0	branched chain fatty acids
2-Ethylbutyrate	1%	33%	6%	0.5% 10-ethyldodecanoic 46 % 12-ethyltetradecanoic 12 % 14-ethylhexadecanoic
2-Methylvalerate	6%	36%	2%	11 % 10-methyltridecanoic 42 % 12-methylpentadecanoic 2 % 14-methylheptadecanoic
Trimethylacetate	1%	51%	10%	13 % 12-dimethyltridecanoic 24 % 14-dimethylpentadecanoic

From our results it appears likely that *B. subtilis* bacteria use branched chain fatty acids in very much the same way as *E. coli* cells need unsaturated (or cyclopropane) fatty acids. In the latter organism Fox has shown that induction of lactose transport seems to depend on exogenous supply of unsaturated fatty acids [16]. We intend to reinvestigate this effect in *B. subtilis* using the new branched chain fatty acid requiring mutant and the citrate transport system. One would expect by analogy with our results of glycerol deprivation that citrate transport can also be induced after deprival of branched chain fatty acids. Possibly biomembranes contain lipid components in such an excess that a shortage in the supply caused by a stop in *de novo* synthesis can be compensated for from this excess.

Acknowledgements

This work was supported by a postdoctoral stipend of the Deutsche Forschungsgemeinschaft to K. Willecke and by Public Health Service Grant CA-Al-11595 from the National Cancer Institute and Institute of Alergy and Infectious Diseases to A.B. Pardee.

References

[1] K: Willecke and A. Pardee, J. Biol. Chem. 246 (1971) 1032.
[2] P. Fortnagel and E. Freese, J. Bacteriol. 95 (1968) 1431.
[3] L. Mindich, J. Mol. Biol. 49 (1970) 415.
[4] C.C. Hsu and C.F. Fox,.J. Bacteriol. 103 (1970) 410.
[5] K. Willecke and L. Mindich, J. Bacteriol. 106 (1971) 514.
[6] L. Mindich, Proc. Nat. Acad. Sci. (U.S.A.) 68 (1971) 420.
[7] J.M. Steim, M.E. Tourtelotte, J.C. Reinert, R.N. McElhaney and R.L. Rader, Proc. Nat. Acad. Sci. (U.S.A.) 63 (1969) 104.
[8] D.M. Engelman, J. Mol. Biol. 47 (1970) 115.
[9] P. Overath, H.U. Schairer and W. Stoffel, Proc. Nat. Acad. Sci. (U.S.A.) 67 (1970) 606.
[10] G. Wilson and C.F. Fox, J. Mol. Biol. 55 (1971) 49.
[11] M.A. Resnick and R.K. Mortimer, J. Bacteriol. 92 (1966) 597.
[12] D.F. Silbert and P.R. Vagelos, Proc. Nat. Acad. Sci. (U.S.A.) 58 (1967) 1579.
[13] T. Kaneda, J. Biol. Chem. 238 (1963) 1222.
[14] K. Willecke and A. Pardee, J. Biol. Chem. 246 (1971) 5264.
[15] L.L.M. van Deenen, in: Progress in the Chemistry of Fats and other Lipids, Vol. VIII, part I, ed. R.T. Holman, (Pergamon Press, 1965), p. I.
[16] C.F. Fox, Proc. Nat. Acad. Sci. (U.S.A.) 63 (1969) 850.

GENETIC ASPECTS OF AMINOACID TRANSPORT IN YEAST

M. GRENSON
Laboratoire de Microbiologie, Faculté des Sciences, Université Libre de Bruxelles and Institut de Recherches du C.E.R.I.A. 1070, Brussels, Belgium

The yeast *Saccharomyces cerevisiae* is a eucaryotic micro-organism which is especially suited for the use of genetics as a tool in the study of transport across biomembranes as well as of membrane biosynthesis.

One of the advantages of this yeast is the possibility of obtaining stable strains of different degrees of ploidy starting from one single haploid cell. The series of isogenic strains used in our laboratory was obtained as follows. From the haploid wild-type strain Σ1278b of mating α, a mating-type mutant was isolated according to Laskowski [1], as described previously [2]. It is known that single mutations at the mating-type locus can convert a strain of mating type α into a strain of mating *a* which is able to conjugate with a α strain, giving rise to diploid clones. Since diploid cells are twice as large as haploid cells, they can be recognized easily under the microscope, and picked up with a micromanipulator. Diploid cells can be grown indefinitely as diploids, or they can be induced to undergo meiosis and sporulation [3]. A diploid cell line which spontaneously appeared in a culture of Σ1278b was isolated and induced to sporulate. Among the haploid segregants, a strain of mating type *a*, strain 3962c, was selected. Similarly, mutations at the mating-type locus and crosses were used to produce triploid and tetraploid cells, all deriving from the same haploid strain Σ1278b [1, 4 and Hilger, unpublished].

The existence of strains which contain 1, 2, 3, or 4 complete sets of chromosomes per cell, and which have the same genetical background makes possible several types of experiments.

1) *Isolation of mutants.* From haploid cells, all kinds of mutants can be isolated (see further).
2) *Dominance-recessivity relationships.* By crossing a haploid mutant with a

haploid wild-type strain, heterozygous diploids are obtained in which it is possible to see whether the mutation is dominant over the wild-type gene or recessive.

3) *Complementation tests.* By crossing two haploid mutants which seem to exhibit the same permeability deficiency, one can see whether the defects are complementary in the diploid.
4) *Chromosome mapping.* As a result of sporulation each diploid cell gives rise to 4 haploid cell lines in which the segregation of genetic markers can be studied, so that mutations can be localized on the chromosomes [5]. As a rule, the point of interest is to know whether genetic markers are closely linked, an information which is obtained easily.
5) *Effect of gene dosage.* It is possible to detect whether the product of a given gene limits an uptake process or an enzymic activity. This can be done in a series of tetraploids containing a different number of normal copies of this gene: 0/4, 1/4, 2/4, 3/4 and 4/4.
6) *Influence of the ratio cell surface to cell volume on uptake.* The cell volume or the cell mass are directly proportional to the degree of ploidy of the yeast: for instance, the volume of a tetraploid cell is exactly fourfold that of a haploid cell. Since the cell surface does not increase that much, the ratio surface to volume is lower in a tetraploid (by 30%). Hence, by comparing the uptake activity in cells of different ploidies, it is possible to study the influence of the ratio surface to volume on uptake [6].

A few applications of these possibilities will be considered below.

Isolation of mutants affected in the uptake of amino acids

Two classical methods were used. One is the selection of mutants resistant to toxic analogues of amino acids. The other method is to start from a strain which requires an aminoacid, histidine for instance, and to isolate mutants which are unable to grow on normal histidine concentrations and need very high histidine concentrations. Once these mutants are obtained, it is possible to study their deficiency and select those which are defective at the level of permeability.

The isolation of permeability mutants makes possible to identify transport systems, i.e. to see how many systems are operating and what is their degree of specificity. For instance, a mutation was selected which strongly depresses L-lysine uptake, whereas the uptake of other amino acids is not affected, even that of arginine, ornithine and histidine [7]. This made possible to identify a very specific lysine transport system. In the same way mutations were selected which specifically affect the uptake of arginine [8], methionine [9], histidine [10] and dicaboxylic amino acids [11] respectively.

The fact that amino acids are taken up in yeast by a number of distinct transport systems or permeases has been confirmed by competition experiments [7–10]. It was shown that the uptake of a given amino acid in normal yeast cells is inhibited only by a few substances which have a closely related structure. The lowest specificity was observed in the case of the arginine permease, the activity of which is inhibited by L-canavanine, L-lysine, L-ornithine, D-arginine (competitive inhibition), and by L-histidine (mixed type of inhibition). However, these inhibitors have a much lower affinity for the recognition site of the arginine permease than L-arginine itself [8].

In addition to the specific amino acid permeases, there is a general amino acid permease, with much lower specificity, which is able to transport all the basic and neutral amino acids [12]. In order to demonstrate the existence of specific amino acid permeases, it was necessary to inhibit the activity of the general amino acid permease. This can be obtained in different ways. In the case of the wild-type strain Σ1278b and its derivatives, the easiest way is to grow the yeast in a culture medium which contains ammonium ions as a source of nitrogen. We shall return to this point later.

Genetic and biochemical analysis of the mutants are affected at the level of uptake shows that they belong to two main classes: those in which the uptake system is affected directly, and those in which uptake is affected indirectly, as a result of the type of regulation of uptake which occurs in yeast. An example illustrating this point will be chosen in the case of pyrimidine uptake were the situation is more schematic and completely analyzed [13]. External cytosine, uracil or uridine can be used as a source of pyrimidine in the wild-type strain Σ1278b. They are taken up by three distinct uptake systems. The uptake of uridine for instance is prevented in two classes of mutants which complement each other genetically. One class of mutants is blocked at the level of the uptake system itself. In the second class of mutants the uridine-kinase activity is missing and the inhibition of uridine uptake is due to the accumulation of an endogenous pool of uridine.

Also in the case of amino acid uptake in yeast the same two classes of mutants might be found, since the same mechanism of regulation of uptake by feedback inhibition is operating. The evidence is as follows. It is known that in bacteria like *Escherichia coli* the internal concentration of exogenous sugars or amino acids seems to be determined by a balance between rates of uptake and exit [14]. When labelled amino acids or sugars are accumulated from the external medium they can be dispaced by adding the same unlabelled substrate to the medium. In yeast, when an amino acid like histidine has been accumulated from the external medium, no exit can be demonstrated, and the accumulated histidine cannot be displaced [10]. However, histidine is still free and intact, and reaches concentrations as high as 0.2 M inside the cell. Hence, the uptake of histidine appears to be a one way process in yeast. What is then the mechanism of regulation of the pool size? If it is a feedback inhibition of an irreversible uptake system, this inhibition should be specific, otherwise it could not have a regulatory function. So, one should be able to inhibit histidine uptake by preloading the cells with histidine, but not by other amino acids. This is what was actually observed [10]. Other experiments [15, and Crabeel and Grenson unpublished] indicate that the specificity of the feedback inhibition of a given amino acid permease is the same as the specificity of its binding site for uptake, or at least very similar.

Until now, we only considered mutants which are affected at the permease level, i.e. at the level of the specific substrate binding site of a transport system. Mutants affected in other steps of transport were isolated on the following basis. If there are steps which are common to several uptake systems, one should be able to select mutants which, as result of a single mutation, have become resistant to several analogues simultaneously. Such mutants were isolated; on the basis of genetical analysis, they belong to several classes. One of the mutations, called *apf*, was further studied [16]. The activity of all the individual amino acid uptake systems is depressed. What is changed is the V_{max} of these systems, but the K_m are unchanged. The mechanism which is affected by this mutation could be energy coupling, but it might also be a cofactor of another kind which is common to several transport systems and limited in amount, like the one considered by Winkler and Wilson [17].

The general amino acid permease and its regulation

In the wild-type strain Σ1278b, the general amino acid permease is not active after growth on ammonium ions as the nitrogen source, but it is active after growth on proline as sole nitrogen source. Nitrogen starvation after growth on ammonia rapidly restores the general amino acid permease activity.

The difficulty of studying the properties of the general amino acid permease is the same as for the specific ones: it is necessary to eliminate (or at least to know exactly the contribution of) the other amino acid permeases.

A mutant was isolated which has lost this general amino acid permease activity [12]. It is called *gap*. It was used to delineate the specificity of the general amino acid permease. Competition experiments confirmed the conclusions obtained by this way.

About the mechanism of ammonia inhibition, we know that it is an inhibition of the activity of the transport system the synthesis of which is not affected. The effect of ammonia is not immediately maximal, and the inhibition is not of a competitive type.

The lack of specificity of the general amino acid permease and its inhibition in the presence of ammonia are in agreement with the view that this uptake system has the catabolic function of supplying the yeast cells with amino acids as sources of nitrogen.

Mutants have been isolated which are resistant to ammonia inhibition and should permit to understand the mechanism of this regulation.

Comparison of uptake in a series of cells with increasing degree of ploidy

The fourfold genome of a tetraploid cell in principle allows it to produce a fourfold amount of each of its constituents, as compared to a haploid cell. However, the increase in cell surface from a haploid to a tetraploid is not fourfold. In fact, if the relative surface to volume ratio in a haploid cell (Σ1278b) is taken as 1, this ratio becomes 0.7 in a tetraploid. If the cell membrane is saturated or nearly saturated with an element of a transport system in a haploid cell, the cell surface will become more and more limiting for its insertion with increasing ploidy. This appears to be the case for the uptake of several substances known to utilize different uptake systems, for instance arginine, lysine and uridine [6]. The uptake rate of these substances per unit of cell mass in the tetraploids is about 70% of what it is in the haploids. The close quantitative correlation which was observed between the reduction of the ratio surface to volume clearly indicates a limitation of space for inserting some constituent of each of these uptake systems into the membrane. Whether this constituent is the specific binding protein remains to be examined. For other substrates, the transport system of which is inducible, this limitation was not observed in uninduced cells.

These observations suggest several possible approaches of the problems of membrane edification and of regulation of the synthesis of membrane constituents.

References

A more detailed bibliography can be found in the cited articles.

[1] W. Laskowski, Z. Naturforsch. 15b (1960) 495.
[2] J. Bechet, M. Grenson and J.M. Wiame, Eur. J. Biochem. 12 (1970) 31.
[3] R.R. Fowell, Nature, 170 (1952) 578.
[4] W. Laskowski, Z. Naturforsch. 17b (1962) 93.

[5] D.C. Hawthorne and R.K. Mortimer, Genetics, 45 (1960) 1085.
[6] C. Hennaut, F. Hilger and M. Grenson, Biochem. Biophys. Res. Commun. 39 (1970) 666.
[7] M. Grenson, Biochim. Biophys. Acta 127 (1966) 339.
[8] M. Grenson, M. Mousset, J.M. Wiame and J. Bechet, Biochim. Biophys. Acta 127 (1966) 325.
[9] J.J. Gits and M. Grenson, Biochim. Biophys. Acta 135 (1967) 507.
[10] M. Crabeel and M. Grenson, Eur. J. Biochem. 14 (1970) 197.
[11] C.R. Joiris and M. Grenson, Arch. Intern. Physiol. Biochim. 77 (1969) 154.
[12] M. Grenson, C. Hou and M. Crabeel, J. Bacteriol. 103 (1970) 770.
[13] M. Grenson, Eur. J. Biochem. 11 (1969) 249.
[14] A. Kepes and G.N. Cohen, in: The Bacteria, eds. Gunsalus and Stanier (Academic Press, New York 1962) Vol. IV, p. 179.
[15] J.J. Gits and M. Grenson, Arch. Intern. Physiol. Biochem. 77 (1969) 153.
[16] M. Grenson and C. Hennaut, J. Bacteriol. 105 (1971) 477.
[17] H.H. Winkler and T.H. Wilson, Biochim. Biophys. Acta 135 (1967) 1030.

GENETIC VARIATION IN MEMBRANE TRANSPORT OF WATER AND ELECTROLYTES

John STEWART
Dartmouth Medical School, Dept. Physiology, Hanover, New Hampshire, USA

Introduction

As has been well illustrated by the two previous speakers, it is now becoming widely recognised that genetics can provide a powerful tool for membrane research in micro-organisms. In this paper I shall suggest that genetics might also make a contribution to research in higher organisms; and in fact that for work on higher organisms, the use of genetic variation as a tool has some additional advantages that are not readily apparent in work on micro-organisms. Higher organisms are distinguished from micro-organisms by their much greater complexity. As a result of this complexity, research on higher organisms is carried out at many different levels – molecular, ultrastructural, cell physiology, organ physiology, behaviour of the whole animal, population studies and ecology. The necessity for specialising in one of these levels means that too often our knowledge is fragmented, and that our understanding at one level is unrelated to our understanding at another. The underlying theme of this paper will be that the particular advantage of using genetic variation as a tool in higher organisms, lies in the power of genetics to integrate our knowledge at different levels of biological organisation.

Sources of genetic variation

An obvious prerequisite for the use of genetics as a tool in membrane research, is to find a suitable source of genetic variation. In micro-organisms it is often possible to produce 'tailor-made' variation by large scale mutation runs followed by automatic mass screening procedures designed to select genetic mutants with exactly the required properties. The effectiveness of these techniques has been largely responsible for the success of genetics as a tool in membrane research ever since the classical work of Jacob and Monod [1] on lactose permeability in *E. coli*. However, for purely practical reasons, such mass screening is not feasible in higher organisms, so that an alternative approach is necessary.

All of the genetic variations in vertebrates that I shall discuss today occurred naturally, having arisen spontaneously. These variations were discovered initially because they affected a relatively superficial, easily measured character in the general area of water and electrolyte metabolism. In order for this variation to be useful for membrane research, it is necessary to follow the initial discovery by an investigation of the specific cause, ultimately in molecular terms, of the original variation.

Interspecific variation

The differences between species, that have been widely studied by comparative physiologists, offer one obvious source of naturally occurring genetic variation. As an example of this interspecific variation, which is potentially of interest with respect to mechanisms of membrane transport, I have chosen the effect of neurohypophysial octapeptide hormones on epithelial sodium and water transport. A very typical experiment on frog skin, taken from the work of Bourguet and Maetz [2], is shown on the left of fig. 1. The upper half of this figure shows the effect of these hormones on water transport, the lower half shows the effect on sodium transport. In this experiment, the hormones clearly had a stimulatory effect on both sodium and water. In fact, these hormones appear to have a parallel effect on sodium and water transport under a wide variety of circumstances, so that it has seemd possible that these two effects might be intimately related to each other. One hypothesis has been that there is a single permeability barrier for both sodium and water, and that the hormones act by increasing the size and/or number of pores in this barrier. At the molecular level, Orloff and Handler [3] have suggested that the effects of neurohypophysial hormones on sodium and water transport might both be mediated by cyclic AMP. However, the study of interspecific variation is sufficient to show that the effects of these hormones on sodium and on water can be dissociated. Thus in the experiment illustrated in the center of fig. 1, Bentley and Heller [4] demonstrated that in a newt preparation the hormone had a stimulatory effect on sodium transport but not on water transport. Conversely, the micropuncture experiment in rats

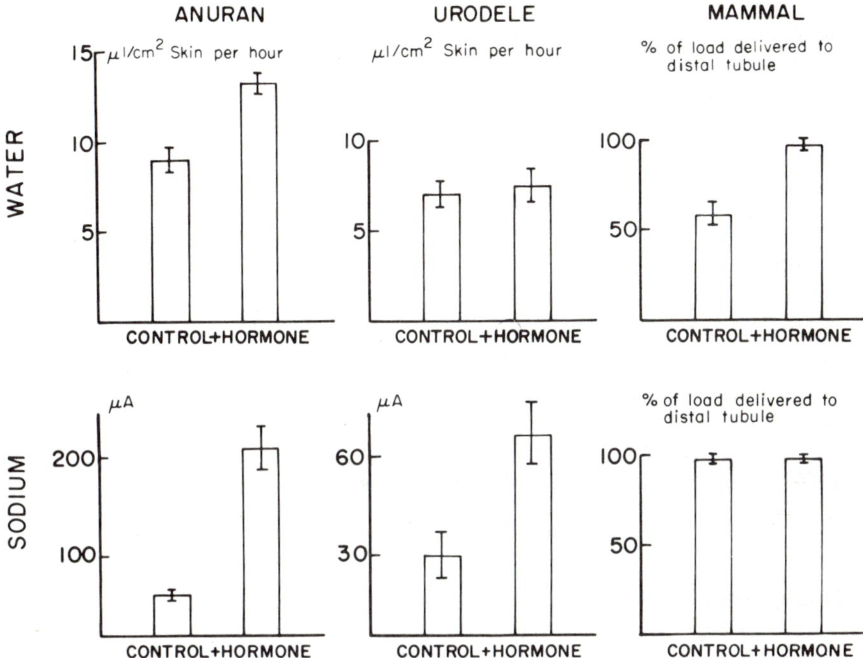

Fig. 1. Effect of neurohypophysial hormones on sodium and water transport.

shown on the right of fig. 1, taken from the work of Schnermann et al. [5], illustrates the well-known fact that in the distal nephron of mammals, vasopressin has an effect on water transport but not on sodium transport.

Of course this independence of water and sodium effects, which creates some complcations for the cyclic AMP theory, can be shown on frog skin without using genetic variation. For example, raising the concentration of calcium can block the hormone effect on sodium but not on water transport [6]. Even in the paper already cited by Bourguet and Maetz [2], it was shown that doses of vasotocin and oxytocin which had similar effects on water transport could have very different effects on sodium transport. The authors concluded that the effects on water and on sodium must have been mediated by different receptor molecules. Now if we could be sure that the frog skin and newt skin in fig. 1 differed only in their water permeability response to hormone, studying the difference between them might give valuable clues about the structures and molecules specifically responsible for hormonal effects on water as opposed to sodium transport. But, of course, frogs and newts actually differ in a great number of characters, and we have no independent means of testing which of those differences might be specifically related to hormonally induced water permeability. This illustrates the limitations of genetic differences between species as a tool in this type of study. Although interspecific variation can often dissociate two variables, and can sometimes even suggest an association, it cannot provide a critical test of a positive causal relationship.

Intraspecific variation

This limitation can be overcome by the use of genetic variation within – rather than between species. I shall explain later in some detail how the observation of genetically segregating generations, in particular the so-called 'backcross experiments' can provide a test of whether a biochemical or ultrastructural difference is related to a difference at the level of gross physiology. However before that I want to describe a few of the many intraspecific variations that are available in mammals.

DI rats (Brattleboro strain)

My first example is the 'Brattleboro' strain of rats. These rats were first discovered because they produced copious amounts of dilute urine [7]. This diabetes insipidus could be corrected by administration of the antidiuretic hormone vasopressin, and subsequent investigations have confirmed that the defect was due to an inability of these rats to produce vasopressin. These rats illustrate some of the advantages of intraspecific genetic variation as a tool. The defect is due to a single autosomal gene locus, so that litter mates can be used as controls; and when compared to possible surgical or experimental manoevers designed to remove vasopressin, the genetic defect is not only convenient but highly specific as well, since the other pituitary hormones are not primarily affected. Incidentally, the precise molecular mechanism whereby the Brattleboro gene causes a failure of neurosecretion, including the absence of neurosecretory granules, poses an interesting but currently unsolved problem [8]. Nevertheless, the advantages offered by the Brattleboro rats are such that they have been widely used as a model for the investigation of many effects of vasopressin, ranging from synthetic organic chemistry to behavioural psychology [7]. In particular, the effects of vasopressin on the ultrastructure of renal distal tubules and collecting ducts have been studied by electron microscopy [9].

DI mice

Severe diabetes insipidus also occurs in a strain of mice named DI. Like the Brattleboro rats, the inability of these mice to concentrate their urine is apparently due to water impermeability of the distal tubule and collecting ducts.

This is demonstrated by fig. 2, taken from the work of Kettyle and Valtin [10]. If the collecting duct is fully permeable to water, the urine should be close to osmotic equilibrium with the surrounding papillary interstitium. This situation is illustrated by control mice from another strain, VII, shown on the left of fig. 2. The urine osmolality is not significantly different from that of the papillary interstitium.

On the other hand, if the urine is substantially less concentrated than the papillary interstitium, this indicates that the collecting duct is relatively impermeable to water. This situation is illustrated by the DI rats, which we know have relatively low water permeability due to an absence of vasopressin. In these rats, on the right of fig. 2, we see that there is substantial osmotic disequilibration between urine and papillary interstitium. The interstitial osmolality is reduced as a secondary effect. The DI mice are shown in the center of fig. 2. It can be seen that they are very similar to the DI rats, which suggests that water permeability is impaired in the DI mice.

However in the mice, in contrast to the rats, the urinary concentrating defect cannot be corrected by exogenous vasopressin administration. The

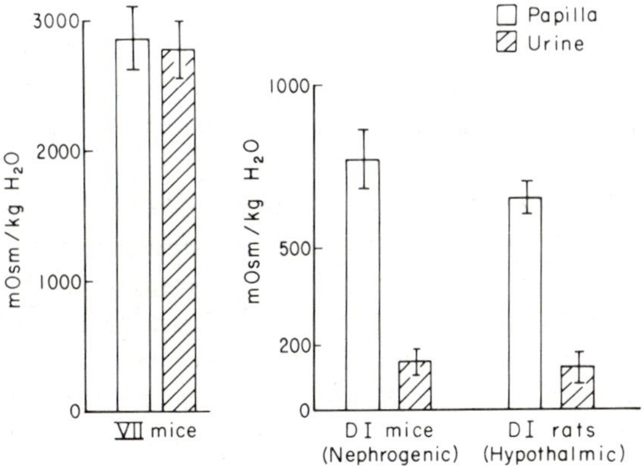

Fig. 2. Osmotic disequilibration between urine and papillary interstitium in diabetes insipidus.

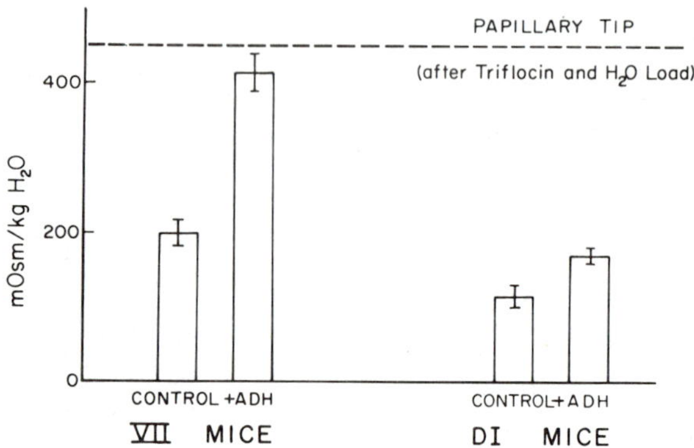

Fig. 3. Sensitivity to ADH at constant interstitial cortico-papillary osmotic gradient.

defect here thus appears to lie in the kidney itself. The experiment illustrated in fig. 3 confirms that DI mice probably have a defect in vasopressin-induced membrane permeability to water. Mice were treated with the diuretic drug TRIFLOCIN, which inhibits sodium reabsorption from ascending loops of Henle and hence abolishes the cortico-papillary osmotic gradient. The mice were simultaneously water loaded to reduce their endogenous vasopressin levels. Under these conditions, a supra-maximal injection of exogenous vasopressin caused a large increase in urinary concentration in VII mice, but only a very slight increase in DI mice. The fact that the DI mice did respond slightly to vasopressin indicates that the membranes had 'receptors' with at least some affinity for vasopressin; but the fact that even after very large doses of vasopressin, the membrane was still relatively water impermeable, as evidenced by osmotic disequilibration between urine and interstitium, indicates that the 'intrinsic activity' of the hormone-receptor complex was probably impaired.

A further indication that receptor affinity is unimpaired in DI mice comes from the work of A.D. Stewart [11], who has shown that the rate of inactivation of vasopressin by kidney slices *in vitro*, a measure of anti-diuretic potency which probably related to binding of vasopressin by specific receptors, was as great in DI mice as in control mice from other strains.

It has been postulated that the action of vasopressin is mediated by cyclic AMP, and an adenyl cyclase from the renal medulla which is specifically activated by vasopressin has been identified (for a list of references, see Dousa et al. [12]. It will be interesting to see if the adenyl cyclase from these DI mice has an altered intrinsic activity, or an altered affinity for activation by vasopressin. But in any case this variation should prove very useful in unravelling the mechanism of action of vasopressin on the water permeability of collecting duct membranes, and in particular in determining the role of adenyl cyclase in this system.

Antidiuretic potencies of arginine- and lysine-vasopressin

If intrinsic activity of the adenyl cyclase is a possibility in the previous case, affinity for vasopressin is a distinct possibility in the next. It has been known for sometime that some members of the Suinae family, in particular domestic pigs, synthesise lysine vasopressin (LVP) rather than the arginine vasopressin (AVP) which is generally found in other mammals [13]. It is also known that whereas in most mammals LVP is distinctly less potent than AVP as an antidiuretic agent, in domestic pigs LVP has almost equal potency with AVP on a molar basis [14]. In view of this apparent coadaptation of porcine kidneys to LVP, it is of considerable interest that Dousa et al. [12] have shown that in pigs the renal medullary adenyl cyclase has greater affinity for LVP than for AVP, whereas in rats, the situation is reversed and the adenyl cyclase has greater affinity for AVP. However, although highly suggestive, it is not legitimate to conclude that the difference in relative affinities for LVP and AVP was actually the cause of the difference in relative potencies of LVP and AVP, since rats and pigs differ in many respects besides their adenyl cyclase affinities.

Formally, any one of these other differences, and not the difference in adenylcyclase affinity, may be the cause of the difference in relative potencies of AVP and LVP. This further illustrates the earlier point that interspecific variation cannot be used to demonstrate positive relationships. As I mentioned before, in order to use genetics to demonstrate a positive relationship, it is necessary to find variation within a single species. Intuitively, one might expect to find the required difference in receptor affinity in 'co-adapted-association' with a different hormone; it was in pigs, that the relatively high receptor affinity for LVP was found. Recently, some wild populations of African warthogs and peccaries have been discovered, in which some individuals make LVP whereas others make AVP [13]. This raised visions of an expedition to Africa with the aim of measuring antidiuretic potencies in wild warthogs! For better or worse, however, it seems that an experimentally more amenable species also makes both AVP and LVP.

A.D. Stewart [15] has described a strain of mice the 'Peru' strain, which synthesise LVP instead of the AVP common to other strains of mice. Moreover, it seems as though Peru mice may have coadapted sensitivity to LVP. The antidiuretic potencies of LVP and AVP have not yet been tested in these mice, but it has been observed that kidney slices from Peru mice inactivate LVP and AVP at about equal rates, whereas in other strains AVP is inactivated much faster than LVP [11]. It will be fascinating to see if the adenyl cyclase in Peru mice does have a relatively high affinity for LVP. If so, it will be possible through the backcross experiments which I shall explain in a moment, to test whether this biochemical difference is actually related to a difference in the relative antidiuretic potencies of AVP and LVP.

Sodium transport and aldosterone

The previous examples have all been concerned with vasopressin and membrane permeability to water. My final example deals with aldosterone and the control of active sodium transport by the kidney. The initial discovery of this variation came from observations on the sodium excretion following waterload in two strains of mice [16]. Fig. 4 shows the total sodium excretion, in μeq, during the 3 hr following a stomach load of water. When given waterload alone, CBA mice excreted more sodium than Peru mice. (This is the same Peru strain which synthesises LVP.) In Peru mice acute administration of 154 μeq of potassium chloride along with the waterload caused a more rapid diuresis and an increase in the total amount of

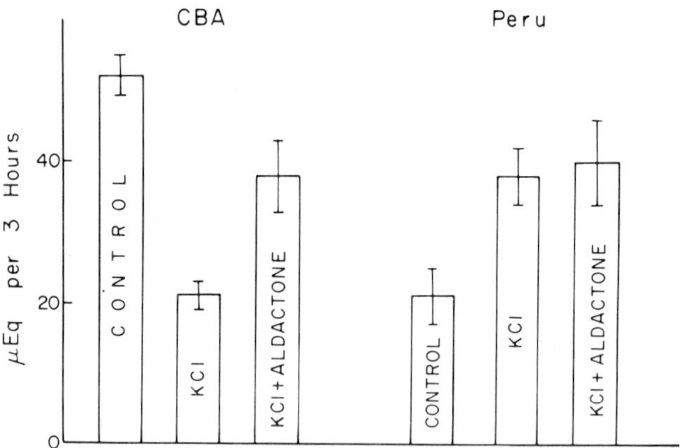

Fig. 4. Sodium excretion following water load (6% body weight).

sodium excreted. However in the CBA strain of mice, the same acute addition of KCl caused a decrease in the sodium excretion. In order to account for this decrease, it was postulated that in CBA mice, the potassium stimulated aldosterone secretion, and that this in turn stimulated sodium reabsorption by the kidney. This hypothesis was strengthened by the observation that in CBA mice, pretreatment with the aldosterone blocking agent aldactone-A returned the sodium excretion almost to the control levels. And, interestingly, the aldactone treatment had virtually no effect on sodium excretion in the Peru mice. This suggested that in Peru mice, either the kidneys were insensitive to aldosterone, or plasma levels of aldosterone were low. The second possibility had to be discarded when we found that plasma levels were similar, and in fact slightly higher in Peru than in CBA mice. The conclusion that the kidneys of Peru mice were relatively insensitive to aldosterone was confirmed more directly by the experiments shown in fig. 5. Plotted here is the total excretion of sodium and potassium during the 3 hr following an acute water load. In CBA mice, adrenalectomy caused a considerable increase in sodium excretion, whereas acute administration of aldosterone to adrenalectomized animals returned the sodium excretion to control values. In Peru mice, on the other hand, adrenalectomy caused no significant increase in sodium excretion, and aldosterone caused no significant decrease, at least under the conditions of these experiments.

In passing it is interesting to note that aldosterone stimulated potassium

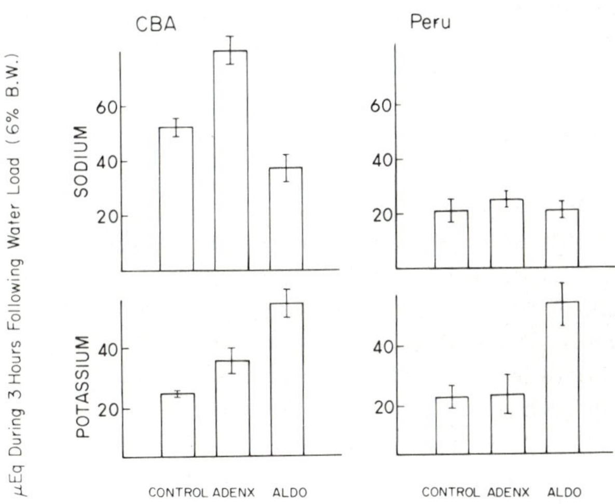

Fig. 5. Effect of aldosterone on sodium and potassium excretion.

secretion not only in CBA mice, where aldosterone presumably had also stimulated sodium reabsorption, but also in Peru mice, where aldosterone had no discernible effect on sodium transport. Thus, genetic variation, by dissociating the effects of aldosterone on sodium and potassium transport, demonstrates that the effect on potassium secretion is not simply a consequence of a primary effect on sodium transport. A similar conclusion can also be drawn without using genetic variation. Fimognari et al. [17] have shown that actinomycin-D can block the effect of aldosterone on sodium excretion, but not on potassium secretion. This last experiment indicates that the effect of aldosterone on sodium transport may be mediated at the transcriptional level. Protein fractions which specifically bind aldosterone and which occur in the cell nucleus have been isolated from renal tissue by the Edelman group [18, 19]. It will be fascinating to see whether any of these specific protein fractions are altered or reduced in Peru mice. If so, this genetic variation should be invaluable in investigations of the mechanism of action of aldosterone.

The point must be made again, however, that simply finding such a difference at the molecular level might suggest but would not *prove* a causal relationship between the molecular difference and the insensitivity of Peru kidneys to aldosterone. CBA and Peru mice differ in many respects at the molecular level, and in formal logic any one of these molecular differences could be responsible for the gross physiological difference. This is the same general problem that I have alluded to throughout this talk, and the time has come to describe the type of genetical breeding experiment that can lead to its resolution.

The 'backcross' experiment

As a specific illustration I shall take my final example. There is a gross physiological difference between the strains CBA and Peru, i.e., the effect of potassium on sodium excretion. We wish to know whether a particular molecular difference between these strains is actually responsible for the physiological difference. The genetical experiment consists of breeding not only first generation hybrids but also seond generation hybrids between the two strains. The results of such an experiment for the single character 'Effect of potassium on sodium excretion', are shown in fig. 6. These are histograms, plotting change in sodium excretion on the abscissa against number of mice on the ordinate. As previously described, potassium caused an increase in sodium excretion in Peru mice, but caused a slight decrease in CBA. The cross be-

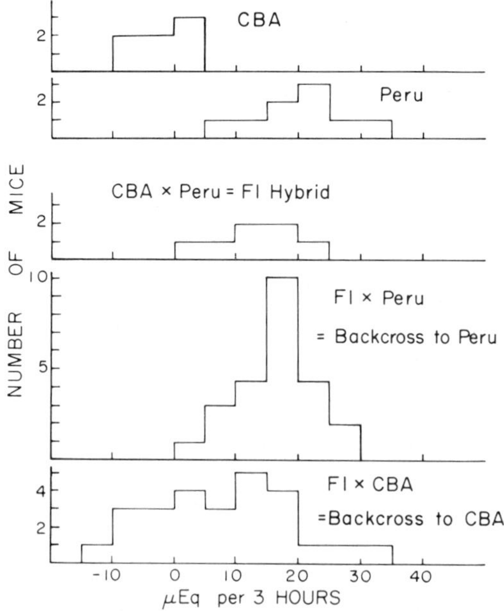

Fig. 6. Effect of KCl on Na^+ excretion in hybrids.

tween CBA and Peru, or so-called F1 hybrids, were similar to the Peru strain, i.e., in genetical parlance, the Peru genes were 'dominant' to CBA.

For reasons which I will not go into here, the most convenient second generation hybrids are the so-called 'backcrosses' of F1 mice to the parental strains [20]. As might be expected, since the F1 and Peru mice are similar to each other, the backcross of F1 mice to the Peru strain produces mice that are similar to both their parents. The important backcross is that of the F1 hybrids to the 'recessive' CBA parent. The offspring are not simply uniformly intermediate between their F1 and CBA parents; instead, they are extremely variable and cover the whole range from the high values of the F1 parent or the low values of the CBA parent. This wide variation is due to the fundamental phenomenon of genetic *segregation,* or 'sorting out' of the genetic factors derived from CBA and Peru, that are contained in the F1 parent.

The important point for our present discussion is that in general, genetic factors which affect unrelated characters will segregate more or less independently from each other in such a backcross. An example of this is shown on the left of fig. 7. In separate experiments it has been shown that the strains CBA and Peru also differ from each other at the molecular level, in hepatic

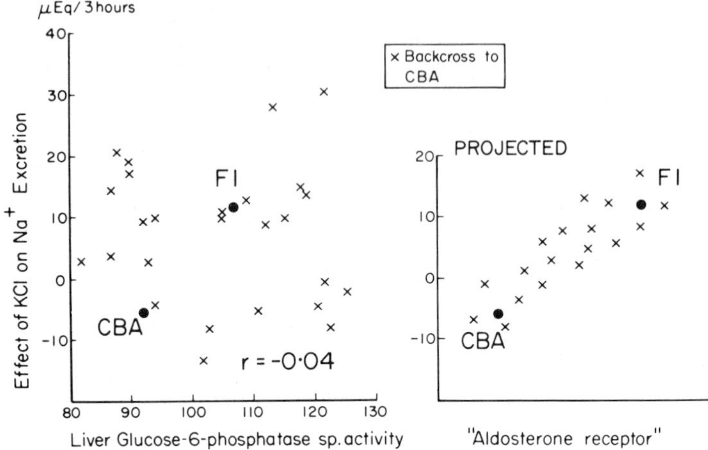

Fig 7. Backcross test of correlation between physiological and molecular levels.

Glucose-6-phosphatase activity. The F1 mice are again similar to the Peru parent, so that in fig. 7 we see that CBA and the F1 hybrids differ from each other both in phosphatase activity, and in the effect of potassium on sodium excretion. One might suspect that the variation in phosphatase activity is caused by different genetic factors than those responsible for the difference in the effect of potassium on sodium excretion; and indeed we see in the graph on the left, where each point represents a single mouse, that in the backcross the genetic factors are segregating independently and variation in glucose-6-phosphatase activity is completely unrelated to variation in the effect of potassium on sodium excretion. Conversely, if backcross variation at the molecular level is highly correlated with a physiological character, then it can be concluded that segregation of the same genetic factors is causing variation in both the molecular and the physiological characters. Indulging for a moment in some wishful thinking, a hypothetical example is shown on the right of fig. 7. If Peru and F1 mice do indeed have an altered or reduced aldosterone receptor in the kidney, and if variation in this receptor correlated well with the effect of potassium on sodium excretion in a backcross experiment, this would constitute strong evidence that the aldosterone receptor was actually involved in mediating the effect of potassium on sodium excretion.

A '*backcross experiment*' of the sort I have just described has certain limitations, and it cannot provide logically conclusive proof that a particular molecular difference is the cause of a particular physiological difference.

Rather than go into these special limitations, however, I would like to propose the positive view that a backcross experiment can provide a critical test of an hypothesis, since it is based on observation of a correlation under conditions where such a correlation may very well be broken down. A 'backcross experiment' can be used to test any relationship between a molecular and a physiological level, as in the example just described, or in the previous examples testing whether variation in adenyl cyclase affinity and/or intrinsic activity is related to membrane permeability and urinary concentrating ability. Moreover, the backcross-experiment is not limited to relating molecular and physiological levels; it can be used to test a relationship between *any* two levels of biological organisation, such as structure and function, or physiology and behaviour. As such, the backcross experiment illustrates my theme that a major advantage of using genetic variation as a tool lies in its power to integrate our knowledge at different levels of biological organisation.

Population genetics and evolution

I would like to conclude by discussing another way in which the study of genetic variation can integrate our understanding of biological organisms. I mentioned at the beginning of this talk that all the genetic variation I was going to describe was not specially produced by artificial mutagenic agents, but occurred naturally. Such naturally occurring variation is worthy of study in its own right, apart from its use as an experimental tool for a number of reasons. The first reason is that its very existence poses something of a problem in population genetics. If we were to make two assumptions: firstly that a given form of a gene can be assigned a definite '*fitness*', irrespective of other genes in the same animal, and secondly that one animal can be assigned a definitive fitness irrespective of other animals in the same population, then it can be shown mathematically that natural selection would lead to the predominance of the 'most fit' form of each gene and the virtual elimination of all other forms. We would thus expect all members of a species to conform uniformly to a single 'ideal' type, with rare exceptions due to recurrent mutation. Now in fact natural populations appear to be far from uniform genetically; the variations I have described in this paper are far from exceptional, and in fact it is not too much to say that we have found genetic variation almost wherever we have looked for it.

The mathematics tells us that it is very unlikely that this ubiquitous variation can exist because it is selectively neutral; on the contrary, the variation is a positive phenomenon which requires an explanation in terms of selective forces which will lead positively to its maintenance. At the moment we can only guess about these selective forces – and in fact we will almost certainly need more studies on the molecular and ultrastructural mechanisms by which these genes cause variation before we can fully understand the action of selective forces. But in general terms, it is clear that the two assumptions on which the mathematical prediction of uniformity was based must be wrong. We have a probable example of the falsity of the first assumption in the co-existence of LVP and AVP in the same species that I have discussed in this paper. It is simply not possible to say that LVP or AVP is 'superior' as an antidiuretic hormone irrespective of whether the renal receptors have higher affinities for LVP or AVP. The falsity of the second assumption, that an individual can be assigned a definite 'fitness' irrespective of other individuals in the same population, is even more speculative but highly intriguing. In a population of social animals, where cooperation, or at least the avoidance of destructive mutual antagonism is important, we might indeed expect that genetic variation which fits different individuals into different ecological niches might be mutually advantageous to all the individuals in the population [21].

These considerations may seem far removed from molecular and ultrastructural aspects of membrane transport. I think, however, that they are in fact related, because the naturally occurring variation which I have been discussing is the raw material on which natural selection acts in order to produce evolutionary change. If it is true that we can fully understand the organisation of present-day organisms only if we also understand how they have arisen through evolution, then it is essential for us to know how and why organisms vary not only between but also within species.

Acknowledgements

This paper has been largely based, both in conception and in many of the experimental results, on the work of the research group on "Endocrine Genetics", supported by the Medical Research Council of Great Britain, and carried out at the Department of Genetics, Milton Road, England. The work was originally inspired by the late S.G. Spickett. I am also grateful to other members of the Endocrine Genetics research group, particularly A.D. Stewart and D. Charlesworth, for making some of their data available to me.

References

[1] F. Jacob and J. Monod, Genetic regulatory mechanisms in the synthesis of proteins. J. Mol. Biol. 3 (1961) 318–356.

[2] J. Bourguet and J. Maetz, Arguments en faveur de l'independance des mecamismes d'action de divers peptides neurohypophysaires sur le flux osmotique d'eau et sur le transport actif de sodium au sein d'un même recepteur. Biochim. Biophys. Acta 52 (1961) 552–565.

[3] J. Orloff and J. Handler, The role of adenosine 3'-5'-phosphate in the action of antidiuretic hormone. Am. J. Med. 42 (1967) 757–768.

[4] P.J. Bentley and H. Heller, The action of neurohypophysial hormones on the water and sodium metabolism of urodele amphibians. J. Physiol. 171 (1964) 434–453.

[5] J. Schnermann, H. Valtin, K. Thurau, W. Nagel, M. Horster, H. Fischbach, M. Wahl and G. Liebau, Micropuncture studies on the influence of antidiuretic hormone on tubular fluid reabsorption in rats with hereditary hypothalamic diabetes insipidus. Pfluegers Eur. J. Physiol. 306 (1969) 103–118.

[6] M.J. Petersen and I.S. Edelman, Calcium inhibition of the action of vasopressin on the urinary bladder of the toad. J. Clin. Invest. 43 (1964) 583–594.

[7] H. Valtin, Hereditary hypothalamic diabetes insipidus in rats (Brattleboro strain). Am. J. Med. 42 (1967) 814–827.

[8] H. Valtin, J. Stewart and H.W. Sokol, Genetic control of the production of neurohypophysial hormones, in: Handbook of Physiology (1972).

[9] C.C. Tisher, R.E. Bulger and H. Valtin, Morphology of renal medulla in water diuresis and vasopressin-induced antidiuresis. Am. J. Physiol. 220 (1971) 87–94.

[10] W.M. Kettyle and H. Valtin, Chemical and dimensional characterisation of the renal countercurrent system in mice. Kidney International 1 (1972) 135–144.

[11] A.D. Stewart, Genetic Variation in Neurohypophysial Hormones in Mice. Ph.D. thesis, University of Cambridge, 1969.

[12] T. Dousa, D. Hechter, I.L. Schwartz and R. Walter, Neurohypophyseal hormone-responsive adenyl cyclase from mammalian kidney. P.N.A.S. 68 (1971) 1693–97.

[13] D. Ferguson and H. Heller, Distribution of neurohypophysial hormones in mammals. J. Physiol. 180 (1976) 846–863.

[14] R.A. Munsick, W.H. Sawyer and H.B. Van Dyke, The antidiuretic potency of arginine and lysine vasopressins in the pig with observations on porcine renal function. Endocrinology 63 (1958) 688–693.

[15] A.D. Stewart, Genetic variation in the neurohypophysial hormones of the mouse *mus musculus.* J. Endocrinol. 51 (1971).

[16] J. Stewart, Diuretic responses to electrolyte loads in four strains of mice. Comp. Biochem. Physiol. 30 (1969) 977–987.

[17] G.M. Fimognari, D.D. Fanestil and I.S. Edelman, Induction of RNA and protein synthesis in the action of aldosterone in the rat. Am. J. Physiol. 213 (1967) 954–962.

[18] T.S. Herman, G.M. Fimognari and I.S. Edelman, Studies on renal aldosterone-binding proteins. J. Biol. Chem. 243 (1968) 3849–56.

[19] G.E. Swanek, L.L.H. Chu and I.S. Edelman, Stereospecific binding of aldosterone to renal chromatin. J. Biol. Chem. 245 (1970) 5382–89.

[20] J. Stewart, Biometrical genetics with one or two loci. I. The choice of a specific genetic model. Heredity 24 (1969) 211–224.

[21] J.M. Thoday, Natural selection and biological progress, in: A Century of Darwin, ed. S.A. Barrett (London, Heinemann (1958) pp. 313–333.

HUMAN HEREDITARY DISORDERS OF MEMBRANE TRANSPORT

J.C. CRAWHALL

Department of Experimental Medicine, McGill University Clinic, Royal Victoria Hospital, Montreal, Quebec, Canada

At the present time the only human disorders of membrane transport that have been described relate either to intestinal transport or to renal transport. It is known that every other cell type in the body has some specific mechanisms for transport of electrolytes and organic substrates, but disorders of these other types have not so far been described. For the purposes of this lecture I will restrict myself mainly to disorders of amino acid transport and certain disorders of sugar transport which might be relevant.

The disorders of amino acid transport in the kidney can be classified into three types (table 1). From the point of view of this discussion, the most important is the first group, the specific disorders of substrate transport. This group has to be clearly delineated from the third group, in which the aminoaciduria which is present is really the result of an overflow phenomenon caused by the high levels of amino acid in the plasma. Because of the phenomenon of competitive inhibition of amino acid transport it is possible for a specific aminoaciduria to occur which is secondary to an overflow aminoaciduria for a different amino acid. In such a case an apparent disorder of amino acid transport is induced by the high levels of another amino acid acting as a competitive inhibitor for the reabsorption sites in the kidney tubule. This group has been called the competitive aminoaciduris.

Numerous disorders of amino acid transport in the kidney have now been described that are clearly not related to an overflow phenomena, and I have attempted to classify some of these in a way which will simplify understanding of the mechanism of the disorder involved. The first of these I have called primary generalized disorders of amino acid transport (table 2). This group is characterized by the fact that a wide range of amino acids is involved, and

Table 1
Etiology of aminoacidurias.

Specific transport defect	Competitive aminoacidurias	Overflow aminoacidurias

Table 2
Generalized disorders of amino acid transport.

Primary
Lowes syndrome (Oculo cerebrorenal syndrome) (ref. [1])
Fanconi syndrome (some cases of, in children and adults) (ref. [2])
Busby syndrome (familial growth retardation, renal aminoaciduria and cor pulmonale) (ref. [3])
Luder-Sheldon syndrome (aminoaciduria with glycosuria) (ref. [4])

Table 3
Generalized disorders of amino acid transport.

Secondary to other hereditary disorders of metabolism
Tyrosinemia (ref. [5])
Hereditary fructose intolerance (ref. [6])
Galactosemia (ref. [7])
Wilsons disease (ref. [8])
Fanconi syndrome with cystinosis (ref. [9])

although the syndromes have been well described the nature of the underlying metabolic or transport defect remains completely unknown. The second group includes generalized disorders which are secondary to identifiable metabolic disorders (table 3). In this group, the nature of the metabolic disorders is understood, at least in part, and the ability of the renal tubule to reabsorb amino acids at a normal rate appears to be impaired as a result of the disorders of metabolism. Finally, there is a third group in which the failure of tubular reabsorption of amino acids is restricted to either single amino acids or well defined groups of amino acids, and a pure transport defect appears to be involved (table 4).

To complete the list of hereditary renal tubular transport defects two other important and well described disorders should be mentioned, that of renal glycosuria and X-linked primary hypophosphatemic rickets (table 5). Of particular interest in the study of disorders of transport are those disorders which appear to affect both intestinal and renal transport simultaneously.

Table 4
Specific disorders of amino acid transport.

Cystinuria (ref. [10])
Iminoglycinuria (ref. [11])
Some basic amino acidurias (familial protein intolerance with dibasic amino aciduria, renal and intestinal) (ref. [12])
Pure cystinuria (ref. [13])
Lysinuria (ref. [14])
Glycinuria (ref. [15])

Table 5
Other renal disorders of transport.

Renal glycosuria (ref. [16])
X-linked primary hypophosphatemic rickets (ref. [17, 18])

Table 6

Intestinal disorders of amino acid transport
Tryptophan malabsorption (blue diaper syndrome) (ref. [19])
Methionine malabsorption (ref. [20])
Disorders of amino acid transport with a combined effect on the intestine and kidney
Hartnup disease (ref. [21])
Hyperdibasic aminoaciduria (ref. [22, 23])
Familial protein intolerance with dibasic aminoaciduria (ref. [12])
Cystinuria (ref. [10])
Intestinal disorders of sugar transport
Glucose-galactose malabsorption (ref. [24])

Some disorders of intestinal transport are listed in table 6. Two disorders of amino acid transport appear to affect specifically the intestinal absorption mechanisms: these are tryptophan malabsorption (the blue diaper syndrome) and methionine malabsorption. In addition, there is a well described disorder of sugar transport, the glucose-galactose malabsorption syndrome, in which the renal transport of sugars appears to be normal. There are also four disorders of amino acid transport: Hartnup disease, Hyperdibasic amino-aciduria, familial protein intolerance with dibasic amino-aciduria, and cystinuria, in which the hereditary defects of transport seem to affect both intestinal and renal transport mechanisms simultaneously.

Table 7
Group transport by the kidney.

1. Monoamino-monocarboxylic amino acids:
   alanine, serine, threonine, valine, leucine, isoleucine, phenylalanine, tyrosine, tryptophan, histidine, methionine
2. Dibasic amino acids:
   lysine, arginine, ornithine, and cystine
3. Dicarboxylic amino acids:
   glutamic and aspartic
4. Imino-acid and glycine:
   proline, hydroxyproline, and glycine

In order to investigate these human disorders of transport it is necessary to appreciate the complexity of the biological systems involved. For instance a large number of investigations which have now been carried out in various biological systems lead us to believe that the mechanisms for amino acid transport can be grouped into four major groups (table 7) [24]. Of these, the second and fourth group are now quite clearly represented as specific amino-acidurias in human subjects. However, the overall mechanism of amino acid transport is vastly more complex than this and can be usefully be divided into four stages. The first is the identification of the amino acid by the specific transport mechanism in the membrane. This must be a four point stereo-chemical identification in order to achieve the known specificity of the transport system for specific groupings. These include the carboxyl group, the amino group, the stereochemical configuration of the alpha carbon atom, and the nature of the terminal group of the molecular chain. Once identification has taken place the membrane transport then proceeds, mediated by what is

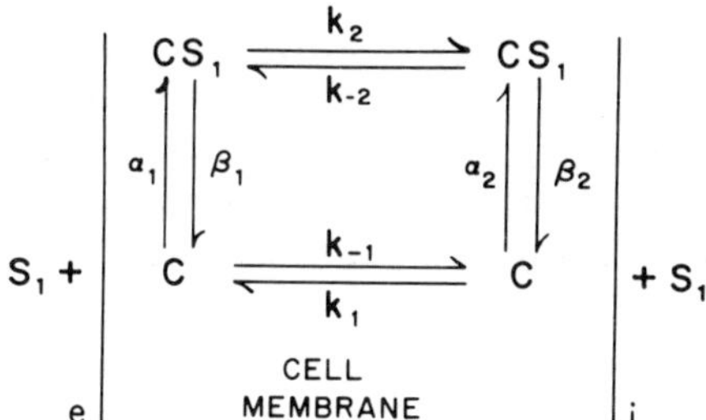

Fig. 1. Schematic mechanism of active membrane transport. e = extracellular phase; i = intracellular phase; S_1 = substrate 1; C = carrier; CS_1 = carrier-substrate complex.

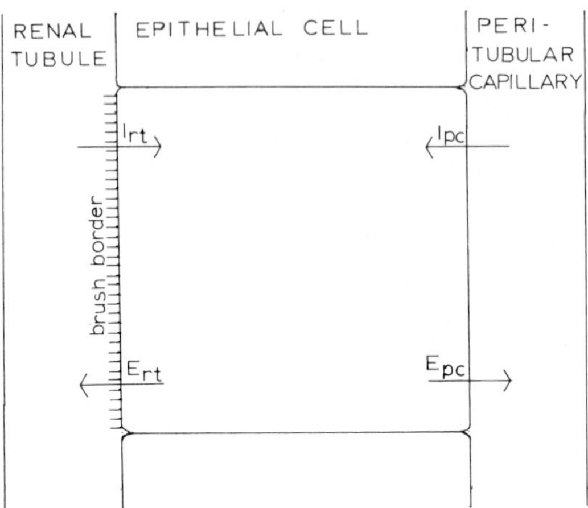

Fig. 2. Possible pathways for epithelial cell transport.

thought to be an active transport mechanism and a carrier protein. This is symbolised in fig. 1. The next stage after membrane transport is that of epithelial cell transport (fig. 2). In order to transfer substrate right across the epithelial cell two separate membranes have to be crossed. Each of these has potential influx and efflux pathways and, hence, it is the overall movement of substrate that has to be studied rather than the individual properties of the membrane. The proximal tubule of the mammalian kidney almost certainly makes use of each of these four pathways under various different circumstances. Finally, even if the mechanisms of membrane and epithelial cell transport could be elucidated, the final result of the transport process is determined either by the counter current mechanism in the proximal renal tubule or by the considerable length of the absorbing intestinal epithelium. I will use the example of cystinuria to discuss one or two of these phenomena.

In the case of the completely recessive type of cystinuria, it has been shown by in vitro transport studies of intestinal mucosal biopsy specimens that there is a complete defect of the active transport of lysine [25]. However, despite the fact that this is an essential amino acid and that up to 2 g of lysine per day is lost in the urine by such patients, they do not demonstrate an obvious defect of protein synthesis or lysine deficiency. Two different types of in vivo study have now been performed on these patients. The first was an intestinal perfusion experiment by Hellier, Perrett and Holdsworth [27], who showed by this technique that the absorption of free lysine in patients with cystinuria was only 11% of that of the mean of the absorption of control subjects. Using an oral lysine load, however, Asatoor et al. [28] found that free lysine was very adequately absorbed in patients with cystinuria. The apparent discrepancy between these two results can be explained by the fact that the first investigators were making a specific study of the absorption of the amino acid over a relatively short length of gut, whereas when the total length of the small intestine was available for absorption of the amino acid, this could apparently occur quite adequately. In addition, Asatoor et al. [28] showed that lysine could also be absorbed by cystinuria patients in combination as a dipeptide.

The complexity of the transport mechanism in the kidney in cystinuria has been the subject of much investigation [10], and one noteworthy observation is that in some patients with cystinuria the apparent renal clearance of cystine is greater than the glomerular filtration rate. This was first commented on by Frimpter et al. [29] and was further investigated by Crawhall et al. [30]. A possible explanation of this phenomenon is based upon the diagram of epithelial cell transport shown in fig. 2. At the present time the transport defect in cystinuria is believed to be along the pathway labelled I_{rt}. It is possible that there is also a slow efflux mechanism (E_{rt}) occurring, which is normally not observed because of the high rate of influx. However, in the genetic defect of cystinuria in which the influx mechanism is absent, it is possible for an amino acid such as cystine to appear in the urine as the result of the amino acid moving in the reverse direction, from the peritubular capillary side to the renal tubular side. The overall result would be the appearance of more amino acid in the urine than was actually filtered through the glomerulus. This complex of transport pathways has made it more difficult to interpret the genetics of transport defects because partial defects of transport such as might be observed in heterozygotes could be obscured by the counter current phenomena in the renal tubule. Despite this however, Harris et al. [31] were able to divide patients with cystinuria into two groups on the basis of the appearance of aminoaciduria in the patients and their siblings. He called the two types completely recessive and incompletely recessive cystinuria. The classification of patients with cystinuria in this way, however, is only satisfactory if one can distinguish clearly between the homozygous and the heterozygous state. In order to make this distinction, Crawhall et al. [32] have taken data on urinary amino acid excretion from a large number of patients with cystinuria and from their relatives, and have treated this data in the statistical manner of canonical variate analysis. On the basis of this data, possible homozygotes for cystinuria can be identified in families in which one propositus has been identified, and the homozygous can be distinguished from the heterozygous state with a reasonable degree of certainty. Another observation apparent from this study was that the expressivity of the heterozygote cystinuria gene is very variable within families, to the extent that there is no clear distinction between the ranges of normality and heterozygosity. The probability that we are looking at more than one genotypic variant may in part explain this. There may also be other minor genetic variants of the transport process comparable to other genetic polymorphisms which would influence the more direct hereditary defect of transport which occurs in cystinuria. Although at least two and possibly three genotypic variants of cystinuria have been described, the possibility of a renal tubular transport defect occurring in subjects who are doubly heterozygous for the two major genotypes also has to be considered. Evidence in support of this possibility has been presented by Rosenberg et al. [33] and more recently by Morin et al. [34].

Some investigators have tackled the problem of identifying heterozygotes for various hereditary defects of transport by determining the tubular absorption maxima of the kidneys of these subjects by means of oral loading with a suitable substrate. Examples of this are the classification of the renal glycosurias on the basis of loading studies by Rosenberg and Elsas [35] and also the loading studies with proline in patients with iminoglycinuria [36].

In conclusion, it can be said that the study of hereditary disorders of transport in human subjects has been particularly illiminating by its demonstration of the grouping of amino acids for transport purposes, and of the differentiation of the transport mechanisms for amino acid, sugars and phosphate re-absorption as separate identifiable functions of the renal tubule under independent genetic control. The overlap between intestinal and renal disorders of transport is also particularly interesting from the genetic point of view in that there must be some stage of the genetic control of the transport mechanisms which is common to both the kidney and the intestine for this phenomenum to take place. On the other hand, disorders of transport can occur separately in these two tissues so there must be other hereditary determinants for transport which are not common to the two tissues. Finally, it is significant that so far no generalized disorders of transport affecting all human cell types have ever been described, although this could be explained by the possibility that such a mutation would be non-viable.

References

[1] L. Hambraeus, G. Pallisgaard and P. Kildeberg, The Lowe syndrome. Observations on the amino acid metabolism in a two year old affected boy. Acta Paediat. Scand. 59 (1970) 631–636.

[2] D.D. Hunt, G. Stearns and J.B. McKinley, Long-term study of a family with Fanconi syndrome without cystinosis (DeToni-Debre-Fanconi syndrome). Amer. J. Med. 40 (1966) 492–510.

[3] P.T. Rowley, P.S. Mueller, D.M. Watkin and L.E. Rosenberg, Familial growth retardation, renal aminoaciduria and cor pulmonale. I. Description of a new syndrome, with case reports. Amer. J. Med. 31 (1961) 187–204.

[4] W. Sheldon, J. Luder and B. Webb, A familial tubular absorption defect of glucose and amino acids. Arch. Dis. Child. 36 (1961) 90–95.

[5] J. Gentz, R. Jagenburg and R. Zetterstrom, Tyrosinemia. J. Pediat. 66 (1965). 670–696.

[6] R. Lindemann, L.R. Gjessing and B. Merton, Amino acid metabolism in hereditary fructosemia. Acta Paediat. Scand. 59 (1970) 141–147.

[7] S.V. Darling and O. Mortensen, Aminoaciduria in galactosaemia. Acta Paediat. Scand. 43 (1954) 337–341.

[8] W.H. Stein, A.G. Bearn and S. Moore, The amino acid content of the blood and urine in Wilson's disease. J. Clin. Invest. 33 (1954) 410–419.

[9] J.C. Crawhall, P.S. Lietman, J.A. Schneider and J.E. Seegmiller, Cystinosis, plasma cystine and cysteine concentrates and the effect of D-penicillamine and dietary treatment. Amer. J. Med. 44 (1968) 330–339.

[10] J.C. Crawhall and R.W.E. Watts, Cystinuria. Amer. J. Med. 45 (1968) 736–755.

[11] C.R. Scriver, Renal tubular transport of proline, hydroxyproline, and glycine, J. Clin. Invest. 47 (1968) 823–835.

[12] J. Perheentupa and J.K. Visakorpi, Protein intolerance with deficient transport of basic aminoacids. Another inborn error of metabolism. Lancet 2 (1965) 813–816.

[13] J. Brodehl, K. Gellissen and S. Kowalewski, Isolierter Defekt der tubularen Cystin-Ruckresorption in einer Familie mit idiopathischem Hypoparathyroidismus. Klin. Wschr. 45 (1967) 38–40.

[14] K. Oyanagi, R. Miura and T. Yamanouchi, Congenital lysinuria: A new inherited transport disorder of dibasic amino acids. J. Pediat. 77 (1970) 259–266.

[15] A. DeVries, S. Kochwa, J. Lazebnik, M. Frank and M. Djaldetti, Glycinuria, an hereditary disorder associated with nephrolithiasis. Amer. J. Med. 23, (1957) 408–415.

[16] L.J. Elsas and L.E. Rosenberg, Familial renal glycosuria: a genetic reappraisal of hexose transport by kidney and intestine. J. Clin. Invest. 48 (1969) 1845–1854.

[17] R.W. Winters, J.B. Graham. T.F. Williams, V.W. McFalls and C.H. Burnett, A genetic study of familial hypophosphatemia and vitamin D resistant rickets with a review of the literature Medicine 37 (1958) 97–142.

[18] C. Arnaud, F. Glorieux and C. Scriver, Serum parathyroid hormone in X-linked hyposphosphatemia. Science 173 (1971) 845–847.

[19] K.M. Drummond, A.F. Michael, R.A. Ulstrom and R.A. Good, The blue diaper syndrome: Familial hypercalcemia with nephrocalcinosis and indicanuria. Amer. J. Med. 37 (1964) 928–948.

[20] C. Hooft, J. Timmermans, J. Snoeck, I. Anterer, W. Uyaert and C. Van Den Hende, Methionine malabsorption syndrome. Amer. J. Paed. 205 (1965) 73–104.

[21] M.D. Milne, M.A. Crawford, C.B. Girao and L.W. Loughridge, The metabolic disorder in Hartnup disease. Quart. J. Med. 29 (1960) 407–421.

[22] D.T. Whelan and C.R. Scriver, Hyperdibasicaminoaciduria; an inherited disorder of amino acid transport. Pediat. Res. 2 (1968) 525–534.

[23] M. Kekomaki, J.K. Visakorpj, J. Perheentupa and L. Saxen, Familial protein intolerance with deficient transport of basic amino acids. Acta Paediat. Scand. 56 (1967) 617–630.

[24] G. Meeuwisse and B. Lindquist, Glucose-galactose malabsorption, studies on the intermediate carbohydrate metabolism. Acta Paediat. Scand. 59 (1970) 74–79.

[25] J.A. Young and B.S. Freedman, Renal tubular transport of amino acids. Clin. Chem. 17 (1971) 245–266.

[26] S. Their, M. Fox, S. Segal and L.E. Rosenberg, Cystinuria: In vitro demonstration of an intestinal transport defect. Science 143 (1964) 482–282.

[27] M.D. Hellier, D. Perrett and C.D. Holdsworth, Dipeptide absorption in cystinuria. Brit. Med. J. iv (1970) 782–783.

[28] A.M. Asatoor, M.R. Crouchman, A.R. Harrison, F.W. Light, L.W. Loughridge, M.D. Milne and A.J. Richards, Intestinal absorption of oligopeptides in cystinuria. Clin. Sci. 41 (1971) 23-33.

[29] G.W. Frimpter, M. Horwith, E. Furth, R.E. Fellows and D.D. Thompson, Insulin and endogenous amino acid renal clearances in cystinuria: Evidence for tubular secretion, J. Clin. Invest. 41 (1962) 281–288.

[30] J.C. Crawhall, E.F. Scowen, C.J. Thompson and R.W.E. Watts, The renal clearance of amino acids in cystinuria. J. Clin. Invest. 46 (1967) 1162–1171.

[31] H. Harris, U. Mittwoch, E.B. Robson and F.L. Warren, Phenotypes and genotypes in cystinuria. Ann. Human. Genet. 20 (1955) 57–91.

[32] J.C. Crawhall, P. Purkiss, R.W.E. Watts and E.P. Young, The excretion of amino acids by cystinuric patients and their relatives. Ann. Hum. Genet. 33 (1969) 149–169.

[33] L.E. Rosenberg. S. Downing, J.L. Durant and S. Segal, Cystinuria: Biochemical evidence for three genetically distinct diseases. J. Clin. Invest. 45 (1966) 365–371.

[34] C.L. Morin, M.W. Thompson, S.H. Jackson and A. Sass-Kortsak, Biochemical and genetic studies in cystinuria: Observations on double heterozygotes of genotype I/II. J. Clin. Invest. 50 (1971) 1961–1976.

[35] L.J. Elsas L.E. Rosenberg, Familial renal glucosuria: a genetic reappraisal of hexose transport by kidney and intestine. J. Clin. Invest. 48 (1969) 1845–1854.

[36] C.R. Scriver, Renal tubular transport of proline, hydroxyproline and glycine. III. Genetic basis for more than one mode of transport in human kidney. J. Clin. Invest. 47 (1968) 823–835.

ENERGY COUPLING IN THE LACTOSE TRANSPORT SYSTEM OF *ESCHERICHIA COLI*

T. Hastings WILSON
Department of Physiology Hardvard Medical School
Boston, Massachusetts U.S.A.

Since the discovery of active transport of lactose by *E. coli* [1] several speculations have been made concerning the role of energy coupling. Two basic hypothesis have been suggested: the first states that energy is required for substrate entry into the cell [2–5] while the second proposes that energy coupling is required to prevent exit, having no effect on the entry step [6–10].

The latter proposal was supported by the inhibitor experiments of Winkler and Wilson [8]. Their model is given in fig. 1. According to this model the external substrate reacts with the membrane transport carrier (the product of the y gene) to give substrate-carrier complex which migrates to the inner surface of the membrane. Energy is couples in such a manner as to reduce the affinity of the carrier for the substrate which is thus ejected into the cytoplasm of the cell. The empty carrier returns to the outside of the membrane where it spontaneously regains its high affinity for substrate.

Important confirmation of this hypothesis might come from the study of a

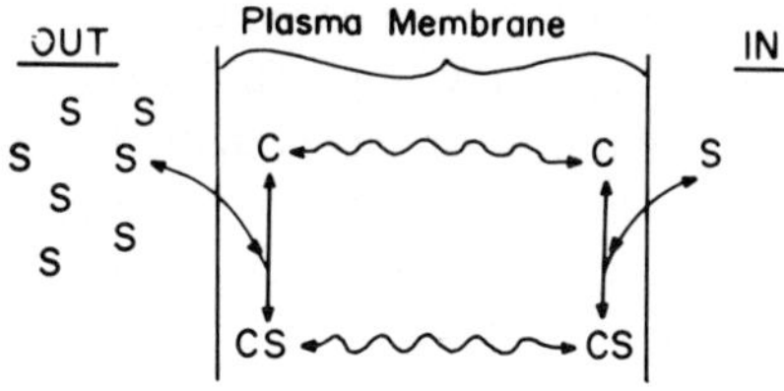

Fig. 1. Model for membrane transport in *E. coli.* S is substrate, C is carrier, and CS is carrier substrate complex.

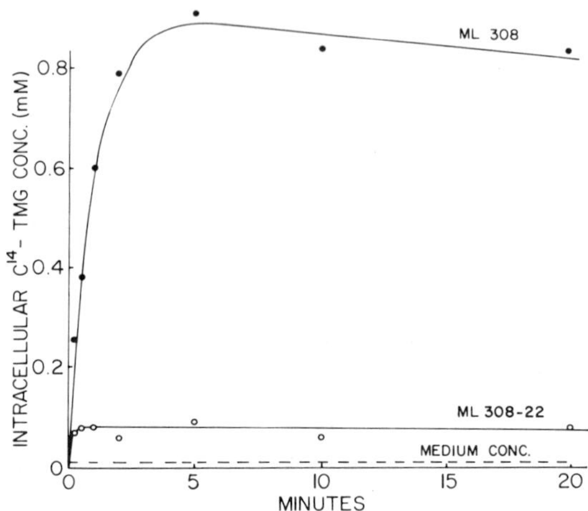

Fig. 2. Accumulation of ^{14}C-TMG by ML 308-22 and ML 308. The cells were suspended to a density of 300 Klett units (No. 42 filter). Aliquots of 2.0 ml were mixed with 2.0 ml ^{14}C-TMG in medium 63 (final concentration of 0.011 mM at 0.08 μc/μM), and the mixture incubated at 25°. Samples (0.5 ml) were withdrawn at the indicated intervals, filtered through Millipore filters, washed, and counted.

mutant energy uncoupled for lactose transport. The predicted property of such a mutant would be its inability to actively accumulate substrate but its retention of the membrane carriers which performed those transport functions not dependent upon the expenditure of energy. Two such mutants have been isolated and their properties studied [11–13].

Such mutants show a severe defect in accumulation of thiogalactosides. Fig. 2 shows that mutant 308-22 has lost 90% of the accumulating capacity of the parent. According to the above hypothesis an energy uncoupled mutant should possess intact membrane carriers. Three assays for carriers were performed. The first assay was one widely used for this system, the rate of entry of o-nitrophenyl-galactoside (ONPG). The lactose analog thiodigalactoside (TDG) possesses a very high affinity for the carrier and is therefore useful as a specific inhibitor of this system. The TDG-inhibited ONPG entry rate was taken as a measure of the lactose transport carriers. Table 1 shows that the mutant possessed more than normal activity of this carrier. Similar results were obtained by two other assays; the counterflow method [14, 15] and TDG binding [16].

If membrane carriers were intact (and energy coupling does not affect entry) then normal growth of the mutant cells on lactose would be predicted.

Table 1
ONPG hydrolysis by ML 308-22 and 308

Cells	Addition	ONPG hydrolyzed (nmoles/min/l × 10^9 cells)
ML 308-22	–	182
ML 308-22	TDG	4.2
ML 308	–	112
ML 308	TDG	4.5

ONPG hydrolysis was measured by incubating 1.0-ml cells at 350 Klett units (2.1 × 10^9 cells/ml) with 3.5 ml medium 63 and 0.5 ml 20 mM ONPG. TDG was added to a concentration of 20 mM, where indicated.

Both parent and mutant showed the same doubling time when grown on 25 mM lactose (table 2) as a sole source of carbon and energy. Thus the carriers are functional for lactose. The mutant, however, possessed a growth defect at very low substrate concentrations: at 0.5 mM the doubling time 63 min and at 0.25 mM no growth occurred. Clearly, accumulation is essential for growth under these conditions.

If entry were not affected in the energy-uncoupled mutant then the exit must be more rapid than normal to account for the inability to accumulate. Cells were preloaded with radioactive TMG and the rate of exit measured. Exit from the mutant was much faster than that from the parental cells [11].

One might legitimately ask whether the rapid exit pathway was non-specific. This was investigated by measuring accumulation of other actively transported compounds. If non-specific leakage were present a defect would

Table 2
Growth of ML 308-22 and ML 308 in lactose at various concentrations

Lactose concentration (mM)	Doubling time (min)	
	ML 308	ML 308-22
25	55	52
0.5	52	63
0.25	56	∞

The inoculum was grown aerobically overnight at 37° in medium 63 containing 2.5 mM lactose. The cell suspension, 0.05 ml, was transferred to sidearm flasks containing 30 ml medium 63 and lactose at the indicated concentrations. The flasks were shaken vigorously at 37° and growth measured with a Klett-Summerson colorimeter (No. 42 filter).

be found in the accumulation of all substrates. The mutant was found to possess normal accumulation of D-fucose, α-methylglucoside, L-proline and α-aminoisobutyric acid. This demonstrates that the mechanism of energy coupling was not shared by these compounds.

It would be of interest to know whether the defect in this mutant was in the membrane carrier protein or in some other protein of the cell. A partial answer was obtained by transducing the lactose operon from the mutant to a lactose-deleted strain. If the uncoupling property were in the lactose operon both the uncoupling property and the lactose genes should be co-transduced by the bacteriaphage Pi in such an experiment. Experiments of this type showed that 13 out of 13 lactose-positive transductants showed the energy-uncoupling property [13]. It is thus probable (although not yet unequivocally proven) that the defect resides in the y gene which codes for the membrane carrier protein. This view is supported by a biochemical abnormality of the membrane carrier of the mutant cells. The mutant was much more suseptable to inhibition of transport by SH inhibitors (N-ethylmaleimide and *p*-chloromercurybenzoate) than was the parent.

In summary, mutants have been isolated with membrane carrier activity but a servere defect in accumulation. The properties of these mutants confirm the view that energy coupling lowers the affinity of the carrier for its substrate on the inner surface of the membrane, having no effect on entry. An interesting physiological function of active transport is illustrated. While energy coupling is not essential for growth on high concentrations of lactose, accumulation is necessary for growth on very low concentrations of substrate.

References

[1] H.V. Richenberg, G.N. Cohen, G. Buttin and J. Monod, Ann. Inst. Pasteur 91 (1956) 829.
[2] A. Kepes, Biochim. Biophys. Acta 40 (1960) 70.
[3] G.A. Scarborough, M.K. Rumley and E.P. Kennedy, Proc. Nat. Acad. Sci. 60 (1968) 951.
[4] D. Schachter, N. Johnson, M.A. Kirkpatrick, Biochem. Biophys. Res. Commun. 25 (1966) 603.
[5] A.L. Koch, Fed. Proc. 28 (1968) 868.
[6] G.N. Cohen and J. Monond, Bacteriol. Rev. 21 (1957) 169.
[7] A.L. Koch, Biochim. Biophys. Acta 79 (1964) 177.
[8] H.H. Winkler and T.H. Wilson, J. Biol. Chem. 241 (1966) 2200.
[9] A. Kepes, J. Membrane Biol. 4 (1971) 87.
[10] C.F. Fox and E.P. Kennedy, Proc. Nat. Acad. Sci. 54 (1965) 891.
[11] P.T.S. Wong, E.R. Kashket and T.H. Wilson, Proc. Nat. Acad. Sci. 65 (1970) 63.

[12] T.H. Wilson, M. Kusch and E.R. Kashket, Biochem. Biophys. Res. Commun. 40 (1970) 1409.
[13] M. Kusch and T.H. Wilson, Fed. Proc. 30 (1971) 313.
[14] A. Kepes, In: the Molecular Basis of Membrane Function, ed. Tosteson, (Prentice-Hall Inc., 1969), p. 358.
[15] P.T.S. Wong and T.H. Wilson, Biochim. Biophys. Acta 196 (1970) 336.
[16] E.P. Kennedy, The Lactose Operon, eds. J.R. Beckwith and D. Zipser (Cold Spring Harbor Laboratory, 1970) p. 49.

Part III

ACTIVE TRANSPORT ACROSS A SINGLE MEMBRANE

INTRODUCTION

D.C. TOSTESON
Department of Physiology and Pharmacology, Duke University Medical Center, Durham, North Carolina, U.S.A.

This session of the Conference deals with the physico-chemical basis of active transport across single membranes. It serves as a bridge between earlier sessions devoted to the turnover and genetic determination of membranes and subsequent sessions which consider active transport across epithelia. The active transport systems in a single membrane are the resultant of the processes considered in the first two sessions. Furthermore, an adequate understanding of active transport across a single plasma membrane is clearly a necessary step in the elucidation of active transport across the two or more plasma membranes arranged in series in complex geometrical patterns in epithelia. Rather than attempt a consideration of the vast subject of active transport of all kinds of substances across single membranes, we chose to restrict attention to the transport of monovalent and divalent cations. Thus, this session anticipates the final section on coupling mechanisms, e.g. the coupling of active transport of Na^+ with movement of sugar or amino acids.

The argument of this session proceeds from a detailed consideration of the distinctive characteristics of active transport of Ca^{2+}, Na^+, and K^+ in well studied membranes to an examination of membrane antigens and enzymes which may be involved in the active transport process. The format consists of four major papers each followed by several shorter papers bearing on the same subject.

The first major paper by W. Hasselbach and M. Makinose deals with Ca^{2+} transport in sarcoplasmic reticulum membranes. They report observations which lead them to the important conclusion that the Ca^{2+} pump in this system can be reserved so that "high-energy" phosphate bonds are produced rather than consumed during the transport process. A short paper by Y. Ogawa also treats some problems related to the mechanism of Ca^{2+} uptake by sarcoplasmic reticulum. A second short paper by E. Carafoli et al. discusses Ca^{2+} uptake in cardiac muscle.

The second major paper by L.J. Mullins describes active transport of Na^+ and K^+ across the squid axon membrane. Because of the relatively large diameter of the squid axon which makes possible internal perfusion or dialysis, this membrane is almost unique in biology in that its ion permeability has been studied simultaneously by both electrical and tracer methods.

The third long paper by J.F. Hoffman concentrates on the use of ^{3}H-ouabain in estimating the number of active transport sites in human and other mammalial red cell membranes. Short papers by L. Bolis et al. and G. Gardos consider further the transport of substances across red cell membranes.

The final long paper of the session by I.M. Glynn and J.C. Ellory discusses the effect of reaction of a membrane antigen with its specific antibody on active $K^+ - Na^+$ transport in LK sheep red cells. The paper makes clear that immunological techniques may be of increasing value in investigations of the molecular basis of membrane transport processes. Short papers by P.J. Garrahan and by A. Rega shed new light on the properties of the K^+ stimulated phosphatase activity in human red cell membranes. Finally, P.L. Jørgensen reports on the impressive progress which he has made in the purification of the $(Na^+ + K^+)$-ATPase isolated from the outer medulla of the mammalian kidney.

THE REVERSAL OF THE SARCOPLASMIC CALCIUM PUMP

W. HASSELBACH and M. MAKINOSE

Max-Planck-Institut, Department of Physiology, Heidelberg, Germany

Calcium transport plays a vital role in a great number of physiological activities. So far as it is known, in all cells the cytoplasmic calcium level is considerably lower than the calcium level in the extracellular fluid. As a consequence, many physiological activities can be turned on and off by the liberation and the removal of small quantities of calcium. One of the most intensively studied systems involved in the regulation of the intracellular culum (SR) in fast striated muscles [1–5]. These membranes offer a great number of acvantages of which only two might be mentioned.
of advantages of which only two might be mentioned.

1. The specific calcium transport activity of these membranes is extremely high.

2. After isolation of the fragmented tubular membrane network, firmly closed vesicles are formed, so that not only the energy yielding reactions but also the transport processes which lead to an accumulation of calcium can be observed.

The accumulation of calcium in the sarcoplasmic vesicles is the result of a sequential process during which high energy phosphate compounds like ATP and other NTP or acetylphosphate and carbamylphosphate are used up [7–9]. Some steps in this reaction sequence have been ascertained.

1. Calcium combines with these membranes at sites of an extremely high calcium affinity. Identical calcium binding curves have been obtained for open as well as for closed SR vesicles and even for vesicles from which the membrane lipids have been removed (fig. 2) [10, 11]. This intrinsic high

Abbreviations: ACP: acetylphosphate; EGTA: ethylenglycol bis (β-aminoethylether) -NN-tetraacetic acid; NTP: nucleosidtriphosphate; prenylamine: N-(3 phenyl-propyl 2)-1-1 diphenylpropyl-3-amine.

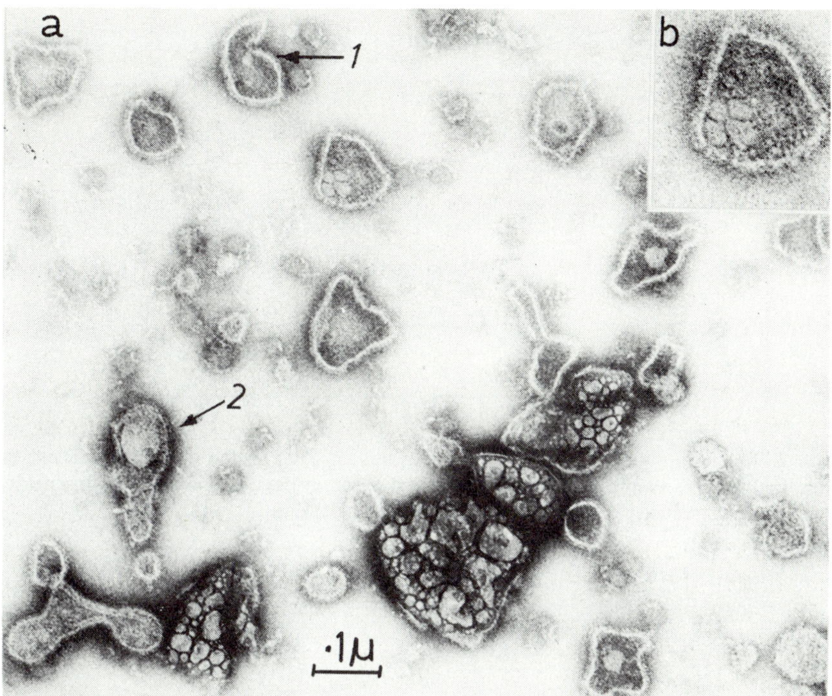

Fig. 1. Calcium loaded vesicles of rabbit fragmented sarcoplasmic reticulum, negatively stained with 1% potassium phosphotungstate after 10 min glutaraldehyde fixation. Vesicular structures of different size and shape can be observed. Some vesicles are completely filled with nodular deposits of cacium oxalate, while others contain smaller precipitates (arrow 1) or show areas of higher density (arrow 2). The surface of most vesicles is covered with fine particles. × 100 000. Inset: × 180 000. Loading conditions: [ATP] = [$MgCl_2$] = [K_2oxalate] = 5 mM; histidine 0.02 M *p*H 7.0, 40 mM KCl, 0.1 mM $CaCl_2$, SR protein 0.02 mg/ml. Incubation time 10 min, T = 20 °C. (see ref. [6]).

calcium affinity neither depends on the presence nor on the absence of magnesium ions. In the same concentration range in which the calcium binding sites are saturated the activation of the calcium transport system takes place. Since in the respective concentration range the calcium transport activity only depends on the calcium concentration in contact with the outer surface of the vesicles one must assume that the reaction sequence is initiated at the outer surface of the vesicles, i.e. the high affinity calcium sites face the outer solution [12].

2. Independent of the discussed high affinity binding site of calcium,

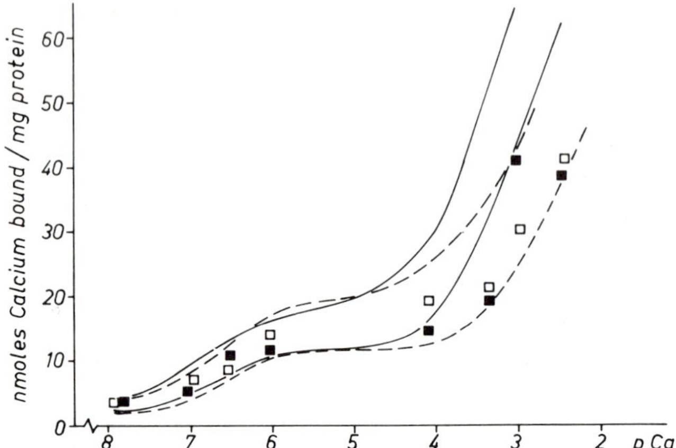

Fig. 2. Calcium binding to various preparations of SR vesicles. = native vesicles; = = modified vesicles (sonicated and oleate treated preparations), ATP absent or present; □ ■ lipid deprived vesicles; □ ATP absent; ■ ATP present. (See ref. [10]).

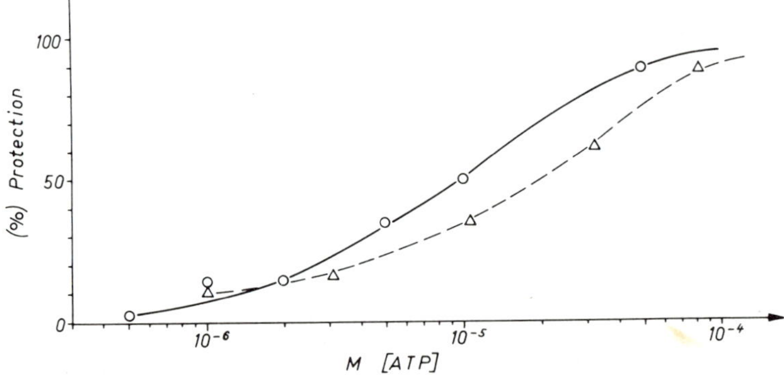

Fig. 3. The effect of ATP on lipid deprived SR vesicles and the calcium dependent ATPase activity of native vesicles. The vesicles were delipidated [13] incubated with NEM in the presence of ATP the concentration of which is plotted on the abscissa. Subsequently the ATPase was reconstituted by the addition of oleate (0.2 mg/mg prot.). The low ATP concentrations were established by pyruvatekinase (EC.2.7.1.40) and phosphoenolpyruvate. △—△ ATPase activity of native vesicles [14].

ATP combines with the SR membranes [3]. The binding curve shown in fig. 3 has been obtained with a membrane preparation reversibly inactivated by enzymatic lipid depletion. The experimental procedure is described in the legend of fig. 2. Magnesium ions are not required for ATP binding. The high affinity of the membranes for ATP found in these static binding experiments agrees quite well with the K_m-value found in experiments in which the dependency of calcium uptake on the concentration of ATP has been studied [14, 15]. Since ATP protects thiol groups located at the outer surface of the membranes [16] it seems reasonable to assume that the binding sites for ATP are directed outward as are the high affinity binding sites for calcium.

3. In the following step which occurs in less than 2 sec the terminal phosphate of ATP or the phosphoryl group of acetylphosphate is transferred to the membranal protein [17–19]. In contrast to calcium and ATP binding, this step requires the presence of calcium and magnesium ions and furthermore presumably the presence of lipids. The phosphorylation yield depends on the phosphate donator in the system. With ATP maximal 5 moles phosphate per 10^6g protein were found, ITP gives 8 moles phosphate per 10^6g protein while a yield of 10–12 moles phosphate per 10^6g protein is obtained with acetylphosphate as P-donator (fig. 4) [17, 20]. All phosphorylated intermediates have been isolated as acylphosphates and in the native protein all phosphoryl groups can be transferred readily to ADP. Calcium ions which activate the phosphoryl transfer reactions are solely those present in the solution outside of the vesicles. High calcium concentrations inside the vesicles which may occur as the result of the accumulation process do not interfere with the phosphorylation reaction [18].

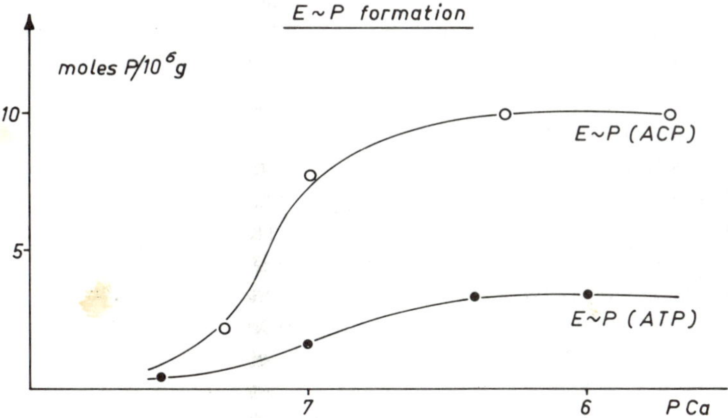

Fig. 4. The activation of phosphoprotein formation by ionized calcium with ATP and acetylphosphate as phosphate donator. Ordinate: phosphoprotein formation (E ~ P) mole/10^6g of prot. Abscissa: free calcium concentration (pCa) adjusted by CaEGTA buffers. (Makinose; Friedman and Makinose, unpublished observations.)

4. The final reactions which can be observed are phosphate liberation and calcium accumulation. Both processes proceed with an undiminished rate only for a longer period of time when calcium precipitating agents (oxalate, phosphate, pyrophosphate) are present in the medium [2, 21]. The prolongation of the accumulation process is required for the simultaneous measurement of phosphate liberation and calcium accumulation. As the result of the precipitation of calcium inside the vesicles, calcium uptake can continue at a low internal free calcium concentration whose level is determined by the solubility product of the respective calcium salt and the concentration of the precipitating agent. In the absence of calcium precipitating agents the fast rising calcium concentration inside counteracts not only the uptake of calcium but also inhibits the liberation of phosphate. At an internal calcium concentration of about 0.5 mM the rate of the two processes is reduced to 50% [12] while the phosphoryl transfer reactions are not effected by the internal calcium concentration. The simultaneous affection of phosphate liberation and calcium accumulation by internal calcium is one example of the close coupling between the two processes. This close coupling exists as long as the SR vesicles are not leaky and under such conditions, for every phosphate liberated two calcium ions are taken up provided that the extra vesicular calcium level is not too low. This transport ratio has been found under numerous conditions. It is most remarkable that this ratio does not change even when the absolute rates of calcium uptake and of phosphate liberation are varied by a factor of 5 to 10. The result does not depend whether such variations have been performed by the use of other nucleosidtriphosphates as energy donators or by the application of transport inhibitors like prenylamine. This observation suggests that the calcium influx can only be counteracted by a neglectibly small passive calcium efflux. This suggestion is supported further by the observation that like the transport ratio, the final inside-outside ratio of the activity products for calcium and oxalate does not depend on the initial rates of calcium influx.

The close coupling between calcium translocation and phosphate liberation during net calcium uptake suggests that the two events are two aspects of one and the same reaction. It seems therefore tempting to suspect that corresponding to the coupling between calcium influx and phosphate liberation a similar coupling between calcium efflux and ATP synthesis may exist. This would mean that the ATP dependent calcium translocation in the sarcoplasmic membrane is a reversible reaction. At first, the occurrence of the suspected backward reaction could be demonstrated under conditions where calcium influx is balanced by a calcium efflux of equal size. This steady state is reached when calcium uptake ceases after the limiting activity ratio

Table 1
Various properties of the SR calcium pump driven by different energy donators and inhibited by prenylamine.

	ATP	UTP	ATP + phenyl-amine (10^{-4} M)
Ca uptake (μmoles/mg·min)	1.40	0.3	0.44
Ca uptake / ATP splitting	2.0	2.0	2.0
Ca_i Ox_i / Ca_o Ox_o	2000	2000	2000
Ca exchange (steady state) μmoles/mg·min	0.12	0.04	0.04
E~P (μmoles/g)	3		3
E + ATP ⇌ EP + ADP (μmoles/min·mg)	3		3

(Ca_i/Ca_o) has been reached [12]. This ratio is maintained as long as ATP is present in the medium. Under the usual conditions the calcium influx and efflux proceeds very slowly. The rate has been determined by measuring the disappearance of tracer amounts of ^{45}calcium added after a constant calcium level in the medium is reached. To reveal the reverse reaction we supplement the usual uptake medium with ADP and ^{32}phosphate. As shown in fig. 5 incorporation starts to occur not until the steady state is attained. It proceeds with a rate of 0.05–0.1 μmoles P/mg prot. min. [22]. The rate of the simultaneously occurring calcium exchange is in the same order of magnitude. The incorporation of ^{32}P into the ATP fraction is abolished when the vesicles are made leaky and it is suppressed by prenylamine. It is affected neither by dinitrophenol nor by azide. These findings promote the search for a net ATP synthesis driven by the net calcium efflux. In order to find the appropriate conditions for this reaction we studied calcium efflux from calcium loaded vesicles. The rate of calcium release from these vesicles in media containing no ATP is much lower than the rate of calcium uptake. Even when the calcium concentration in the media is drastically lowered by EGTA the release rate does not increase. The rate of release is limited by the low calcium permeability of the membrane because the calcium appears at once in the medium when the membranes are destroyed [23]. Since the rate of calcium release under these conditions is even smaller than the activity of the calcium inde-

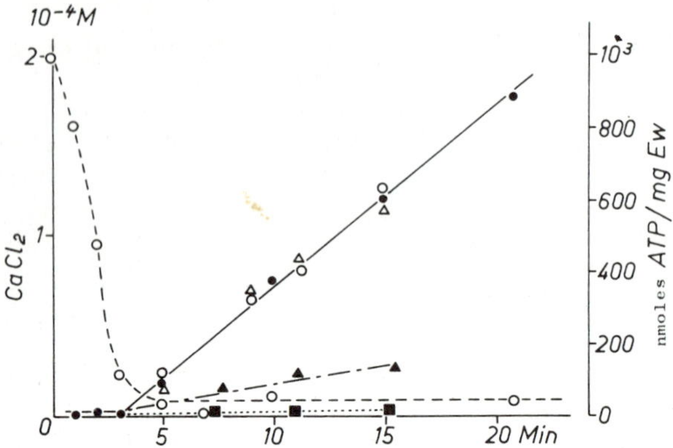

Fig. 5. The incorporation of inorganic phosphate into the ADP fraction after the cessation of net Ca uptake. – – – attainment of a constant calcium level; ––– ^{32}P-ATP formation; Additions: ○ none; △ 1 m azide; ● 0.2 mM DNP; ▲ 0.1 mM prenylamine; ■ vesicles treated with ether.

pendent basal ATPase it seemed impossible to demonstrate the occurrence of a net ATP synthesis.

However, to our surprise we found that a much faster calcium efflux can be induced from intact vesicles. It requires the combined action of ADP, phosphate, and magnesium [24]. The experiments are depicted in fig. 6. The

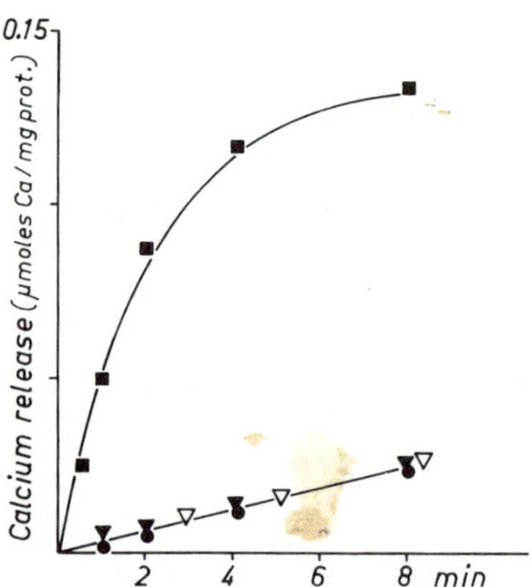

Fig. 6. The effect of phosphate, ADP and magnesium on the release of calcium from Ca loaded vesicles. ■ complete release medium contained 20 mM phosphate, 2 mM ADP, 5 mM $MgCl_2$. One of the essential components was omitted. ▼ no ADP, ● no phosphate, ▽ no magnesium.

vesicles were loaded with calcium oxalate using ITP or acetylphosphate instead of ATP as energy donator. ITP or acetylphosphate are used to prevent the accumulation of ADP in the medium. When the energy donator present in the uptake medium is used up, the suspension of calcium loaded vesicles is diluted by different releasing solutions containing KCl, Mg, EGTA and in addition inorganic phosphate and ADP. Fig. 6 shows that the efflux is enhanced by more than 20 fold in the medium containing both phosphate and ADP. The efflux is optimally activated when the medium contains 0.05 mM ADP and about 3 mM phosphate. The enhancement of the calcium efflux produced by phosphate and ADP is most effectively reduced by the lack of magnesium or low concentrations of calcium ions in the medium. The effective concentrations of calcium are identical with those that activate the calcium uptake. For a number of different activating phosphate concentrations an inhibition constant of 0.2 μM was obtained. These findings exclude a simple diffusion mechanism for calcium efflux and suggest that the mem-

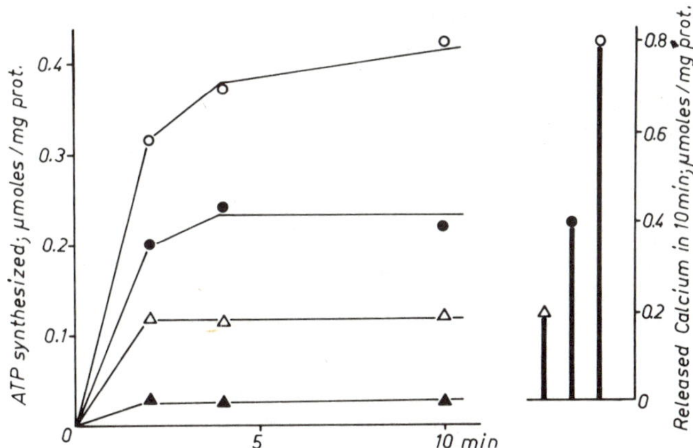

Fig. 7. Calcium release from calcium phosphate loaded vesicles and net formation of ^{32}P-ATP. Left ordinate: ATP synthesis; right ordinate: calcium released. The amount of calcium released during 10 min is identical with the amount of calcium stored. During an incubation period of 15 min in media containing 2 mM acetylphosphate, 7 mM $MgCl_2$, 40 mM KCl, 0.1 M glucose, 20 mM Na orthophosphate (^{32}P) and 0.5 mg/ml of vesicular protein at pH 7.0, the vesicles (one mg of protein) were loaded with different amounts of calcium: ○ 800 nmoles, ● 400 nmoles, △ 200 nmoles and ▲ no calcium. Subsequently the calcium release was started by the addition of 2 mM ADP, hexokinase (0.02 mg/ml) and 1 mM to 2 mM EGTA at time 0. The amount of EGTA was at least 5 times higher than the total calcium in the system. The synthesis of ATP was measured as glucose-6-phosphate (^{32}P) formation (left ordinate).

brane components involved in the active uptake of calcium also take part in the efflux of calcium. In fact, under the conditions where a fast calcium release occurs, the net outward movement of calcium is stoichiometrically connected with the formation of ATP [25]. Fig. 7 illustrates that for two calcium ions that leave the vesicles one phosphate group is transferred to ADP. The rate of ATP formation reaches approximately 20% of the initial rate of ATP consumption during active calcium uptake. As under the conditions of steady state P-incorporation, net ATP formation disappears when the vesicles are made permeable for calcium. The net formation of ATP is inhibited by prenylamine like the steady state P-incorporation.

The constant relationship between the amount of calcium stored and the amount of ATP formed, shows that ATP synthesis does not depend on the calcium load. Therefore, we have to assume that the ratio of calcium inside over calcium outside does not fall below a critical level until nearly all calcium is released. Inside the vesicles the calcium activity remains constant because calcium is replenished from the calcium precipitates and therefore the decline of the ratio Ca_i/Ca_o is only the result of the slowly rising calcium concentration in the solution outside.

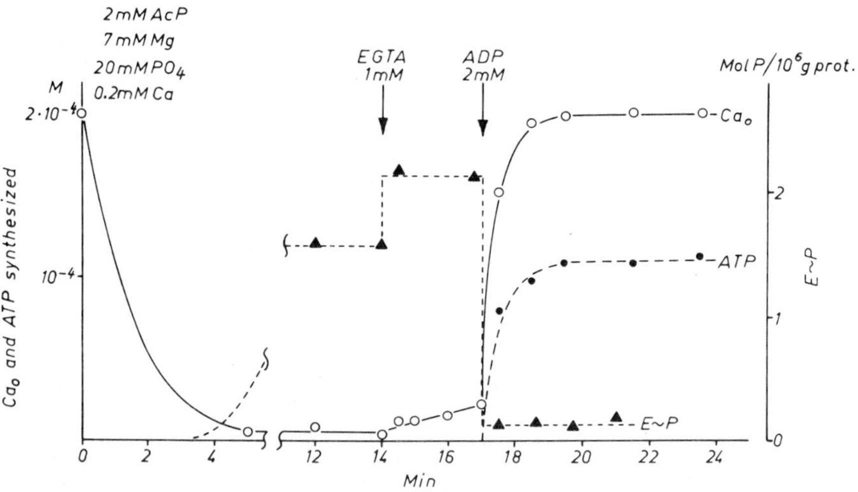

Fig. 8. The incorporation of inorganic phosphate into the SR membranes and the transfer of the phosphoryl group to ADP during calcium release. The incubation medium is identical with that of fig. 7. EP-formation was measured as described by Makinose (1970). Glucose-6-phosphate formed after the addition of ADP was measured after chromatographic isolation. ○—–—○ calcium concentration in the solution (left ordinate) ▲– –▲ phosphoprotein E ~ P (right ordinate) ●– –● ATP formation (left ordinate)

The minimal energy which has to be provided for ATP formation by the calcium gradient under the experimental conditions can be estimated to be about 2000 calories. This value is so low because the ATP level is kept low by the hexokinase reaction which exert a considerable energy pull. Since the translocation of two calcium ions givės rise to the incorporation of one phosphate, the ratio of calium inside to calcium outside of about 20 is sufficiently high to provide the required energy.

As described for the forward reaction the back reaction also gives rise to the formation of a phosphorylated intermediate in the vesicular membranes (fig. 8). The incorporation of inorganic phosphate (^{32}P) can be observed after the cessation of net calcium uptake, presumably after the calcium gradient has reached a threshold value. The phosphorylation yield can be enlarged further when the ratio Ca_i/Ca_o is increased by the addition of EGTA to the system. ADP takes over readily the phosphoryl group from the membranal protein which has been characterized to be an acylphosphate as it is the intermediate formed by ATP (Makinose, unpublished results). Unlike the synthesis of ATP by the mitochondria or chloroplasts the calcium efflux

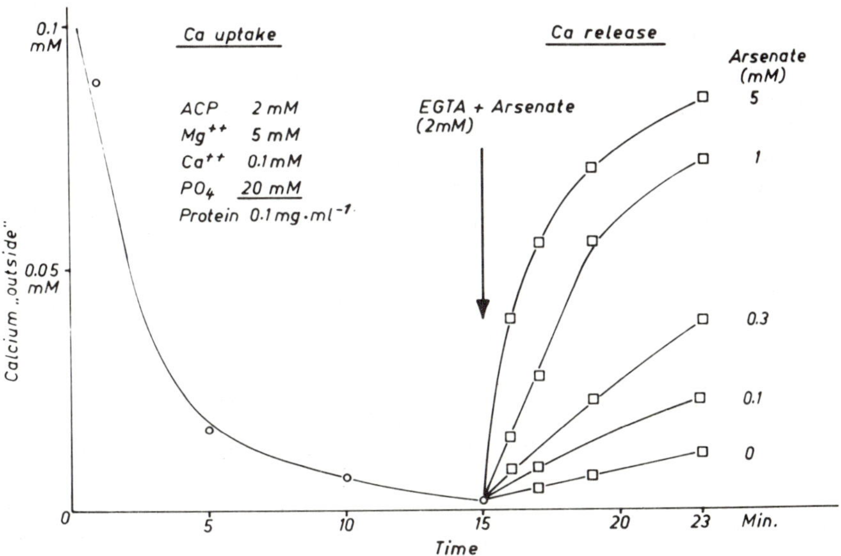

Fig. 9. The arsenate induced calcium release from calcium loaded vesicles and its inhibition by phosphate. The vesicles were loaded with calcium phosphate using acetylphosphate as energy donator in the presence of 5 mM Mg^{++} . The calcium release is induced by the addition of arsenate and EGTA.

dependent ATP formation of the SR vesicles is affected neither by olygomysin nor dinitrophenol and CCP. However, one classical uncoupler of oxidative phosphorylation proves to be effective also at the sarcoplasmic vesicles, namely arsenate. In the presence of magnesium ions, arsenate induces a fast calcium release from calcium loaded vesicles which reaches its maximal rate at a concentration of 5 mM (fig. 9). In contrast to the release induced by phosphate, the presence of ADP is not required for the effect of arsenate. Moreover, phosphate as well as ADP counteract the effect of arsenate. If calcium efflux is induced by a combination of phosphate, arsenate and ADP it depends on the proportion of the three reagents whether or not uncoupling of ATP formation from calcium efflux does occur. While high concentration of phosphate and ADP prevent uncoupling completely, complete uncoupling occurs when the concentration of phosphate and ADP are kept low in the system (fig. 10) (Hasselbach, Makinose, unpubl. results).

The described results demonstrate that the calcium translocation across the sarcoplasmic membranes is reversibly connected with the phosphoryl transfer reaction giving rise to a splitting of ATP when calcium moves inward and synthesis of ATP when calcium moves outward.

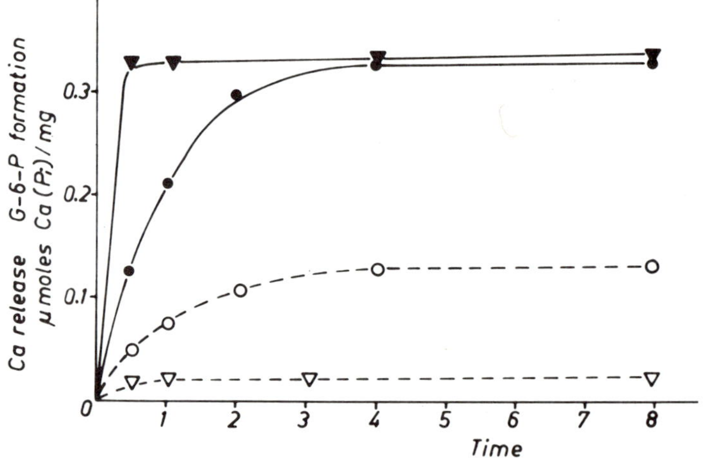

Fig. 10. The dissociation of calcium efflux and ATP formation produced by arsenate. The vesicles were loaded in media containing 2 mM acetylphosphate, 1 mM phosphate, 1 mM oxalate, 0.5 mM $CaCl_2$, 40 mM KCl, 0.1 M glucose, 0.5 m vesic. protein/ml. The calcium release was induced either by the addition of EGTA and ADP or EGTA, ADP and arsenate together with hexokinase (EG 2.7.1.1) 0.02 mg/ml spec. activity 140 U/mg. Final concentrations: EGTA 2 mM, ADP 0.05 mM, arsenate 5 mM. ●—● calcium release without arsenate; ▲—▲ calcium release with arsenate; ○––○ P-incorporation without arsenate; ▽––▽ P-incorporation with arsenate.

References

[1] S. Ebashi and F. Lipman, J. Cell. Biol. 17 (1962) 389.
[2] W. Hasselbach and M. Makinose, Biochem. Z. 333 (1961) 518.
[3] W. Hasselbach, in: O. Meyerhof Festschrift (Springer, Berlin, 1971).
[4] A. Martonosi and R. Feretos, J. biol. Chem. 239 (1964) 659.
[5] A. Weber, R. Herz and I. Reiss, Biochem. Z. 345 (1966) 329.
[6] B. Agostini and W. Hasselbach, Histochemie 28 (1971) 55.
[7] M. Makinose and R. The, Biochem. Z. 343 (1965) 383.
[8] L. de Meis, J. biol. Chem. 244 (1969) 3733.
[9] A. Pucell and A. Martonosi, J. biol. Chem. 246 (1971) 3389.
[10] W. Fiehn and A. Migala, Eur. J. Biochem. 20 (1971) 245.
[11] J. Chevallier and R.A. Butow, Biochemistry 10 (1971) 273.
[12] M. Makinose and W. Hasselbach, Biochem. Z. 343 (1965) 360.
[13] W. Fiehn and W. Hasselbach, Eur. J. Biochem. 13 (1970) 510.
[14] L. de Meis, J. biol. Chem. 246 (1971) 4764.
[15] G. Inesi, J. Goodman and S. Watanabe, J. biol. Chem. 242. (1967) 4637.
[16] W. Hasselbach and L.G. Elfvin, J. Ultrastruct. Res. 17 (1967) 598.
[17] M. Makinose, 2. Int. Biophys. Congress Wien (1966) 276.
[18] M. Makinose, European J. Biochem. 10 (1969) 74.
[19] T. Yamamoto and Y Nonomura, J. Biochem. (Tokyo) 62 (1967) 557.
[20] Z. Friedman and M. Makinose, FEBS Letters 11 (1970) 69.
[21] W. Hasselbach, and M. Makinose, Biochem. Z. 339 (1963) 94.
[22] M. Makinose, FEBS Letters 12 (1971) 269.
[23] W. Hasselbach, Biochem. Intracell. Struct. ed. by L. Wojtcak, W. Drabikowski, H. Strzelecka-Golaszewska (PWN Warszawa, 1969).
[24] B. Barlogie, W. Hasselbach and M. Makinose, FEBS Letters 12 (1971) 267.
[25] M. Makinose and W. Hasselbach, FEBS Letters 12 (1971) 271.
[26] W. Hasselbach and M. Makinose, Biochem. Z. 3331 (1969) 351.

A COMMENT ON THE MECHANISM OF CA-UPTAKE OF THE SARCOPLASMIC RETICULUM

Yasuo OGAWA
Department of Pharmacology, Faculty of Medicine, University of Tokyo, Tokyo

As regards the Ca-uptake of the sarcoplasmic reticulum (SR) *, two interesting but somewhat puzzling observations have been reported. First, there is a discrepancy between the concentration of ATP which gives maximum amount of Ca-uptake and that which gives maximum rate of Ca-uptake [1] (fig. 1). Secondly, not only ATP and other nucleoside triphosphates [2], but also some other phosphate compounds, such as AcP [3] and pNPP [4], can induce the Ca-uptake of SR. The present communication is intended to find out a common aspect of these observations.

All experiments were performed using fragmented frog SR which was prepared according to the method described in the previous paper [1]. The binding constant of SR for Ca is expressed by the reciprocal of the concentration of Ca^{2+} which gives half-maximum amount of Ca-uptake. In the absence of oxalate, the curve representing the relationship between the amount of Ca-uptake and the concentration of Ca^{2+} [1] can well be fitted to the calculated curve of Michaelis-Menten type (in the presence of oxalate, however, the curve becomes much steeper and cannot be expressed by a curve based on a simple kinetics; moreover, the use of oxalate masks the difference in the activities between the fresh and deteriorated preparations [5]).

The binding constants of SR for Ca were determined at the intermediate concentrations of ATP (0.3–30 μM) where the rate of Ca-uptake did not reach the maximum but the amount of Ca-uptake did. As shown in fig. 2, the

* The following abbreviations are used: SR, sarcoplasmic reticulum; E∿P, phosphorylated intermediate of SR; PK, pyruvate kinase; PEP, phosphoenolpyruvate; AMPCPOP, α, β-methylene adenosine triphosphate; AMPOPCP, β, γ-methylene adenosine triphosphate; AcP, acetylphosphate; CarbP, carbamylphosphate; pNPP, *p*-nitrophenylphosphate.

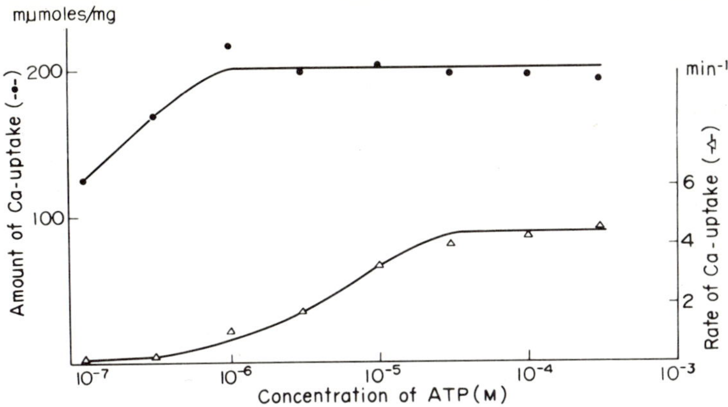

Fig. 1. The relation between the Ca-uptake activity of SR and the concentrations of ATP [1].

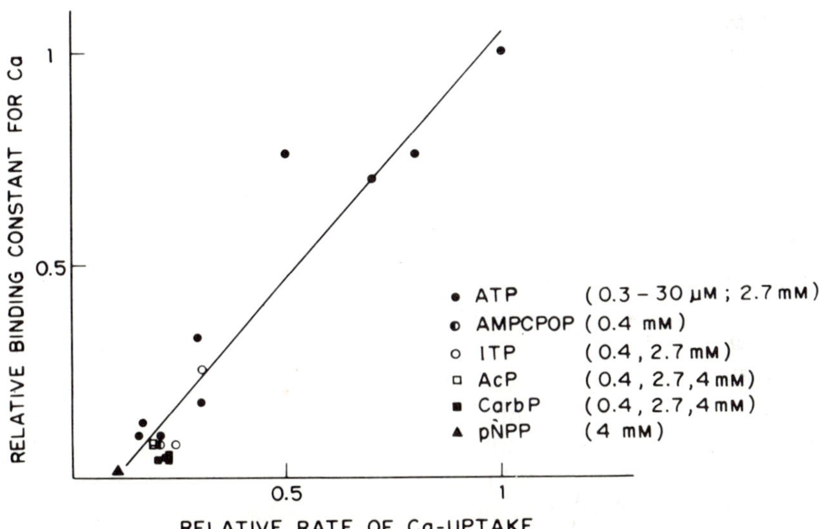

Fig. 2. The relation between the rate of Ca-uptake and the binding constant of SR for Ca [9]. When the concentrations of ATP were 0.3–30 μM, ATP regenerating system (PEP + PK) was used. In the case of the substrates other than ATP, the concentrations to give the maximum amount and rate of Ca-uptake were used. The binding constant of SR for Ca was determined by centrifugation methods. The amount of Ca accumulated was determined by using radioactive ^{45}Ca or atomic absorption. The unit binding constant in this figure corresponded to $3\times 10^6\ M^{-1}$. The rate was determined by dual-wave spectrophotometric method using murexide. The reciprocal of the time required for reaching half-maximum capacity was taken as the rate of Ca-uptake. The unit rate in this figure corresponded to 0.13–0.14 sec^{-1}. Temperature was 15°C.

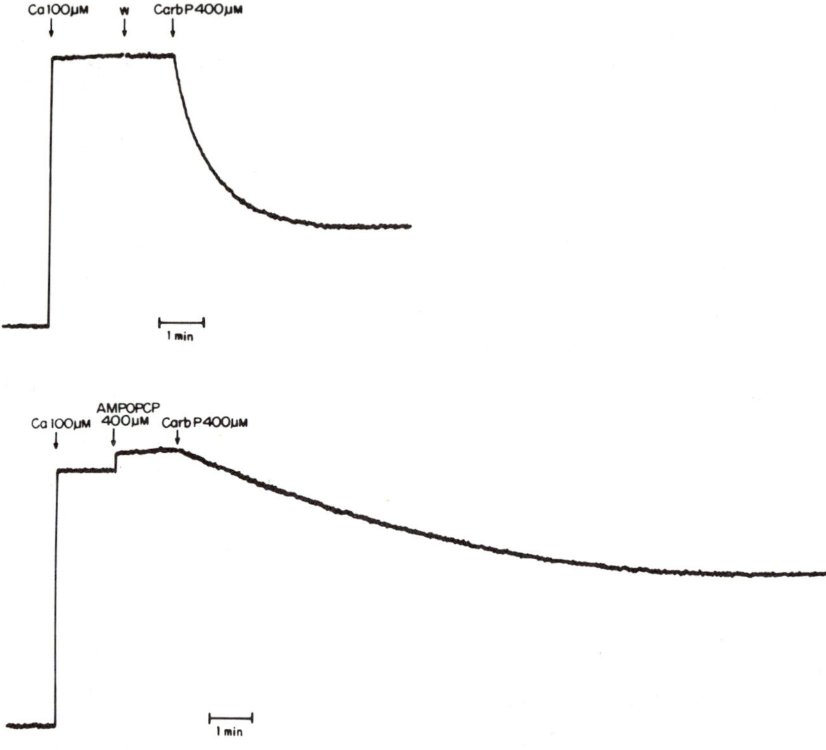

Fig. 3. Effect of AMPOPCP on the Ca-uptake by CarbP [9].

binding constants of SR for Ca are linearly related to the rates of Ca-uptake.

The maximum amounts of Ca-uptake by the substrates other than ATP, including AMPCPOP and CarbP (the effectiveness of CarbP was also reported by Pucell and Martonosi [6]), were not so much different from that of ATP *. However, the maximum rates of Ca-uptake induced by these substrates were much slower than that by ATP. Corresponding to these slower responses, much lower binding constants for Ca were demonstrated. If we plot the binding constants for Ca against the rates of Ca-uptake with these substrates, they fall on the line for ATP (fig. 2). The binding constant was independent of temperature for each substrate.

* The ratios of Ca-uptake to extra-hydrolysis were as follows: with lower concentrations than 1 μM of ATP, or AMPCPOP, the ratio is about 1; with CarbP or ITP, near 2; with higher concentrations of ATP, far below 1.

AMPOPCP itself cannot induce the Ca-uptake of SR and cannot be hydrolyzed by the ATPase of SR. It showed, however, an inhibitory effect on the Ca-uptake with CarbP (fig. 3). The effect was more remarkable on the rate than on the amount of Ca-uptake (however, the apparent binding constant was not reduced but seemed to be rather slightly increased, since the decrease in the amount of Ca-uptake was found only at higher concentrations of Ca^{2+}). This is also the cases with other substrates. With low concentrations of ATP, AMPOPCP exerted a marked depressing effect on the rate of Ca-uptake at its early stage. AMPOPCP inhibited ^{14}C-ATP binding to SR and E~P formation by CarbP. These observations together with other evidence may support that the sites for all substrates are identical.

In addition to the above common feature, however, the Ca-uptake with ATP in the presence of AMPOPCP showed another feature which was not shared by other substrates, i.e., the inhibitory phase induced by AMPOPCP was followed by an acceleratory phase (fig. 4). Corresponding to this acceleration, the binding constant for Ca increased from 3×10^5 M^{-1} to $2.2. \times 10^6$ M^{-1}.

It is reasonably assumed that E~Ps formed by different kinds of substrates should be identical, since the sites for substrates are identical. Therefore, if E~P be solely responsible for the Ca-uptake, the binding constants for Ca with various substrates should be the same. However, this is not the case as shown in fig. 2. This indicates that the enzyme-substrate-complex would play a crucial role in the processes of Ca-uptake. It should also be mentioned that the amount of E~P shows the same dependence on the concentration of Ca^{2+} as the amount of Ca-uptake [7, 8], and that the amount of ^{14}C-ATP

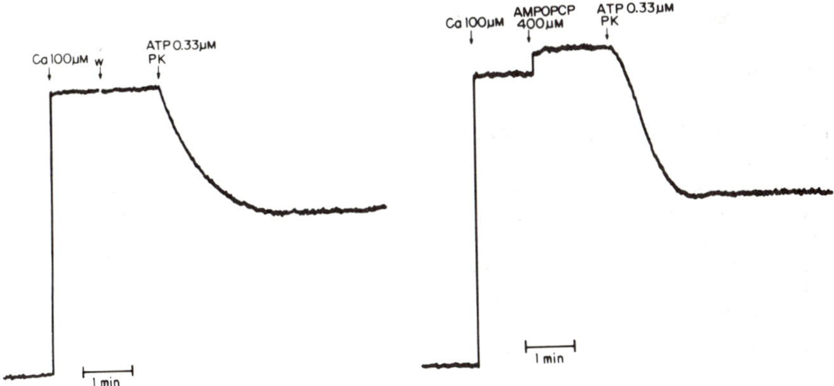

Fig. 4. Effect of AMPOPCP on the Ca-uptake by ATP [9].

binding showed the mirror image of the dependence of the amount of Ca-uptake on the concentration of Ca^{2+} (unpublished). In view of these findings, the following scheme can be inferred:

$$E + X{\sim}P \rightleftharpoons E \cdot X{\sim}P \qquad (1)$$

$$Ca_o^{2+} + E \cdot X{\sim}P \rightleftharpoons Ca \cdot E \cdot X{\sim}P \qquad (2)$$

$$Ca \cdot E \cdot X{\sim}P \rightleftharpoons Ca \cdot E{\sim}P + X$$

$$Ca \cdot E{\sim}P \rightarrow E \cdot Ca + P$$

$$E \cdot Ca \rightleftharpoons E + Ca_i^{2+}$$

where X~P represents substrates and suffixes i and o represent Ca inside and outside the vesicles, respectively. There is a possibility that a more quantitative analysis of the reactions represented by eq. (1) and (2) would explain the observations noticed at the beginning of this article. Further investigation along this line is now under progress.

Acknowledgement

The author wishes to express his cordial thanks to Prof. S. Ebashi and Dr. M. Endo for their advice and discussions. He is also indebted to Dr. K. Ikeda, Dept. Medical Electronics, for his construction of electrical devices for dual-wave spectrophotometer. This work was supported in part by the research grants of the U.S. Public Health Service, AM-04810, Muscular Dystrophy Association of America, Inc., the Ministry of Health and Welfare, Japan, No. 216, the Iatrochemical Foundation, Tokyo, and Toray Science Foundation.

References

[1] Y. Ogawa, J. Biochem. 67 (1970) 667.
[2] A. Martonosi and R. Feretos, J. Biol. Chem. 239 (1964) 648.
[3] LeDeMeis, J. Biol. Chem. 244 (1969) 3733.
[4] G. Inesi, Science, 171 (1971) 901.
[5] G.D. Baird and S.V. Perry, Biochem. J. 77 (1960) 262.
[6] A. Pucell and A. Martonosi, J. Biol. Chem. 246 (1971) 3389.
[7] T. Yamamoto and Y. Tonomura, J. Biochem. 62 (1968) 558.
[8] M. Makinose, Europ. J. Biochem. 10 (1969) 74.
[9] Y. Ogawa, J. Biochem., in press (1971).

MITOCHONDRIAL Ca^{2+} UPTAKE AND HEART RELAXATION

Ernesto CARAFOLI, Roberta TIOZZO, Carlo S. ROSSI
and
Giovanni LUGLI
Istituto di Patologia Generale, Universitá di Modena, Italy

1. Introduction

It has been known for many years that Ca^{2+} is the essential agent in the coupling of excitation to contraction in muscle [1]. Contraction occurs when the Ca^{2+} concentration in the sarcoplasm rises above 0.1 μM. When its concentration falls below this level, relaxation ensues [2–6]. It is generally accepted that the movements of Ca^{2+} linked to the contraction and relaxation of skeletal muscle are under the control of sarcoplasmic reticulum [7–10], since its vesicles are capable of taking up Ca^{2+} from the medium very rapidly and with optimal efficiency at concentrations of Ca^{2+} similar to those present in the muscle cell [11] and can lower the concentration of Ca^{2+} in the reaction medium below the level necessary for relaxation [12]. The energy necessary for uptake of Ca^{2+} by the fragments of the sarcoplasmic reticulum is derived from the splitting of ATP. It has also been demonstrated that sarcoplasmic reticulum from skeletal muscle can prevent the contraction of isolated myofibrils by substracting the necessary Ca^{2+} from the medium.

Whereas the role of sarcoplasmic reticulum in skeletal muscle is generally accepted, its role in cardiac muscle is far less clear, and has been questioned by a number of authors [13–15]. The rationale for the questioning rests on the well known differences in dimensions and duration of the contraction in cardiac and skeletal muscle cells, and on the finding that the ability to take up Ca^{2+} is extremely low in heart sarcoplasmic reticulum [16].

In addition, sarcoplasmic reticulum is poorly represented in the cardiac cell, and even if it were very active in the transport of Ca^{2+}, it would most likely be insufficient to move the amount of Ca^{2+} necessary for contraction and relaxation.

By contrast, mitochondria are very abundant in heart cells, are situated in close connection to the myofibrils, are capable of taking up Ca^{2+} very rapidly from media containing very low Ca^{2+} concentrations [17, 18] and can lower the concentration of Ca^{2+} in the medium well below the level necessary to induce relaxation [19]. In addition, recent studies of Patriarca and Carafoli [15], confirmed by Horn et al. [20] have shown that $^{45}Ca^{2+}$ injected to the rats labels more extensively heart mitochondria than the other organelles of the heart cells. The specific activity of mitochondrial Ca^{2+} is far higher than that of Ca^{2+} in the sarcoplasmic reticulum, suggesting that mitochondrial Ca^{2+} is in rapid equilibrium with Ca^{2+} in the cytosol and/or the extracellular spaces. Observations by Carafoli et al. [21] on rat diaphragm, and of Horn et al. [20] on rat heart have also shown that mitochondria control the entrance of Ca^{2+} into the muscle cell, and its exit from it.

In considering a role of mitochondria in the contraction and relaxation of heart, two factors are important. One is the ability to eject Ca^{2+} at a rate compatible with the known speed of contraction, a problem which is pertinent to the sarcoplasmic reticulum as well, since no agents have so far been found which induce rapid release of Ca^{2+} from isolated sarcoplasmic reticulum. A series of agents (including quinine, quinidine, chloropromazine, caffeine, ryanodine, local anesthetics and zinc ions) [22, 23] induce release of Ca^{2+}, but at a slow rate. Release of Ca^{2+} from mitochondria is on the other hand induced by a variety of agents, including uncouplers, respiratory inhibitors, anaerobiosis [24]. The release is certainly very rapid, but no attempts have so far been made to relate it to the velocity of contraction.

The second factor to be considered is the ability to relax isolated myofibrils. As mentioned, sarcoplasmic reticulum from skeletal muscle can prevent the contraction of myofibrils *in vitro*; heart mitochondria were also tested [25, 26] and found to be much less effective than sarcoplasmic reticulum. In some of the studies mentioned, however, mitochondria were tested under conditions which were optimal for the uptake of Ca^{2+} by the sarcoplasmic reticulum, but which have later been found to be completely unsuitable for the uptake of Ca^{2+} by mitochondria. We have made a study of the proper experimental conditions, and have found that mitochondria, when incubated in appropriate media, indeed show a high capacity for relaxation of isolated myofibrils.

2. Materials and methods

Rabbit muscle myofibrils were prepared according to Perry and Grey [27]. Mitochondria were prepared from rabbit and rat heart, and from rat liver. Liver mitochondria were prepared with a conventional 0.25 M sucrose procedure. For the preparation of heart mitochondria the animals were rapidly killed, the hearts were quickly removed and placed in ice-cold medium containing 100 mM KCl, 10 mM Tris–Cl, *p*H 7.4, and 1% bovine serum albumin (BSA). The hearts were squeezed with filter paper, chopped into small pieces, and homogenized with 12 volumes of the KCl-Tris-BSA medium in a Polytron homogenizer (mod. PT 10), for 10 sec with the setting at 5.5. Mitochondria were isolated from homogenates with a conventional differential centrifugation procedure and washed once.

The protein concentration of myofibrils and mitochondria was determined with a biuret reaction. The uptake of Ca^{2+} by mitochondria was measured in a medium containing 100 mM NaCl, 10 mM Tris-Cl, pH 7.4, 5 mM Na-glutamate, 5 mM Na-malate, 4 mM $MgCl_2$, and 3 mg of mitochondrial protein in a final volume of 2 ml, at 25°. At the desired times, aliquots of the suspension were withdrawn and filtered through Millipore membranes. The filters were washed very rapidly with cold medium, dried, and counted in a Packard mod. 2010 Scintillation Counter. The uptake of Ca^{2+} was determined by comparison with appropriate standards. ATPase activity was measured in the same medium containing 5.0 mM Na-ATP and the amounts of mitochondrial and/or myofibrillar protein specified in the legend for table 1. Temperature, 30°. After 15 min of incubation, 2 ml cold 20% trichloroacetic acid were added, and the denatured protein was removed by filtration. On

Table 1

ATPase activity of mitochondria from rat liver, rat heart, rabbit heart. The reaction medium contained 100 mM NaCl, 10 mM Tris-Cl, pH 7.4, 4 mM $MgCl_2$, 5 mM Na-glutamate, 5 mM Na-malate, 5 mM Na-ATP, and the following amounts of mitochondrial protein: 3 to 4.8 mg (rat liver), 3 mg (rat heart), 3 mg (rabbit heart). The final volume was 2.0 ml, the temperature was 25°, and the incubation time was 10 min. In brackets the number of experiments.

Mitochondria from	ATPase activity (μMoles Pi/mg protein/10 min)
Rat liver	0.07 (2)
Rat heart	4.26 (2)
Rabbit heart	0.49 (7)

aliquots of the clear filtrate, phosphate was determined with the method of Fiske and Subbarow [28].

The contraction of isolated myofibrils was evaluated by following the clarification of the suspension medium as the contraction progressed in graduated, conical test tubes.

3. Results and conclusions

As fig. 1 shows, heart mitochondria take up Ca^{2+} rapidly and efficiently when suspended in an appropriate medium containing 10 μM Ca^{2+}. The uptake slows down to a plateau when about 60% of the total Ca^{2+} added has been taken up, that is, when the extramitochondrial total Ca^{2+} concentration has been lowered to about 5 μM. Fig. 1 also shows that the mucopolysaccharide stain ruthenium red completely blocks the uptake of Ca^{2+} at the very low concentration of 25 μM, in agreement with the recent findings of Moore [29] on liver mitochondria. It is important that ruthenium red blocks the binding of Ca^{2+} to mitochondria even at the very low Ca^{2+} concentration used; since mitochondria bind some Ca^{2+} also in the absence of metabolism [30], when very low concentrations of Ca^{2+} are employed, the traditional inhibitors of active uptake become less efficient. Ruthenium red, by contrast, blocks both active and energy-independent binding of Ca^{2+} [31]; it is thus

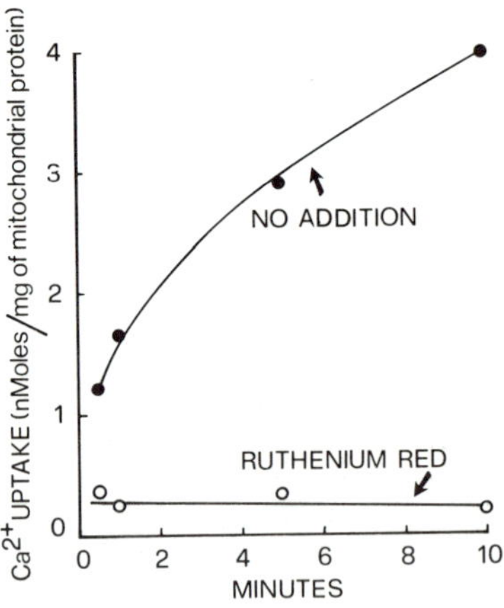

Fig. 1. Effect of ruthenium red on Ca^{2+} uptake by heart mitochondria. Technical details are found in the Methods. Mitochondrial concentration, 5 mg of protein; ruthenium red concentration, 25 μM.

the only agent known capable of suppressing all types of Ca^{2+} interaction with mitochondria, a fact that makes it a very useful tool in assessing the role of mitochondrial Ca^{2+} transport (and binding) in heart relaxation.

One of the parameters of relaxing activity most frequently used is the ability to inhibit the Mg^{2+}-stimulated myofibrillar ATPase. The measurement can be carried out without complications using sarcoplasmic reticulum, which has a very negligible ATPase as compared to that of the myofibrils. Also liver mitochondria have a very low Mg^{2+}-stimulated ATPase, and can thus be conveniently used as models in studies of relaxation. By contrast, the Mg^{2+}-stimulated ATPase activity is very high in heart mitochondria, making them hardly suitable to study the inhibition of myofibrillar ATPase. We have however found that rabbit heart mitochondria are rather exceptional as compared to mitochondria from other heart tissues, since they have a low Mg^{2+}-stimulated ATPase (table 1), almost comparable to that of liver. We have thus used rabbit heart mitochondria, and have found that they inhibit the ATPase of isolated myofibrils by about 50% (liver mitochondria inhibit somewhat more). More important, the inhibition of ATPase is completely removed by ruthenium red, which has no effect on the ATPase of mitochondria. This shows the relationship between the block of Ca^{2+} uptake by mitochondria and the inhibition of myofibrillar ATPase.

An additional demonstration of the ability of mitochondria to relax myofibrils is presented in the experiment of fig. 2. After a few minutes in the

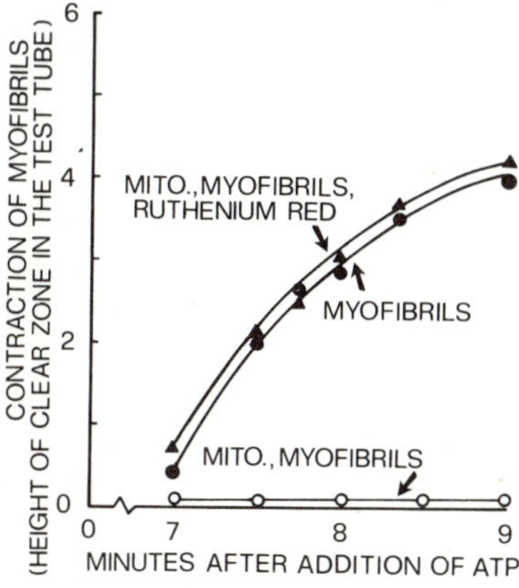

Fig. 2. Effect of ruthenium red on the relaxing activity of heart mitochondria. Conditions as in table 1, including the concentration of mitochondria and myofibrils. However, the final volume was scaled up to 5 ml.

Table 2

Relaxing activity of mitochondria. Each tube contained 100 mM NaCl, 10 mM Tris-Cl, *p*H 7.4, 4 mM $MgCl_2$, 5 mM Na-glutamate, 5 mM Na-malate, 5 mM Na-ATP, 2.2 mg myofibrillar protein, or/and 3 mg mitochondrial protein in a final volume of 2.0 ml. Temperature, 25°, incubation time, 10 min.

Mitochondria from	Inhibitor	ATPase (μmoles Pi)			Relaxing activity (% inhibition of myofibrillar ATPase)
		Myo-fibrils	Mito-chondria	Myofibrils plus mitochondria	
Rat liver	none	2.68	0.38	1.08	74
Rabbit heart	none	3.49	2.69	4.30	54
Rabbit heart	25 μM Ru-nium red	3.45	2.66	5.81	9

medium used for the experiments of tables 1 and 2 myofibrils contract very efficiently, as judged from the clarification of the top portion of the test tube. The contraction is prevented by mitochondria, and is restored if mitochondria are added together with ruthenium red.

It seems thus clear that heart mitochondria are capable of relaxing myofibrils *in vitro*. However, they must be tested in media which are suitable for their activity. As mentioned in the Introduction, it is very likely that some of the previous negative conclusions on the relaxing activity of mitochondria were due to the use of media which were optimal for the uptake of Ca^{2+} by the sarcoplasmic reticulum, but where unsuitable for the study of mitochondria.

More detailed studies on the relationship between mitochondrial Ca^{2+} uptake and heart relaxation are now under way. Also under way are studies of the velocity of Ca^{2+} release from mitochondria under the influence of a variety of conditions.

Acknowledgments

This research has been supported by the National Research Council of Italy (contract No. 6902109–115. 3453. 0) The excellent technical assistance of Mrs. D. Ceccarelli-Stanzani is gratefully acknowledged.

References

[1] L.V. Heilbrunn and F.H. Wiercinski, J. Cell. Comp. Physiol. 29 (1947) 15.
[2] A. Weber and R. Herz, J. Biol. Chem. 238 (1963) 599.
[3] R.J. Podolsky and L.L Constantin, Federation Proc. 23 (1964) 933.
[4] W. Hasselbach, Progr. Biophys. Mol. Biol. 14 (1964) 167.
[5] A. Sandow, Pharmacol. Rev. 17 (1965) 265.
[6] A.M. Katz, Physiol. Rev. 50 (1970) 63.
[7] W. Hasselbach and M. Makinose, Biochem. Z. 333 (1961) 518.
[8] S. Ebashi, Progr. Theoret. Phys. (Kyoto), Suppl. 17 (1961) 35.
[9] A. Weber, R. Herz and I. Reiss, J. Gen. Physiol. 46 (1963) 599.
[10] U. Muscatello, E. Andersson Cedergren, and G.F. Azzone, Biochim. Biophys. Acta, 63 (1962) 55.
[11] R. Ohnishi and S. Ebashi, J. Biochem. 55 (1964) 599.
[12] W. Hasselbach and M. Makinose, Biochem. Z. 339 (1963) 94.
[13] M.C.O. Fehmers, Intracellular Ca^{2+} en de werking van de hartspier, Ph. D. Thesis, Univ. of Amsterdam (Hollandia offsetdrukkerij, Amsterdam, 1968).
[14] N. Haugaard, E.S. Haugaard, N.H. Lee and R.S. Horn, Federation Proc. 28 (1969) 1657.
[15] P. Patriarca and E. Carafoli, J. Cell. Physiol. 72 (1968) 29.
[16] J. Gergely, D. Pragay, A.F. Scholz, J.C. Seidel, F.A. Sreter and M.M. Thompson, Molec. Biol. of Muscle Contraction, BBA Library 9 (1965) 145.
[17] A.L. Lehninger, E. Carafoli and C.S. Rossi, Advances in Enzymol. 30 (1967) 259.
[18] G.P. Brierley, E. Murer and E. Bachmann, Arch. Biochem. Biophysis, 109 (1964) 89.
[19] B. Chance, A. Azzi, I.Y. Lee, C.P. Lee and L. Mela, in: Mitochondria, Structure and Function, eds. L. Ernster and Z. Drahota (Academic Press, New York, 1969) 233.
[20] R.S. Horn, A. Fyhn and N. Haugaard, Biochim. Biophys. Acta, 226 (1971) 459.
[21] E. Carafoli, P. Patriarca and C.S. Rossi, J. Cell. Physiol. 74 (1969) 17.
[22] A.S. Fairhurst and W. Hasselbach, Europ. J. Biochem. 13 (1970) 504.
[23] A.P. Carvalho, J. Gen. Physiol. 52 (1968) 622.
[24] Z. Drahota, E. Carafoli, C.S. Rossi, R.L. Gamble and A.L. Lehninger, J. Biol. Chem. 240 (1965) 2712.
[25] A. Weber, R. Herz and I. Reiss, Federation Proc. 23 (1964) 896.
[26] L. Lorand, Federation Proc. 23 (1964) 905.
[27] S.V. Perry and T.C. Grey, Biochem. J. 64 (1956) 184.
[28] C.H. Fiske and Y. Subbarow, J. Biol. Chem. 66 (1925) 375.
[29] C. Moore, Biochem. Biophys. Res. Commun. 42 (1971) 298.
[30] C. Rossi, A. Azzi and G.F. Azzone, J. Biol. Chem. 242 (1967) 951.
[31] F.D. Vasington, P. Gazzotti, R. Tiozzo and E. Carafoli, Biochim. Biophys. Acta 256 (1972) 43.

ACTIVE TRANSPORT OF NA$^+$ AND K$^+$ ACROSS THE SQUID AXON MEMBRANE*

L.J. MULLINS

Department of Biophysics, University of Maryland, School of Medicine, Baltimore, Maryland, U.S.A.

Introduction

The squid giant axon has been used for Na/K transport studies for a somewhat shorter time than have red blood cells and muscle. When one considers the relative difficulty of obtaining a reliable supply of squid compared with a supply of red cells or of muscle fibers, as well as the greater effort and expense required, it is clear that investigators choosing to work on squid axons must have compelling reasons for doing so. The most obvious advantage of using squid axons is that in principle it is possible to control solute concentrations on both sides of the membrane and therefore to study transport under conditions of fixed internal substrate and electrolyte concentration. It is difficult to see how this might be possible, even in principle, with red cells or with 100 μ muscle fibers, given the mechanical properties of ordinary materials. A second advantage of using squid axons is that one is not studying a population of cells of variable age (and therefore presumably variable membrane properties), but rather an area of membrane about 0.2 cm^2 that is isolated from the end regions of the axon.

While there can be little doubt that the Na/K transport systems of red cells, muscle, and squid axons are very similar in their modes of operation, it is the differences between these systems that may be helpful in telling us more about the mechanism of active transport. For example, it is not clear why red cells have a low and rather constant Na/K coupling of 3:2, while both muscle and nerve have higher and readily variable coupling ratios. Other

* Aided by grants from the U.S. National Institute of Neurological Diseases and Stroke (NS 05846) and from the U.S. National Science Foundation (GB 8279).

notable differences between these different cells is the manner in which their Na efflux responds to changes in $[K]_o$ and $[Na]_i$. These points will be emphasized in the discussion of the squid axon that follows.

Technical improvements in the methods for measurement of ion fluxes in squid axons have been continuous. Originally, the loading of radioactive isotopes into axons depended on pretreating the axon with radioactive seawater followed by washing in inactive seawater. Subsequent studies [1–2] showed that such treatments, at least for K, selectively loaded the Schwann cells with isotope. The specific activity inside the axon can seldom be raised to more than 10% of that of the seawater, given the relatively few hours available for loading, while the Schwann cells can be brought to isotopic equilibrium in a very much shorter time. Microinjection, originally used by Grundfest, Kao and Altamirano [3] for electrophysiological observations on the effects of injected substances, was developed by Caldwell et al. [4] into an extremely useful technique for introducing both isotopes and substrates into axons and much of the information we have on the behavior of the Na pump has resulted from the application of this technique. With the discovery [5] that axons can be perfused with a simple salt solution and still retain their electrophysiological properties, the way appeared clear for a major improvement in technique for the study of active transport since in principle it ought to have been possible to introduce substrates and maintain a defined set of substrate and ion concentrations in the axon on a time-independent basis. With regard to the maintenance of a substrate-dependent Na efflux, the results from internal perfusion experiments have not been entirely satisfactory; either the fluxes are abnormally high [6] or are much too high in the absence of a substrate [7]. Fluxes of Na do appear to be stable in perfused squid axons when F^- is a major constituent of the perfusion fluid but this substance irreversibly inactivates membrane ATPase, so that a sensitivity of Na efflux to ATP can hardly be expected. Another technique for the study of Na transport in squid axons is that of internal dialysis, in which a porous glass dialysis tube is inserted into the axon and dialyzable solutes in the dialysis fluid are exchanged with solutes in the axon [8]. With this technique, Na efflux is stable with time, is of normal magnitude when physiological concentrations of ATP are included in the dialysis fluid, and in the absence of substrates it agrees rather closely with the value calculated from flux ratio considerations. The main purpose of the following discussion is to examine some of the results that have recently been obtained with squid axons with a view to seeing what sorts of models of active transport will accommodate these newer experimental findings.

Much experimental work carried out over the past decade has led to the following sorts of conclusions regarding active transport:

1) it is a coupled movement of Na out of and K into cells
2) it is dependent on ATP or on some substrate readily converted to ATP
3) the enzyme, membrane ATPase, is both the system that converts ATP into transport work and the pump that moves the ions involved
4) ouabain, or other cardiac glycosides are specific inhibitors of both transport and membrane ATPase.

Using techniques such as internal dialysis, it has proved possible to examine these conclusions somewhat more rigorously than was previously possible. The results obtained suggest that active transport may not be as simple as one might have thought, and that some of the conclusions listed above do not actually lend themselves readily to some of the more conventional schemes that have been proposed.

In part, this difficulty stems from the increasingly complex phenomena that have been observed in studies of Na-K transport, and in part difficulties have become apparent when well known processes have been observed in greater detail than was formerly possible. The phenomena of transport can be classified into substrate or inhibitor dependent fluxes of Na or K:

A. The "classical" Na efflux.
   1. Requires $[ATP]_i$
   2. Proportional to $[Na]_i$
   3. Stimulated by $[K]_o$ or $[Na]_o$
   4. Insensitive to $[ADP]_i$ and $[Ca]_o$
   5. Inhibited by $[P_i]_i$ and $[Mg]_i$
   6. Glycoside inhibits but also initiates a new Na efflux

B. The glycoside dependent Na efflux [9]
   1. Independent of $[ATP]_i$ or $[K]_o$
   2. Dependent on $[Na]_i$
   3. Requires ouabain, strophanthidin or similar compounds

C. The ATP dependent K influx
   1. Requires $[ATP]_i$
   2. Proportional to $[K]_o$
   3. Stimulated by $[Na]_i$
   4. Inhibited by $[ADP]_i$
   5. Inhibited by glycoside
   6. Insensitive to $[Na]_o$, $[Li]_o$, or $[Choline]_o$

D. The Ca dependent Na efflux [10–11]
   1. Independent of glycoside or $[K]_o$
   2. Dependent on $[ATP]_i$ and $[Na]_i$

3. Stimulated by $[Li]_o$
4. Dependent on $[Ca]_o$

E. The ATP dependent Na influx [9]

1. Independent of glycoside or $[K]_o$
2. Dependent on $[ATP]_i$ and $[Na]_i$

Flux (A) is that usually considered when active transport of Na is meant; point (A.1.) was not entirely settled by the injection experiments of Caldwell et al. [4] which showed that the injection of either ATP or ArgP would restore Na efflux but that only ArgP would yield a Na efflux that was decreased when $[K]_o$ of seawater was lowered. A test of the relative efficacy of ArgP and ATP in dialyzed axons in promoting Na efflux [12] showed that ArgP was virtually inactive while ATP gave a normal Na efflux. Requirement (A.3.) is a well-known one and one that is usually ascribed to a coupling between inward K transport and Na efflux. However, as I shall show later, this supposed coupling can vary over such wide limits than an obligatory coupling of Na to K becomes somewhat implausible. It may, at this point, be

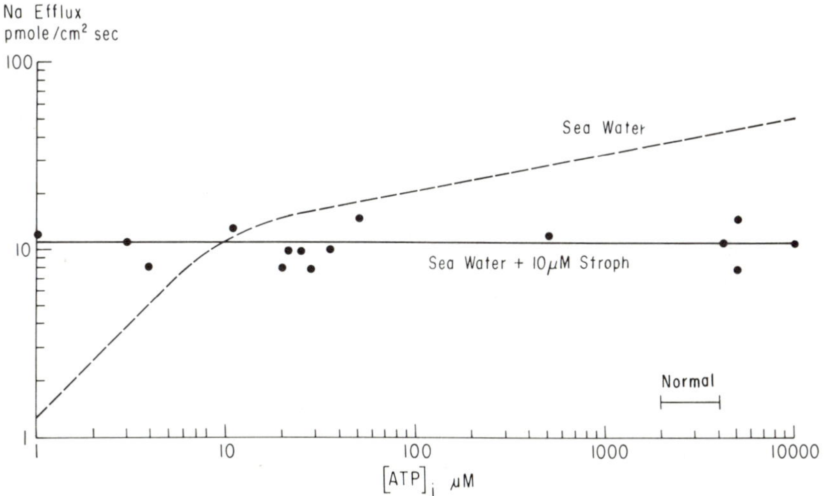

Fig. 1. Na efflux is plotted as a function of [ATP] supplied to the axoplasm by dialysis. The broken line is Na efflux into seawater and the solid line and experimental points refer to seawater + 10 mM strophanthidin. Na efflux is apparently stimulated by this glycoside at low $[ATP]_i$ and inhibited at high $[ATP]_i$. The measurements show that Na efflux in glycoside is independent of [ATP]. The bracketed horizontal line labelled "normal", represents the experimentally measured range of [ATP] in fresh axoplasm. Plot uses double logarithmic scales.

safer to assume that external K has a role in promoting Na transport which is independent of K transport. The finding (A.5.) is a new one [13] and may give important clues as to the mechanism of transport. Although glycosides have been frequently used to define active transport, the finding (A.6.) [9] shows that it is impossible for such a definition of transport to be an accurate one. This fact is emphasized by point (B.2.) which categorizes a separate Na efflux that is glycoside dependent. The difference in substrate dependence for Na fluxes (A) and (B) is shown as fig. 1 (data of [9]). This flux has not been studied extensively enough to characterize it further; it differs from (D) in that the Na efflux is ATP independent. The fluxes in (C) describe what is usually called the K pump; (C.6.) lists the effect of Na-free solutions – these have a large effect on Na efflux but none on K influx.

What evidence there is suggests that fluxes (A), (B) and (C) may be brought about by mechanisms that are closely related, while (D) and (E) may be brought about by another mechanism. Further discussion must depend on the way in which one defines the mechanisms responsible for metabolism-dependent ion fluxes.

Definitions of active transport

Active transport has been defined as that which is involved in a net movement of material from a region of lower to one of higher electrochemical potential. That is, the definition is concerned entirely with the uphill movement of solute. It is easy to imagine, however, that transport processes involving the expenditure of energy might also move material in a downhill direction. Such processes might play a physiological role in vastly accelerating the movement of solutes into the cell even though active transport in the sense of the uphill movement was not involved. It would seem therefore that a more appropriate definition of active transport might relate the amount of metabolic substrate used to the quantity of solute transported across the membrane without regard for whether the work performed was with or against an electrochemical potential gradient.

Another difficulty with a simple definition of active transport becomes apparent when coupled ion transport processes are considered. For example, there are circumstances in which the movement of sodium ion via active transport may be uphill, while the allegedly coupled movement of potassium ion in the opposite direction may involve transporting the ion against no electrochemical potential gradient or even moving it downhill. Under these circumstances, it is difficult to see whether the work involved in moving sodium against a gradient can be partially compensated by the coupled ion movement in the opposite direction or whether this is an unrealistic assumption to make. Clearly the answer to such a problem lies in a complete understanding of the mechanism of pumping and entirely consistent definitions of active transport can only come when the actual mechanism is much better understood.

Transport studies must usually rely on an operational definition of active transport and a variety of inhibitors has been used in order to furnish a means of testing whether a particular ion flux is or is not sensitive to the substrate for active transport processes, ATP. For tissues which do not have a prominent glycolytic system, cyanide is a useful inhibitor because it will rather promptly reduce the cellular ATP level to tens of micromolar and in such tissues sodium efflux is observed to fall to low levels. Efflux does not fall to the value predicted by flux ratio considerations but then given the small residual ATP concentration which is observed in the tissue it is to be expected that some sodium transport will still be energized. More specific inhibitors of sodium transport such as ouabain and related cardiac glycosides might be expected to be even more useful because *in vitro* studies have shown that these glycosides inhibit membrane ATPase while physiological studies show that cardiac glycosides reduce sodium efflux. Another method of operationally defining active transport is to say that when the ATP concentration inside a cell is increased from zero to any particular value, the change in ionic fluxes which follows this step-change in ATP concentration is a measure of the transport that is taking place. A difficulty with this definition is that, in principle, we can recognize two sorts of effects which ATP might have; it might induce a permeability change in the membrane by virtue of its binding to some membrane component, or it might bring about the active transport which we wish to define. It may prove exceedingly difficult to separate these two hypothetical effects of ATP so that a definition of active transport based on a change in ATP concentration may be a somewhat inperfect one. In turn, a further improvement in the definition of active transport would be to have it specify the quantity of substrate (ATP) that is split per quantity of ion transported; this would allow the definition to relate the chemical free energy utilized in ATP hydrolysis to a definite quantity of ion transported and if one found, for example, that an ion flux occurred under conditions where no ATP was hydrolyzed but where ATP was necessary, then it would be justifiable to assign another role to the effect of ATP on the system.

The difficulties with inhibitors used to define active transport can be summarized by a detailed consideration of the findings with cardiac glycosides. The following results [9] show how difficult it is to work out an entirely

Effect of glycoside on Na fluxes

	ATP absent		ATP present	
	Control	Glycoside	Control	Glycoside
Na influx	40 *f* *	no effect	increased	no effect
Na efflux	1.2 *f*	12 *f*	increased	12 *f*

* pmole/cm² sec.

consistent scheme for transport based on the action of cardiac glycosides. The fact that glycoside has no effect on Na influx when ATP is absent must mean that this substance does not affect Na permeability, while the fact that Na efflux is the same in the presence or absence of ATP plus glycoside must mean that this efflux is ATP independent. Sodium influx is increased by ATP but this change is glycoside insensitive. We might consider two separate explanations for this effect. Either the increase in Na influx is a permeability change induced by ATP, or the Na influx is a carrier movement that happens to be glycoside insensitive. We cannot rule out a permeability change mechanism on the basis of looking at Na efflux since this is known to be highly sensitive to ATP; we can, however, examine the Na efflux in the presence of strophanthidin and find that this movement is independent of internal [ATP]. If ATP were to produce a permeability change leading to an increased Na influx, the simplest explanation would be that this would be independent of glycoside and that efflux as well as influx would be affected. Measurements of Na efflux in the presence of ATP and glycoside show that Na efflux is the same whether or not ATP is present (although admittedly the increment in efflux to be observed would be only a few pmole/cm^2 sec) so that the best provisional conclusion is that the increment in Na influx with ATP present is a carrier mediated process that is glycoside insensitive.

The examples cited above were chosen to illustrate the difficulty inherent in using a glycoside to define active transport fluxes. Further details relating to mechanism will be given later in this paper, but at present it seems that the distinction between purely passive and active fluxes may be a somewhat blurred one. This is inferred because the large sodium efflux induced by ouabain in an axon that has a very low $[ATP]_i$ must have some energy source since it is clearly a movement against a large electrochemical gradient. The only energy source available under the experimental circumstances cited would appear to be the sodium concentration gradient itself so that the allegedly purely passive sodium entry into the cell could, in principle, supply the necessary energy to allow for the ouabain induced sodium efflux. In turn one infers that the inward sodium movement must be through some kind of carrier system because otherwise it is difficult to see how this purely dissipative ion movement could contribute energy for the outward movement of sodium.

Given the present capability for flux measurement, it appears that an operational definition of active transport may be: the change in flux that is produced by a change in the $[ATP]_i$ from zero to a particular value. This definition does not distinguish between downhill and uphill movements of an ion, nor does it separate possible permeability effects of ATP from carrier mediated ion movement. More importantly, it rather arbitrarily separates ion movement that may be coupled into separate ion pumps. I shall argue later that there is no evidence that ATP induces permeability changes in the membrane for either Na or K, and that evidence for the coupling of ionic fluxes is equivocal. It appears, therefore, that the provisional definition of a pump in terms of the ATP dependence of its ionic flux may be a useful one.

The coupling of ion transport fluxes

While the original idea of Dean [14] in suggesting a Na pump was that this was a device that would maintain a non-equilibrium distribution of Na^+ across the cell membrane, it was noticed rather early that $[K]_o$ stimulated the loss of Na from muscle [15] and that removing $[K]_o$ from the solution bathing a nerve fiber reduced Na efflux. It seemed reasonable, therefore, to suppose that K entry was coupled to Na efflux and an early suggestion was that this coupling was 1:1; that is, that 1 K^+ was transported inward for each Na^+ transported outward. This assumption was convenient for studies then taking place on the control of the resting potential of muscle fibers because such a pumping mechanism was itself electroneutral and could not contribute membrane current. Subsequent work on ion fluxes has shown rather convincingly, however, that in general Na efflux is larger than K influx so that the original transport mechanism became more complicated than it might first have seemed. In the squid axon, the number of Na extruded per K taken in can be varied from greater than 3:1 to 1:1 depending on $[Na]_i$ [16,17].

These experimental findings make it at least worthwhile questioning whether a coupled ion pump (even if the coupling ratio is allowed to vary) correctly describes the active transport system. One might, for example, consider separate Na and K pumps, operating from a common energy source. The conventional carrier scheme is shown below:

$$(1)\ nNa_i + X_i \leftrightarrow (Na_nX)_i \rightarrow (Na_nX)_o \leftrightarrow nNa_o + X_o$$
$$(2)\ X_o \rightarrow Y_o$$
$$(3)\ Y_o + mK \leftrightarrow (K_mY)_o \rightarrow (K_mY)_i \leftrightarrow mK_i + Y_i$$
$$(4)\ Y_i \rightarrow X_i$$

Separate pumps would merely require the deletion of reactions (2) and (4) so that X is solely a Na carrier and Y is a K carrier; interconversion would not be possible, and X_o, Y_i would return to X_i, Y_o unloaded.

At first sight, such a scheme might seem highly unpromising in view of the evidence that $[K]_o$ affects Na efflux, that both K influx and Na efflux are reduced by glycosides, and that membrane ATPases require Na and K as activators for ATP hydrolysis. There are, however, some difficulties in reconciling the conventional carrier scheme with experimental findings. In particular, we might consider the K-free effect in squid axons. When an axon is transferred to K-free seawater, Na efflux falls to between 1/2 to 1/3 of control values. If, on the other hand, $[ATP]_i$ is reduced to levels of 1 μM, Na efflux will fall to 1/30 of control efflux [9]. Clearly K-free seawater does not stop an ATP-dependent Na extrusion. It has been suggested [11] that the [K] outside the membrane is not zero under K-free conditions because K leakage from the axon and Schwann cells makes $[K]_o$ of the order of 1 mM. This is a reasonable suggestion but it does not help much in explaining the Na efflux under K-free conditions because K influx is linearly proportional to $[K]_o$ over the concentration range 1–20 mM so that if Na efflux under K-free conditions is ½ of control, then for a K_o of 1 mM, instead of the normal 10 mM of seawater, K influx will be 1/10 of control or the coupling ratio will have increased 5-fold. In an axon with normal $[Na]_i$ and $[ATP]_i$ the ratio would have been 3, so that under these altered conditions it would now be 15. This is stretching the carrier concept beyond its elastic limit!

It might be suggested that, under K-free conditions, Na efflux becomes coupled not to K influx but to Na influx. Unfortunately, the effect of Na-free conditions outside the squid axon is to increase Na efflux and this effect has been explained rather satisfactorily by Baker et al. [11] as a competition between Na_o and K_o for sites outside the axon. Under this scheme, an axon is less K-free when choline$^+$ is the external cation because the K^+ leaking from axon and Schwann cells are more able to activate external K sites. Such an arrangement makes it impossible to apply Na-free solutions as a test for the extent to which Na:Na exchange takes place. Such a hypothetical Na:Na exchange would require that Na influx increase under K-free conditions but such evidence as there is shows that Na influx is not altered by K-free seawater, nor is there a glycoside sensitive component to Na influx in axons with a normal $[ATP]_i$ and phosphoarginine. These experimental findings show that it is difficult to test for Na:Na exchange; it is possible that there is a carrier mediated Na influx which is normally uncoupled and that K-free conditions allow the coupling of this flux to Na efflux. The resulting ion movement must be postulated to be glycoside insensitive and hardly helps support the idea of a Na:K coupled pump. A glycoside insensitive, Na_o-dependent Na efflux has been observed in muscle [18] so that the idea of such a form of transport is not unique.

To summarize the effect of nominally K-free solutions on Na efflux from squid axons, one can only say that the residual Na efflux appears uncoupled to the movement of any appreciable quantity of Na or K inward. External [K] has a large effect on Na efflux but one whose concentration dependence is different from that on K transport.

Much the same set of experimental findings is obtained if $[K]_o$ is held constant and $[Na]_i$ varied. The findings with respect to the ATP-dependent fluxes are [17]:

1) Na efflux is linearly dependent on $[Na]_i$; K influx is linearly dependent on $[K]_o$.
2) K influx is a non-linear function of $[Na]_i$ and appears half saturated at $[Na]_i \sim 10$ mM.
3) K influx is unaffected by the replacement of $[Na]_o$ by $[choline]_o$.

These results suggest that K^+ and Na^+ on the outside do not compete for the K pump as otherwise one would expect an altered rate of K pumping under Na-free conditions. Both $[Na]_i$ and $[K]_o$ affect the rate of pumping of K^+ and Na^+ but they do so in a manner that is not consistent with a simple coupling of Na efflux with K influx. Allowing the coupling ratio to be set by both $[K]_o$ and $[Na]_i$ is a device for retaining the notion of coupling, but it is hardly helpful in understanding the mechanism of pumping.

At very low $[K]_o$, an increment in $[K]_o$ promotes far more Na extrusion than K uptake, so that the apparent coupling ratio is very high. On the other hand, at $[K]_o$ 10–20 mM, an increment in $[K]_o$ promotes only a small increment in Na efflux but yields the same increment in K influx as at low $[K]_o$. This leads to a lower coupling ratio, although the limiting ratio for saturating $[K]_o$ would appear to be greater than 3 Na/K.

The rather complicated relationships between the influence of $[K]_o$ and $[Na]_i$ on the Na and K fluxes at an [ATP] of 3–4 mM are summarized in fig. 2. This shows that the pumped ion is linearly related to its own concentration, but that the ion on the opposite side of the membrane has a non-linear effect on the rate of pumping. One of the simplest interpretations of such experimental results is that K_o is an activator of Na efflux and Na_i is an

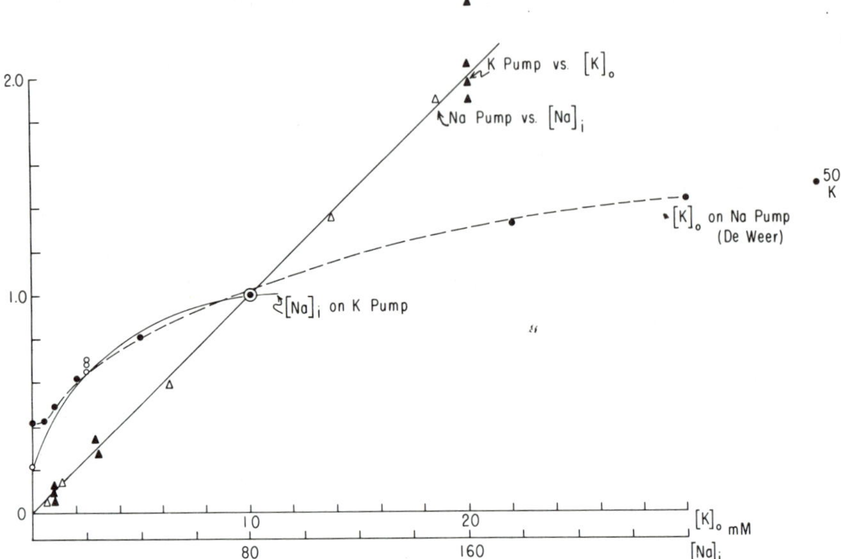

Fig. 2. Ordinate is Na efflux or K influx relative to that at 10 mM $[K]_o$ seawater with $[Na]_i$ 80 mM, $[ATP]_i$ 3 mM. The abscissa is either $[K]_o$ in Na seawater or $[Na]_i$. Open circles show the effect of $[Na]_i$ on K influx from 10 K (Na) seawater, solid circles the effect of $[K]_o$ in Na seawater on Na efflux (at constant $[Na]_i$). Solid triangles give the effect of $[K]_o$ on the ATP dependent K influx while open circles show the effect of $[Na]_i$ on Na efflux (at constant $[K]_o$).

activator of K influx but that ion transport and ion activation of transport are separate effects.

As Na and K are known to be activators of membrane ATPase, these observations suggest that "activation" of transport by Na_i and K_o relates to an improved energy supply to the ion pumps but that it is possible to distinguish between pumping and its energy supply.

Another mode of coupled ion movement which has received a great deal of attention is Na:Na exchange. The concept of this sort of interchange arose from a suggestion of Ussing that it might explain a Na efflux from muscle which appeared too large to be compatible with the energy sources available. The efflux estimate on which Ussing based this calculation proved to be erroneous – muscle has a Na efflux that makes only modest demands on the available energy sources. Meanwhile, the term "exchange diffusion" has remained as a concept and is frequently invoked to explain various sorts of measured fluxes.

There is the usual sort of difficulty in defining this coupled ion movement. The original idea was that a 1:1 carrier-mediated Na interchange took place which did not require ATP. A definition of exchange diffusion which has often been used is the reduction of Na efflux when $[Na]_o$ is replaced with another cation such as Li^+ or Choline$^+$. This test assumes that $[Na]_o$ has no effect on Na efflux other than providing Na for Na:Na interchange and it assumes further that the ion used to replace Na (e.g. Li) has no effect of its own.

That neither of these assumptions is likely to be generally true became apparent first in studies on frog sartorius muscle [19] in which the effect of replacing Na in Ringer with Li in some cases led to a decrease in Na efflux, while in others produced no change in Na efflux. It proved possible to show that replacing Na with Li outside could, in Na-loaded muscles, lead to an *increase* in Na efflux. Other studies [20] have shown that Li, in addition to whatever other effects it may have, has a K-like action that stimulates Na efflux. The effects of Li in increasing or decreasing Na efflux proved to be related to $[Na]_i$; a low $[Na]_i$ favored a decrease in Na efflux, while a high $[Na]_i$ had the effect of producing an increased Na efflux into Li Ringer solution. A further complication was that the decrease in Na efflux observed in Li Ringer was not sensitive to ouabain [18] suggesting possibly that the effect was not related to the Na carrier involved in Na:K pumping. A difficulty with the muscle fiber, however, is that one may be dealing with a complex compartment situation and the fluxes observed into Li solutions may be a reflection of this complexity rather than of a true Na:Na exchange.

For the squid axon, there can be little doubt that compartmentalization of internal Na is nonexistent; the experimental findings with Na-free solutions are, however, quite different from those of muscle. The usual effect of Li seawater on a squid axon is to increase Na efflux although the effect may not be marked in all axons. In any case, one seldom observes a decrease in Na efflux in Na-free solutions. The effect of such solutions on Na efflux has been analyzed in great detail [11] and a convincing case has been made for the idea that Na and K compete for sites on the outside of the membrane which activate Na pumping. Because K is much more potent in activating Na efflux than is Na, the effect of Na-free solutions is to make the effective [K] in seawater much greater. If there were a component of the Na efflux that were cut off by removing Na_o, one could argue that this situation is present but undetected by the method used. The maximum Na extrusion at very high $[K]_o$ appears, however, to be the same in Na-containing or Na-free seawater so that it appears unlikely that there is any Na:Na exchange in normal axons.

Further confirmation of this view is found in the absence of an effect of strophanthidin on Na influx into such axons [17].

A recent finding is that the effects of Na-free and K-free solutions on the Na efflux from squid axons are reciprocal phenomena in the sense that if an axon shows a large decrease in Na efflux on applying K-free solutions, it will be relatively insensitive to external Na, while axons relatively insensitive to external K will require Na to support maximum Na extrusion [21, 13]. One might think that this is a demonstration that some Na efflux is coupled Na/K and some Na/Na. Were this true, however, in normal axons there should then be a glycoside sensitive Na influx, and this remains undetected. One is, instead, led to the view that some of the pump elements in the membrane are K activated while others are Na activated. One can, of course, define a good pumping system as one in which K activation is the sole mode of operation but this merely avoids consideration of the problem of how the specificity requirement for external cation activation of the Na pump is set.

A variety of experimental treatments will abolish the K sensitivity of the squid axon; these include treatment with CN (and the reinjection of ATP) or with alkaline DNP [11] and the injection of Arg or ADP into normal axons [13]. Treatment with alkaline DNP (which reduced ArgP to low levels but left ATP unaltered) is also known to induce a glycoside sensitive Na influx in the axon and it has been suggested that all of these experimental conditions which lead to an interference with metabolism also lead to a loss of K sensitivity and a glycoside sensitive Na influx which is Na:Na exchange. ATP dependent K influx is inhibited about 50% by an equimolar mixture of ADP and ATP [17] while injected P_i is without effect on the K sensitivity of the Na efflux [13], so that a summary of what might be happening when the [ADP] of axoplasm is increased, is that the K influx is inhibited and an ATP dependent Na influx replaces it, while K sensitivity may be lost and replaced by an $[Na]_o$ sensitivity. This is equivalent to saying that Na/K pumping changes to Na/Na exchange.

Before accepting this scheme, however, it is prudent to ask whether the experimental findings might be explained in quite another way. In view of the suspicion that external ions may activate a mechanism without necessarily being transported by it, one may ask whether in fact the loss of the K-free effect produced by CN or other treatments is really different from the low K-free effect shown spontaneously by some fresh axons. A known effect of CN on axons which has recently been described [22] is a great enhancement of the Ca efflux. Presumably this observation reflects the discharge of Ca accumulated in the mitochondria into the axoplasm and its subsequent removal from the axoplasm via a membrane pump. Therefore, at least in principle, it is possible that agents which interfere with metabolism may cause the release of material from stores in the axoplasm and that the release of this material affects the Na pump to the extent of reducing its K sensitivity. If this is so, then ADP may be a substance which changes the specificity of the K pump such that it pumps Na, and we do not have an example of Na:Na exchange under the experimental conditions described above. Such a scheme would have the advantage of making the loss of the K-free effect the same for normal and CN treated axons; i.e. it would be dependent on the release of material from the mitochondria. It would also separate from the K-free effect the effects of ADP on the K pump. Fortunately, an experimental test of this proposal is possible; if with ADP, the K pump becomes a Na pump, but the rest of the system remains unchanged, then ATP hydrolysis will take place in connection with Na/Na exchange. If, however, ATP is required in only a catalytic role, then ATP hydrolysis will not take place.

A sodium influx pump

While attention has focussed mainly on active transport of Na out of and K into the cell, evidence of several sorts is accumulating that there may be a carrier mediated inward movement of Na which is not connected with the Na/K transport system. Evidence for the independence of this Na transport from Na/K transport comes from measurements of Na influx in the presence of strophanthidin; these show that Na influx is unaffected by this inhibition both when $[ATP]_i$ is normal [17] and when $[ATP]_i$ is of the order of 2 μM (Brinley and Mullins, unpublished). Sodium influx is, however, sensitive to ATP and an increased influx of about 25 f results from adding physiological concentrations of ATP to axons first dialyzed free of ATP.

Lithium, when used as a replacement for Na in seawater, induces a large increase in Na efflux which is ouabain insensitive [10]. Since ouabain, in Na seawater, induces a Na efflux of 10–12 f, the observations with Li seawater suggest that this ion induces a Na efflux of 20–24 f and that half of this increased efflux is $[Ca]_o$ and $[ATP]_i$ dependent. Although the effect of Na substitutes on Na efflux in the presence of glycosides has not been systematically studied, Sjodin and Beauge [21] have shown that in the presence of strophanthidin, Na efflux in $Tris^+$ seawater is lower than in Na seawater while Baker et al. [10] show that $Choline^+$ seawater gives about the same Na efflux as Na seawater in the presence of ouabain.

Other evidence that makes a Na influx pump seem more plausible has been provided by Baker et al. [10] who described a ouabain insensitive, $[Ca]_o$

dependent Na efflux. This system is an exceedingly complex one so that any attempt to summarize it will be necessarily inaccurate. It does appear, however, that Na and Ca compete for entry and that in seawater with a normal $[Na]_o$, Ca entry is negligible. The sensitivity of the Na efflux to [Ca] is also very small in seawater but a large $[Ca]_o$ sensitive Na efflux is apparent in Li seawater.

An interesting example of a glycoside insensitive Na efflux is the finding [9] that when $[ATP]_i$ in dialyzed axons is ~ 1 μM and Na efflux is ~ 1 *f*, the application of strophanthidin increases Na efflux 10 fold; as Na influx is not affected, this cannot be ascribed to a permeability change and it seems necessary to suppose that the movement of Na down its electrochemical gradient becomes coupled to Na efflux in the presence of glycoside. Since, under the conditions of the experiment, ATP was present inside at negligible concentrations, it appears that some of the Na influx measured in the absence of ATP is carrier mediated and is capable of being coupled to Na efflux by glycoside. The efflux of Na in the presence of strophanthidin is independent of $[ATP]_i$ so that it is clear that glycoside abolishes the Na/K pump and adds a new component to the Na efflux.

The effect of strophanthidin on ATP-free axons in promoting a Na efflux differs from the effect of $[Li]_o$ in promoting a Na efflux mainly in that the Li-stimulated Na efflux requires ATP and a high $[Na]_i$. It may be that a high internal Na is necessary for the activation of a Na dependent ATPase. The observations clearly suggest that the ouabain insensitive fraction of membrane ATPase preparations may be of more than trivial interest.

The influx of Na in a carrier mediated mode is also suggested by the observations of Blaustein and Hodgkin [22] that Ca efflux from squid axons is very likely coupled to Na entry; they suggest that a carrier mediated Na entry could provide the energy for the extrusion of Ca^{2+}. Again, the Ca^{2+} efflux is insensitive to glycoside and the scheme proposed is very similar to that suggested above for the effect of glycoside on Na efflux in the absence of ATP. While Li_o was most effective in promoting Ca influx and Na efflux, Na_o was most effective in promoting Ca efflux suggesting that if the same system is involved for both movements, then the nature of the external cation partly determines whether Na or Ca will be extruded. Other factors are $[Na]_i$ and $[ATP]_i$.

To summarize the status of the Na influx pump, it is clear that it is at present poorly defined but may have the following properties with respect to influx.

1) glycoside independent
2) $[ATP]_i$ dependent

3) $[Na]_i$ dependent
4) a competition between Ca and Na for entry
5) relative activating effect of external cations for Na efflux: Li > Na, Choline > Tris.

Membrane ATPase

There is a great deal of recent experimental evidence to persuade one that there are two forms of enzyme membrane ATPase, i.e. E_1 and E_2. Such evidence is summarized by Post et al. [23] and consists of showing that $E_1 \sim P$ is a form which allows terminal P exchange between ATP and ADP. It is also clear that $E_1 \sim P$ and $E_2 \sim P$ show differing sensitivities to ADP and K (with respect to the loss of P by the enzyme). It seems likely, therefore, that the change from E_1 to E_2 is a conformational change in the enzyme which might be associated with ion pumping (presumably of Na), while the reverse transition $E_2 \rightarrow E_1$ might be associated with K pumping. What is actually known, however, is that Na and K are activators of the ATP hydrolysis reaction sequence and an alternate view of membrane ATPase is that it supplies energy to pump ions but that the pumps are separate molecular entities.

A choice between these alternatives is not possible at present, but as was shown earlier in this paper, the requirements for activation of pumping and the requirements for the actual transport of ions often appear to be different with respect to both [Na] and [K]. Another method of analyzing the action of the enzyme in regulating ion transport is to consider the effects of the products of ATP hydrolysis on transport. Since the energy available from ATP hydrolysis is dependent on the term $[ATP]/[ADP]\ [P_i]$, one might expect that varying [ADP] or $[P_i]$ inside the membrane would be equivalent operations. Experiments show, however, that Na efflux is largely independent of [ADP] [9] but that it is reduced by P_i [13]. That fraction of K influx which is dependent on ATP is inhibited by [ADP] [17] while the K-free effect in squid axons is not influenced by large increases in $[P_i]$ [13]. An increase in internal [ADP] can be expected to promote the formation of E·ATP so that one hypothesis for explaining why ADP has an inhibitory effect on K influx but not on Na efflux would be that the Na efflux is a different one from that existing before the addition of ADP, and that this new Na efflux is Na:Na exchange. This scheme rather plausibly explains why Na:Na exchange requires ATP, and why K transport is inhibited – it suggests that ATP hydrolysis is not required for Na:Na exchange. However, this is a point on which we have no data. There is one difficulty with this proposal.

ADP ought to be a powerful inhibitor of ATP hydrolysis if an appreciable fraction of the enzyme is to be stabilized as E·ATP; this does not seem to be the case.

A second scheme for the action of ADP is to suppose that it allows a reaction such as the following to proceed:

$$E_1 + ATP \leftrightarrow E_1 \cdot ATP \leftrightarrow E_1{\sim}P + ADP$$
$$E_1{\sim}P \rightarrow E_1 + P_i$$

This would be a Na-activated ATPase, and would in effect be a short circuit on the usual cyclic scheme involving E_1 and E_2. This arrangement would avoid the necessity for supposing that ADP is highly inhibitory to ATP hydrolysis but would require that a different pathway be utilized.

Finally, it should be recognized that if enzyme and ion pumps are separate molecular entities (or functionally separate parts of the same molecule), it is possible that ADP may have separate effects on the pump and on its energy source.

Ion specificity in transport

The Na pump appears to have a high degree of specificity for Na; there is no evidence of transport of other ions (in the squid axon system). The inclusion of Li at a concentration 8 times that of Na has shown no effect on Na extrusion [9] so that competition for either transport or activation would appear minimal. Much the same sort of specificity is exhibited by membrane ATPase, which has an absolute requirement for Na. Such a finding argues somewhat in favor of an identity between ATPase activating and transport sites. The site on the outside of the membrane has been called the "not Na" site by Post et al. [23] since its specificity for a particular monovalent cation is not great. All cations from Li to Cs show some activation of this site in the ATPase and K, Rb, and Cs are clearly transported inward by a pump that is usually called the K pump. The activation of Na extrusion by Rb and especially Cs is considerably less than for K and transport is usually less, as well. Unfortunately, experimental information is insufficient to show whether the apparent coupling ratios of Na/Cs differ greatly from those for Na/K over a range of external concentrations.

What is significant about the ion specificity of both the membrane ATPase and transport is the extreme specificity of the Na site and the distinctly lesser specificity of the K site. The K site resembles the sort of specificity found with certain cyclic polypeptides which have been introduced into artificial lipid bilayers to confer upon them K-RB specificity, that is, a specificity that depends on a steric property of the ion. The Na specificity is different; it does not confuse Na-like ions (such as Li) with Na and suggests very strongly that it depends on the electronic structure of Na rather than on the steric properties of the ion. Thus, one infers that the chemical nature of the pumps and activating sites for Na and K are intrinsically different. Such a conclusion makes it somewhat implausible that a Na binding site can be transformed (by some conformational rearrangement of a protein) into a K binding site if the mechanisms for ion selectivity at the two sites are intrinsically different.

The activation of ion transport

Recent developments in ion transport studies using squid axons have shown that various cations applied to the outside of the membrane can have a complicated set of effects on transport. Many of these effects can be described as the activating effect of external cations on transport.

1) K activation. This is the most familiar effect. The effect of $[K]_o$ on Na efflux is half maximal at a concentration less than 1 mM in Na-free solutions. Na competes for sites and at seawater [Na], the required [K] for half maximal Na efflux is about 10 mM.
2) Na activation. Even when the competition of Na for K sites is allowed for, Na has an activating effect on Na efflux which varies from axon to axon such that the sum of Na efflux activated by Na and by K is roughly constant.
3) Ca activation. In Na-free seawater, there is a Na efflux which depends on $[Ca]_o$. Na competes at the Ca activating site so that in Na seawater there is very little Ca dependent Na efflux. $[Li]_o$ appears to promote the Ca dependent Na efflux to a greater extent than simple Na-free conditions would suggest.
4) Ion independent Na efflux. There is an appreciable Na efflux into isotonic dextrose solutions. Part of this efflux is undoubtedly K activated Na efflux even though the solution is nominally K-free but it seems unlikely that all the Na efflux is so activated. A difficulty with experiments of this sort lies in deciding on a suitable reference substance from which activating effects are to be measured. Both choline and Mg-dextrose have been used but not enough work has been done to be sure that choline has no activating properties.

The activation of K transport inward appears simpler, although this may only reflect the fact that fewer experiments have been done on K influx than on Na efflux. At any rate, the Na concentration inside an axon has an important activating effect on K transport and one that is half saturated at about 10 mM Na. Since the Na concentration of squid axons as usually studied is often around 100 mM, the K pumping system is fully activated in most experiments. Changes in external [Na], which have large effects on the K activation of Na efflux, have no effect on K transport so that the Na/K competition previously noted does not affect the K transport system.

Given the complexity of the observations of ion activation of transport and the competition among ions for activating sites, it is clear that ion transport cannot be explained in the rather simplistic terms of a pump mechanism that moves 3 Na outward and 1 K inward per transport cycle. The best that can be done at present is to describe the observed transport as unambiguously as possible and to consider criteria that seem presently valid for deciding whether a particular transport system is related to another system or is, in fact, a separate entity.

Summary

The contribution of the squid axon to active transport studies has been mainly that of showing that the energized movements of Na and K across its membrane are rather more complicated than present schemes of active transport envisage. Much more experimental work will be required before the phenomena are adequately understood and the following scheme can only be considered a more provisional one for representing transport phenomena. Fig. 3 shows membrane ATPase as supplying energy to Na and K pumps;

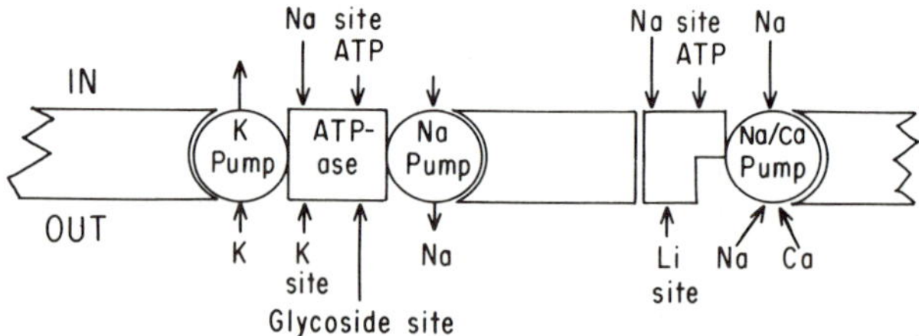

Fig. 3. A possible arrangement whereby Na^+ and K^+ activation of membrane ATPase and pumping of Na^+ and K^+ are separated. The features illustrated are: activating sites for Na^+ and K^+ as well as pumping sites. Both Na and K pumps derive energy from the same source but actual pumping rates depend both on the degree of activation and ion concentration at the pump site.

there are separate activating sites for the enzyme and transport sites for the pumps in order to accommodate the differing concentration dependencies of these functions. The Na/Ca pump is shown with an enzyme for ATP utilization. The enzyme differs from membrane ATPase in that it has no glycoside site, and it is Li activated on the outside. While information on this system is still meager, it is intended to accommodate the following experimental information: a glycoside insensitive, ATP dependent Na influx that is Na_i activated, a competition between Na and Ca for pumped entry, and a glycoside insensitive, ATP sensitive Na efflux.

The ATP independent Na efflux observed in the presence of glycoside is, rather clearly, not a property of the Na/Ca pump but is the result of transforming membrane ATPase E into E-gly, where this is a form of the enzyme that occurs in the presence of ouabain and related substances. This change can be most easily visualized by supposing that the K pump is stopped while the Na pump is allowed to cycle freely and to carry cations inward and Na outward. Details of the activation and competition mechanisms involved in ion pumping remain to be worked out but it is quite possible that in the end the system will become even more complicated than it now appears.

References

[1] P.C. Caldwell and R.D. Keynes, The permeability of the squid giant axon to radioactive potassium and chloride ions. J. Physiol. (London) 154 (1960) 177.

[2] R.A. Sjodin and L.J. Mullins, Tracer and non-tracer potassium fluxes in squid giant axons and the effects of changes in external potassium concentration. J. Gen. Physiol. 50 (1967) 533.

[3] H. Grundfest, C.Y. Kao and M. Altamirano, Bioelectric effects of ions microinjected into the giant axon of *Loligo*, J. Gen. Physiol. 38 (1954) 245.

[4] P.C. Caldwell, A.L. Hodgkin, R.D. Keynes and T.I. Shaw, Partial inhibition of the active transport of cations in the giant axons of *Loligo*. J. Physiol. (London) 152 (1960) 591.

[5] P.F. Baker, A.L. Hodgkin and T.I. Shaw, Replacement of the axoplasm of giant nerve fibres with artificial solutions. J. Physiol. (London) 164 (1962) 330.

[6] T.I. Shaw, Cation movement in perfused giant axons. J. Physiol. (London) 182 (1966) 209.

[7] M. Canessa-Fischer, F. Zambrano and E. Rojas, The loss and recovery of the sodium pump in perfused giant axons. J. Gen. Physiol. 51 (5, Pt. 2) (1968) 162s.

[8] F.J. Brinley, Jr. and L.J. Mullins, Sodium extrusion by externally dialyzed squid axons. J. Gen. Physiol. 50 (1967) 2303.

[9] F.J. Brinley, Jr. and L.J. Mullins, Sodium fluxes in internally dialyzed squid axons. J. Gen. Physiol. 52 (1968) 181.

[10] P.F. Baker, M.P. Blaustein, A.L. Hodgkin and R.A. Steinhardt, The influence of calcium on sodium efflux in squid axons. J. Physiol. (London) 200 (1969) 431.
[11] P.F. Baker, M.P. Blaustein, R.D. Keynes, J. Manil, T.I. Shaw and R.A. Steinhardt, The ouabain-sensitive fluxes of sodium and potassium in squid giant axons. J. Physiol. (London) 200 (1969) 459.
[12] L.J. Mullins and F.J. Brinley, Jr., Some factors influencing sodium extrusion by internally dialyzed squid axons. J. Gen. Physiol. 50 (1967) 2333.
[13] P. De Weer, Effects of intracellular adenosine-5′-diphosphate and orthophosphate on the sensitivity of sodium efflux from squid axons to external sodium and potassium. J. Gen. Physiol. 56 (1970) 583.
[14] R.B. Dean, Theories of electrolyte equilibrium in muscle. Biol. Symp. 3 (1941) 331.
[15] H.B. Steinbach, Na and K in frog muscle. J. Biol. Chem. 133 (1940) 695.
[16] R.A. Sjodin and L.A. Beaugé, Coupling and selectivity of sodium and potassium transport in squid giant axons. J. Gen. Physiol. 51 ((1968) 152S.
[17] L.J. Mullins and F.J. Brinley, Jr., Potassium fluxes in dialyzed squid axons. J. Gen. Physiol. 53 (1969) 704.
[18] P. Horowicz and C. Gerber, Effect of external potassium and strophanthidin on sodium fluxes in frog striated muscle. J. Gen. Physiol. 48 (1965) 489.
[19] R.D. Keynes and R.C. Swan, The effect of external sodium concentration on the sodium fluxes in frog skeletal muscle. J. Physiol. (London) 147 (1959) 591.
[20] L.A. Beauge and R.A. Sjodin, The dual effect of lithium ions on sodium efflux in skeletal muscle. J. Gen. Physiol. 52 (1968) 408.
[21] R.A. Sjodin and L.A. Beaugé, The influence of potassium- and sodium-free solutions on sodium efflux from squid giant axons. J. Gen. Physiol. 54 (1969) 664.
[22] M.P. Blaustein and A.L. Hodgkin, The effect of cyanide on the efflux of calcium from squid axons. J. Physiol. (London) 200 (1969) 497.
[23] R.L. Post, S. Kume, T. Tobin, B. Orcutt and A.K. Sen, Flexibility of an active center in sodium-plus-potassium adenosine triphosphatase. J. Gen. Physiol. 54 (1969) 306S.

SIDEDNESS OF THE RED CELL Na:K PUMP

Joseph F. HOFFMAN
Department of Physiology, Yale University School of Medicine, New Haven, Connecticut, U.S.A.

It is implicit in the concept of active transport that pumps, *per se*, possess or display both functional and structural asymmetry. This is so not only because such transport is oriented, but in order to direct the flow of material uphill there must be an obligatory linkage with metabolism [1]. The anisotropic nature of the underlying molecular device follows directly from thermodynamic analysis of the coupling of vectorial transport to metabolic (scalar) reactions [2]. Therefore the study of the sidedness of a pump should provide at least in principle useful insight into its molecular mechanism. In this paper I will consider only the Na:K pump of red blood cells since this system appears optimal at the moment for defining with the greatest specificity those features which can be assigned to one or the other side of the membrane.

The general features of the side dependence of the Na:K pump are outlined in fig. 1. The pump operates to move K in and Na out against their respective electrochemical potential gradients. In addition, the pump is evidently tightly coupled [3–5] in the sense that neither ion can be transported alone, establishing the simultaneous requirement for K on the outside and Na on the inside of the membrane [6]. The proximate source of energy for the pump is intracellular ATP and the products of the reaction remain within the cell interior [7, 8]. One consequence of the pump's utilization of ATP is that instead of measuring the appropriate fluxes the activity of the pump can effectively be followed by measuring (Na + K)-dependent ATPase activity [9]. This approach has been exploited in recent years [10] to characterize a number of membrane preparations (particle or microsomal) even though the transport capacity cannot be evaluated because the membrane integrity has been lost as a result of isolation. Obviously the permeability barrier must remain intact (or perhaps reconstituted or fabricated), distinguishing inside

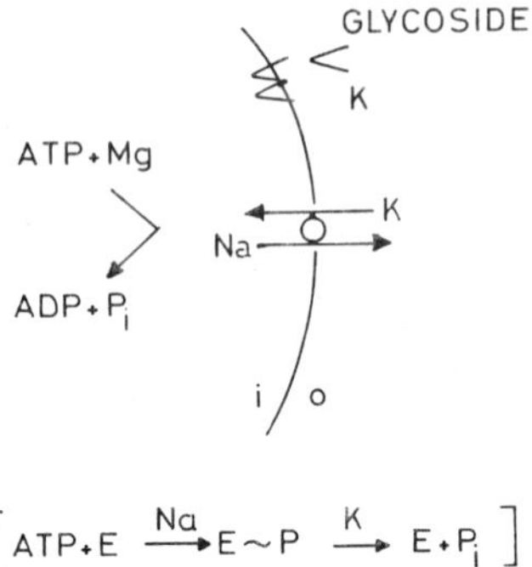

Fig. 1. Basic features of the Na:K pump mechanism. The membrane is indicated by the vertical line separating the inside (i) from the outside (o). The active transport system resides within the membrane pumping, at the expense of intracellular ATP, Na and K in the indicated directions. The Na:K pump is inhibited, non-competitively, by cardiac glycosides which bind to the outside of the membrane. Glycoside binding is antagonized by external K.

from outside, for the determination of the function and orientation of any membrane component especially with regard to the mechanism of translocation of transported ions. On the other hand, considerable detail is available concerning the sequence of reactions associated with the utilization of ATP by the Na,K-ATPase, based on studies dealing with the incorporation and release of ^{32}P from gamma-labeled ^{32}P-ATP [11–13]. The overall reaction of the pump can thus be summarized by the equation:

$$\mathrm{ATP} + \mathrm{E} \xrightarrow{\mathrm{Na}} \mathrm{E} \sim \mathrm{P} \xrightarrow{\mathrm{K}} \mathrm{E} + \mathrm{P_i} + \mathrm{ADP}$$

where E denotes the pump as enzyme and E $\sim$ P, a phosphorylated form in which Na acts to promote phosphorylation and K, dephosphorylation, of the pump complex, in conjunction with different postulated conformational states of E $\sim$ P. It should be recognized that the molecular mechanism of translocation and its relation to transphosphorylation is still obscure even in the light of evidence that reversible transformation of different forms of E $\sim$ P might be associated with particular types of cation exchange [14,15]. Mention of these partial reactions is made here (with no attempt at a thorough account) in order to emphasize asymmetry in the transport cycle in terms of possible time-dependent spatial states of the process.

While the foregoing has dealt with the polarity of the pump and its normal reactants and substrates we now shift to consider the mode of action of various inhibitors with regard to their use as probes for the sidedness of the pump. From the different ways in which the pump can be inhibited we will only be concerned with those which act directly on the pump such as cardiotonic steroids [16], sulfhydryl reagents [9] and Ca [8]. Thus inhibitors which work indirectly by interference with the supply of ATP, for instance, starvation, or substrate competition (e.g., glucose + hexokinase) or metabolic inhibitors like iodoacetate, are of limited interest in the present context with the possible exception of fluoride [8]. Fluoride may act directly on the pump since it inhibits in the presence of ATP [8] perhaps as consequence of its interaction with Mg and perhaps from the inside but the data available are insufficient to specify its mechanism or site of action.

Cardiotonic glycosides (and their aglycones), in being potent and specific inhibitors [16, 17], have proved useful in at least two different but related ways in defining membrane sidedness of the pump complex. The first is concerned with the localization of the particular binding sites for these inhibitors and the second, with the determination of various factors which govern their binding.

The sites to which cardiac glycosides bind have been localized to the outside surface of the membrane, as indicated in fig. 1, on the basis that these steroids are inactive when placed inside [17, 18] and that their binding is antagonized directly by external K [19]. In addition, since cardiac glycosides appear to inhibit the Na:K pump by an allosteric mechanism [18], it follows that glycosides bind to sites which are spatially distinct and therefore some distance away from the loci involved with the inward carriage of K. Nevertheless, the glycoside binding site must be a component part of the pump complex since glycosides bound to the pump before solubilization remain attached and inhibiting afterwards [20]. While the number of glycoside binding sites per cell can be measured directly using, for instance, tritiated ouabain, the important parameter to establish has been the number of glycosides bound to pump-associated sites [21, 22]. Thus, from studies concerned with relating ouabain binding to pump inhibition, the number of molecules needed for complete inactivation averages about 250 per cell for human red cells [23]. This number represents an upper limit since not all of the nonspecific binding may have been taken into account [23]. If the number of glycoside binding sites counts the number of pumps, that is, one site one pump, then some 6000 ions would be pumped per site per minute and the surface density of pumps would approximate two per square micron of surface and comprise in sum about one part in 10^4 of the total membrane mass (see [20]).

Since glycoside binding by the membrane is known to be influenced or controlled by a variety of factors, it is possible, by study of the side dependence of the various factors involved, to specify other asymmetries of the Na:K pump than those mentioned before. Determinations of this type of sidedness necessarily depend upon the use of a system, such as red cells, where the inside–outside relationship is known and amenable to experimental manipulation. Thus, while many of the different determinants of glycoside binding have been worked out on microsomal [11, 24–30] or permeable red cell ghost [21] preparations, the side dependencies of some of these factors (e.g. nucleotides, Na and P_i) have been evaluated in intact red cells or resealed red cell ghosts [31]. It should be understood that in all cases referred to, the sites to which the glycosides bind have been shown to be specific in terms of the number bound and in the sense that the binding results in inhibition of either the Na:K pump or the Na, K-ATPase.

Thus, the glycoside binding that is promoted by nucleoside triphosphates (NTP), such as ATP, requires NTP on the inside as opposed to the outside of the membrane. This conclusion is based on studies utilizing metabolically depleted reconstituted ghosts since it was found that glycoside binding to these ghosts only occurs in the presence of appropriate substrates [31]. For instance, it can be shown that glycosides will bind to the membrane once the metabolic intermediates (including NTP) are regenerated intracellularly by incubation with adenosine (which penetrates to the interior). This result indicates that whatever substrates promote glycoside binding to the outside must be present on the inside. It is likely that the substances involved are in fact NTP since they are the only compounds regenerated by this procedure which are also known to support glycoside binding. Another but more direct type of experiment leading to the same conclusion has also been carried out on the reconstituted ghost system in which specific substrates were incorporated inside the ghosts during hemolysis (and then washed) prior to assay for glycoside binding. Thus incorporated ATP or UTP (the only NTP tested in this manner) were found to be equally effective in promoting glycoside binding. On the other hand, ghosts reconstituted without incorporated substrates failed to bind glycosides when either ATP or UTP was present only on the outside, in the external medium. The fact that essentially any NTP promotes glycoside binding (as determined using the permeable ghost preparation referred to before) contrasts sharply with the nucleotide requirement for running the Na:K pump. The pump utilizes almost exclusively ATP and only purine nucleotides (the pyrimidine nucleotides are inactive) but glycoside binding is promoted equally well by ATP, GTP, ITP, UTP, CTP and dATP (and also by many of their diphosphates) down to concentrations as low as 1 μM [8, 21]. These results imply that the binding of glycosides to the pump

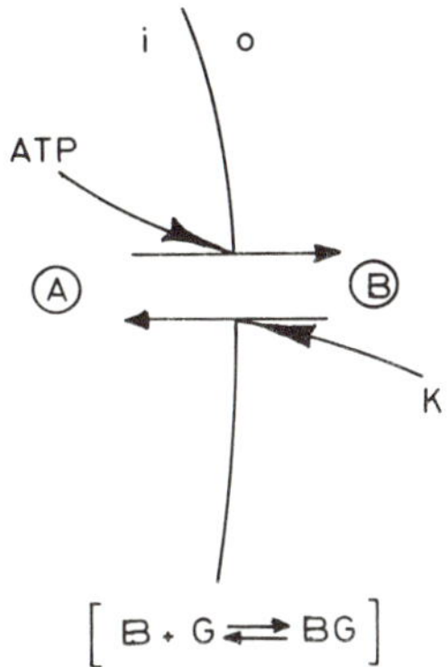

Fig. 2. A model showing the side dependencies of the membrane associated with the binding of glycosides to the pump. Different conformational states of the binding site are represented by A and B but glycosides (G) can only combine with the B form. Once BG is formed the complex in red cells is almost irreversible meaning that the equilibrium constant for the equation is very large (of the order 10^{11} for ouabain). ATP on the inside favors the formation of B; K on the outside favors the A form. Although other NTP will substitute for ATP in promoting glycoside binding, ATP is indicated here since it is the physiological substrate of the pump. See text for further details.

complex does not require the formation of any phosphorylated intermediate. It should be added, by the way, that glycoside binding requires the presence of Mg regardless of the binding promoter and therefore parallels the pump's requirement for Mg.

A model depicting the membrane asymmetries involved with the interaction of cardiac glycosides with their specific membrane sites is indicated in fig. 2. The binding sites are seen to exist in two membrane conformations, A and B, but glycoside (G) can combine only with the B form. When BG is formed the pump is inactivated. Since this model involving different conformational states of the pump complex is based on the fact that ATP on the *inside* promotes the binding of glycoside to the outside, ATP may act to shift the $A \rightleftharpoons B$ equilibrium to the right or to increase the rate of interconversion between the two forms; external K would act to antagonize these changes promoting in effect a shift to the left. The basis for the transition from A to B would be the adsorption of substrate (NTP) to its active site thereby inducing a change in the conformation of the pump complex. Since glycoside binding can occur without phosphorylation the conformational change associated with the B form (whether or not glycoside is bound) must represent only a partial rearrangement of the pump components which would ordinarily be connected with the different stages of transphosphorylation underlying the translocation cycle. (It should be mentioned that in the absence of promoters like ATP, and with the set of the equilibrium shifted to the left, there must be an oscillation between forms A and B which occurs spontaneously [driven by thermal energy] since in this circumstance glycoside will bind to the membrane given sufficient time and/or concentration but in terms of the present discussion these effects are negligible.) Clearly, models of the pump based on sequential stages of different conformational states of its molecular components (e.g. [21, 32]) can rather simply be extended to accommodate the requirements of glycoside binding as well as providing for the fact that glycosides presumably inhibit the pump under normal (physiological) conditions by preventing its dephosphorylation (see fig. 1).

Previous work has shown that the NTP dependent binding of glycosides can be influenced by the presence of Na but the effect of Na on red cell ghosts is evidently different than it is on the various microsomal preparations referred to before. Thus, while Na seems to be an obligatory requirement for glycoside binding in the microsomal preparations, Na appears to control the rate rather than the extent of glycoside binding in red cells. Since our particular concern is not with the differences between these different preparations but with membrane asymmetries we will consider only the side dependence of the effects of Na in red cell ghosts. It turns out that Na will affect the rate of glycoside binding from either side of the membrane whether or not Na is present on the opposite side. Analyses of the different effects of Na were carried out using resealed ghosts in which the ionic composition of the internal as well as the external medium could not only be defined but maintained almost constant during the course of glycoside binding [31]. Thus ghosts were studied in which the levels of Na or K on both sides of the membrane were varied independently with choline added to keep the total isotonic concentration (Na + K + choline) constant. Considering first glycoside binding that is promoted by incorporated ATP and Mg, it was found as also previously reported [33] that the rate of binding, at constant internal composition, increased as the concentration of Na in the external medium increased. Since in this situation the effectiveness of Na is decreased as external K is increased it appears that external Na acts indirectly on glycoside binding by decreasing the ability of K to antagonize the binding of glycoside. In terms of the model presented in fig. 2 this type of competition between Na and K increases the rate of glycoside binding by preventing K from decreasing the availability of the B form. However, the most dramatic effects of Na have been obtained not by external but by internal Na. Thus, as internal Na is increased, under conditions where the internal K and external Na and K are held constant, the rate of glycoside binding is found to be *decreased*. The concentrations of internal Na necessary to inhibit glycoside binding by 50 percent is approximately 20 mM, the same as the apparent K_m of the activation of the Na:K pump by internal Na. These results, on the one hand, provide further evidence dissociating glycoside binding from phosphorylation while, on the other hand, they imply in terms of the model presented in fig. 2, that Na only interacts with the conformation of the pump complex associated with the A form when it initiates the formation of the phosphoenzyme as a requisite for its translocation.

While the results just reviewed were obtained using incorporated ATP and Mg to promote glycoside binding, another system to consider is the glycoside binding that is promoted by orthophosphate (P_i) and Mg [26]. The effect of Na in this system is again complex and depends upon which side of the membrane Na is presented to [31]. In addition, the sidedness of the effect of P_i in promoting glycoside binding is also of interest but this is complcated by the fact that ghosts, like intact cells, are rather permeable to P_i. This difficulty can be overcome, however, by treating ghosts with agents which will inhibit their permeability to P_i but which will not alter their capacity to bind glycosides or to pump Na and K. Thus the P_i permeability of resealed ghosts pretreated with SITS* [34] is reduced to such an extent that P_i can be effectively restricted to one or the other side of the membrane during the time which glycosides bind [31]. In this way it was first established that P_i promotes glycoside binding when it is present on the inside as opposed to the outside of resealed ghosts. Next, the effects of internal and external Na on glycoside binding were controlled and evaluated on resealed ghosts containing P_i in a fashion completely analogous to that described previously on ghosts containing ATP. It was found as before that increasing the concentration of external Na increased the rate of glycoside binding by competition with external K and that as the concentration of internal Na is increased the rate of glycoside binding is *decreased* [31]. While the side dependent effects of Na action appear similar in the two systems, the details of the mechanisms which underly and promote glycoside binding would presumably be different since the interaction of glycosides and P_i appears mutual, at least in microsomal preparations, in the sense that the binding of glycosides promotes the incorporation of P_i [11, 26]. In terms of the model presented in fig. 2 this would mean that the B form can result from different conformational sets of the pump complex induced in different ways, which may (E–P) or may not (E–NTP) involve phosphorylation.

It is of interest now to consider the effects of different sulfhydryl reagents

* 4-acetamido-4′-isothiocyano-stilbene-2,2′-disulfonic acid.

Table 1

The differential effects of prior exposure to different sulfhydryl reagents on transport parameters of red cell membranes. Hemoglobin-free ghosts were frozen-thawed and exposed for 60 min at 37°C, *p*H 7.5, either to 1 mM ethacrynic acid (ETHA) or to 5 mM *N*-ethylmaleimide (NEM) or to 0.15 mM *p*-chloromercuriphenyl sulfonic acid (PCMBS) in a solution containing (mM): NaCl (40) + $MgCl_2$ (1.25) + EDTA (0.25) + Tris (10). The ghosts were then washed and suspended in this solution prior to testing either glycoside binding or Na,K-ATPase activity. Glycoside binding was accessed after incubation at 37°C for 30 min in the presence of 10^{-7} M [^{3}H]-ouabain and 2 mM ATP (see ref. [21]). Na,K-ATPase was measured after the addition of 2 mM ATP and 20 mM KCl to the above solution, by following the rate of P_i formation in the presence and absence of 10^{-4} M ouabain. The effects of ETHA [35] and NEM [9, 36, 37] on phosphorylation were determined by others on Na.K-ATPase prepared from kidney [35], red cells [36], brain [9] or the electric organ of the electric eel [37]. Thus, the incorporation of ^{32}P from [γ-^{32}P]-ATP in the presence of Na was used to estimate the level of phosphoenzyme. The asterisk refers to Na:K pump activity rather than to Na,K-ATPase since intact cells were used. The parenthesis here indicates that the inhibition of the pump is reversible upon incubation with a reducing agent such as dithiothreitol [40]. See text for further details.

Sulfhydryl reagent		Ouabain binding	Na,K-ATPase	Phospho-enzyme
ETHA (ghosts)		normal	inhibited	inhibited
NEM (ghosts)		prevented	inhibited	normal
PCMBS	ghosts	prevented	inhibited	unknown
	cells	normal	(inhibited)*	unknown

on glycoside binding since their use has disclosed other asymmetries associated with the Na:K pump complex. The effects of three different reagents are summarized in table 1. It is apparent that ethacrynic acid and *N*-ethylmaleimide (NEM) both inhibit the Na, K-ATPase but that ethacrynic acid inhibits phosphorylation without affecting glycoside binding, in contrast to NEM, which inhibits glycoside binding without affecting phosphorylation. Thus the differential action of these two reagents provides further support for the idea that phosphorylation is not a necessary requirement for glycoside binding. On the other hand, under other conditions these reagents have been useful in discriminating between the various phosphorylated forms of the pump complex which can bind glycosides (see [38, 39]). But more pertinent to the present discussion is the possibility that the differences in the effects of ethacrynic acid and NEM on glycoside binding are due to the fact that the two reagents may act on different sides of the membrane. If it is assumed that ethacrynic acid acts only from the outside of the membrane (and the avail-

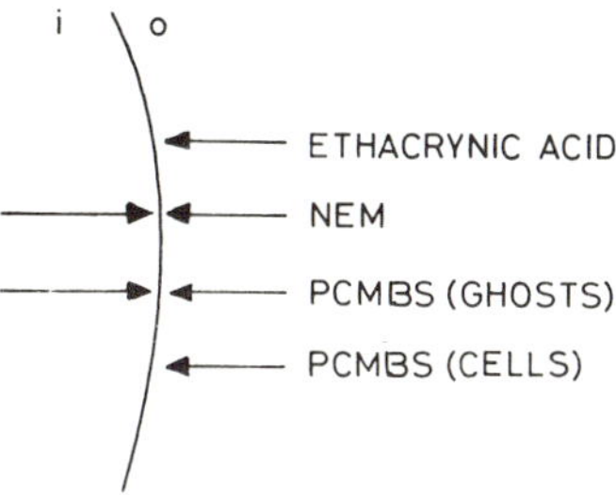

Fig. 3. Membrane sidedness and the action of sulfhydryl reagents on glycoside binding as outlined in table 1. Ethacrynic acid appears to interact with the membrane only on the outside (o) while *N*-ethylmaleimide (NEM) interacts from the inside (i) as well. Depending upon the permeability of the membrane PCMBS (*p*-chloromercuriphenyl sulfonic acid) can interact either from the outside or on both sides. See text for further discussion.

able evidence favors this view since its effects on the inhibition of the Na:K pump in intact red cells can be directly antagonized by external K and since it appears to act on the Na, K-ATPase in ghosts the same as it acts on the pump in intact cells) and that NEM acts from the inside as well as from the outside of the membrane, as indicated in fig. 3, then it can be supposed that the Na,K-ATPase can be inhibited by either reagent acting from the outside but that glycoside binding can be prevented only by NEM acting from the inside. This interpretation can presumably be tested by using the sulfhydryl reagent, PCMBS (*p*-chloromercuriphenyl sulfonic acid). On the basis that the intact cell is known to be impermeable to PCMBS [40] its sidedness of action can be defined operationally as the difference in its effects on intact cells as compared to ghosts. This is to say that in intact cells the effect of PCMBS is limited to the outside while in ghosts PCMBS has access to both sides of the membrane (fig. 3). As shown in table 1, PCMBS inhibits the Na,K-ATPase and prevents glycoside binding in ghosts in parallel with the effects of NEM. In contrast, it is apparent that PCMBS has no effect on glycoside binding to intact cells although it inhibits the Na:K pump (as Na,K-ATPase) and therefore simulates the effect of ethacrynic acid in ghosts. However, the inhibition of the Na:K pump (Na,K-ATPase) by PCMBS in intact cells can be reversed by incubation with a reducing agent such as dithiothreitol [40] but the effects of PCMBS in ghosts like NEM appear to be irreversible. Thus, the results obtained with the three sulfhydryl reagents indicate that there are at least two classes of sulfhydryl groups associated with the Na:K pump complex and that the two classes are located on opposite sides of the membrane.

Reaction with the outside groups alone is sufficient to cause inhibition of the Na:K pump (or Na,K-ATPase) but it is only by interference with the class associated with the inside that glycoside binding can be prevented.

Brief mention should be made of the membrane asymmetry with regard to the effects of Ca as a known inhibitor of the Na,K-ATPase [8]. In either intact cells or reconstituted ghosts the Na:K pump is unaffected by the presence or absence of Ca in the external medium. On the other hand, the Na:K pump as studied in reconstituted ghosts containing ATP and Mg is inhibited by incorporated Ca in complete parallel to its inhibition of the Na,K-ATPase [8]. In addition, the glycoside binding that is promoted by either NTP or P_i as discussed eariler is not affected by Ca if Mg is present but Ca cannot replace Mg in the binding reaction [21].

Finally, another type of membrane sidedness should be considered which is connected with the mechanism of action of an antibody (anti-L) which is known to stimulate the Na:K pump activity in low potassium (LK) type sheep [41] and goat [42] red cells. While the situation is complicated anti-L appears to exert its effects both by increasing the total number of pumps [43, 44] and by changing the apparent affinities of all the pumps for Na and K [43–46]. Both of these effects are necessary in order to account for the total increase in activity whether measured in terms of the Na:K pump or the Na,K-ATPase. However, for the purposes of the present discussion, it is the side dependence of the membrane associated with the change in affinities which is of special interest. Thus it has been shown in both sheep and goat red cells that anti-L acts on the outside to change the apparent affinities of the pumps on the inside but not on the outside. Although the mechanisms involved in these effects are not known it appears at least in LK goat red cells [44] that K on the inside is required before bound anti-L can act to change the transport parameters.

Acknowledgement

Portions of the work reported in this paper were supported by the National Institutes of Health, Grants PHS-09906 and AM-05644 and by the National Science Foundation, Grant GB-18924.

References

[1] T. Rosenberg, in: Active Transport and Secretion. Symposia Society Exptl. Biol. No. 8 (Academic Press, New York, 1954) p. 331.

[2] A. Katchalsky, in: Permeability and Function of Biological Membranes, eds. L. Bolis, A. Katchalsky, R.D. Keynes, W.R. Loewenstein and B.A. Pethica (North-Holland Publishing Co., Amsterdam, 1970) p. 20.
[3] E.J. Harris and M. Maizels, J. Physiol. 113 (1951) 506.
[4] I.M. Glynn, J. Physiol. 134 (1956) 278.
[5] R.L. Post and P.C. Jolly, Biochim. Biophys. Acta 25 (1957) 118.
[6] I.M. Glynn, J. Physiol. 160 (1962) 18P.
[7] G. Gardos, Acta Physiol. Hung. 6 (1954) 191.
[8] J.F. Hoffman, Circulation 26 (1962) 1201.
[9] J.C. Skou, Physiol. Rev. 45 (1965) 596.
[10] S.L. Bonting, in: Membranes and Ion Transport, ed. E.E. Bittar (J. Wiley and Sons, New York, 1970) Volume 1, p. 257.
[11] R.W. Alberts, G.J. Koval and G.J. Siegel, Molec. Pharmacol. 4 (1968) 324.
[12] R.L. Post, S. Kume, T. Tobin, B. Orcutt and A.K. Sen, J. Gen. Physiol. 54 (1969) 306s.
[13] R. Blostein, J. Biol. Chem. 243 (1968) 1957.
[14] I.M. Glynn, J.F. Hoffman and V.L. Lew, Phil. Trans. Roy. Soc. (London) B, 262 (1971) 91.
[15] I.M. Glynn and J.F. Hoffman, J. Physiol. 218 (1971) 239.
[16] H.J. Schatzmann, Helv. Physiol. Pharmacol. Acta 11 (1953) 346.
[17] P.C. Caldwell and R.D. Keynes, J. Physiol. 148 (1959) 8P.
[18] J.F. Hoffman, Am. J. Med. 41 (1966) 666.
[19] I.M. Glynn, J. Physiol. 136 (1957) 148.
[20] P.B. Dunham and J.F. Hoffman, Proc. National Acad. Sci. (Wash., D.C.) 66 (1970) 936.
[21] J.F. Hoffman, J. Gen. Physiol. 54 (1969) 343s.
[22] J.F. Hoffman and C.J. Ingram, in: Metabolism and Membrane Permeability of Erythrocytes and Thrombocytes, eds. E. Deutsch, E. Gerlach and K. Moser (Georg Thieme Verlag, Stuttgart, 1968) 420.
[23] C.J. Ingram, Ph.D. Dissertion (Yale University, 1971).
[24] H. Matsui and A. Schwartz, Biochim. Biophys. Acta 151 (1968) 655.
[25] A. Schwartz, H. Matsui and A.H. Laughter, Science, 160 (1968) 323.
[26] G.E. Lindenmayer, A.H. Laughter and A. Schwartz, Arch. Biochem. Biophys. 127 (1968) 187.
[27] A.K. Sen, T. Tobin and R.L. Post, J. Biol. Chem. 244 (1969) 6596.
[28] J.C. Skou and C. Hilberg, Biochim. Biophys. Acta 185 (1969) 198.
[29] P.F. Baker and J.S. Willis, Nature, 226 (1970) 521.
[30] O. Hansen, Biochim. Biophys. Acta 233 (1971) 122.
[31] H. Bodemann, C.J. Ingram and J.F. Hoffman, unpublished studies.
[32] J.F. Hoffman, in: Biophysics of Physiological and Pharmacological Actions, ed. A.M. Shanes (Am. Assoc. for the Advancement of Sc., Wash., D.C., 1961) p. 3.
[33] H.J. Schatzmann, Biochim. Biophys. Acta 94 (1965) 89.
[34] P.A. Knauf and A. Rothstein, J. Gen. Physiol. 58 (1971) 190.
[35] S.P. Bannerjee, V.K. Khanna and A.K. Sen, Biochem. Pharmacol. 20 (1971) 1649.
[36] R. Blostein and V.K. Burt, Biochim. Biophys. Acta 241 (1971) 68.
[37] S. Fahn, M.R. Hurley, G.J. Koval and R.W. Alberts, J. Biol. Chem. 241 (1966) 1890.

[38] S.P. Bannerjee, S.M.E. Wong, V.K. Khanna and A.K. Sen, Molec. Pharmacol. 8 (1972) 8.
[39] S.P. Bannerjee, S.M.E. Wong and A.K. Sen, Molec. Pharmacol. 8 (1972) 18.
[40] P.J. Garrahan and A.F. Rega, J. Physiol. 193 (1967) 459.
[41] J.C. Ellory and E.M. Tucker, Nature, 222 (1969) 477.
[42] J.C. Ellory and E.M. Tucker, Biochim. Biophys. Acta 219 (1970) 160.
[43] P.K. Lauf, B.A. Rasmusen, P.G. Hoffman, P.B. Dunham, P.B. Cook, M.L. Parmelee and D.C. Tosteson, J. Membrane Biol. 3 (1970) 1.
[44] J.C. Ellory, J.R. Sachs, P.B. Dunham and J.F. Hoffman, Symposium on Passive Permeability of Cell Membranes, in: Biomembranes, ed. L.A. Manson (Plenum Press, 1972) in the press.
[45] J.C. Ellory, I.M. Glynn, V.L. Lew and E.M. Tucker, J. Physiol. 217 (1971) 61P.
[46] R. Blostein, P.K. Lauf and D.C. Tosteson, Biochim. Biophys. Acta 249 (1971) 623.

MONOSACCHARIDE PERMEABILITY IN BROWN TROUT *SALMO TRUTTA* L. ERYTHROCYTES

L.BOLIS and P. LULY
Istituto di Fisiologia Generale, Università di Roma, Italy

1. Introduction

Monosaccharide transport (hexoses and pentoses) has been studied largely in human red blood cells [1–4] and it is now clear that in human erythrocytes the transport does not follow Fick's law, but shows characteristics indicating the presence of a specific transport system in the membrane.

Monosaccharide transport has been studied also in few animal species: Kozawa [5] demonstrated that the erythrocytes of Macacus rhesus are similar in behaviour, as to monosaccharide transport, to human red blood cells. Wilbrandt [6] has shown a low degree of permeability of hexoses in rabbit, rat and dog red blood cells, and in 1955 Morgan et al. [7] have suggested that the penetration of glucose in rabbit erythrocytes shows a behaviour which is in favour of the presence of a specific transport system in the membrane. Widdas [8] has shown a high degree of permeability for glucose in the case of foetal red blood cells from pig, rabbit, guinea-pig, sheep and the rates of penetration were comparable with that of adult human red cells. Faust and Parpart [9] concluded from experiments that monosaccharides (hexoses and pentoses) penetrate into red cells of Ground-hog by free diffusion.

Concerning red cell permeability in lower vertebrates, Bolis et al. [10] demonstrated that D(+) glucose does not penetrate red blood cells from brown trout *Salmo trutta* L.

The aim of this work is to present further data on monosaccharide penetration in trout erythrocytes.

2. Material and methods

Blood samples were collected by heart puncture from the living fish to avoid any possible action of anesthetics on cell permeability. Red cells were washed three times (1:20 v:v) with a physiological saline buffered with Na/Na_2 phosphate buffer (pH 7.4), the overall osmolarity was 340 mOsm i.e. equal to trout plasma osmolar concentration.

The monosaccharides used were: D(+)mannose, D(+)galactose, D(+)glucose, D(–)ribose, L(+)arabinose and D(+)xylose.

Washed cells were resuspended in buffered saline containing 150 mM or 300 mM sugar at 18° or 30° unless otherwise stated. Sugar penetration in trout red cells was observed during 4 hr using an osmotic method according to Wilbrandt [11].

Isotope countertransport and isotope exchange experiments* were also performed. The technique of isotope countertransport consists in equilibrating cells in a 1:1 (cell volume:medium volume) suspension with cold substrate in high and labelled substrate in low concentration, followed by dilution with medium containing only low concentration of labelled substrate. By the dilution a gradient for the cold substrate is established which induces, if the transport uses a carrier, a countertransport of the labelled substrate.

The technique of isotope exchange consists in equilibrating cells with cold and labelled substrate, but then diluting with medium containing cold substrate and no labelled substrate. Then a gradient for the labelled substrate induces outward movement of the labelled substrate which follows first order kinetics and whose rate in a carrier system is decreased in the presence of cold substrate to a degree depending on the concentration of the cold substrate and its Michaelis constant K_m.

3. Results and discussion

As indicated in fig. 1, D(+)mannose, D(+)galactose and D(+)glucose behave in the same way: a slow increase of the osmotically active cell content, that within four hours does not reach equilibration, might indicate either slow entry of sugar or else, possibly Na entry. Previous experiments with labelled glucose yielded no indication of glucose penetration [10].

As to pentose, it was found that D(–)ribose, L(+)arabinose and D(+)xy-

* Isotope experiments have been performed in the Department of Pharmacology, University of Bern.

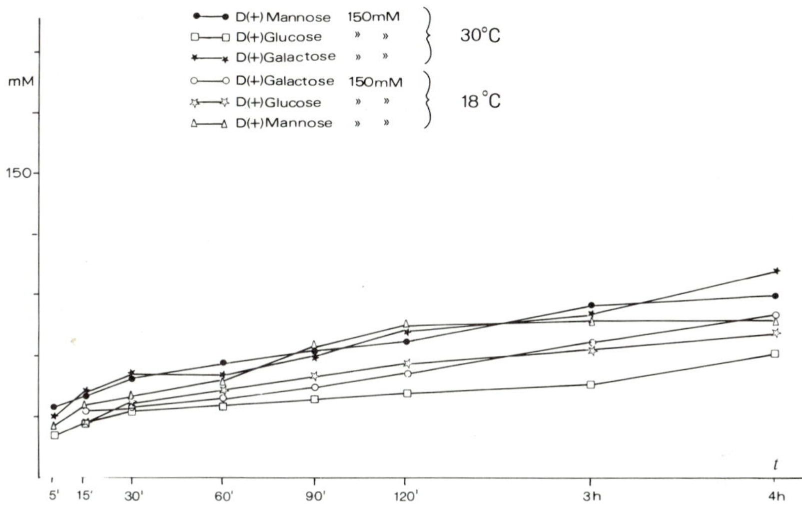

Fig. 1. Hexose penetration in *Salmo gairdneri* Rich. erythrocytes.

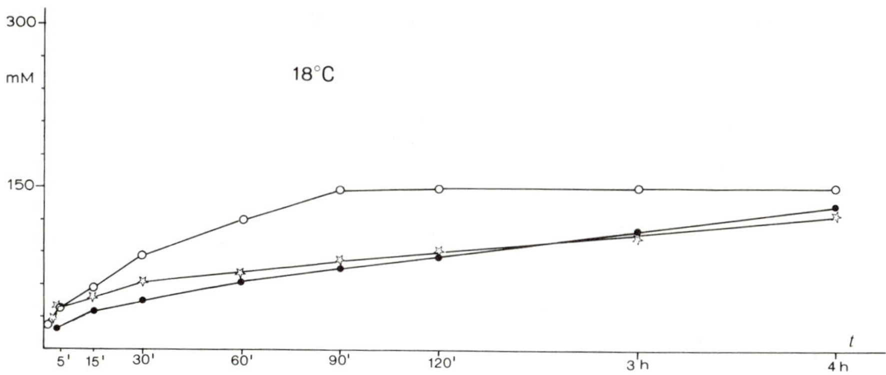

Fig. 2a. L(+)arabinose, D(+)xylose, D(−)ribose (150 mM) penetration in *Salmo gairdneri* Rich. erythrocytes. T. 18° C.

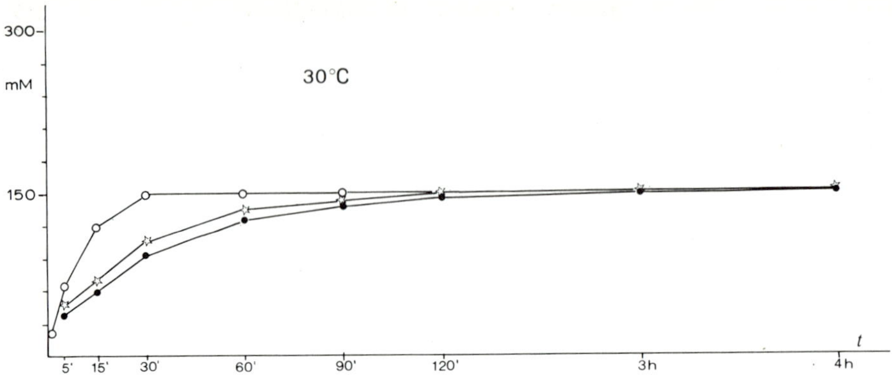

Fig. 2b. L(+)arabinose, D(+)xylose, D(–)ribose (150 mM) penetration in *Salmo gairdneri* Rich. erythrocytes. T. 30° C.

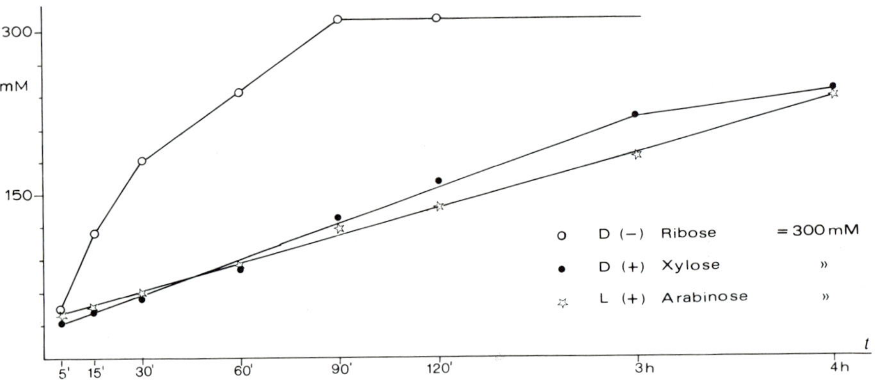

Fig. 3a. L(+)arabinose, D(+)xylose, D(–)ribose (300 mM) penetration in *Salmo gairdneri* Rich. erythrocytes. T. 18° C.

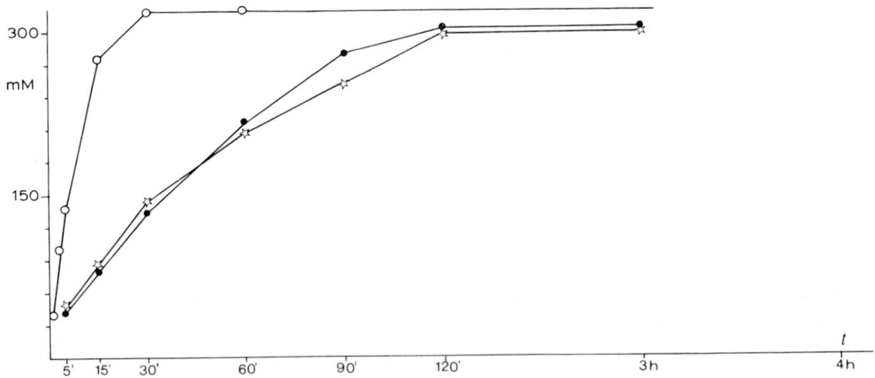

Fig. 3b. L(+)arabinose, D(+)xylose, D(–)ribose (300 mM) penetration in *Salmo gairdneri* Rich. erythrocytes. T. 30° C.

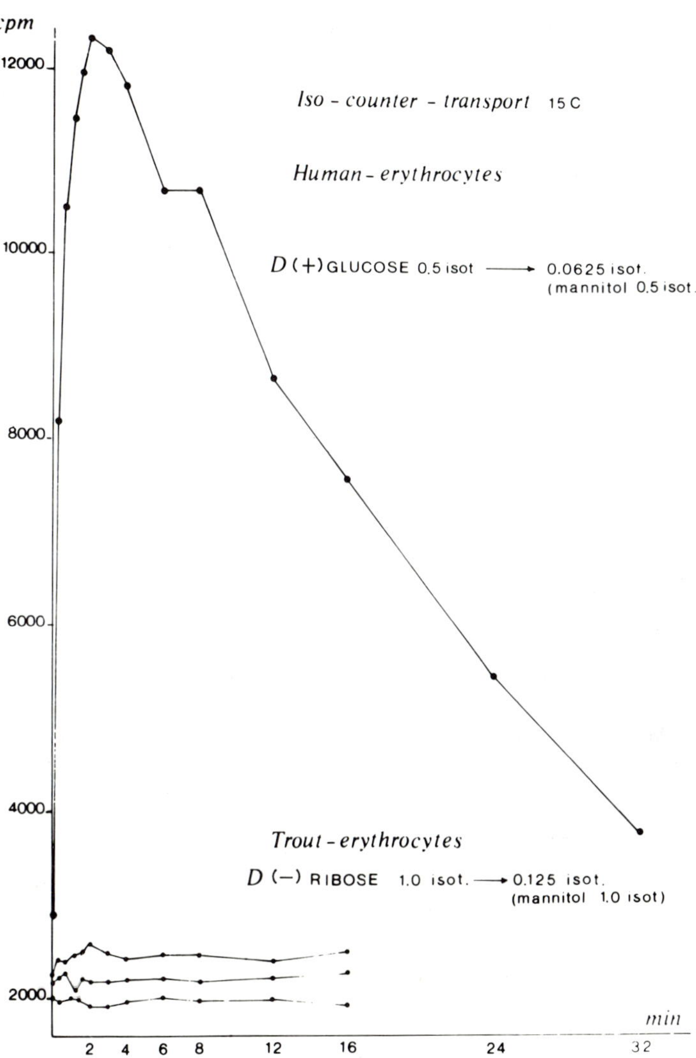

Fig. 4. D(+)glucose iso-countertransport in human erythrocytes and D(–)ribose in *Salmo gairdneri* Rich. erythrocytes. [1 isot. = 300 mM.]

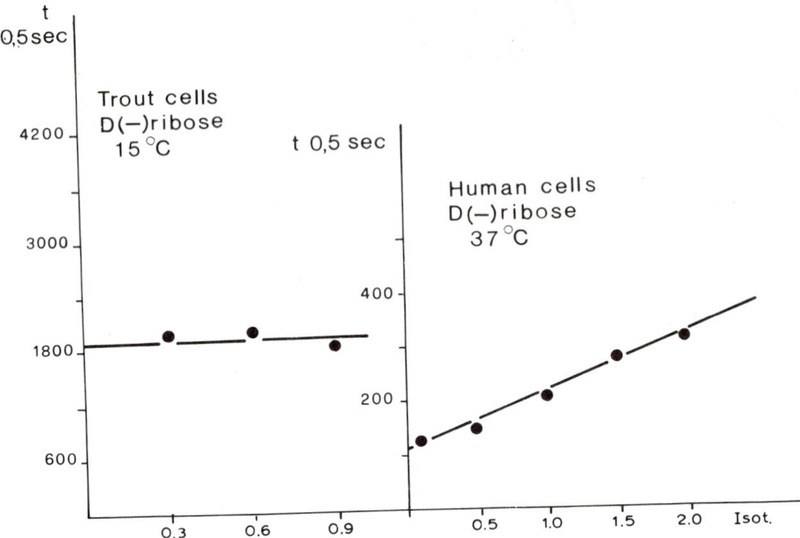

Fig. 5. Isotope exchange for D(–)ribose in human and *Salmo gairdneri* Rich. erythrocytes. [1 isot. = 300 mM.]

lose do penetrate with different rates, as shown in figs. 2a/b and 3a/b. The highest rate of permeation is shown by ribose; the influence of concentration on pentose penetration is clear. The temperature coefficient (Q_{10}) (18–30°C) appears to be: for ribose 3.07, for xylose 2.56 and for arabinose 2.66.

In order to test the possibility of a carrier transport for D(–)ribose in trout red cells, two different sets of experiments were carried out. Iso-countertransport experiments have been performed for glucose in human and for ribose in trout red cells: fig. 4 gives a clear positive evidence for human red cells and glucose, and no evidence for trout red cells and ribose. D(–)ribose has a very low affinity for human red blood cells, excluding marked iso-countertransport. Isotope exchange experiments with D(–)ribose established that cold substrate concentration has no effect on the rate of labelled substrate movement in trout red blood cells, whereas there is a clear positive effect in human red cells (fig. 5).

Thus both these experiments fail to provide evidence for a carrier mediated diffusion for ribose in trout red blood cells.

Acknowledgments

We wish to express our gratitude to Professor W. Wilbrandt for his suggestions and for many stimulating discussions.

References

[1] W. Wilbrandt and T. Rosenberg, The concept of carrier transport and its corollaries in pharmacology. Pharmacol. Rev. 13 (1961) 109.

[2] P.G. LeFevre, Rate and affinity in human red blood cell sugar transport. Am. J. Physiol. 293 (1962) 286.

[3] D.M. Miller, Monosaccharide transport in erythrocytes, in: Red Cell Membrane Structure and Function, eds. G.A. Jamieson and T. Greenwalt (Lippincott Co., Philadelphia and Toronto, 1969) p. 240.

[4] R.G. Faust, Monosaccharide penetration into human red blood cells by an altered diffusion mechanism. J. Cell Comp. Physiol. 56 (1960) 103.

[5] S. Kozawa, Beiträge zum arteigenen Verhalten der roten Blutkorperchen. III. Artdifferenzen in der Durchlässigkeit der roten Blutkorperchen. Biochem. Z. 60 (1914) 231.

[6] W. Wilbrandt, Die Permeabilität der roten Blutkorperchen für einfache Zucker, Pflügers Arch. 241 (1938) 302.

[7] H.E. Morgan, C.F. Kalman, R.L. Post and C.R. Park, Kinetics of glucose transport across red cell membrane. Fed. Proc. 14 (1955) 103.

[8] W.F. Widdas, Hexose permeability of foetal erythrocytes. J. Physiol. 127 (1955) 318.

[9] R.G. Faust and A.K. Parpart, Permeability studies on the red blood cell of the Ground-hog *Marmota monax*. J. Cell Comp. Physiol. 57 (1961) 1.

[10] L. Bolis, P. Luly and V. Baroncelli, D(+)glucose permeability in brown trout *Salmo trutta* L. erythrocytes. J. Fish Biol. 3 (1971) 273.

[11] W. Wilbrandt, Handbuch der Physiologisch und Pathologisch-chemischen Analyse (Springer-Verlag, Berlin, 1955) 10 Aufl. b II, 49–70.

STUDIES ON THE INHIBITION OF ACTIVE K^+-TRANSPORT IN HUMAN ERYTHROCYTES

G. GÁRDOS

Department of Cell Metabolism, National Institute of Haematology and Blood Transfusion, Budapest, Hungary

I should like to contribute to the problem of the inhibition of active K^+-transport in human erythrocytes by a few own observations. It was conspicuous in our experiments aiming ^{42}K-accumulation that if adenosine was used in the control tubes as substrate a slow net K^+ outflow occurred as a rule, while in the presence of inosine as a substrate K^+ equilibrium was not disturbed. The rate of K^+ outflow in the former case was dependent on the adenosine concentration and seemed to be the consequence of NH_3 liberation from the nucleoside, a process catalyzed by the erythrocyte enzyme adenosine deaminase. NH_4Cl in equimolar concentration to the NH_3 liberated from adenosine inhibited ^{42}K accumulation to the same extent (table 1).

The table clearly demonstrates that while inosine at 1–20 mM final con-

Table 1
The effect of inosine, adenosine and NH_4Cl on the ^{42}K influx of human erythrocytes at 37°C.

Concentration (mM)	Inosine		Adenosine		NH_4Cl	
	K^+ influx (mEq/ l cell/hr)	Inhibition (%)	K^+ influx (mEq/ l cell/hr)	Inhibition (%)	K^+ influx (mEq/ l cell/hr)	Inhibition (%)
0	3.56		3.54		3.52	
1	3.58	0	2.98	16	3.30	6
5	3.57	0	2.06	42	1.98	44
10	3.30	7	1.09	69	1.35	62
20	3.05	14	0.79	78	0.80	77

Table 2
The effect of antihistamines on the ^{42}K influx of human erythrocytes at 37°C.

	K^+ influx (meq/l cell/hr)	Inhibition (%)
Control	3.74	
1 mM Tripelennamine	2.64	29.5
1 mM Phenindamine tartarate	2.52	32.6
1 mM Chloropyramine	2.46	34.0
1 mM Promethazine	2.22	40.5

centration has no effect on ^{42}K influx, a considerable inhibition approximating 80% at 20 mM final concentration is caused by adenosine and NH_4Cl.

A similar net K^+ outflow was elicited when antihistamines were added to erythrocyte suspensions. From among the tested drugs Promethazine was the most effective in this respect, followed by Chloropyramine and then by Phenindamine tartarate and Tripelennamine (table 2).

^{42}K influx values indicated a marked inhibition of the active K^+ accumulation. At 1 mM antihistamine concentration the extent of the inhibition amounted to 30–40%. Calculations based on the accelerated net K^+ flux in the presence of antihistamines refer to the fact that these compounds affect not only the active ion movement but also the rate of the leak ion transport.

The presented findings suggest that active K^+ movement is inhibited by a part but not all of the NH_2-group containing compounds. Histamine and hydrazine, namely, enhancing leak ion movements under appropriate conditions [1, 2], are completely ineffective with respect to the active ion transport. As to the mechanism of action of the reported active, NH_2-group containing compounds we still have to confine ourselves to assumptions. It is likely that the clue will be found in the inhibition of the ouabain-sensitive ATPase system. This assumption, however, needs further experimental confirmation.

References

[1] H. Passow, in: Cell Interface Reactions, ed. H.D. Brown (Scholar's Library, New York, 1963) p. 57.

[2] G. Gárdos and I. Szász, Acta Biochim. Biophys. Acad. Sci. Hung. 3 (1968) 13.

STIMULATION OF A SODIUM PUMP BY AN ANTIBODY THAT INCREASES THE APPARENT AFFINITY FOR SODIUM IONS OF THE SODIUM-LOADING SITES*

I.M. GLYNN

Physiological Laboratory, University of Cambridge, England

and

J.C. ELLORY

Agricultural Research Council Institute of Animal Physiology, Babraham, Cambridge, England

Sheep, goats, cattle, buffalo and 'possums are peculiar among mammals in that some varieties have red cells rich in K and others red cells poor in K and rich in Na [1–5]. In sheep, at any rate, the difference is determined by a single gene pair with low K (LK) dominant to high K (HK) [6]. A few years ago Ellory and Tucker [7], working at the A.R.C. Institute of Animal Physiology just outside Cambridge, made the surprising observation that if you took LK sheep red cells and injected them into an HK sheep, the sheep produced a serum – anti-L serum – which had the remarkable property of increasing the activity of the Na pump in LK red cells that were exposed to it. Both the pump and the associated ATPase activity were stimulated between 3-fold and 8-fold. These effects of anti-L serum were confirmed by Lauf, Rasmusen, Hoffman, Dunham, Cook, Parmelee and Tosteson [8] at Duke University, and work both at Duke and at Cambridge showed that the action of the antibody was accompanied by an increase in the number of ouabain-

* The last part of Dr. Glynn's lecture described work, done in collaboration with Dr. J.F. Hoffman, on the nucleotide requirements for Na-Na exchange by the Na pump. This part of the lecture has not been reproduced here since a full account of that work has now been published (J. Physiol. Lond. 218 (1971) 239).

binding sites on the surface of the red cells, as though the antibody were unmasking latent pump sites [9, 10]. The magnitude of the increase in ouabain binding was, however, much too small to account for the increased pumping activity if the unmasked sites behaved identically with those active in the untreated cells. It is possible that the estimates of ouabain binding sites were confused by non-specific binding, but in any event it seemed worthwhile to look for other effects of the antibody on the pump. Recently Lauf et al. [10] suggested that the antibody acted by increasing the selective affinity for Na ions of the pump sites facing towards the inside of the cell, i.e. the Na-loading sites. The evidence on which this suggestion is based is shown in fig. 1, which is taken from their paper. Lauf et al. used the parachlormercuribenzoate method of Garrahan and Rega [11] to prepare LK sheep red cells containing different proportions of Na and K, and then they measured K influx from a medium which contained enough K to saturate the external pump sites, i.e. the K-loading sites. The curves in fig. 1 show the relation between K influx by the pump and the internal ion concentrations. The upper curve is for cells treated with anti-L serum; the lower curve, which serves as a control, is for cells treated similarly but with a non-immune serum. There is no doubt that the anti-L serum shifted the curve to the left, but it is arguable whether this implies that there was an increase in the selective affinity for Na i.e. the ability of the sites to discriminate between Na and K.

Because the sum of the intracellular concentrations of Na and K was held constant, as we move to the right along the x-axis K decreases as Na increases, and the sharp rise in the curves at the right hand side of the figure is associated with a much bigger proportional change in K concentration than in Na concentration. Let us suppose that K ions compete with Na ions for the Na-loading sites, and, to keep the argument simple, suppose also that competition is at a single site. If the system follows Michaelis-Menten kinetics we may write

$$v = \frac{V_{max} s}{K_m(1 + i/K_i) + s}$$

where s is the substrate concentration, here the concentration of Na, and i is the inhibitor concentration, by hypothesis the concentration of K. If the sum of the concentrations of Na and K is held constant, this may be written

$$v = \frac{V_{max} s}{K_m[1 + (T-s)/K_i] + s}$$

where $T = i + s$, and solving this equation for v we get an equation of the form

$$v = \alpha s/(\beta + s)$$

where

$$\alpha = V_{max}/(1 - K_m/K_i) \quad \text{and} \quad \beta = \frac{K_m(1 + T/K_i)}{1 - K_m/K_i}$$

If the affinity for the inhibitor is less than the affinity for the substrate, then α and β are both positive and a plot of v against s gives a rectangular hyperbola passing through the origin with the horizontal asymptote at α in the upper right hand quadrant and the vertical asymptote at $-\beta$ in the lower left-hand quadrant. Such a curve does not look much like the curves in fig. 1. But if the affinity for the inhibitor is greater than the affinity for the substrate, than α and β will both be negative, and the theoretical curve is similar in shape but swung round through 180° so that the real part of the curve, i.e. the part corresponding to positive concentrations of substrate and inhibitor, is upwardly inflected. The question we must ask, then, is: can the data of Lauf et al. be fitted by curves of this form? Fig. 2 shows that they can and, furthermore, that it is possible to fit the data for both control and anti-L

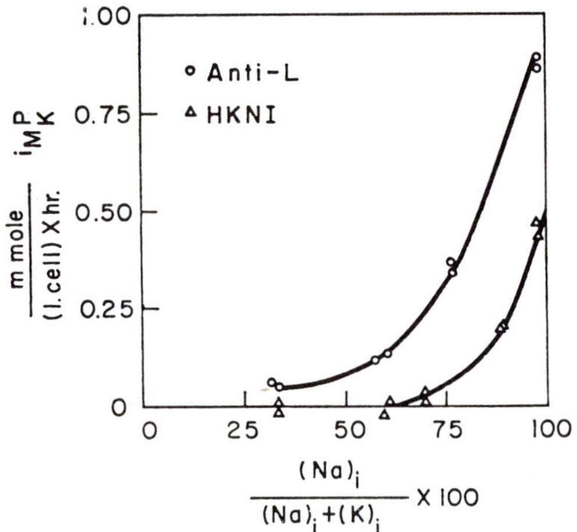

Fig. 1. The effect of intracellular Na and K concentrations on K^+-pump flux in LK sheep red cells that had been treated with anti-L serum or non-immune (HK-NI) serum. The incubation medium contained 5 mM-K and 130 mM-Na. (Reproduced from Lauf et al. [10].)

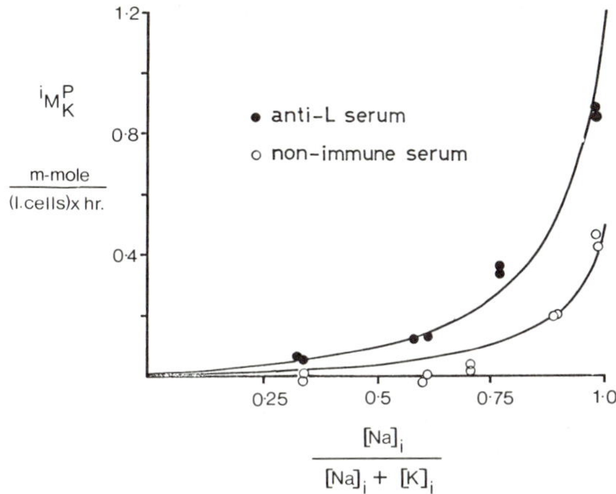

Fig. 2. The experimental points are from Lauf et al. [10], and the curves are theoretical curves derived from the Michaelis-Menten equation for competitive inhibition when the sum of the concentrations of substrate and inhibitor is held constant (see text). The two curves differ only in the value assigned to α. Note that the fit is tolerably good except in the region where the experimental results are least reliable because the ouabain-sensitive K influx was only a small fraction of the total K influx.

treated cells tolerably well by altering only V_{max}: it is not necessary to assume that the affinity for Na or for K is changed by the antibody.

At this stage, we thought that Dr. Lauf and his colleagues were probably wrong; but it would of course have been extremely embarrassing for me to come to this conference and, with Dr. Tosteson, who is one of those colleagues, in the Chair, say that their conclusions were wrong; so we had to go on until we showed that they were right.

An obvious approach was to vary the internal Na concentration while the K concentration was held constant; but this is difficult to do. Since the cell contents must be kept isotonic, if the Na concentration is to be reduced the Na must be replaced by an inert cation like choline. Unfortunately the Garrahan and Rega method for changing the cation composition of red cells works well for Na and K but rather badly for choline. We therefore gave up the idea of using intact cells and decided to work with fragmented red cell ghosts, measuring ouabain-sensitive ATPase activity instead of ouabain-sensitive K influx. With a broken membrane preparation it is, of course, not possible to have different solutions in contact with the outer and inner faces of the membrane, but by working only under conditions in which the external pump sites were saturated with K we were able to assume that any effects

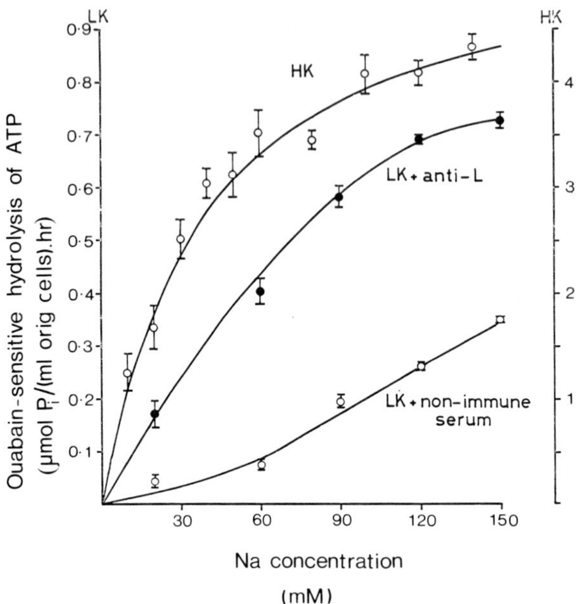

Fig. 3. The effect of anti-L serum on the stimulation by Na of the ouabain-sensitive ATPase activity of ghosts prepared from LK goat red cells. For comparison, the behaviour of ghosts from HK goat cells is also shown. The K concentration was 10 mM. The vertical bars show ± 1 S.E.M. Note that the scale of the ordinate is different for the HK ghosts.

we observed were brought about by changes in the cation concentrations at the inner face of the membrane. We used ghosts prepared from LK goat red cells rather than from sheep cells because the goat cells are similarly affected by sheep anti-L serum [12], and a much larger fraction of their membrane ATPase activity is inhibited by ouabain – 70% rather than 20% under optimal conditions.

Figure 3 shows the results of an experiment in which ouabain-sensitive ATPase activity was measured in media containing different concentrations of Na and 10 mM-K. Treatment with anti-L serum caused a marked increase in the sensitivity of the ATPase to Na; indeed, in their response to Na the anti-L-treated cells behaved much more like HK cells, whose behaviour is also shown in the figure. Four other similar experiments gave similar results. It seems that Lauf et al. are right after all, and that anti-L serum really does increase the apparent affinity for Na of the Na-loading sites of the pump.

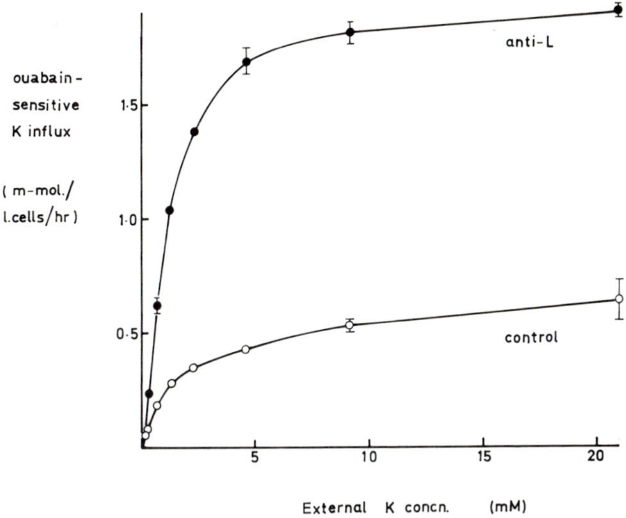

Fig. 4. An unpublished experiment by P.B. Dunham and J.C. Ellory showing the relation between external K concentration and ouabain-sensitive K influx for LK goat red cells that had been treated with anti-L serum or with a non-immune serum. The vertical bars show ± 1 S.E.M.

It is interesting that this increase in affinity for Na of the Na-loading sites is not accompanied by a change in the affinity for K of the outward facing pump sites. Fig. 4 shows the magnitude of the ouabain-sensitive K influx when control cells and anti-L-treated cells were incubated in media containing different concentrations of K. It is clear that the sensitivity to K in the medium is not appreciably affected by treatment with the antibody. Similar results have been found with sheep red cells [9, 10].

Stimulation of enzymes by specific antibodies

The experiment of Ellory and Tucker [7] which demonstrated the stimulating effect of anti-L serum on the Na pump in LK sheep red cells was prompted by an earlier observation of Tucker [13] that the K level in the red cells of LK lambs was high at birth, and fell gradually over the first six weeks of life with a time course similar to that of the rise in the level of M antigen in the red cells of HK lambs. This observation led her to suggest that the product of an alternative allele to the gene controlling the M antigen (later detected serologically and called 'L') might inhibit the pump, and to wonder whether an antibody to this product would stimulate the pump by reversing the inhibition. Although the stimulation her hypothesis predicted has been found to occur, the hypothesis must be reconciled with three more recent discoveries. (i) The pump in reticulocytes from LK sheep is not LK in character even though serological tests show that L antigen is present [14]. (ii) Serological tests fail to detect L antigen in LK goat red cells [12]. (iii) After LK sheep red cells are treated with trypsin, the pump is not stimulated by anti-L serum though the serum still lyses the cells if complement is present [15]. The fact that goat cells, in which L antigen cannot be detected serologically, absorb something from anti-L serum which, when later eluted from the cells, is able to stimulate the pump in LK cells from both sheep and goats, yet which is not lytic for goat cells and only weakly lytic for sheep cells, suggests that the anti-L serum contains antibodies to two different antigens (L and L_p) of which only one (L_p) is present in the goat cells [12].

If this is correct, it is possible that stimulation of the pump by anti-L serum could result from the reversal of an inhibitory effect of the L_p antigen. Neutralizing the effect of an inhibitory protein is, however, only one way in which antibodies can stimulate enzymes, and over the last few years several examples of stimulation by antibodies have been described in which the enzyme appears to be directly affected by the antibody. Since this work is to be found mainly in the immunological literature, which is not always widely read by membrane physiologists, it may be helpful to review it briefly here.

The most dramatic example of the stimulation of an enzyme by a specific antibody involves a defective β-galactosidase found in an *E. coli* mutant studied by Rotman and Celada [16]. Rotman and Celada showed that interaction of normal *E. coli* β-galactosidase with its specific antibody, produced in rabbits, did not significantly affect the enzymic activity. When the defective enzyme from the mutant was allowed to react with the antibody to the normal enzyme, however, the enzymic activity, which had been very low, was increased by a factor of between 500 and 1000. Further experiments by Celada, Strom and Bodlund [17] showed that stimulation of the defective enzyme by the antibody followed first-order kinetics even when the molecules of antigen and of antibody were present in similar numbers and when, therefore, second-order kinetics might have been expected. The half-time for the stimulation was about 45 min, which is very long compared with the 500 milliseconds taken for precipitation of the enzyme by the antibody under appropriate conditions (unpublished observations of Antonini and Celada, quoted by Celada et al.). Celada et al. explain both the first-order kinetics and the discrepancy between the times by supposing that the rate-limiting step in stimulation is a slow reaction following the combination of enzyme and antibody and probably involving a configurational change.

Less dramatic examples of stimulation of enzymes were described several years earlier by Cowie, Spiegelman, Roberts and Duerksen [18], working with a ribosome-bound β-galactosidase from *E. coli*, by Okada, Ikenaka, Yagura and Yamamura [19], working with an amylase from *Aspergillus oryzae*, and by both Richmond [20] and Pollock [21] working with penicillinases. Richmond found a four-fold stimulation of a penicillinase from *Staphylococcus aureus* unaccompanied by any change in the affinity of the enzyme for its substrate (private communication quoted by Pollock, 1964). Pollock, working with *B. licheniformis* and methicillin as substrate, found a ten-fold increase in V_{max} but the affinity of the enzyme for its substrate was greatly reduced – by a factor of about 100 – so that at low substrate concentrations the antibody was inhibitory. Further subtleties appeared when penicillinases from a series of mutants of *B. licheniformis* were treated with a rabbit anti-serum to the wild-type enzyme [22]. The mutant enzymes differed from each other, and from the wild-type enzyme, both in the maximum rates of hydrolysis that they catalysed and in their relative affinities for different substrates. The anti-serum altered both the V_{max} and the K_m for each substrate, sometimes very strikingly – the methicillinase activity of one mutant was increased 70-fold. In general, the direction of the alterations was such as to reduce the difference between mutant and wild-type activity, but the attractive notion that, in combining with the antibody to the normal enzyme, the abnormal enzyme was somehow forced into the normal configuration was not supported by experiments with an anti-serum to a mutant enzyme. This anti-serum did not make the wild-type enzyme behave like the mutant enzyme, as might have been expected, but had effects rather like those of the anti-serum to the wild-type enzyme.

All the above examples are from micro organisms, but Cinader and his colleagues have shown that anti-sera from rabbits injected with bovine ribonuclease contain activating as well as inhibitory antibodies [23, 24]. Separation on DEAE Sephadex gave a number of fractions, of which one behaved as if it contained only activating antibody. The activating effect was detected when cyclic cytidylic or uridylic acids were used as substrate; with ribonucleic acid – the normal substrate – no activation was observed but the enzyme was less susceptible to inhibition by inhibitory antibodies. The suggested explanation of this protective effect is that the inhibitory antibody finds its binding sites on the enzyme already occupied by the activating antibody, or conceivably by some other antibody in the activating DEAE Sephadex fraction which does not itself activate or inhibit. The curious lack of activation when the enzyme acts on its normal substrate was not found in some earlier experiments of Cinader [25] in which one of a series of six rabbits injected with a polyalanine polytyrosine derivative of ribonuclease produced antibodies which activated both the normal and the modified ribonuclease, whether ribonucleic acid or cyclic cytidylic or uridylic acid was the substrate.

Most of those who have discussed the mechanism of stimulation of enzymes by antibodies suggest that configurational changes are responsible for the increased enzymic activity. This is probably correct, but it must be admitted that in no case is there any direct evidence for it.

Inhibition of the Na pump in LK goat red cells by intracellular K

We saw earlier that the data of Lauf et al. [10] suggest that in LK sheep red cells intracellular K ions are strongly inhibitory. To investigate this inhibition further we measured the ouabain-sensitive ATPase activity of ghosts from LK goat red cells as a function of Na concentration with K fixed at either 10 mM or 50 mM. Even the lower K concentration was sufficient to saturate the outward facing sites of the pump (see fig. 4), so again it is fair to

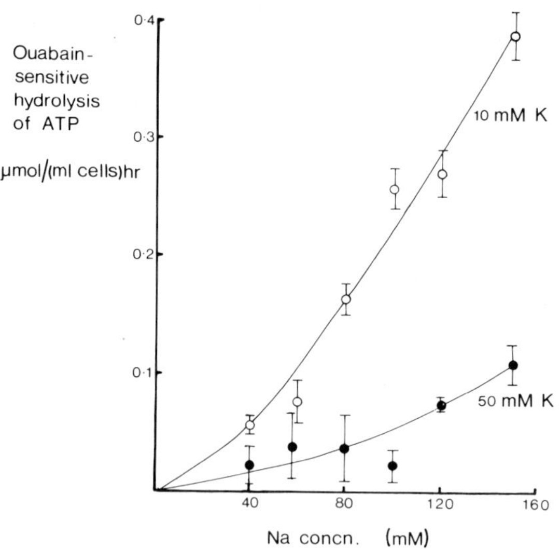

Fig. 5. An experiment showing the powerful inhibitory effect of K ions on the ouabain-sensitive ATPase activity of ghosts from LK goat red cells. The vertical bars show ± 1 S.E.M.

Table 1
Inhibition by K of the ouabain-sensitive hydrolysis of ATP by ghosts prepared from LK goat red cells

Concn. in medium		Ouabain-sensitive hydrolysis of ATP (mmole/(l·orig. cells·hr)	Hydrolysis in high K / Hydrolysis in low K	Minimum possible value of K_m/K_i assuming competition at:	
Na (mM)	K (mM)			1 site	3 sites
150	10	0.388 ± 0.020	0.281	87	3.2
150	40	0.109 ± 0.016			
120	10	0.270 ± 0.020	0.278	77	2.6
120	40	0.075 ± 0.006			

assume that the effects we observed were produced at the inner surface. The results of this experiment, which are illustrated in fig. 5, make it clear that K ions inhibit strongly, but the experiment was rather unsatisfactory because there was a good deal of scatter particularly at the low Na concentrations. In further experiments we made more careful measurements of ouabain-sensitive

Table 2
The effect of pretreatment with anti-L serum on the inhibition by K of the ouabain-sensitive hydrolysis of ATP by ghosts prepared from LK goat red cells

	Concn. in medium		Ouabain-sensitive hydrolysis of ATP (mmole/(l·orig. cells·hr)	Hydrolysis in high K / Hydrolysis in low K	Minimum possible value of K_m/K_i assuming competition at:	
	Na (mM)	K (mM)			1 site	3 sites
Control cells	120	10	0.309 ± 0.011	0.275	87	2.6
	120	40	0.085 ± 0.014			
	90	10	0.227 ± 0.006	0.101*	impossible	2.1*
	90	70	0.023 ± 0.011			
Anti-L treated cells	120	10	0.755 ± 0.006	0.435	9.1	1.4
	120	40	0.328 ± 0.012			
	90	10	0.581 ± 0.014	0.234	10.8	1.0
	90	70	0.165 ± 0.010			

* Figures unreliable because of the relatively large experimental error in the estimate of hydrolysis in the 70 mM-K medium.

ATPase activity at two different K concentrations at just one or two Na concentrations. The results of experiments of this kind are summarized in table 1 and 2, but before they are examined something must be said about their analysis.

Ideally, one would like to get an idea of the absolute magnitudes of the apparent affinities of the Na-loading sites for both Na and K. Unfortunately, because the curves of activity against Na concentration show no tendency to saturate even at the highest concentrations we can reach without making the solutions hypertonic, we are not able to get any estimate of V_{max}. What we can do, however, is to compare the rates of ouabain-sensitive hydrolysis of ATP under two different sets of conditions; then, if we take the ratio of those hydrolysis rates, V_{max} cancels from our equations. Suppose, for example, we measure the rate of ouabain-sensitive hydrolysis at two different K concentrations, i_1 and i_2, at the same Na concentration s. If Na and K compete then the fraction of sites occupied by Na will be $s/[K_m(1 + i_1/K_i) + s]$ when the K concentration is i_1, and $s/[K_m(1 + i_2/K_i) + s]$ when the K concentration is i_2.

If each pump has only a single Na-loading site, the ratio (R) of hydrolysis rates will be the same as the ratio (Z) of the number of sites loaded with Na, and we can therefore write

$$R_{\text{(single site)}} = Z = \frac{K_m + K_m i_2/K_i + s}{K_m + K_m i_1/K_i + s}.$$

If each pump has three Na-loading sites each of which must be loaded with Na for the pump to work, the ratio of hydrolysis rates will be given by the cube of the ratio of the number of sites carrying Na – assuming for simplicity that all the sites are identical. We cannot solve these equations to get values for K_m and K_i, nor can we get a value for the ratio K_m/K_i, that is to say the ratio of the affinities for K and for Na. But for any given value of Z the ratio K_m/K_i must increase as K_m tends to zero. We therefore obtain a minimum figure for K_m/K_i by putting K_m equal to zero and solving the equation. For the single site case:

$$\left(\frac{K_m}{K_i}\right)_{min} = \frac{s(1-R)}{i_1 R - i_2}$$

and for the case where three sites have to be occupied by sodium simultaneously:

$$\left(\frac{K_m}{K_i}\right)_{min} = \frac{s(1-\sqrt[3]{R})}{i_1\sqrt[3]{R} - i_2}.$$

If the three sites are not identical, the expression just calculated gives a minimum value of the affinity ratio at the site at which potassium competes most effectively. (It would, of course, be possible to get a complete solution of the Michaelis equation if we had accurate data under three sets of conditions; but since the number of sites at which competition occurs is unknown, and it is unlikely that the sites are identical, more detailed investigation has not seemed worthwhile.)

Table 1 shows the results of an experiment in which ouabain-sensitive hydrolysis of ATP was measured at two different K concentrations – 10 mM and 40 mM – with the Na concentration fixed at either 150 mM or 120 mM. The ratio of hydrolysis rates has been calculated for each pair of results, and the last two columns in the table give the minimum values for the affinity ratio assuming competition at one site or competition at three sites. It is clear that if competition is at one site the preference for K over Na has to be very marked indeed, and even if competition is at three sites K must be preferred to Na by a factor which cannot be less than about 3. These calculations assume, of course, that the K ions inhibit by competing with Na ions for the Na-loading sites, and they also assume that the combination of Na ions with the sites is sufficiently rapid compared with subsequent reactions for equilibrium kinetics to be applicable.

Table 2 shows the results of a similar experiment in which we have also looked at the effect of anti-L serum. For the control cells the results were very similar to those of table 1. For the cells treated with anti-L serum, the ratio of hydrolysis rates was less and, correspondingly, the minimum values for the affinity ratio are also less, though even for the three-site case the affinity for K must be at least as great as the affinity for Na. This effect of anti-L serum is consistent with our earlier finding that anti-L serum increased the apparent affinity for Na. What we cannot say is whether the increase in the apparent affinity for Na is the result of an increase in real affinity for Na or a decrease in the affinity for K, or both.

The very powerful inhibition of the Na pump by intracellular K demonstrated in these experiments is unusual and seems to be a kind of behaviour peculiar to LK sheep and goat red cells. Unless the intracellular K concentration is kept low, the Na pump will be inhibited, so that the level of K rather than the level of Na or the supply of energy is likely to be the factor controlling the rate of Na pumping in these cells. It would be interesting to know whether an effect of this kind is at all responsible for the low level of K in any other cells, e.g. cat and dog red cells.

We do not, of course, know how the K ions inhibit, but competition with Na ions seems much the most likely explanation; and if that is the explanation, then, as we have seen, we are led to the further conclusion that the inward facing sites – the Na-loading sites – have a higher affinity for K than for Na. That is a bizarre conclusion but not an impossible one. For however much the internal sites prefer K, the pump will still work as a Na pump provided that the machinery operates only when the sites are loaded with Na.

For the time being, the conclusion that in some cells the Na-loading sites prefer K to Na may be more amusing than important; but it would be very satisfying if one could relate this peculiarity in ion selectivity to a particular molecular structure. When methods of purifying the pump ATPase have been further improved it will be tempting to look for evidence of amino-acid substitution, though if Dunham and Hoffman [26] are right in thinking that there are only half a dozen pump sites per LK cell the protein chemists are going to have to work quite hard to get material for their analyses. Without looking so far ahead, the notion that the Na-loading sites prefer K may be helpful by reminding us that not everything depends on Eisenman's 11 sequences. The enormous success of those sequences at accounting for ion preferences in biological as well as in inanimate systems (see Diamond and Wright [27]) sometimes makes us forget that they are concerned only with affinities. But reactivity can matter more than affinity. Most of us are dark haired because, though "gentlemen prefer blondes" [28], "gentlemen marry brunettes" [29].

Acknowledgements

We are grateful to Professor W.A.H. Rushton, F.R.S., Dr. V.L. Lew, Dr. P.J. Garrahan and Dr. E.M. Tucker for helpful discussion, and to Dr. Tucker also for supplying the anti-L serum.

References

[1] J.V. Evans, Nature (London) 174 (1954) 931.
[2] J.V. Evans and A.T. Phillipson, J. Physiol. (London) 139 (1957) 87.
[3] M.D. Pandey and A. Roy, Current Sci. 9 (1968) 256.
[4] J.C. Ellory and E.M. Tucker, J. Agr. Sci. 74 (1970) 595.
[5] E. Baker and W.J. Simmonds, Biochim. Biophys. Acta 126 (1966) 492.
[6] J.V. Evans, J.W.B. King, B.L. Cohen, H. Harris and F.L. Warren, Nature (London) 178 (1956) 849.
[7] J.C. Ellory and E.M. Tucker, Nature (London) 22 (1969) 477.

[8] P.K. Lauf, B.A. Rasmusen, P.G. Hoffman, P.B. Dunham, P.H. Cook, M.L. Parmelee and D.C. Tosteson, Third Int. Biophys. Cong. of the Int. Union for Pure and Applied Biophys., Cambridge, Mass. (1969) Abstracts p. 71.
[9] J.C. Ellory and E.M. Tucker, in: Permeability and Function of Biological Membranes, eds. L. Bolis, A. Katchalsky, R.D. Keynes, W.R. Loewenstein and B.A. Pethica (North-Holland, Amsterdam, 1970) p. 120.
[10] P.K. Lauf, B.A. Rasmusen, P.G. Hoffman, P.B. Dunham, P.H. Cook, M.L. Parmelee and D.C. Tosteson, J. Membrane Biol. 3 (1970) 1.
[11] P.J. Garrahan and A.F. Rega, J. Physiol. (London) 193 (1967) 459.
[12] J.C. Ellory and E.M. Tucker, Biochim. Biophys. Acta 219 (1970) 160.
[13] E.M. Tucker, Proc. 11th European Conference on Animal Blood Groups and Biochemical Polymorphism, Warsaw (1968) p. 483.
[14] J.C. Ellory and E.M. Tucker, J. Physiol. (London) 208 (1970) 18P.
[15] P.K. Lauf, M.L. Parmelee, J.J. Snyder and D.C. Tosteson, J. Membrane Biol. 4 (1971) 52.
[16] M.B. Rotman and F. Celada, Proc. Nat. Acad. Sci. 60 (1968) 660.
[17] F. Celada, R. Strom and K. Bodlund, Antibody mediated activation of defective β-galactosidase (AMEF). Characteristics of binding and activation processes, in: The Lactose Operon, eds. J. Beckwith and D. Zipzer (Cold Spring Harbor Laboratory publishing house, 1970) pp. 291–303.
[18] D.B. Cowie, S. Spiegelman, R.B. Roberts and J.D. Duerksen, Proc. Nat. Acad. Sci. 47 (1961) 114.
[19] Y. Okada, T. Ikenaka, T. Yagura and Y. Yamamura, J. Biochem. (Tokyo) 54 (1963) 101.
[20] M.H. Richmond, Biochem. J. 88 (1963) 452.
[21] M.R. Pollock, Immunology 7 (1964) 707.
[22] M.R. Pollock, J. Fleming and S. Petrie, in: Antibodies to Biologically Active Molecules, ed. B. Cinader (Pergamon, New York, 1966) p. 139.
[23] T. Suzuki, H. Pelichová and B. Cinader, J. Immuno. 103 (1969) 1366.
[24] H. Pelichová, T. Suzuki and B. Cinader, J. Immunol. 104 (1970) 195.
[25] B. Cinader, in: Antibodies to Biologically Active Molecules, ed. B. Cinader (Pergamon, New York, 1967) p. 85.
[26] P.B. Dunham and J.F. Hoffman, Fed. Proc. 28 (1969) 339.
[27] J. Diamond and E.M. Wright, Ann Rev. Physiol. 31 (1969) 581.
[28] A. Loos, Gentlemen Prefer Blondes (Liveright, New York, 1925).
[29] A. Loos, But – Gentlemen Marry Brunettes (Liveright, New York, 1928).

EFFECTS OF A SUBSTRATE OF THE MEMBRANE PHOSPHATASE ON CATION MOVEMENTS IN HUMAN RED CELLS

P.J. GARRAHAN and A.F. REGA

Departamento de Quimica Biologica, Facultad de Farmacia y Bioquimica, Universidad de Buenos Aires, Buenos Aires, Argentina

Current evidence suggests that hydrolysis of ATP by the cell membrane cation transport system involves a sodium-dependent phosphorylation of an intermediate followed by a potassium-activated release of inorganic phosphate (see Glynn [1]). It has been suggested that the K-activated and glycoside-sensitive phosphatase activity found in most cell membranes is the expression of the ability of the cation transport system to hydrolyze other phosphate esters apart from the phosphorylated intermediate formed from ATP [1].

The red cell membrane is highly permeable to *p*-nitrophenylphosphate (*p*-NPP) one of the substrates of the phosphatase [2]. This property is illustrated on fig. 1 which shows that when red cells are incubated in solutions containing from 17 to 70 mM *p*-NPP, intracellular *p*-NPP builds up in cell water reaching with a half time of 3 min a steady value of about one fourth the extracellular concentration. This distribution ratio is to be expected for a divalent anion in Donnan equilibrium. It is clear therefore that *p*-NPP added to the incubation media of intact red cells will have quick access to the active centre of the membrane phosphatase which is located at the inner surface of the cell membrane [2]. This finding allowed us to test whether active sodium and rubidium movements in human red cells are modified by *p*-NPP in a way consistent with the proposed role of the phosphatase in the active transport system. The use of rubidium seems justified since it is as effective as potassium for both, active transport [3] and membrane phosphatase activity [4]. In all experiments 70 mM *p*-NPP was used in order to obtain, at equilibrium,

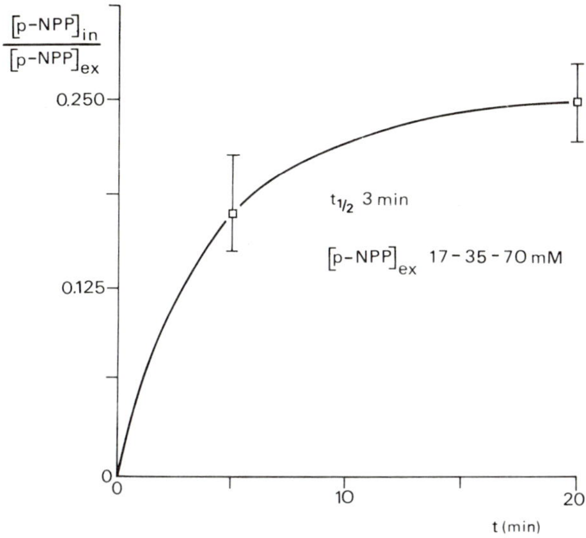

Fig. 1. *p*-Nitrophenylphosphate uptake in red cells incubated in media containing different amounts of *p*-NPP. For the composition of the solutions and other details see legends to tables.

an intracellular concentration of *p*-NPP near the K_m of the membrane phosphatase.

Table 1 shows that when *p*-NPP is added to ATP-free cells there is an increase in both Na and Rb fluxes but that this increase is not prevented by ouabain in concentrations high enough as to fully inhibit both active transport and membrane phosphatase activity. The lack of effect of *p*-NPP does not seem to depend on the free energy necessary to drive active transport since no ouabain-sensistive sodium efflux is measurable even when the adverse gradients for transport are cancelled (table 1, expt. 2). These results are at variance with earlier observations by Askari and Rao [5] and suggest that under our conditions hydrolysis of *p*-NPP is unable to energize active transport.

Being *p*-NPP unabie to energize active transport it follows that *p*-NPP should inhibit this phenomenon in ATP-containing cells if it is hydrolyzed through the pathway used by the phosphorylated intermediate formed from ATP. In agreement with this prediction table 2 shows that partial inhibition of ouabain-sensitive cation fluxes can be achieved with 70 mM extracellular *p*-NPP. The inhibitory effect of *p*-NPP does not seem to be due to the accumulation of the products of its hydrolysis. In fact even though at the end of the incubation the cells treated with *p*-NPP had 4 m-mole/litre cells more

Table 1
Effect of *p*-NPP on sodium and rubidium fluxes in depleted human red cells

Expt. no.	Depletion time (hr)	*p*-NPP concentration (mM)	Fluxes		
			Control	Control + ouabain (mmole/l cell/hr)	Ouabain-sensitive
1	24	0	Na efflux 0.54	0.54	0.00
		70	Na efflux 1.26	1.29	−0.03
		0	Rb efflux 1.78	1.61	0.17
		70	Rb efflux 3.72	3.74	−0.02
2	24	0	Na efflux 0.85	0.82	0.03
		70	Na efflux 2.36	2.41	−0.05

Cells were depleted by incubation at 37°C in a medium containing (mM): NaCl 40, KCl 100, $MgCl_2$ 1, Tris-PO_4H_3 2.5, Tris-HCl 70, bovine serum albumin 5% w/v, Na penicillin G 100 000 U/100 ml, L streptomycin sulfate 0.1 g/100 ml, *p*H 7.8 at 37°C. Sodium efflux was measured using ^{22}NaCl during a 20 min incubation at 37°C in media containing (mH): KCl 20, NaCl 10, $MgCl_2$ 1, Choline chloride 90, Tris-PO_4H_3 2.5, Tris-HCl 70, *p*H 7.8 at 37°C. In experiment No. 2 all the choline chloride and NaCl was replaced by an equivalent amount of KCl. Rubidium influx was measured using ^{86}RbCl in media containing (mM): RbCl 20, NaCl 10, $MgCl_2$ 1, choline chloride 90, Tris-PO_4H_3 2.5, Tris-HCl 70, *p*H 7.8 at 37°C. When present, 70 mM *p*-NPP (Tris salt) replaced all the Tris-HCl. Ouabain-sensitive fluxes are the differences between the fluxes in the presence and in the absence of 10^{-3} M ouabain. Each experiment was performed simultaneously on the same batch of cells.

Table 2
Effect of *p*-NPP on ouabain-sensitive cation fluxes in ATP-containing human red cells

Expt. no.	Depletion time (hr)	Flux	Ouabain-sensitive (mmole/l cell/hr)		Inhibition by *p*-NPP (%)
			0 mM *p*-NPP	70 mM *p*-NPP	
1	0	Na efflux	1.18	0.92	22
	4	Na efflux	0.41	0.16	61
2	4	Na efflux	1.07	0.55	49
	4	Rb influx	0.70	0.28	60

Each experiment was performed simultaneously on the same batch of cell. Other conditions as in table 1.

Table 3
Effect of *p*-NPP on ouabain-sensitive sodium efflux in human red cells in the presence and in the absence of ethacrynic acid

Additions	Ouabain-sensitive Na efflux (mmole/l cell/hr)	Inhibition by *p*-NPP (%)	Final intracellular *p*-NPP concentration (mmole/l cell water)
None	0.66	–	–
70 mM *p*-NPP	0.29	56	17
1 mM Ethacrynic Acid	0.68	–	–
70 mM *p*-NPP + 1 mM Ethacrynic Acid	0.58	15	6

Cells were depleted during 4 h. Intracellular *p*-NPP was estimated at the end of the incubation time from the amount of nitrophenol released, after 2 h treatment at 100°C with 2 N HCl, by the trichloroacetic acid-soluble fraction of a pellet of red cells. The concentration of *p*-NPP in the intracellular water was calculated assuming a cell water content of 65% v/v. The concentration of *p*-NPP measured in the control experiment is very close to that to be expected for a divalent anion assuming Donnan equilibrium. Other experimental conditions were similar to those in table 1.

inorganic phosphate than the *p*-NPP free cells, control experiments showed that when this increase whas prevented by 15 mM inosine, inhibition by *p*-NPP was unmodified. Furthermore neither inorganic phosphate nor nitrophenol in concentrations similar to those released from *p*-NPP had any effect on cation fluxes.

Table 2 also shows that the inhibitory effect of *p*-NPP is strongly enhanced if the cells are preincubated in a substrate-free medium suggesting that inhibition by *p*-NPP depends on the intracellular concentration of ATP. This result is to be expected if *p*-NPP inhibited active transport by combining at the active centre of the phosphatase since it is known that ATP lowers the apparent affinity of the phosphatase for *p*-NPP [6].

Ethacrynic acid significantly reduces the rate of entry of *p*-NPP into red cells [7]. Results in table 3 make clear that: 1) under the conditions tested, ethacrynic acid has little effect on the ouabain sensitive sodium efflux in the absence of *p*-NPP, ii) the reduction in intracellular *p*-NPP by ethacrynic acid is associated with a proportional decrease in the inhibitory effect of *p*-NPP. This result suggests that the site for the inhibitory effect of *p*-NPP is located, like the active site of the phosphatase, at the inner surface of the cell membrane. Table 3 also shows that 17 mM *p*-NPP reduces by about one half the ouabain sensistive sodium efflux. This effect is near to the one expected from the interaction of *p*-NPP with the membrane phosphatase whose K_m for *p*-NPP in the presence of 0.25 mM ATP is around 22 mM [8].

Acknowledgements

Part of the results presented here have been published elsewhere [9]. We acknowledge grants from the Consejo Nacional de Investigaciones Científicas y Técnicas (C.N.I.C.T.) Argentina. The authors are established investigators from the C.N.I.C.T.

References

[1] I.M. Glynn, Brit. Med. Bull. 24 (1968) 165.
[2] P.J. Garrahan, M.I. Pouchan and A.F. Rega, J. Physiol. 202 (1969) 305.
[3] R. Whittam and M.E. Ager, Biochem. J. 93 (1964) 337.
[4] A.M. Vigliocco, (personal communication).
[5] A. Askari and S.N. Rao, Biochem. Biophys, Res. Commun. 36 (1969) 631.
[6] P.J. Garrahan, M.I. Pouchan and A.F. Rega, J. Membrane Biol. 3 (1970) 26.
[7] P.J. Garrahan and A.F. Rega (unpublished data).
[8] M.I. Pouchan, P.J. Garrahan and A.F. Rega, Biochim. Biophys. Acta 173 (1969) 151.
[9] P.J. Garrahan and A.F. Rega, Nature (New Biology) 232 (1971) 24.

ASYMMETRICAL PROPERTIES OF THE RED CELL MEMBRANE K-ACTIVATED PHOSPHATASE

A.F. REGA and P.J. GARRAHAN
*Departamento de Quimica Biologica,
Facultad de Farmacia y Bioquimica,
Universidad de Buenos Aires, Argentina*

In recent years, considerable evidence has been accumulated suggesting that the K-activated and ouabain-sensitive phosphatase activity present in most cell membranes is a part of the (Na + K)-ATPase system. This report concerns the results of experiments performed to test the asymmetrical properties of the K-activated phosphatase from human red cells.

If the participation of the K-activated phosphatase in the (Na + K)-ATPase system is assumed to be true, it can be predicted that the activation of the phosphatase by K will take place, as in the (Na + K)-ATPase system, only when K is at the external surface of the cell membrane. To test this point we prepared reconstituted ghosts [1] with different ionic composition. The ghosts where incubated at 37° in isotonic media containing *p*-nitrophenylphosphate (*p*-NPP). Phosphatase activity was estimated from the amount of nitrophenol released from the ghosts. We had previously found that the red cell membrane is highly permeable to *p*-NPP, thus *p*-NPP added to the incubation media will have quick access to the active centre of the phosphatase which is located on the inner surface of the cell membrane [2].

Results in fig. 1 show that the ouabain-sensitive phosphatase activity of K-rich ghosts is largely abolished when the ghosts are transferred from a medium containing K to a K-free medium. Although this result shows that external K is nessesary for full activation it does not rule out the possibility that K is required at both sides of the membrane.

To test this point, reconstituted ghosts free of K(Na-containing ghosts) where assayed for ouabain-sensitive phosphatase activity. A large ouabain-sensity activity can be detected when K-free ghosts are suspended in a K-rich medium. The lack of effect of intracellular K is further shown by the fact

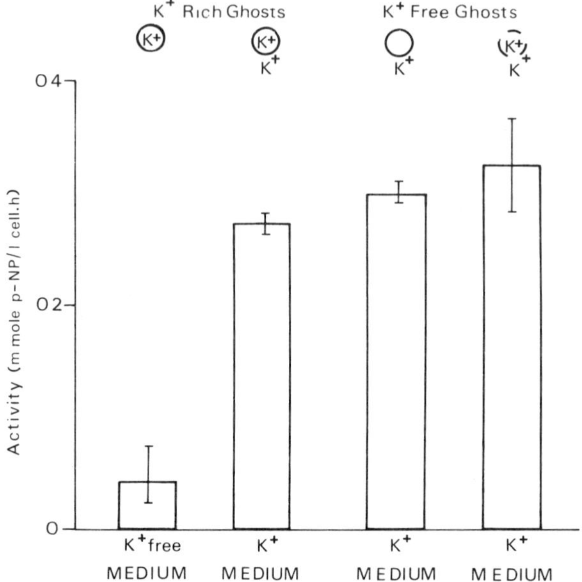

Fig. 1. Ouabain-sensitive phosphatase activity in reconstituted ghosts and in ghosts fragmented by freezing and thawing (first colum from the right). The composition of the "K rich ghosts" before sealing was (mmole/liter) : KCl 50; $MgCl_2$ 5; NaCl 10; Tris-HCl 85, *p*H 7.8. The composition of the "K free ghosts" before sealing was (mmole/liter): NaCl 110; $MgCl_2$ 5; Tris-HCl 40, *p*H 7.8. The "K Medium" contained (mmole/liter): KCl 50; choline-Cl 90; $MgCl_2$ 10; Tris-HCl 10 (*p*H 7.8). The "K free medium" was similar to the "K medium" except that sodium replaced all the potassium. *p*-NPP (6.3 mmole/liter) was present as substrate. The ghosts suspensions were incubated at 37°C for 30 min. In all experiments no less than 60 percent of the ghosts recovered their low permeability to cations. The vertical lines represent the range of three experiments.

that the activity of the same ghosts remains unchanged after freezing and thawing, although K should now have access to both sides of the membrane due to disruption of the permeability barrier.

We have shown elsewere [3] that in the presence of ATP, adequate amounts of Na lower the concentration of K giving half-maximal activation of the membrane phosphatase. This effect has been related to the participation of the K-activated phosphatase in the overall (Na + K)-ATPase reaction. If this were true from the known asymmetric requirements of the (Na + K)-ATPase system, the ATP-dependent effect of Na should require both ATP and Na at the inner surface of the cell membrane. This prediction was tested

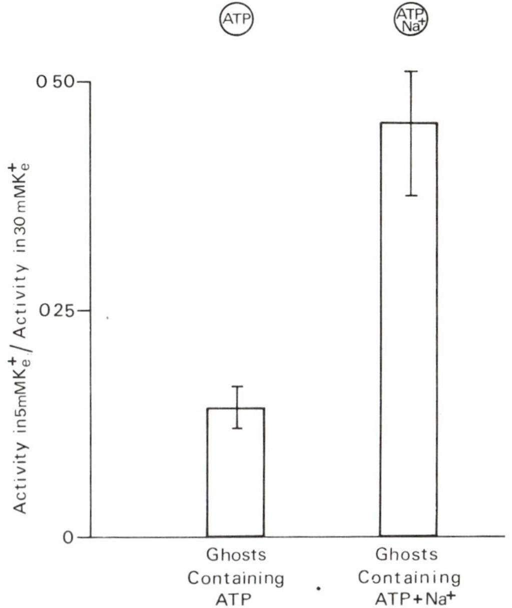

Fig. 2. K-dependent phosphatase activity in resealed ghosts containing either ATP or ATP plus Na. The composition of the ghosts containing ATP before sealing was (mmole/liter): ATP 0.75; choline-Cl 110; $MgCl_2$ 5; Tris-HCl 40; *p*H 7.8. In the ghosts containing ATP plus Na, 20 mmole of NaCl replaced an equivalent amount of choline-Cl. The 30 mM K medium contained (mmole/liter): KCl 30; choline-Cl 80; $MgCl_2$ 10; Tris-HCl 40; *p*H 7.8. In the 5 mM K medium, 25 mM KCl was replaced by an equivalent amount of choline-Cl. The K-dependent phosphatase activity is the difference between the activities in the above-mentioned media and the activities in a medium in which all the KCl was replaced by choline-Cl. For each experiment, the ghosts containing ATP and the ghosts containing ATP plus Na were obtained from a single hemolyzate. The vertical lines represent the range of three experiments.

measuring K-dependent phosphatase activity in ghosts containing either ATP or ATP plus Na, suspended in media with either 5 or 30 mM K.

Results in fig. 2 show that addition of intracellular Na to ATP-containing ghosts leads to a large increase in the ratio of activities in the 5 mM K medium to the activity in the 30 mM K medium. Since it is known that Na does not change the maximum rate of K-dependent phosphatase activity in the presence of ATP [4] the observed increase in the ratio must be ascribed to an increase in the apparent affinity for K induced by ATP and Na from the internal surface of the membrane.

The results presented here suggest that the membrane phosphatase has a definite orientation within the membrane structure. The similitude of the asymmetrical requirements of the membrane phosphatase with those of the (Na + K)-ATPase system strongly support the idea that both enzymic activities are related.

Acknowledgements

A full report of the results presented here has been published elsewhere [5]. We acknowledge grants from the Consejo Nacional de Investigaciones Cientificas y Tecnicas (C.N.I.C.T.) Argentina,. The authors are established investigators from the C.N.I.C.T.

References

[1] S. Lepke and H. Passow, J. Gen Physiol. 51 (1968) 3655.
[2] P. Garrahan, M.I. Puchan and A.F. Rega, J. Physiol. 202 (1969) 305.
[3] A.F. Rega, P.J. Garrahan and M.I. Pouchan, Biochim. Biophys. Acta 150 (1968) 742.
[4] P.J. Garrahan, M.I. Puchan and A.F. Rega, J. Membrane Biol. 3 (1970) 26.
[5] A.F. Rega, P.J. Garrahan and M.I. Pouchan, J. Membrane Biol. 3 (1970) 14.

PURIFICATION OF (Na^+ + K^+)-ATPase FROM THE OUTER MEDULLA OF MAMMALIAN KIDNEY

Peter Leth JØRGENSEN
Institute of Physiology, University of Aarhus, Aarhus, Denmark

1. Introduction

In the further study of the molecular basis of the active transport of cations it is important to have access to pure preparations of (Na^+ + K^+)-ATPase [1]. Using pure preparations, information about the molecular structure of the enzyme system and about its localization in the plasma membrane can be obtained. Such information is necessary for the interpretation of kinetic studies and of tracer studies of the intermediary steps in the enzyme reaction [2]. Also studies of the biosynthesis of the enzyme system, of possible conformational changes during the reaction sequence and of the kinetics of the binding of substrate and activators to the system require pure or at least highly concentrated preparations.

In the outer medulla of mammalian kidneys the predominant histological feature is the broad limb of the loop of Henle, which is known to have a large capacity for reabsorption of sodium against a steep electrochemical gradient [3]. In this zone of rat [4] and rabbit kidney [5], the activity of (Na^+ + K^+)-ATPase is 2–4 fold higher than the activity in the outer cortical zone and about 8 fold higher than in the inner medulla or papilla. Using material from the outer medulla of rat [4] and rabbit kidneys [5], preparations with a relatively high specific activity of (Na^+ + K^+)-AtPase can be obtained by a simple cycle of differential centrifugations. An effort was therefore made to purify further enzyme from this source.

2. The source of the tissue

Fig. 1 gives a survey of the procedure used for preparation of highly active $(Na^+ + K^+)$-ATPase from the outer medulla of rabbit kidney. In the present

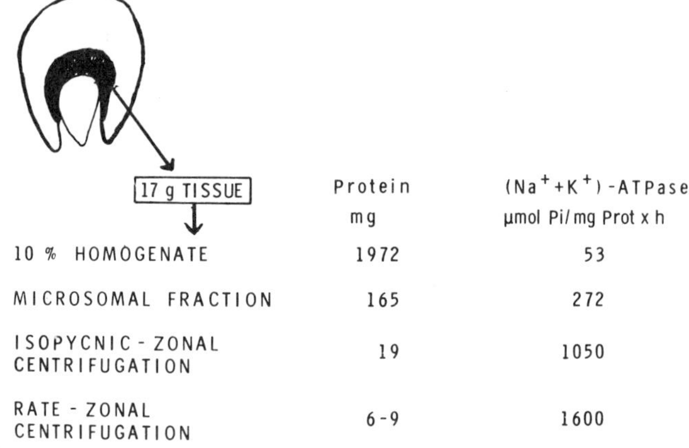

17 g TISSUE	Protein mg	$(Na^+ + K^+)$-ATPase µmol Pi/mg Prot x h
10 % HOMOGENATE	1972	53
MICROSOMAL FRACTION	165	272
ISOPYCNIC - ZONAL CENTRIFUGATION	19	1050
RATE - ZONAL CENTRIFUGATION	6-9	1600

Fig. 1. A summary of the procedure used to obtain highly active $(Na^+ + K^+)$-ATPase from the outer medulla of rabbit kidney. The results are given for a single preparative run using the Ti-15 Beckmann zonal rotor for the isopycnic-zonal centrifugation. The activity of $(Na^+ + K^+)$-ATPase in the homogenate and in the microsomal fraction was measured after incubation with deoxycholate [7]. The microsomal fraction was incubated for 45 min at 20° with 0.6 mg deoxycholate per ml, 2 mM EDTA, 25 mM imidazole, 3% (w/v) sucrose, *p*H 7.0 (20°) in a total volume of 200 ml. This medium was used as a sample for the isopycnic-zonal centrifugation in the Ti-15 zonal rotor [9]. Enzyme and protein analysis were performed as described before [7]. The rate-zonal centrifugation is described in the text.

experiments rabbit kidneys are used. Recent studies show that dog kidney may be used [6] and in our experience pig kidney is also a good source.

3. Incubation of the preparations with detergents

Prior to the isopycnic-zonal centrifugation the microsomal fraction is incubated with a low concentration of sodium deoxycholate. This leads to an increase in the specific activity of $(Na^+ + K^+)$-ATPase from 45 to 270 µmoles Pi per mg protein per hr if the conditions for activation are optimal with respect to temparature, *p*H and concentrations of protein and detergent. A procedure for evaluation of the conditions for maximum activation by deoxycholate has been described [7].

A detailed analysis of the effect of the detergent [7] shows that the existence of micelles in the solution is necessary for the activation. The (Na^+ + K^+)-ATPase activity remains associated with membrane structures after the zonal centrifugation, whereas inactive protein or lipoprotein is removed and rendered soluble by the detergent. Tracer studies show that only insignificant amounts of deoxycholate are bound to the membranes. The data suggest that the activation is due to exposure of latent enzyme sites in the preparation. It is known that the plasma membrane can be transformed into vesicular structures during the homogenization procedure [8]. The (Na^+ + K^+)-ATPase requires an effect of K^+ from one side of the membrane, in the intact cell the outside, and an effect of Na^+, and of Mg^{++} and ATP from the other side, the inside of the membrane [1]. The activation of (Na^+ + K^+)-ATPase can therefore be due to an increase in the permeability of the membranes or to opening of the vesicles resulting in free access of substrate and activators to their respective sites on the membrane.

4. Fractionation by isopycnic-zonal gradient centrifugation

In a series of experiments with the zonal rotor it was found that the particles with which (Na^+ + K^+)-ATPase is associated have a well defined equilibrium density in the zonal gradient, whereas their sedimentation rate varies from 100 to 2000 Svedberg units, suggesting a wide variation in size of these particles. Consequently an isopycnic-zonal gradient centrifugation was chosen as the first step in the fractionation [9].

Fig. 2 shows an isopycnic-zonal gradient centrifugation of the particulate fraction, which contains almost all of the (Na^+ + K^+)-ATPase activity in the homogenate [10]. It is seen that a large part of the (Na^+ + K^+)-ATPase in the tissue from the outer medulla can be collected in a well defined band at a position in the sucrose gradient (1.14–1.15 g/ml), where the contamination by other subcellular structures is relatively low. The specific activity of (Na^+ + K^+)-ATPase in the fractions at the peak is 900–1200 μmoles Pi per mg protein per hr [9] and the yield is 2.5 mg protein per g tissue. The yield of the highly active preparation is lower (1.5 mg protein per g tissue) when the microsomal fraction is used as a sample for the isopycnic-zonal centrifugation, but the output in a single centrifugation is higher. At present the microsomal fraction is therefore used as starting material in the routine preparations.

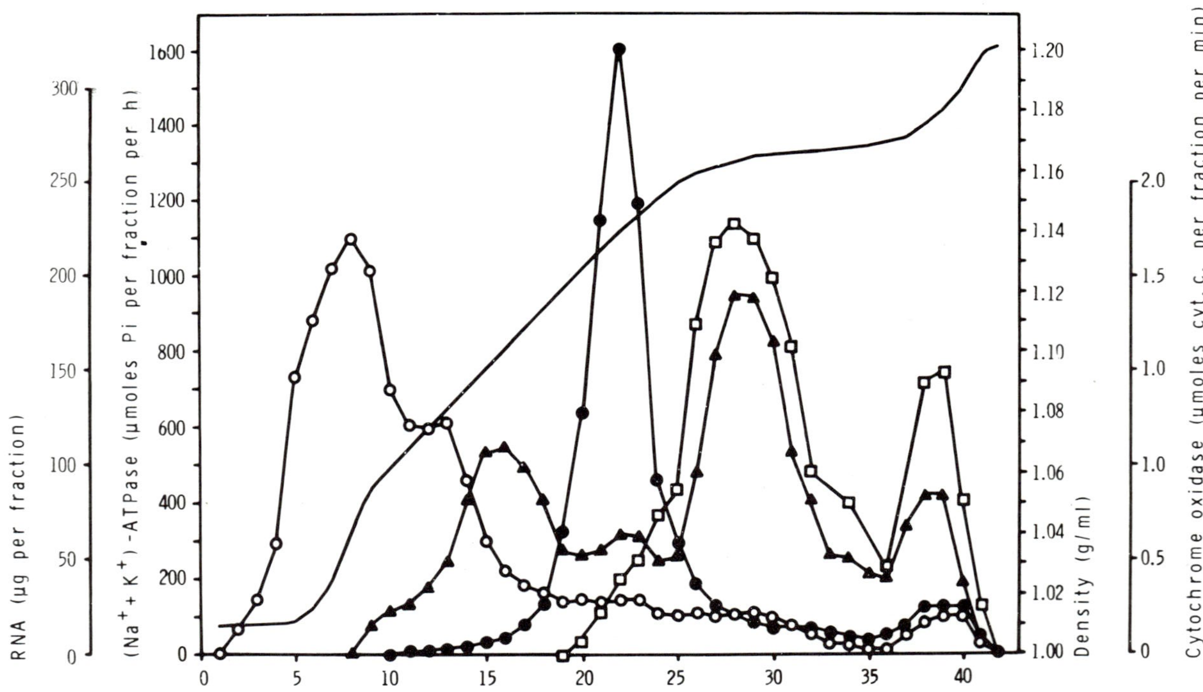

Fig. 2. The distribution of ($Na^+ + K^+$)-ATPase (●–●), RNA (○–○). NADH-cytochrome *c* reductase (▲–▲), and cytochrome oxidase (□–□) in the fractions obtained by isopycnic-zonal centrifugation of the particulate fraction. (——), density of the fractions. The sample contained 72 mg of protein and was prepared as described before [9] from 1.2 g of medullary tissue. Centrifugation at t 40 000 ıev/min for 90 min ($\omega^2 t$ = 10.4 × 10^{10}).

5. Further separation of ($Na^+ + K^+$)-ATPase by rate-zonal centrifugation

The preparation obtained by the isopycnic-zonal centrifugation is contaminated by mitochondria as indicated by the presence of cytochrome oxidase. The mitochondrial fragments and the particles containing the ($Na^+ + K^+$)-ATPase can be separated by rate-zonal centrifugation in a swinging bucket rotor using a gradient ranging from 10% to 30% (w/v) sucrose [9].

Figs. 3–4 illustrate how the rate separation can be performed in the zonal rotor. After the initial rate-zonal centrifugation at a low centrifugal force, the ($Na^+ + K^+$)-ATPase is distributed on 8–10 fractions (fig. 2) and recovery of the enzyme becomes a problem. To solve this, the heavy end of the gradient, containing the mitochondrial fragments is pumped out and replaced by 45% (w/v) sucrose while the rotor is spinning at 3000 rev/min. The contents of the rotor are now recentrifuged at a high centrifugal force and the enzyme is

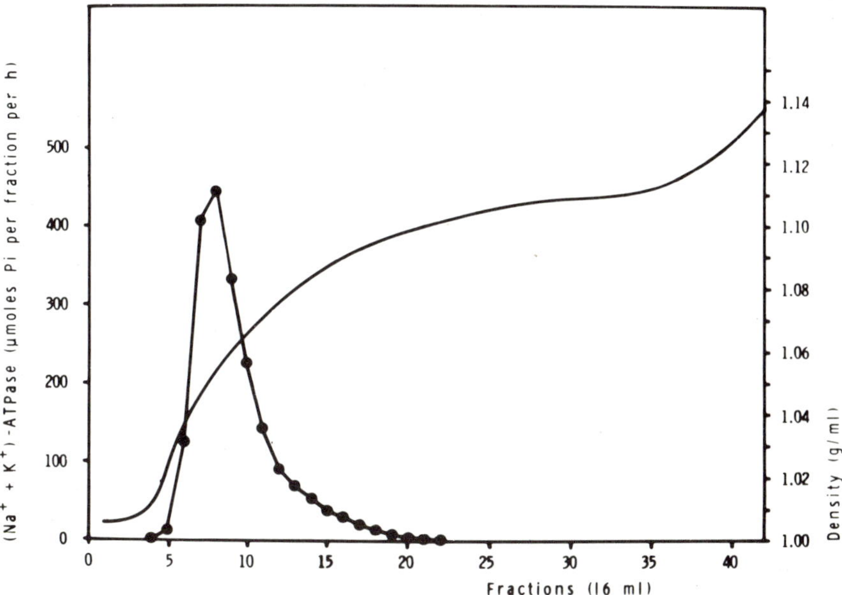

Fig. 3. The distribution of ($Na^+ + K^+$)-ATPase after rate-zonal centrifugation in a gradient ranging from 10% to 30% (w/v) sucrose. The sample was prepared by sedimentation of fractions at the peak of ($Na^+ + K^+$)-ATPase after isopycnic-zonal centrifugation of the microsomal fraction [9]. Centrifugation was for 30 min at 20 000 rev/min ($\omega^2 t = 8.9 \times 10^9$) in the Beckmann Ti-14 rotor. The details of the procedure were as described before [9].

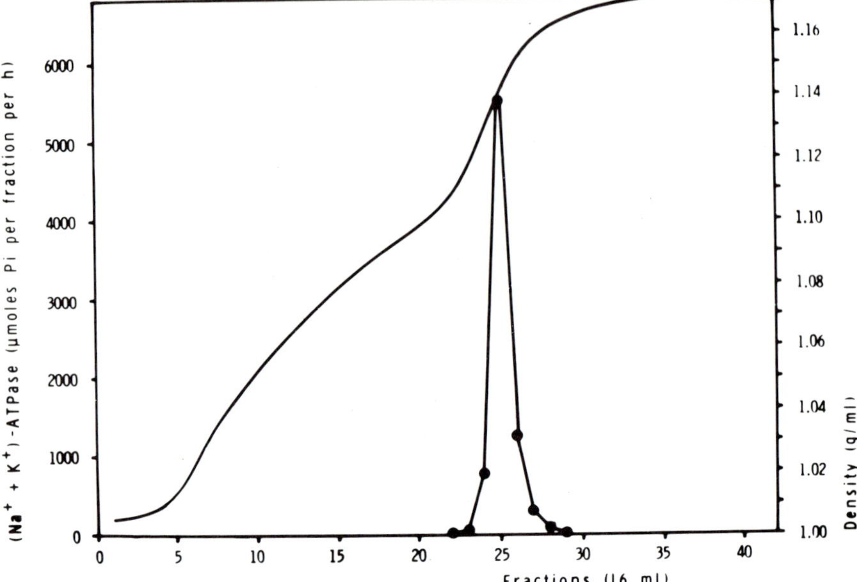

Fig. 4. The distribution of $(Na^+ + K^+)$-ATPase after a combined rate-isopycnic-zonal centrifugation. After an initial centrifugation as described in fig. 3, the rotor was slowed down to 3000 rev/min. 300 ml of the gradient was pumped out through the periphery of the zonal rotor and was replaced by 45% (w/v) sucrose. The rotor was accelerated again and centrifuged for 90 min at 48 000 rev/min ($\omega^2 t = 14.9 \times 10^{10}$).

recovered in one or two fractions. In the preparation obtained by this combined rate-isopycnic-zonal centrifugation, the specific activity of $(Na^+ + K^+)$-ATPase is 1300 to 1500 µmoles Pi per mg protein per hr, and the activity of Mg^{2+}-ATPase is 20–60 µmoles Pi per mg protein per hr.

6. The nature of the preparation

Electron microscopy of the highly active preparations has shown that the particles have the morphological characteristics of plasma membranes [11].

The content of cholesterol and phospholipid is shown in table 1. It is seen that the cholesterol/phospholipid molar ratio is increased to 0.7 in a preparation with a specific a tivity of 1378 µmoles Pi per mg protein per hr. This also suggests that fragments of plasma membranes are isolated [8]. In agreement with the observations on red cell ghosts [15] extraction of lyophilized prepa-

Table 1

The content of phospholipid and cholesterol in the preparations from the outer medulla of rabbit kidney. I is a microsomal fraction. II is obtained by isopycniczonal centrifugation and III by rate-zonal centrifugation. Average values ± S.E. of 4 preparations are shown. Extracts of the fractions were obtained by Folchs method [12]. Phospholipid was determined as inorganic phosphate by the method of Bartlett [13] and cholesterol was estimated using the Tschugaeff reaction in the modification of Hanel and Dam [14].

	($Na^+ + K^+$)-ATPase (μmol Pi/ mg Pr × hr)	Cholesterol (μg/mg protein)	Phospholipid μg/mg protein)	Cholesterol/ phospholipid molar ratio
I	268 ± 12 (4)	135 ± 10 (4)	523 ± 48 (4)	0.53 ± 0.03 (4)
II	934 ± 21 (4)	203 ± 10 (4)	676 ± 12 (4)	0.61 ± 0.03 (4)
III	1378 ± 56 (4)	235 ± 22 (4)	668 ± 37 (4)	0.71 ± 0.09 (4)

rations with dry ether removes 89% of the cholesterol and 34% of the phospholipid. This procedure reduces the activity of ($Na^+ + K^+$)-ATPase only by 10–20%.

Electrophoresis in SDS-polyacrylamide gels [16] show that the large number of protein fractions in the microsomal fraction is reduced considerably by the isopycnic-zonal centrifugation (fig. 5). After rate-zonal centrifugation only two prominent bands are left, one with a molecular weight close to 100 000, the other 50 000 to 60 000. In preparations with a specific activity of 1500–1600, the polypeptide with a molecular weight of about 100 000 forms about two thirds of the total protein.

The high specific activity of ($Na^+ + K^+$)-ATPase in this preparation of membranes can be due either to a high molecular activity of the enzyme or to a high density of enzyme sites per unit membrane area. The first possibility is unlikely because the molecular activity is within the range reported before this enzyme system and because the number of ouabain-binding sites increases in paralles with the specific activity during the preparation [7]. The data therefore suggest that membranes with a high density of ($Na^+ + K^+$)-ATPase or Na^+ pump sites have been isolated. It will be of interest to learn how the relationship is between these membrane fragments and the intact plasma membrane of the cells in the broad limb of Henles loop.

7. Other approaches to purification of ($Na^+ + K^+$)-ATPase

In the present procedure a tissue with a large capacity for active transport of sodium and with a high content of ($Na^+ + K^+$)-ATPase has been selected.

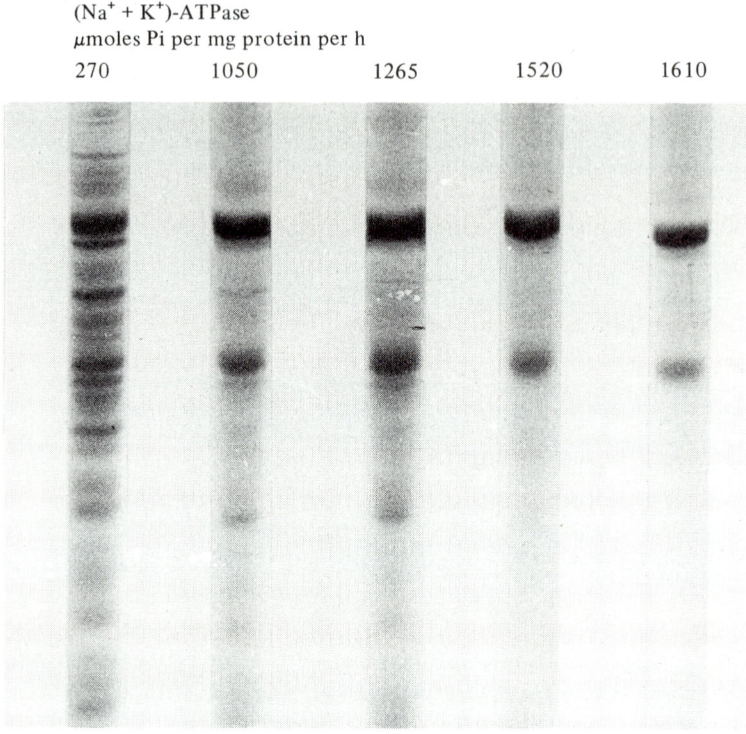

Fig. 5. Electrophoresis in SDS-polyacrilamide gels of the preparations obtained from the outer medulla of rabbit kidney. From the left the preparations are; 1: a microsomal fraction, 2 and 3, preparations obtained by isopycnic-zonal centrifugation, 4 and 5: preparations obtained by rate-zonal centrifugation. The samples contained 20 μg to 40 μg protein. Preparation of the samples and the electrophoresis were performed as described by Weber and Osborn [16].

The subcellular fractions are treated with low concentrations of detergent which do not inactivate $(Na^+ + K^+)$-ATPase. The detergent removes inactive protein and lipid, and membranes with a high density of enzyme sites are isolated by particle separation in inert media. The purity of the preparation with a specific activity of 1500 μmoles Pi per mg protein per hr is estimated to be 50–60% [9].

Soluble preparations of $(Na^+ + K^+)$-ATPase have been obtained by treatment of subcellular fractions from brain and kidney with high concentrations of deoxycholate [17] or Lubrol [18, 19]. After extraction the soluble proteins are fractionated by zonal centrifugation [18], gel chromatography [18,

19] and isoelectric focusing [19]. The most active soluble preparation of ($Na^+ + K^+$)-ATPase has a specific activity of 513 μmoles Pi per mg protein per hr [19].

Uesugi et al. [18] used zonal centrifugation and ammonium sulfate fractionation of Lubrol-extracts from brain and obtained a precipitate with a specific activity of 450–750 μmoles Pi per mg protein per hr. The purity of this preparation was estimated to be about 50%, i.e. nearly the same value as calculated for the present preparation in which the specific activity of ($Na^+ + K^+$)-ATPase is about two times higher. There are two possible explanations for this discrepancy. Either, that the molecular activity of the enzyme extracted with Lubrol is about half that of the membrane-bound enzyme from the outer medulla, or that the Lubrol-extracted preparation contains inactive enzyme units. This problem can be solved only by a critical evaluation of the methods used for determination of the number of enzyme units in the preparations.

References

[1] J.C. Skou, Physiol. Rev. 45 (1965) 596.
[2] J.C. Skou, in: The Molecular Basis of Membrane Function, ed. D.C. Tosteson (Prentice-Hall, New Jersey, 1969) p. 455.
[3] C.W. Gottschalk, Physiologist, 4 (1961) 35.
[4] P.L. Jorgensen, Acta Physiol. Scand. 74 (1968) 11A.
[5] P.L. Jorgensen and J.C. Skou, Biochem. Biophys. Res. Commun. 37 (1969) 39.
[6] J. Kyte, J. Biol. Chem. 246 (1971) 4157.
[7] P.L. Jorgensen and J.C. Skou, Biochim. Biophys. Acta, 233 (1971) 366.
[8] R. Coleman and J.B. Finean, Biochim. Biophys. Acta, 125 (1966) 197.
[9] P.L. Jorgensen, J.C. Skou and L.P. Solomonson, Biochim. Biophys. Acta, 233 (1971) 381.
[10] P.L. Jorgensen, Biochim. Biophys. Acta, 192 (1969) 326.
[11] P.L. Jorgensen and A.B. Maunsbach, in preparation.
[12] J. Folch, M. Lees and G.H. Sloane-Stanley, J. Biol. Chem. 226 (1957) 497.
[13] G.R. Bartlett, J. Biol. Chem. 234 (1955) 466.
[14] H.K. Hanel and H. Dam, Acta Chem. Scand. 9 (1955) 677.
[15] B. Roelofsen, J. de Gier and L.L.M. van Deenen, J. Cellular Comp. Physiol. 63 (1964) 233.
[16] K. Weber and M. Osborn, J. Biol. Chem. 244 (1969) 4406.
[17] D.W. Towle and J.H. Copenhaver, Biochim. Biophys. Acta, 203 (1970) 124.
[18] S. Uesugi, N.C. Dulak, J.F. Dixon, T.D. Hexum, J.L. Dahl, J.F. Perdue and L.E. Hokin, J. Biol. Chem. 246 (1971) 531.
[19] A. Atkinson, A.D. Gatenby and A.G. Lowe, Nature New Biology 233 (1971) 145.

CHLORIDE TRANSPORT IN HUMAN RED CELLS

M. DALMARK, R.B. GUNN, D.C. TOSTESON and J.O. WIETH

Department of Biophysics, University of Copenhagen, Copenhagen, Denmark, and the Department of Physiology and Pharmacology, Duke University Medical Center, Durham, North Carolina, U.S.A.

The purpose of this paper is to reconsider the mechanism responsible for the pronounced permeability of erythrocytes to small inorganic anions. Tosteson [1] showed in 1959 that the self exchange of chloride isotopes at room temperature is a rapid process with a half-time of 0.2 sec. However, it has never been settled whether the rapid chloride exchange occurs by the permeation of free chloride ions, or whether it occurs by the translocation of a neutral complex of a chloride ion with a carrier molecule, which is positively charged in the unloaded state. The second possibility is supported by the findings of an Arrhenius activation energy of 33 kcal/mole for the exchange of chloride ions across the human red cell membrane [2], an unexpected high membrane resistance in *amphiuma* red cells [3], and a *p*H dependence of chloride exchange [4] opposite to that predicted by the fixed charge membrane hypothesis. Together with the evidence that net movements of chloride may be slow – and sometimes rate-limiting for salt transfer [5, 6] – these findings suggest a membrane with an anion carrier, perhaps a lipophilic anion exchange molecule, which transports inorganic anions in much the same way as do amines in artificial membranes [7, 8].

Fig. 1 is an Arrhenius diagram illustrating the pronounced temperature-dependence of chloride exchange. The calculated activation energy of chloride exchange is 33.2 kcal/mole (S.D. 1.2). Fig. 2 shows that both the rate constant and the unidirectional chloride flux, as measured by $^{36}Cl^-$ efflux under conditions of anion equilibrium [4], do not show the increases with increasing hydrogen ion concentration [9], which have been found for divalent anions between *p*H 6.8 and 8.

Aromatic anions appear to reduce chloride transport by non-competitive

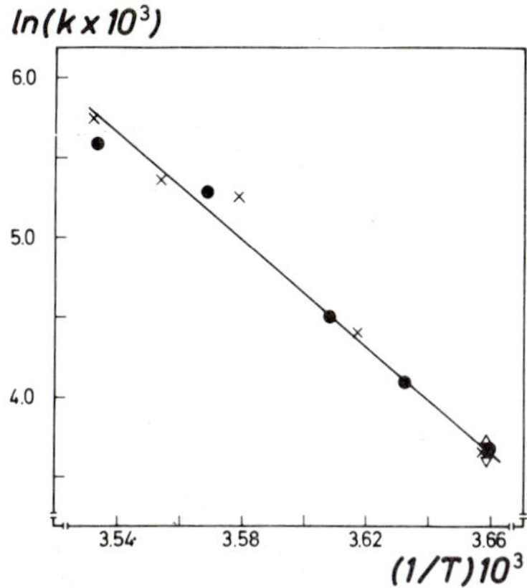

Fig. 1. Arrhenius diagram of the rate of chloride exchange vs temperature between 0 and 10°C [2]. The ordinate is the natural logarithm of ($k \times 10^3$), k being the rate constant of chloride exchange (sec^{-1}), the abscissa is 10^3 times the reciprocal absolute temperature ($1/T$). The slope to the regression line correspond to an Arrhenius activation energy of 33.2 kcal/mole. The rate of chloride exchange was determined by measurements of $^{36}Cl^-$ efflux from red cells under equilibrium conditions where no net flux of chloride occurs across the red cell membrane. The medium with which the cells were equilibrated contained (mM): sodium 142, potassium 3.7, calcium 1.5, magnesium 1, chloride 126.5, bicarbonate 22, phosphate 1.1, and glucose 5. The pH was 7.40 at each experiment.

inhibition. Fig. 3 shows that the high activation energy of chloride exchange is not changed significantly in the presence of the potent aromatic inhibitors picrate and trinitro-cresolate. The similarity between the anion transport in red cells and the anion transport by anion-exchanging amines dissolved in apolar media is underlined by the potent inhibitory action of picrate on both systems [7].

A carrier mediated anion transport may show a greater selectivity than free diffusion in a network of fixed charges where selectivity is based mainly on valence. A carrier molecule which forms complexes with anions may even show a pronounced selectivity between anions with the same hydrated radii and the same mobility in aqueous solutions. This pronounced selectivity has been demonstrated in red blood cells. Thus, the rate of iodide exchange is much less than that of chloride exchange both at room temperature [1] and, as we have found recently, at 0 °C.

The saturation of chloride flux with increasing chloride concentration is an important piece of evidence which would support a carrier model for monovalent inorganic anion transport. Current investigations of red cells in which

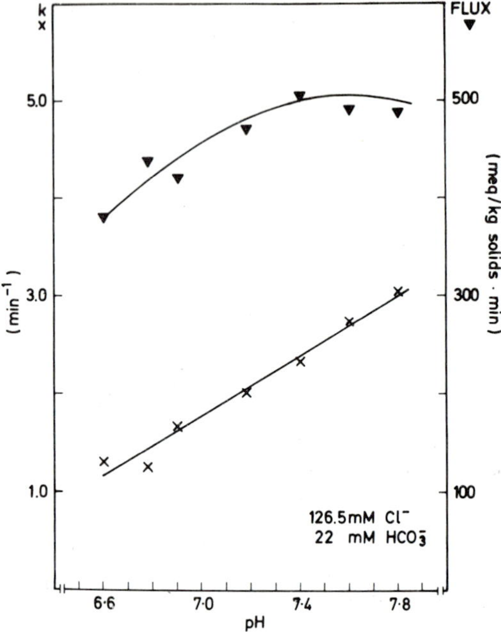

Fig. 2. The *p*H dependence of chloride equilibrium exchange at 0°C [4]. The rate coefficient of chloride exchange × k (min^{-1}) was determined as described in the legend of fig. 1. The chloride flux ▼ was calculated in meq/(kg solids · min) from the rate coefficient of chloride exchange and the intracellular chloride concentration. The variations of *p*H were obtained by titration with carbon dioxide. The bicarbonate buffered electrolyte medium exployed is described in the legend of fig. 1.

the intracellular chloride concentration is varied between 50 to 300 mmoles/l cell water show that the chloride equilibrium exchange flux is constant (about 0.8 mole/kg cell solids per minute in bicarbonate free media at 0 °C and *p*H 7.40), when chloride is the only diffusible anion present in the system. The variations of intracellular chloride concentrations were obtained by a variety of methods: shrinkage or swelling the cells by varying the tonicity of the medium, decreasing the chloride of the cells by increasing the content of 2,3 diphosphoglycerate and increasing the chloride content by equilibrating the cells with various concentrations of ammonium chloride before the flux determinations.

In conclusion, evidence is accumulating that the major mechanism of transport of chloride and other inorganic anions across the human red cell membrane is via a carrier mechanism and not by free diffusion. Further work

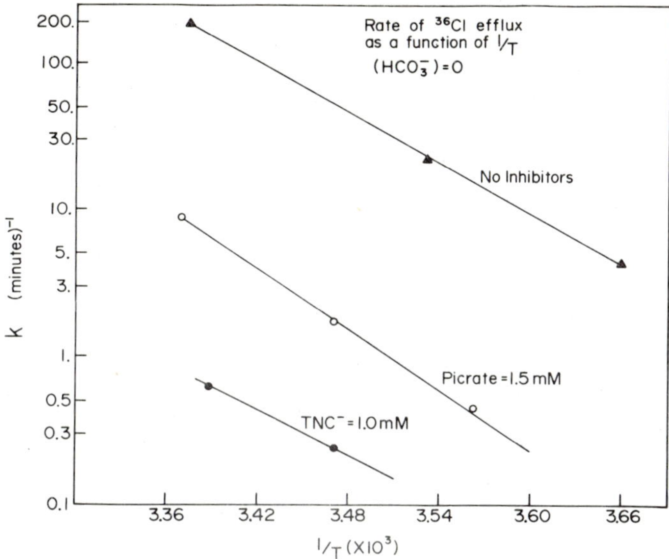

Fig. 3. Arrhenius diagram of the rate of chloride exchange vs temperature in the range 0 to 23°C [10]. The ordinate shows the rate coefficient of chloride exchange (k) on a logarithmic scale, the abscissa is 10^3 times the reciprocal absolute temperature ($1/T$). The rate coefficients of chloride exchange on the absence of inhibitors and in the presence of 1.5 mM picrate, and of 1.0 mM trinitrocresolate (TNC^-) were determined as described in the legend of fig. 1. The bicarbonate free media employed for these experiments were buffered with glycylglycine (27 mM) at pH 7.4 and contained NaCl 150 mM.

is necessary before we can make a reliable estimate of the rate of ionic chloride transport in this system. These considerations have profound implications regarding the relationship between the electrical potential difference and the chloride concentration ratio across the red cell membrane during transient states immediately after changes in the ionic composition of the fluid in which the cells are suspended. However, since there is no evidence of active transport of chloride, the calculation of the membrane potential from the equilibrium distribution of chloride ions between the water phases of cells and medium under steady state conditions, is not precluded by the above considerations.

This work was supported in part by grant HE 12157 from the National Heart and Lung Institute, N.I.H., Bethesda, Md., U.S.A.

References

[1] D.C. Tosteson, Acta physiol. scand. 46 (1959) 19.
[2] M. Dalmark and J.O. Wieth, Biochim. Biophys. Acta 219 (1970) 525.
[3] U. Lassen, in: Oxygen Affinity of Hemoglobin and Red Cell Acid-base Status. Alfred Benzon Symposium IV (Munksgaard, Copenhagen 1972) p. 291.
[4] M. Dalmark, in: Oxygen Affinity of Hemoglobin and Red Cell Acid-base Status. Alfred Benzon Symposium IV (Munksgaard, Copenhagen 1972) p. 320.
[5] A. Scarpa, A. Cecchetto and G.F. Azzone, Biochim. Biophys. Acta 219 (1970) 179.
[6] F.R. Hunter, J. gen. Physiol. 51 (1968) 579.
[7] G. Shean and K. Sollner, Ann. N.Y. Acad. Sci. 137 (1966) 759.
[8] J. Gutknecht and D.C. Tosteson, J. gen. Physiol. 55 (1970) 359.
[9] H. Passow, in: Molecular Basis of Membrane Function, D.C. Tosteson, ed. (Prentice Hall, Englewood Cliffs, 1969) p. 319.
[10] R. Gunn, J.O. Wieth and D.C. Tosteson, Federation Proc. 30 (1971) 314 Abs.

PART IV

TRANSPORT THROUGH EPITHELIA

INTRODUCTION

H.H. USSING
Institute of Biological Chemistry, University of Copenhagen, Denmark

In the last few years there has been a mounting interest in the study of transport through epithelia. Considering the biological importance of epithelia this is not surprising. Most workers in the field agree that the processes underlying active and passive transport across epithelia are basically related to those which are operative in individual cells. However, the arrangement of epithelial cells to form a sheet endows the whole structure with new potentialities with respect to coupling between transports which are not so apparent in individual cells. Some of these complications can be traced back to the presence of slits and pores between the cells, presenting diffusion paths with concentration gradients, bulk flow, etc. These phenomena are highly interesting in their own right, but since this meeting is devoted to transport through cell membranes it would be appropriate to concentrate our efforts mainly on phenomena which can be treated as stemming from transport phenomena in the cell membranes of the epithelia. Indeed it has been found profitable to treat many phenomena in epithelia on the basis of the so-called two membrane hypothesis. In a wider sense the hypothesis simply states that the outward and inward facing membranes of an epithelium have different properties with respect to active and passive transports, and that the transport polarity of the epithelium stems from the different behaviour of these two membranes. The hypothesis was advanced in an explicit form some 15 years ago by Koefoed-Johnsen and myself to explain the responses of the frog skin potential to ionic substitutions on the two sides of the skin.

The hypothesis assumes that a continuous layer of epithelial cells separates the outside and inside bathing solutions. It is implicit that the cells are sealed so tightly together that the whole assembly acts as if it were one giant cell whose permeability is determined solely by the outward and inward facing membranes.

The hypothesis assumes that the outward facing membrane is selectively, but passively permeable to sodium (and lithium) but rather tight to potassium. Furthermore, it has a non-specific permeability to anions, notably small ones like chloride. The inward facing membrane is assumed to be more like say, a nerve membrane. It is permeable to potassium and small anions, but rather tight to passive sodium movements. However, it possesses a cation pump which expels sodium from the cell interior, possibly in exchange for potassium.

The two membrane hypothesis has been applied rather successfully to many vertebrate epithelia. It is interesting however, that every single statement of the hypothesis has been contested, notably with respect to their validity for the frog skin. I may be biased, but it seems to me that after a period of suffering, the hypothesis is now gaining strength again. Judging from the program and from what I have heard in the corridors, we shall get evidence both for and against the hypothesis at today's session.

As for myself, I should like to mention only some experiments performed in our institute by my collaborator Robert Nielsen, experiments which seem to fit in nicely in the hypothesis.

One of the crucial points in the hypothesis is the assumption that the outer cell membrane is rather tight to potassium whereas the inner one is leaky to potassium. This means that even if the pump were a sodium-potassiumexchange pump (whether pumping the two ions in the ratio one to one or any other ratio), one does not get an outward net transport of potassium but only an inward transport of sodium. Thus if one were able to punch holes for potassium in the outer membrane, the dosium-potassium pump should manifest itself, giving rise to outward active transport of potassium as well as inward transport of sodium. Now it is a well known fact that amphotericin B and related polyenes are able to produce rather non-selective "holes" in certain cell membranes. Lichthenstein and Leaf have made use of this effect to increase the sodium transport in toad bladders. Robert Nielsen now studied the effect of this drug on the potassium transport in isolated short circuited frog skins and found clear evidence that the drug induced a sustained active transport of potassium (and rubidium) from the inside to the outside solution. The transport is ouabain-sensitive and behaves as if it were operated by the ordinary cation pump. Obviously amphotericin might activate a dormant potassium pump, but as a working hypothesis it may be permissible to assume that the amphotericin unmasks one of the properties of the normal cation pumping system.

Even if one were to uphold the sodium-potassium exchange pump as the main pumping device in the frog skin and a lot of other epithelia, one should not forget that other pumps exist. This has become particularly apparent in recent studies on insects of which we shall hear more today. Insects seem to possess cation pumps which are totally different from the "common or garden variety" of vertebrates. The interesting problem presents itself whether it is possible to develop a kind of two membrane hypothesis for the potassium pumping insect epithelia. Anion pumps in different tissues may also be arranged to fit two membrane patterns.

With the hope that we shall have an exciting day I shall now leave the floor to the first speaker.

THE ROLE OF IONIC EXCHANGERS AND PUMPS IN TRANSEPITHELIAL SODIUM AND CHLORIDE TRANSPORT ACROSS FROG SKIN

F. GARCIA-ROMEU and J. EHRENFELD
Laboratoire de Physiologie Cellulaire,
Faculté des Sciences, 06, Nice

1. Introduction

Studies on transepithelial ionic transport mechanisms in Amphibia have resulted in two different conceptions of the relationship between the transport of chloride and that of sodium. The short-circuited isolated skin of most frog species when separating two Ringer solutions, only transports sodium actively, the chloride flux being purely passive [1–3]. As the isolated skin became the major tool in the study of transepithelial ionic mechanisms, it was natural to look upon it as truly representative of transport behaviour. The "traditional" view that the chloride simply followed the sodium without a pumping mechanism of its own was thus established. The *in vivo* skin, however, was shown to be capable of absorbing sodium and chloride independently by exchange with endogenous ions [4–7]. An individual pumping mechanism for chloride must therefore exist. Further data support this conclusion. Jorgensen, Levi and Zerahn [8] showed that chloride transport across the *in vivo* skin is active, and Zadunaisky, Candia and Chiarandini [9] reported that the *in vitro* skin of *Leptodactylus ocellatus* transports chloride actively even when the two surfaces of the skin are bathed in Ringer at a physiological concentration. Furthermore, the species which show no active chloride transport when the external Ringer is at a normal concentration, do transport chloride actively when the external sodium chloride concentration is around 2 mM [10, 11].

The following conclusions regarding the amphibian skin can be drawn from the above studies:

1) Sodium is actively transported.
2) Under certain experimental conditions chloride can be actively transported.
3) Under certain conditions sodium and chloride are exchanged against endogenous ions.

It has not been proved that chloride is always actively transported under physiological conditions, but it would seem to be extremely likely. Neither has it been proved that exchanges involving endogenous ions always occur, and the relationship between these mechanisms and those of the ionic pumps remains obscure. Some of these questions will be discussed below.

2. Independent versus linked transport of sodium and cloride

Considering the Na^+/H^+ and Cl^-/HCO_3^- exchange mechanisms [6], which we consider to be different from the active transport pumps (see below), the following questions will be considered:

1) Is all the sodium and chloride absorbed exchanged against the endogenous H^+ and HCO_3^- ions respectively, or
2) does this type of exchange only function to equilibrate the charges at times when sodium and chloride are being pumped at unequal rates?

Unfortunately, the decision as to which of these two possibilities is correct has to depend on indirect experimental approaches since the endogenous H^+ and HCO_3^- interact and cannot accumulate in the external solution. Thus any measurements of exchanges in a sodium chloride solution would suggest the second possibility whichever process was involved. In effect, only excess bicarbonate, or excess hydrogen, corresponding to an excess of pumped chloride or sodium respectively would remain in the external solution. On the other hand, a good relationship between total absorbed sodium or chloride and the respective excreted endogenous ions can be established if each ion absorption is studied independently, this being possible when the ion in question is accompanied by an impermeant ion [6]. But while it is permissible to conclude from experiments using impermeant ions that Na^+/H^+ and Cl^-/HCO_3^- exchanges can occur, the question as to whether they also occur when sodium chloride is present in the external solution remains unanswered.

When the transepithelial ionic fluxes of frogs preadapted to sodium chloride solutions buffered at pH 7.0 are measured *in vivo* in the same solutions, a strict correlation between sodium and chloride fluxes is seen (fig. 1). The slopes of the chloride influx and net flux as a function of sodium influx and net flux suggest that a strict coupling exists between the two pumping mecha-

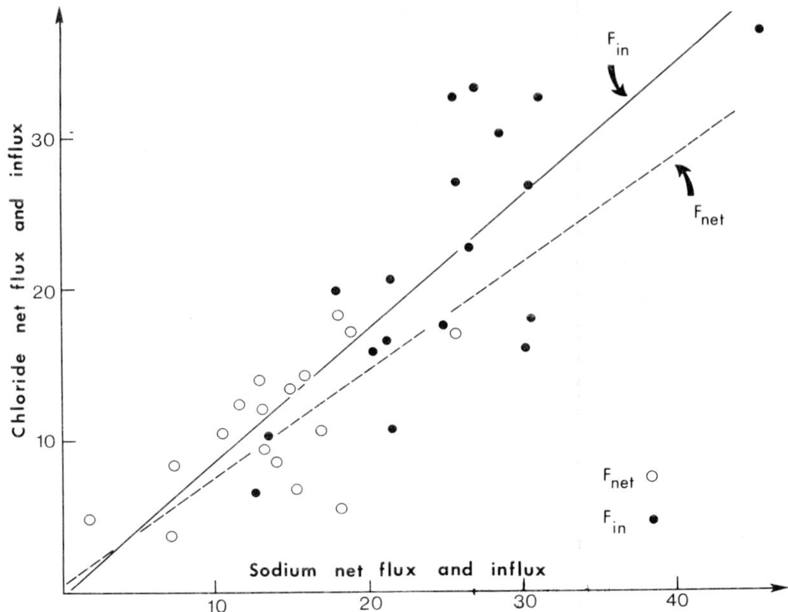

Fig. 1. Transepithelial influx and net flux of chloride as a function of the influx and net flux of sodium respectively in *Rana esculenta in vivo*. Influx of chloride = –0.35 + 0.88 sodium influx. Net flux of chloride = 0.51 + 0.71 sodium net flux. Each point represents an individual experiment. Fluxes measured in a NaCl solution (1750 μeq/1; *p*H 7).

nisms. Under these conditions additional Na^+/H^+ and Cl^-/HCO_3^- exchange mechanisms would seem unnecessary. But if neither bicarbonate nor hydrogen were involved in the chloride and sodium absorption, a carbonic anhydrase inhibitor such as acetazolamide should have no effect on the transepithelial ionic transports (unless the inhibitor acts directly on the pumps, which is unlikely, at any rate for the sodium pump, as will be seen later). Fig. 2 shows the effect on the chloride and sodium unidirectional and net fluxes of an injection of Diamox (dose 10^{-4} M/1000 g). It can be seen that there is a considerable reduction of the influx of both ions and an inversion of the net fluxes which become negative. Table 1 records two experiments in which Diamox was injected into two frogs preadapted to *p*H 7.0 and their fluxes measured at *p*H 6.0; table 2 gives the results of two similar experiments, but with fluxes measured at *p*H 7.0. In both series, during the first three hours after Diamox injection, the chloride and sodium influxes were inhibited by approximately 50%, while the outfluxes increased by 100%.

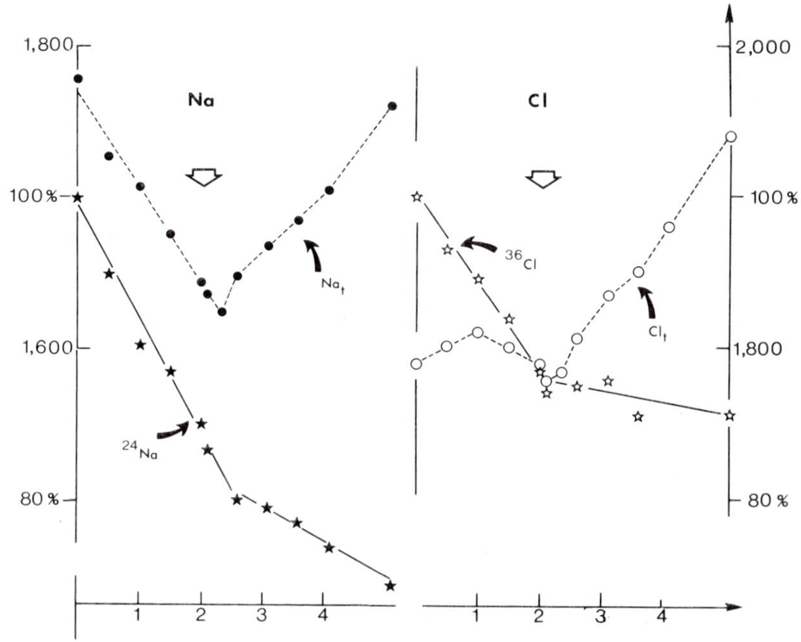

Fig. 2. Effect of intraperitoneal injection of Diamox (10^{-4} M/1000 g) on the sodium (left) and chloride (right) fluxes of *R.esculenta.* The empty arrows indicate the injection of the inhibitor. Ordinates: Na^+ and Cl^- concentrations (μeq/1) and radioactivity (per cent of initial radioactivity) in external medium. Abcissa: time in hours.

These figures should only be taken as a general indication since the course of inhibition varied from one individual to another.

Sleggers and Moons [12] suggest that the inhibitory effect of Diamox on the sodium excretion of the sweat glands is due to cellular acidosis directly inhibiting the sodium pump. In the frog skin however, Diamox acts on the Na^+/H^+ exchange rather than on the sodium pump, as the following facts show:

a) Diamox does not inhibit sodium transport in the *in vitro* skin [13, 11],
b) Diamox inhibition of sodium absorption is accompanied by a simultaneous inhibition of hydrogen excretion (see fig. 3). This simultaneous inhibition of sodium absorption and hydrogen excretion was found in eight out of the nine experiments carried out.

As we have just seen Diamox does not inhibit sodium net flux in the short-circuited isolated skin; it does, however, inhibit hydrogen excretion in

Table 1

Inhibition of transepithelial sodium and chloride fluxes by Diamox in *Rana esculenta in vivo*. Frogs were preadapted in a NaCl solution of 1,750 μeq/l at *p*H 7. Fluxes were measured in a NaCl solution of same concentration but at *p*H 6. Diamox was injected intraperitoneally at a dose of 10^{-4} M; volume injected: 25 μl/100 g. Fluxes are expressed in μeq/hr/100 g.

	Frog B						Frog C					
	Na^+			Cl^-			Na^+			Cl^-		
	F_{in}	F_{out}	F_n	F_{in}	F_{out}	F_n	F_{in}	F_{out}	F_n	F_{in}	F_{out}	F_n
Control period	8.8	2.2	+6.7	8.1	2.0	+6.1	11.9	10.2	+ 1.7	20.6	15.7	+ 4.9
Diamox 1 hr	7.0	6.4	+0.6	4.3	6.6	−2.3	12.6	20.9	− 8.3	14.7	22.6	− 7.9
Diamox 2 hr	4.8	4.8	0.0	5.0	7.2	−2.2	9.3	20.9	−11.6	0.0	10.1	−10.1
Diamox 3 hr	4.3	7.0	−2.7	3.8	8.6	−4.8	5.1	10.0	− 4.9	11.4	19.8	− 8.3

F_{in}, F_{out}, F_n: influx, outflux and net flux respectively.

Table 2

Inhibition of transepithelial sodium and chloride fluxes by Diamox in *Rana esculenta in vivo*. Frogs were preadapted in a NaCl solution of 1,750 μeq/l at *p*H 7. Fluxes were measured in the same solution and are expressed in μeq/hr/100 g. Diamox was injected intraperitoneally at a dose of 10^{-4} M; volume injected: 25 μl/100 g.

	Frog G						Frog H					
	Na^+			Cl^-			Na^+			Cl^-		
	F_{in}	F_{out}	F_n	F_{in}	F_{out}	F_n	F_{in}	F_{out}	F_n	F_{in}	F_{out}	F_n
Control period	14.7	5.5	+9.3	18.4	14.3	+ 4.1	36.0	7.4	+28.6	18.3	5.5	+12.9
Diamox 1 hr	3.3	7.6	−4.3	3.7	13.7	−10.1	21.8	10.3	+11.4	14.3	15.2	− 0.9
Diamox 2 hr	4.0	8.7	−4.7	7.1	14.8	− 7.7	15.0	6.6	+ 8.4	11.5	9.8	+ 1.7
Diamox 3 hr	7.0	14.2	−7.2	0.0	10.2	−10.2	14.5	5.7	+ 8.8	10.7	10.7	0.0

F_{in}, F_{out}, F_n: influx, outflux and net flux respectively.

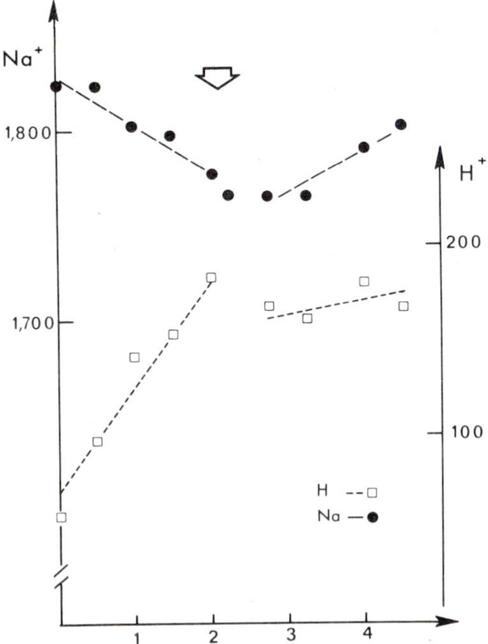

Fig. 3. Inhibition of Na^+ absorption and H^+ excretion by Diamox (Dose injected: 10^{-4} M/1000 g). External solution: Na_2SO_4 at pH 6. Ordinates: Na^+ and H^+ external concentrations in μeq/l. Abscissa: time in hours.

the isolated skin [13]. We are thus faced with the apparently contradictory situation that Diamox inhibits both sodium absorption and hydrogen excretion *in vivo* but only hydrogen excretion *in vitro.* The only possible conclusion is that hydrogen excretion and sodium absorption are no longer linked in the short-circuited skin. This in fact has been shown in the isolated skin by Emilio et al. [13] who found no correlation between hydrogen excretion and sodium net flux, measured at 30 min intervals by the short-circuit current. In *Rana ridibunda,* the species used by Emilio et al. [13] the short-circuit current is a measure of sodium net flux [14] as in other *Rana* species. In other words, during short-circuiting the net transfer of charges is brought about only by the sodium transport mechanism, hydrogen net transfer not occurring under these conditions. Also, it is known that the sodium influx/outflux ratio increases considerably under short-circuiting [2, 15]. As a consequence of these two phenomena, namely the absence of hydrogen net flux and the increase of sodium net flux, it can be seen that there is no relation between

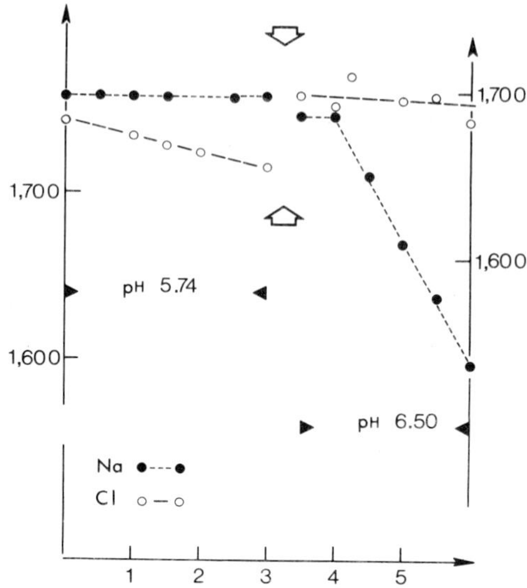

Fig. 4. The effect of the alkalization of the external solution on the net flux of sodium in *R.esculenta*; for comparison, the net flux of chloride is also given. Frog preadapted to Na_2SO_4 solution of *p*H 5.7–5.9; net fluxes first measured in a NaCl solution of *p*H 5.7; at arrows, the external solution was replaced by a NaCl solution of *p*H 6.5. Ordinates: Na^+ and Cl^- external concentrations in μeq/l. Abscissa: time in hours.

the fluxes of these two ions in the short-circuited skin. It has therefore been proposed [7] that hydrogen transport towards the mucous face is not a necessary condition for net sodium transport. The sodium transport mechanism, in fact, must be the motive force of the cationic exchange system, and the hydrogen excretion must serve to equilibrate the charges, in other words, to act as a physiological short circuit for the sodium transport. If the charges are equilibrated by an external circuit, as in the technique of Ussing and Zerahn, the linking of the two ionic fluxes disappears and only the sodium continues to be transported. We have no data concerning the effect of acetazolamide on the HCO_3^- excretion by the frog skin, but the effect of the inhibitor on sodium and chloride absorption and hydrogen excretion would be difficult to explain if, as has been stated, the skin contains no carbonic anhydrase [16]. This enzyme has recently been shown to occur in the urinary bladder of the fresh water turtle *Pseudemys* [17] although previously it was also thought to be absent in this tissue.

A further indication of the constant occurrence of a Na^+/H^+ exchange in the *in vivo* skin should be mentioned. Fig. 4 shows the effect on the net

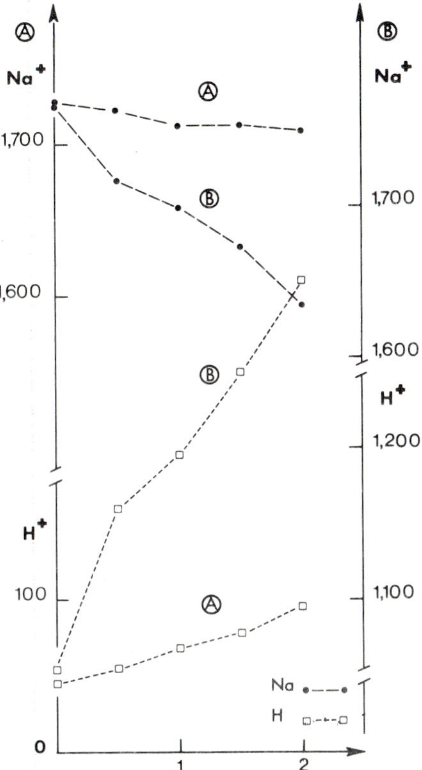

Fig. 5. The effect of sudden changes of *p*H on Na^+ absorption and H^+ excretion in *R.esculenta.* Frogs A and B were preadapted to distilled water at *p*H 6. The fluxes of frog A were measured in a Na_2SO_4 solution of *p*H 6. Fluxes of frog B were measured in Na_2SO_4 solution of *p*H 7. This solution was buffered with imidazol 2 mM + H_2SO_4. Ordinates: Na^+ and H^+ concentrations in external solution in μeq/l. In abscissa: time in hours.

sodium flux of a sudden increase of external *p*H in *R. esculenta in vivo.* This frog was preadapted to a Na_2SO_4 solution at a *p*H of 5.7–5.9. Net fluxes were first measured in NaCl at the same *p*H as the preadaptation solution; then, the external solution was changed to a NaCl solution of *p*H 6.5. The sodium net flux passed from 0.12 to 23.99 μeq/h/100 g. A similar effect has been described in *R. Clamitans in vivo* by Motais and Schmidt-Nielsen [18]. In order to follow the hydrogen excretion, comparable experiments were made in Na_2SO_4 solutions. Fig. 5 illustrates two such experiments. Frogs were preadapted in distilled water at *p*H 6; sodium and hydrogen fluxes of frog A were measured in Na_2SO_4 solution at the same *p*H as the solution of preadaptation, whereas the fluxes of frog B were measured in a Na_2SO_4 solution of *p*H 7.0. It can be seen that the alkalisation of the external medium with respect to the preadaptation solution results in very considerable increases of both sodium absorption and hydrogen excretion.

The data presented above would seem to suggest that, even in sodium chloride solutions, most if not all of the sodium and chloride absorbed must be a result of hydrogen and bicarbonate exchanges.

3. Some characteristics of the chloride transport mechanism

Martin and Curran [10] have shown that active chloride transport is maintained in the isolated short-circuited skin when the chloride concentration bathing the external face is of the order of 2 mM. Under these conditions Erlij [11] found that Diamox inhibited active chloride transport without affecting sodium transport. As has been mentioned above, Diamox also inhibits chloride transport *in vivo.* An important difference between chloride and sodium transports is thus apparent, for Diamox only inhibits the sodium transport *in vivo* whereas it acts both *in vivo* and in the *in vitro* short-circuited skin on the chloride transport. An obligatory and strongly linked Cl^-/HCO_3^- exchange would explain this difference. In fact, it seems that in *in vitro* skin, if HCO_3^- were no longer available for exchange, chloride transport would stop; unlike that of sodium it would not be maintained only by equilibrating the charges by an external circuit.

The evolution of the unidirectional chloride fluxes in the presence of external bicarbonate also suggest a compulsory linkage of chloride and bicarbonate. A bicarbonate concentration of 4 mM in a 1.740 μeq NaCl solution buffered at *p*H 7.0 does not change the chloride influx relative to a previous control period, but it does significantly increase the outflux (table 4). This bicarbonate effect appears paradoxical because it concerns the *trans* flux of chloride but not the *cis* flux. It would seem that a transport inversion takes place, the bicarbonate being transported inwards in exchange for chloride, the outflux of which therefore increases. The lack of effect on the chloride influx indicates that the chloride transport mechanism is far from saturation and that at the external face affinity for chloride is higher than for bicarbonate. The ratio between the bicarbonate and chloride concentrations of the external solutions in the above experiments was approximately 2.

The effect of bicarbonate on the skin can be likened to its action on chloride transport in the stomach. Imamura [19] showed that when the

Table 3
Effect of bicarbonate on chloride fluxes in *R. esculenta in vivo*. Bicarbonate was added to the external bath (final concentration 4 mM/1). Frogs were preadapted in NaCl solution (1 750 μeq/1) buffered with Imidazol + H_2SO_4 at pH 7. Fluxes were measured in the same solution and are expressed as μeq/h/100g (mean ± standard deviation of the mean).

n=4	Control period	Bicarbonate period	Difference *
F_{in}	17.62 ± 0.88	17.47 ± 0.56	−0.15 ± 0.36
F_{out}	7.52 ± 2.01	10.92 ± 2.15	+3.40 ± 0.83 **
F_n	+10.11 ± 1.62	+6.55 ± 2.22	−3.56 ± 0.80 **

n: number of animals.
F_{in}, F_{out} and F_n: influx, outflux and net flux respectively.
* Mean difference of paired data ± standard error of the mean.
** $0.02 < P < 0.05$.

bicarbonate concentration of the secretory solution is raised to 72 mM, the chloride flux from the nutrient to the secretory solution is increased and that in the opposite direction diminished.

4. Correlations between the unidirectional fluxes of each ion

The effect of external bicarbonate on the chloride outflux suggests that this flux passes by the transport mechanism. Measurements of unidirectional chloride and sodium fluxes show a very significant correlation between influx and outflux in each of these ions (figs. 6 and 7) when made at a constant external NaCl concentration of 1 750 μeq. When the external NaCl concentration varies, causing variations of the influx, Linderholm [15] and Kirschner [20] have shown that in the isolated skin the sodium outflux is a function of the influx. Kirschner [20] proposed that the sodium outflux is in series and not in parallel with the sodium pump, in other words, that outward moving sodium ions can interact with the carrier molecules. Indeed, since influx and outflux are correlated, the outflux can hardly be considered as a simple leak independent of the pumping mechanisms.

The correlation of the two fluxes could be explained by an exchange diffusion mechanism [21] linked to the active transport pumps. Exchange diffusion of Na^+ linked to the sodium pump has been shown to occur in the human erythrocyte [22]. Recently an exchange diffusion of sodium passing by the active transport pump has also been demonstrated in the gill of the

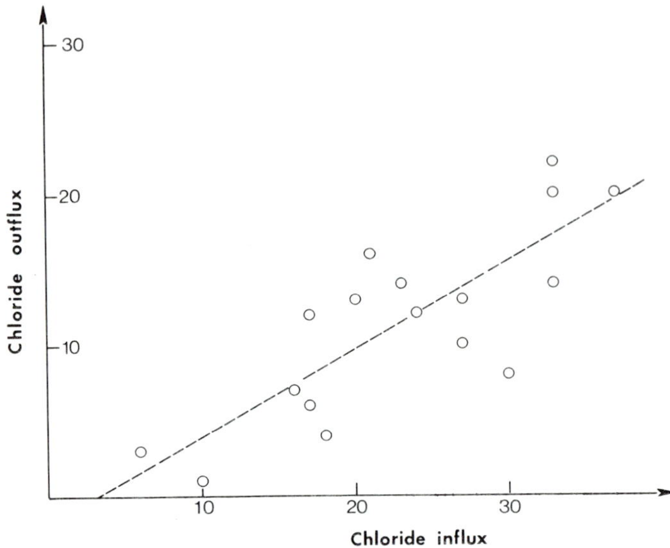

Fig. 6. Transepithelial chloride outflux in relation to chloride influx in *R.esculenta in vivo.* Frogs preadapted in a NaCl solution (1,750 μeq/l; *p*H 7); fluxes measured in the same solution. Each point represents an individual experiment. Chloride outflux = −1.93 + 0.58. Chloride influx ($r = +0.80$; $0.001 < P < 0.01$).

european eel in sea water (Motais and Isaia, personal communication). Nevertheless, although the hypothesis of chloride and sodium exchange diffusions linked to the respective pumps is conceivable, no real proof of the occurrence of these can yet be given. Furthermore, the exchange diffusion concept is based on a precise model at the molecular level [21], but other models, such as that of Kirschner [20] can be made to explain the same phenomena. The important fact to retain as a basis for further considerations is the linkage between influx and outflux.

Rotunno et al. [23] have recently developed a technique for measuring the sodium influx across the external barrier of the epithelium. They showed that when the external face of the isolated skin is bathed in Ringer containing 115 mM sodium chloride, the sodium flux from the external solution into the epithelium is much higher than the transepithelial net flux. From this they conclude that in order to penetrate into the transporting compartment, sodium must interact with sites constituting a mechanism of obligatory exchange diffusion. When the external face is bathed in 1 mM sodium chloride,

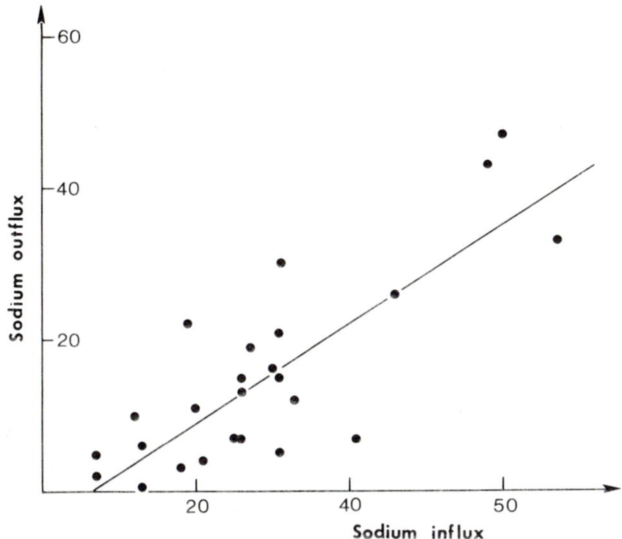

Fig. 7. Transepithelial sodium outflux in relation to sodium influx in *R. esculenta in vivo*. Same experimental conditions as in fig. 6. Sodium outflux = −4.23 + 0.66. Sodium influx ($r = +0.84$; $P < 0.001$).

a concentration near to that in our experiments, however, the influx into the epithelium and the transepithelial flux are approximately equal [23] and thus no exchange diffusion need be postulated. These results cannot be compared with ours since the fluxes measured by Rotunno et al. are those across the external barrier while ours represent the passage from the transport compartment across the internal barrier; in fact, in view of the length of our experiments, the specific activity of the transport compartment must be considered as being in equilibrium with the external solution, flux measurements between the two cannot therefore be made.

5. Conclusions

If exchanges with endogenous ions are normally part of the transport mechanisms, the question of their relationship to the active transport pumps remains to be settled. The mechanism of sodium transport is the better known and it is from data referring to this that interpretation is possible. We have seen that Diamox does not inhibit the sodium pump in the isolated short-circuited skin [13, 11]. Its effect on *in vivo* sodium and hydrogen transports show that it acts on the Na^+/H^+ exchange mechanism. From these facts two levels in the sodium transport mechanism can be differentiated: one at an ionic exchanger and the other at the active transport pump which is inhibited by ouabain. Rotunno et al. [23] have recently shown that sodium entry into the epithelium is not ouabain-sensitive, a fact supporting the hypothesis of Koefoed-Johnsen and Ussing [24] that sodium enters the skin passively, and the report of Farquhar and Palade [25] that ATP-ase is absent on the external border of the skin. Consequently, it is reasonable to suppose that a cationic exchanger is situated at the entrance to the transport compartment where it would assure the equilibrium of the charges and the selectivity of the passive sodium influx into the epithelium. It has in fact been shown that penetration of sodium into the epithelium is a saturable process [26] for which other ions, such as potassium [23] and lithium [26] may enter into competition. From all the above studies and also from those of Nutbourne [27] on the production of streaming currents by hydrostatic pressure, and Pesente [28, 29] (see Garcia-Romeu [7]) on cationic selectivity changes caused by *p*H variations, the presence of fixed charges in the path of sodium through the epithelium would seem very probable. The model proposed by Cereijido and Rotunno [30] (see also [31–33, 23]) as a result of other lines of evidence also accords a very significant role to fixed charges in transepuithelial sodium transport; we feel that this is the most important proposition of Cereijido and Rotunno's model.

Far less is known about the chloride transport mechanism. As we have noted above, the Cl^-/HCO_3^- exchange seems to be much more compulsory than the Na^+/H^+ exchange, but we have no data to show whether chloride transport is brought about by a double mechanism of exchanger and pump or whether the two are one and the same process.

The independence of sodium and chloride transport mechanisms through the epithelium may signify that two transport compartments exist, but we feel that this possibility is unlikely. The correlation between the two fluxes (fig. 1) indicates a mechanism governing both ionic transports and which, under normal conditions keeps their differences minimal. This regulatory mechanism must be situated inside, or at the edge of the transport compartment common to the two ions. This single transport compartment would have separate entrances for chloride and sodium. Further evidence that the two ions enter by different paths is given by the action of amiloride which inhibits sodium entry into the transport compartment and its transepithelial transport [34–37], but which has no effect on active chloride transport [11].

Nothing is known about the mechanisms linking the sodium and chloride fluxes. That ouabain inhibits the active transports of both sodium and chloride across the isolated skin of *Leptodactylus ocellatus* [9] may indicate that the linkage occurs at the level of the active transport pumps.

This investigation was supported with a grant (no. 70-7-2518) from the Direction Générale à la Recherche Scientifique et Technique.

References

[1] V. Koefoed-Johnsen, H. Levi and H.H. Ussing, The mode of passage of chloride ions through the isolated frog skin. Acta Physiol. Scand. 25 (1952) 150–63.

[2] H.H. Ussing and K. Zerahn, Active transport of sodium as the source of electric current in the short-circuited isolated frog skin. Acta Physiol. Scand. 23 (1951) 110–27.

[3] H. Linderholm, The electrical potential across isolated frog skin and its dependence on the permeability of the skin to chloride ions. Acta Physiol. Scand. 28 (1953) 211–17.

[4] A. Krogh, Osmotic regulation in the frog (R. esculenta) by active absorption of chloride ions. Skand. Arch. Physiol. 76 (1937) 60–74.

[5] A. Krogh, The active absorption of ions in some freshwater animals. Z. vergl. Physiol. 25 (1938) 335–50.

[6] F. Garcia-Romeu, A. Salibian and A. Pezzani-Hernandez, The nature of the *in vivo* sodium and chloride uptake mechanisms through the epithelium of the chilean frog *Calyptocephalella gayi* (Dum. et Bibr., 1841). J. Gen. Physiol. 53 (1969) 816–35.

[7] F. Garcia-Romeu, Anionic and cationic exchange mechanism in skin of anurans, with special reference to Leptodactylidae in vivo. Phil. Trans. Roy. Soc. (London) B 262 (1971) 163–74.

[8] C.B. Jorgensen, H. Levi and K. Zerahn, On active uptake of sodium and chloride ions in anurans. Acta Physiol. Scand. 30 (1954) 178–90.

[9] J.A. Zadunaisky, O.A. Candia and D.J. Chiarandini, The origin of the short-circuit current in the isolated skin of the South American frog *Leptodactylus ocellatus.* J. Gen. Physiol. 47 (1963) 393–402.

[10] D.W. Martin and P.F. Curran, Reversed potentials in isolated frog skin. II. Active transport of chloride. J. Cell Comp. Physiol. 67 (1966) 367–74.

[11] D. Erlij, Salt transport across isolated frog skin. Phil. Trans. Roy. Soc. (London) B. 262 (1971) 153–61.

[12] J.F.G. Sleggers and W.M. Moons, Effect of acetazolamide on the chloride shift and the sodium pump in secretory cells. Nature 220 (1968) 181–82.

[13] M.G. Emilio, M.M. Machado and H.P. Menano, The production of a hydrogen ion gradient across the isolated frog skin. Quantitative aspects and the effect of acetazolamide. Biochim. Biophys. Acta 203 (1970) 394–409.

[14] K.T. Gil Ferreira, Anionic dependence of sodium transport in the frog skin. Biochim. Biophys. Acta 150 (1968) 587–98.

[15] H. Linderholm, Active transport of ions through frog skin with special reference to the action of certain diuretics. Acta Physiol. Scand. 27 (1952) Sup. 97.

[16] T.H. Maren, Carbonic anhydrase: chemistry, physiology and inhibition. Physiol. Rev. 47 (1967) 595–781.

[17] W.N. Scott, Y.E. Shamoo and W.A. Brodsky, Carbonic anhydrase content of turtle urinary bladder mucosal cell. Biochim. Biophys. Acta 219 (1970) 248–50.

[18] R. Motais and B. Schmidt-Nielsen, In vivo ionic exchange through the skin of the frog *Rana clamitans*. Bull. MDIBL. 10 (1970) 151–6.

[19] A. Imanura, The effects of carbon dioxide and bicarbonate on chloride fluxes across frog gastric mucosa. Biochim. Biophys. Acta 196 (1970) 245–53.

[20] L.B. Kirschner, On the mechanism of active sodium transport across the frog skin. J. Cell. Comp. Physiol. 45 (1955) 61–87.

[21] H.H. Ussing, Interpretation of the exchange of radiosodium in isolated muscle. Nature 160 (1947) 262–63.

[22] P.J. Garrahan and I.M. Glynn, The behaviour of the sodium pump in red cells in the absence of external potassium. J. Physiol. (London) 192 (1967) 159–74.

[23] C.A. Rotunno, F.A. Vilallonga, M. Fernandez and M. Cereijido, The penetration of sodium into the epithelium of the frog skin. J. Gen. Physiol. 55 (1970) 716–35.

[24] V. Koefoed-Johnsen and H.H. Ussing, The nature of the frog skin potential. Acta Physiol. Scand. 42 (1958) 298–308.

[25] M.G. Farquhar and G.E. Palade, Adenosine triphosphatase in amphibian epidermis. J. Cell Biol. 30 (1966) 359–79.

[26] T.U.L. Biber and P.F. Curran, Direct measurement of uptake of sodium at the outer surface of the frog skin. J. Gen. Physiol. 56 (1970) 83–99.

[27] D.M. Nutbourne, The effect of small hydrostatic pressure gradients on the rate of active sodium transport across isolated living frog-skin membranes. J. Physiol. (London) 195 (1968) 1–18.

[28] L. Pesente, Effetti della concentrazione idrogenionica sulla permeazione di alcuni ioni monovalenti attraverso la pelle di rana. Bol. Soc. Ital. Biol. Sper. 45 (1969) 1161–64.

[29] L. Pesente, Analisi degli effetti della concentrazione idrogenionica sulla permeazione di alcuni ioni monovalenti. Bol. Soc. Ital. Biol. Sper. 45 (1969) 1164–67.

[30] M. Cereijido and C.A. Rotunno, Fluxes and distribution of sodium in frog skin. A new model. J. Gen. Physiol. 51 (1968) 280s–89s.

[31] M. Cereijido, I. Reisin and C.A. Rotunno, The effect of sodium concentration on the content and distribution of sodium in the frog skin. J. Physiol. (London) 196 (1968) 237–53.

[32] M. Cereijido and N. Fraidenraich, Surface *vs.* transcellular route in the transport of sodium across epithelial membranes. Horizons in Surface Science: Biological Applications, eds. L. Prince and D.F. Sears (Appleton-Century-Crofts, New York, 1970).

[33] M. Cereijido and C.A. Rotunno, The effect of antidiuretic hormone on Na movement across frog skin. J. Physiol. (London) 213 (1971) 119–33.

[34] P.J. Bentley, Amiloride: a potent inhibitor of sodium transport across the toad bladder. J. Physiol. (London) 195 (1968) 317–30.

[35] J. Crabbe and N. Ehrlich, Amiloride and the mode of action of aldosterone on sodium transport across toad bladder and skin. Pfluegers Arch. 304 (1968) 284–96.

[36] A. Dorge and W. Nagel, Effect of amiloride on sodium transport in frog skin. II. Sodium transport pool and unidirectional fluxes. Pfluegers Arch. 321 (1970) 91–101.

[37] W. Nagel and A. Dorge, Effect of amiloride on sodium transport of frog skin. I. Action on intracellular sodium content. Pfluegers Arch. 317 (1970) 84–92.

ON THE EVALUATION OF FLUXES ACROSS THE OUTER BORDER OF THE EPITHELIUM

M. CEREIJIDO, J.H. MORENO, E. RODRÍGUEZ BOULAN
and C.A. ROTUNNO

Departamento de Fisicoquímica, Universidad de Buenos Aires y Centro de Investigaciones Médicas Albert Einstein, Buenos Aires

One of the key features in the movement of substances across epithelial membranes is the penetration across the outer border. This is the place where the steapest gradients occur and where most of the electrical resistance is found. In order to study the mechanism of this penetration in detail we [1] have developed a technique to evaluate unidirectional fluxes from the outer solutions to the epithelium (J^i_{12}). It can be demonstrated that, since the flux of tracer across the whole skin equilibrates in a few minutes, the measurement of J^i_{12} has to be done in a fraction of a minute. To overcome this difficulty we proceed as follows.

The abdominal frog skin is mounted as a flat sheet between two Lucite chambers (fig. 1). The exposed part of the skin has a rectangular cross-section. After a preincubation period with non-radioactive Ringers, the chambers are emptied and a dual simultaneous infusion pump injects Ringer solution containing tracer into the outer chamber and non-radioactive Ringer into the inner chamber. The solutions are injected steadily into the bottom of the chambers and the level rises homogeneously in 48 sec. When the solution reaches the upper border and the exposure is interrupted, the lowest part of the membrane had been exposed for the whole time of the injection (48 sec). All other parts have been exposed for a fraction of 48 sec. This fraction is proportional to the distance to the floor of the chamber. Once the solutions reach the upper border the exposure is interrupted and the outer face of the skin is washed with a jet of isotonic sucrose solution at zero °C delivered with a squeeze bottle during 2 sec (15–20 ml). [This procedure constitutes a

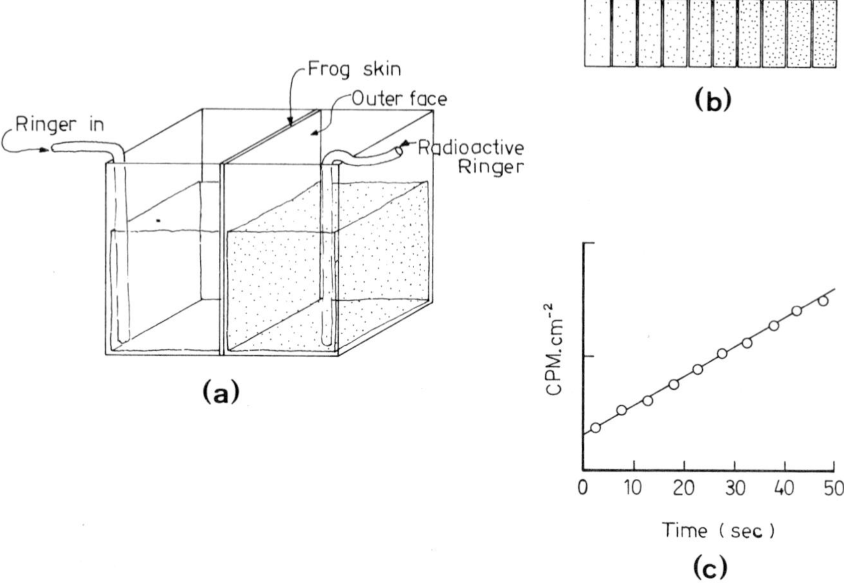

Fig. 1. Schematized representation of the technique to measure J^i_{12}. a: the level of the radioactive (right) and non-radioactive (left) Ringer rises and reaches the top in 48 sec. b: a central strip of the exposed skin is cut into several transverse sections. The one on the left corresponds to the upper part of the chamber, and the one on the right corresponds to the lower border. c: the activity (or else the amount of substance i) per square centimeter in each sample is plotted vs time. The slope gives the value of J^i_{12}.

modification of the method originally described by Rotunno et al. [1] since in their case the rinsing was reduced to plunging the skin deep into a beaker with cold sucrose solution for 1 sec.] The skin is then immersed into a thermos flask containing liquid nitrogen (−196 °C). The central part of the skin is cut into several transverse sections, each representing a certain uptake time of tracer. The samples are analyzed for their concentration of tracer and a kinetic curve, showing rhw uptake of Na as a function of time, is plotted. The slope of the curve is used to calculate the unidirectional flux from the outer solution to the epithelium (J^i_{12}).

As pointed out by Erlij and Smith [2] the existence of an Extracellular Space (ECS) on the outside might introduce a serious error in the evaluation of J^i_{12}. This stems from the fact that by dividing the amount of substance (i) taken up by the skin over a short period of time, by the time of uptake, one includes in the resulting flux the amount of substance trapped in the ECS, thus obtaining an overestimated value of J^i_{12}. Also, since the substance i does not reach directly the "outer barrier" of the epithelium, but has to travel first across the ECS, the delay in reaching the outer barrier might constitute another source of error. Let us therefore discuss how would this ECS distort the evaluation of J^i_{12} as studied with the technique described above.

Kidder, Cereijido and Curran [3] (see also [4] and [5]) have studied the time course of the voltage change that follows a step change in the concentration of Na in the outer bathing solution. They have shown that the new steady state voltage is completed in about 1/5 of a second, most of which is accounted for the delay in crossing the unstirred layer. This time is much shorter than 4.8 sec which is the period covered by the first sample and the error that it would introduce may be disregarded.

A more important source of error might be constituted by the *amount* of tracer of substance i which could remain trapped in the ECS after the rinsing. If this space was not rinsed, each square centimeter of skin would contain (using Ringers with 115 mM Na):

$$1.15 \times 10^{-7} \text{ mole } \mu l^{-1} \times 0.67 \ \mu l = 7.7 \times 10^{-8} \text{ mole.}$$

This amount is even greater than 4.0×10^{-8} mole, which is the amount that penetrates into a square centimeter of epithelium in 48 sec with a flux of 3 μmole h^{-1} cm^{-2} (a typical value). Therefore, if after the exposure to ^{22}Na, one only blots the skin (i.e. does not rinse), the variations of the amount of ^{22}Na contained in each sample of skin, and the ^{22}Na adsorbed to its outer surface will not allow an accurate measurement of the slope. Also, in a skin standing vertically, when the loading solution is sucked out of the chamber at the end of the exposure to ^{22}Na, the solution wetting the outer side drains downwards. Its contribution to the measured uptake becomes proportionally larger as the samples approach the floor of the chamber. If the rinsing were not sufficient, this would constitute a serious of error, as it would modify not only intercept but the *slope* of the curve as well. Although the intercept in the experiments performed with the original technique had erratic values, they usually were about 2×10^{-9} mole cm^{-2}, indicating that most of the ECS was removed by the rinsing. However, the observation that the amount trapped could also modify the slope of the uptake curve led us to adopt a more rigorous way of rinsing, and the jet washing procedure described above was adopted. We shall now illustrate the results obtained with this procedure.

Biber and Curran [6] have found that J^{Na}_{12} appears to be composed by two fractions: one that saturates at high concentrations of Na and which is inhibited by Li, and another one that increases linearly with the concentration

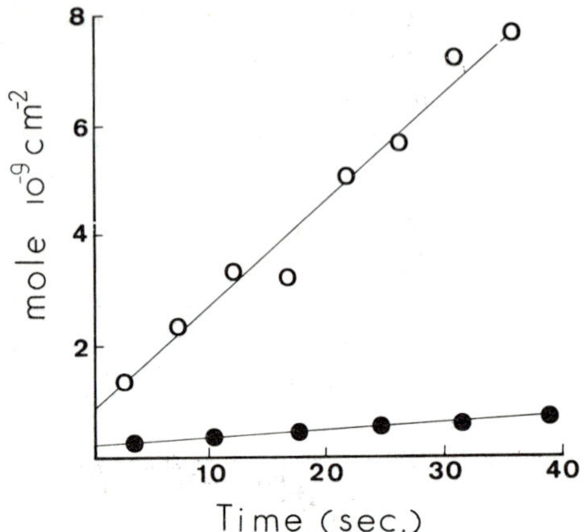

Fig. 2. J_{12}^{Na} from a control Ringer containing 5 mM Na (○) and in the presence of 60 mM Li (●).

of Na and which is not affected by Li. Erlij and Smith [2] have later identified the linear fraction with the amount of Na trapped in the ECS, and the saturing fraction with the one that penetrates into the epithelium. Fig. 2 shows two experiments at 5 mM Na: one control (open circles) and one in the

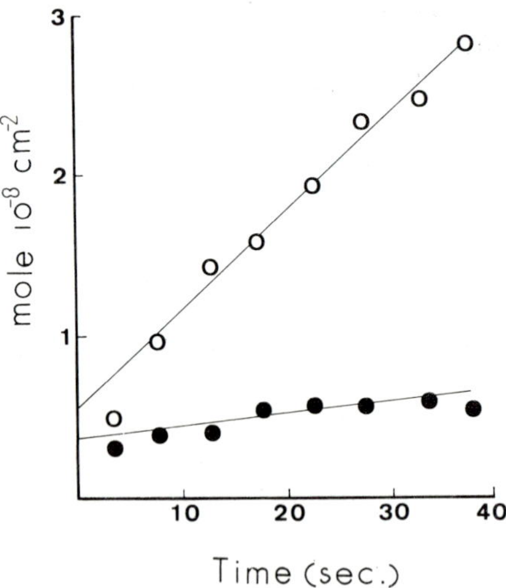

Fig. 3. J_{12}^{Na} from a control Ringer containing 115 mM Na (○) and in the presence of 5 × 10^{-5} M Amiloride (●).

presence of 60 mM Li (full circles). It can be observed that the flux J_{12}^{Na} is almost completely abolished by Li. It may be concluded that the amount of Na taken up does not represent Na in the ECS and that the amount which might not be rinsed is very little. This agrees with the observation that Na in the ECS washes toward the outer solution in a fraction of a second [4, 5].

Since the amount of Na which penetrates in the ECS is a linear function of the concentration of Na in the bathing solution, the error in the estimation of J_{12}^{Na} that it could introduce would be more noticeable at high concentrations of Na [2]. Fig. 3 (open circles) illustrates a measurement of J_{12}^{Na} in Ringers with 115 mM Na. The value of J_{12}^{Na} is 2.73 μmole h^{-1} cm^{-2} which is close to 2.47 μmole h^{-1} cm^{-2}, the value of the influx across the whole skin [7]. This value of J_{12}^{Na} is considerably smaller than the ones reported before from this laboratory [1, 8], indicating that the original procedure of plunging of the skin into cold solution for less than a second was not sufficient to avoid errors arising from the ^{22}Na remaining in the ECS. Fig. 3 (full circles) illustrates also the effect of Amiloride (5×10^{-5} M). Notice that this substance, which is thought to stop the penetration of Na across the outer border of the epithelium [9] does not modify appreciably the intercept but drastically inhibits the flux.

One may conclude that the amount of Na which is used to compute J_{12}^{Na} with the technique described above is not contained in a ECS but behind a sensitive barrier (sensitive, at least, to Li and to Amiloride).

M.C. and C.A.R. are Career Investigators from the National Research Council of Argentina (C.N.I.C.T.). J.M. and E.R.B. are Research fellows from the C.N.I.C.T.

References

[1] C.A. Rotunno, F.A. Vilallonga, M. Fernandez and M. Cereijido, J. Gen. Physiol. 55 (1970) 716.
[2] D. Erlij and M.W. Smith, J. Physiol. 213 (1971) 33.
[3] G.W. Kidder, M. Cereijido and P.F. Curran, Amer. J. Physiol. 207 (1964) 935.
[4] B. Lindemann and U. Thorns, Science 158 (1967) 1473.
[5] J. Dainty and C.R. House, J. Physiol. 182 (1966) 66.
[6] T.U.L. Biber and P.F. Curran, J. Gen. Physiol. 56 (1970) 83.
[7] J.A. Zadunaisky, O.A. Candia and D. Chiarandini, J. Gen. Physiol. 47 (1963) 393.
[8] M. Cereijido and C.A. Rotunno, J. Physiol. 213 (1971) 119.
[9] J. Eigler and J. Crabbe, in: Renal Transport and diuretics, eds. Thurau and Jahrmarker (Springer-Verlag, Berlin, 1969).

RESISTANCE RESPONSE OF FROG SKIN TO BRIEF AND LONG LASTING CHANGES OF $(Na)_0$ AND $(K)_0$*

U. GEBHARDT, W. FUCHS and B. LINDEMANN
Abt. Membranforschung an Epithelien,
2nd Department of Physiology, 665 Homburg/Saar, Germany

1. Introduction

Na ions which are to be absorbed from pond water by epithelial cells of frog skin have to pass a number of series "layers" of organic material which cover the apical cell membranes of the outermost layer of transporting cells and serve for chemical and mechanical protection. There is first a thin layer of fibrillar mucous, then a layer of dense protein about 5 μ thick – the partly cornified cells of the *str. corneum* – then another layer of mucous [1, 2]. Hidden below this, finally, is a Na-selective membrane, sometimes called "outer resistive membrane" or "outer barrier", which mainly determines the total electrical resistance of the epithelium [3–5]. This, at least, is a position of the Na-selective membrane which is currently favoured by many authors (reviewed in [6]). According to their views, the Na-selective membrane is the apical plasmalemma of the outer cell layer of the *str. granulosum*, the "first replacement cell layer", which is very likely to outermost layer of transporting cells [7, 8, 6].

The cells are linked together sideways by tight junctions of a type known as zonula occludens. This linkage makes the apical membranes of individual cells into one structurally almost continuous [1] ion-selective barrier stretching all around the frog. The linkage is not perfect, however, in that the membranes can be shunted by extracellular pathways [7] to various extents. In this paper, the term "Na-selective membrane" refers to the apical plasmalemma of the outer layer of cells of the *str. granulosum*, while "outer barrier" will mean this membrane together with its extracellular shunt-pathways.

* Supported through "Sonderforschungsbereich 38".

We have studied the Na-selective membrane of frog skin *in vitro*, by observing its voltage and resistance changes in response to concentration steps in the outside bathing solution. The interpretation of such experiments would be more difficult if secondary changes of cellular composition had taken place. To minimize such changes, we have used a rapid-flow chamber, which permits outside concentrations to change so fast and so briefly that all electrical responses to concentration steps will be responses only of the Na-selective membrane, its protective layers and, possibly, its extracellular shunt pathways.

2. Method

Abdominal skin of *Rana esculenta* is mounted such that 1 cm^2 of its outer surface lines the lower wall of a flow channel (fig. 1), and is in a steady state

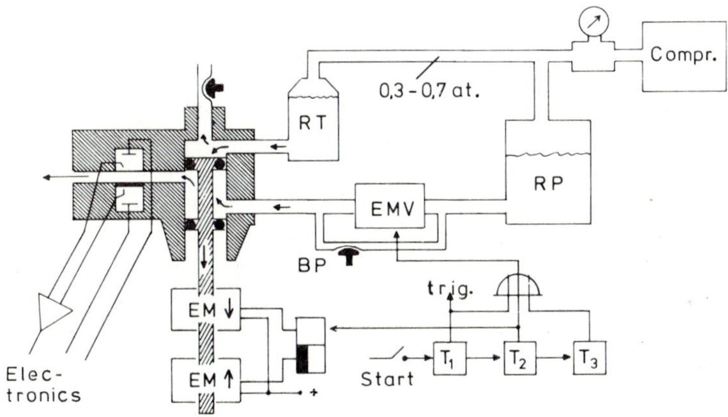

Fig. 1A. Scheme of fast-flow arrangement. Lucite chamber hatched. RP: reservoir for preexposure fluid (P-solution); RT: reservoir for test-fluid (T-solution). BP: bypass to adjust slow flow of P-solution. EMV: electromagnetic valve to change speed of flow of P-solution; EM: electromagnets for switching between P- and T-solutions.

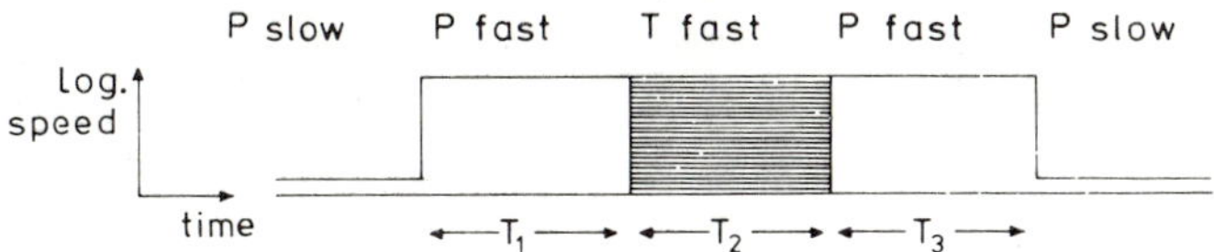

Fig. 1B. Flow program.

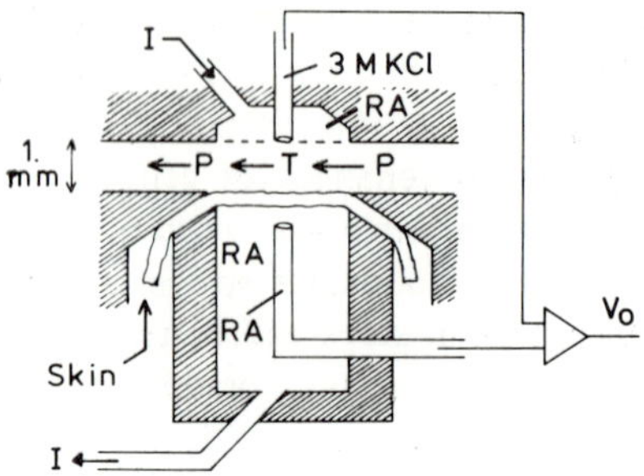

Fig. 2A. Chamber and electrode arrangement. 1 cm^2 of skin was glued with its inner surface to Sulfate-Ringer-Agar (RA) contained in the lower half-chamber, using a small amount of Histo-acryl-N-Blau (B. Braun-Melsungen). The rim of the exposed skin area was not "crushed" between lucite parts, but gently touched by smooth lucite surfaces which were covered with a thin coat of silicon grease. The outer voltage electrode contained fresh 3 M KCl – agar to minimize diffusion potentials. The outer current electrode's opening was K – or Na – Ringer-agar.

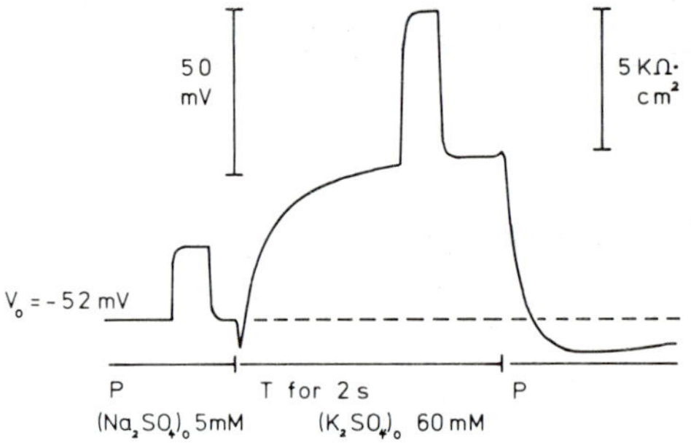

Fig. 2B. Original recording, showing voltage and resistance response to a P-T-P sequence. Abdominal skin of Rana esculenta at room temperature. All outside solutions used in this study contained a background-concentration of 1 mM $CaSO_4$ and 5 mM Tris-SO_4 buffer, *p*H 6.0. In most cases, the voltage response was faster than shown here.

with solution "P", flowing slowly through this channel. When we press a "start" switch, the solution is first accelerated to a flow of 40 ml/sec going through 1.5 × 0.1 cm of rectangular cross-sectional area. After a predetermined time t_1, the solution is replaced by a test solution "T" flowing at the same rate for a period t_2. The chamber allows mounting of ion exchange membranes in place of frog skin. Exposure times to test solutions can be chosen between 50 ms and several seconds. The changeover time between solutions – measured optically or by means of changes in solution resistivity

– was found to be about 25 ms. In this period, pressure transients measured in the center of the membrane area with a "Pitran" transistor were less than 0.02 atm.

The electrode arrangement is shown in fig. 2A. To record changes of membrane small signal resistance (R) without interference from changes in resistivity of the switched outer bathing solutions, we used our chopped-current clamp method [9]. Chopped current pulses of $10\,\mu$ A/cm^2 of equivalent current density were used. Chopping frequency was 10 kHz, pulse (traun) duration 0.1 to 0.2 s [9]. A first pulse was passed through the skin in the inward direction directly before each change of solution, a second pulse of equal density and duration after the change of solution, when the skin voltage had become constant again (fig. 2B). Voltage was sampled with chopping frequency in the current-free intervals of 50 μs. The time course of the sampled transepithelial voltage was stored in a dynamic shift register (transient recorder 802, Biomation), and plotted on an XY recorder. The deflection of skin voltage (5–50 mV) caused by a current pulse was taken to be proportional to the skin's small signal slope resistance R (measured in the vicinity of the resting voltage). Differences in this deflection, which occurred within 2 s after a change of the outer solution, are thought to be caused by changes of membrane – or shunt-resistance of the "outer barrier".

3. Time course of voltage response

The response of a Permaplex C 20 cation exchange membrane to an increase of KCl-concentration was found to be complete within 30–40 ms. In contrast, the response of frog skin to changes of $(Na)_0$ (substituting K) was much slower. In a typical experiment the outer solution contained predominantly K, which was briefly replaced by Na for a period of 400–500 ms while the anion concentration (poorly penetrating gluconate) was maintained. The delay with which the membrane polarized in response to the cation substitution, indicated that it took $t_{1/2} = 80$ ms to make the increase of Na-concentration at the surface of the responding membrane half complete. Fitting the delay curve to Crank's diffusion equation for unstirred layers [10], we computed a nominal thickness of 10.3 μ for an idealized, homogenous unstirred layer with a single diffusion coefficient of 0.5×10^{-5} cm^2/s. It has to be kept in mind, however, that the real unstirred layer constains sublayers of organic material, very likely including the *str. corneum,* which contribute about half of the total thickness. Diffusion coefficients will hardly be equal in these sublayers. For this reason a localization of the responding membrane based on delay data is difficult if not impossible. When decreasing $(Na)_0$ by switching back from T- to P-solution, we find larger delays ($t_{1/2}$ = 140 ms, δ = 13.7 μ). Thus, the skin shows *up-down asymmetry*, indicating that the Na-concentration at the outer surface of the Na-selective membrane can rise faster than it can drop (or that the membrane can respond to an increase of the Na-concentration at its surface faster than to a decrease). There was less delay in skins that had just shed their *str. corneum* and in skins where the *str. corneum* had been treated with a Papain-Cystein solution.

The initial voltage change was much faster, when – instead of substituting $(K)_0$ by Na – $(Na)_0$ was stepped up or down together with the anion concentration (Na-gluconate "step" or Na_2SO_4 "step"). The increase in initial speed observed under these conditions could be caused by Donnan potentials or diffusion potentials set up in front of the Na-selective membrane, because

1. there was less delay, as expected from structures closer to the surface
2. the fast component preceded the change of membrane resistance, which (under some conditions) occurs after a change of $(Na)_0$. This is in contrast to substitution experiments, where the fractional voltage change did not precede the fractional resistance change. And
3. the fast voltage response was less rapid when the Na step change was done in the presence of a large and constant background of electrolyte (for instance Tris-gluconate). Such a background would shortcircuit diffusion potentials.

The secondary, slow part of the voltage transient observed with a Na-gluconate step showed much more up-down asymmetry than was observed with a Na-K substitution. Keeping the osmolarity constant with sucrose proved to be of very little effect [11]. The voltage response was usually 95% complete within 0.3–1.5 s (except in cases where test solutions contained neither Na nor K). Thus, the exposure time could be limited to 2 s.

4. Concentration dependence of skin resistance: hysteresis of R-c functions

In preliminary experiments it was noted, that the electrical resistance of many skin preparations was – in the range 20–100 mM – independent of $(Na)_0$ or $(K)_0$. In other preparation, where a dependence remained, a pronounced hysteresis was often found: when after systematically increasing concentrations they were decreased again, the resistance-concentration function was found to have shifted to lower resistance values. This happened even when poorly penetrating anions like gluconate or sulfate were used on both sides of the skin. The hysteresis behaviour has forced us to distinguish resistance changes caused by *brief* concentration changes from those observed after *long time* exposure to different concentrations. Less hysteresis was found when exposure to test solutions was brief [12, 13]. In subsequent experiments we found quite reproducible, hysteresis-free resistance-concentration functions when using minimal exposure times in our fast-flow chamber. Curves obtained in this way may be called *instantaneous R-c functions.* They were found to be different when the outer surface of the epithelium was preequilibrated with different P-solutions (containing only the poorly penetrating anions mentioned above). For instance, the skin was first equilibrated with solution P_1 and its instantaneous R-c function was measured with one set of test solutions. Afterwards, the skin was equilibrated with solution P_2 and a different instantaneous R-c function was found *using the same set of test solutions.* After shifting back to P_1, the R-c function obtained first could be reproduced.

5. Some instantaneous R-c functions of frog skin

While reproducibility of data obtained from one preparation was improved by the use of brief exposure, we still found large variations between different skins. We believe that in this case there is little point in averaging results of many preparations as long as different skins cannot be studied in the same phase of their moulting cycle. Instead, we have tried to do all important comparisons with the same preparation, i.e. to expose each skin to as many solutions as its survival time would permit.

Fig. 3A shows instantaneous R-(K_2SO_4) functions obtained from one epithelium which was preexposed with P-solutions of 50 mM Na_2SO_4, 50 mM K_2SO_4, 5 mM Na_2SO_4 and 5 mM K_2SO_4. Each preexposure period lasted 30 min. The figure shows, that the Rc-curves depend strongly on the medium, with which the outer surface was in contact before anyone step change of concentration. Absolute R-values and slopes of the R-log c_0 plot were with K containing P-solutions larger than with Na–P-solutions of equal molarity. Furthermore, R-values and slopes obtained with 5 mM P-solutions were larger than those obtained with 50 mM P-solutions. In this preparation, R was independent of $(K)_0$ over the large range of concentrations used, if the preexposure medium was of high Na-concentration.

The instantaneous R-(Na_2SO_4) functions of fig. 3B were obtained from the same epithelium using the same P-solutions. The four curves have almost the same general shape: with increasing $(Na)_0$, R was lowered down to a limiting value. There was a tendency of R to increase again at large $(Na)_0$, particularly when the P-solution was of low concentration.

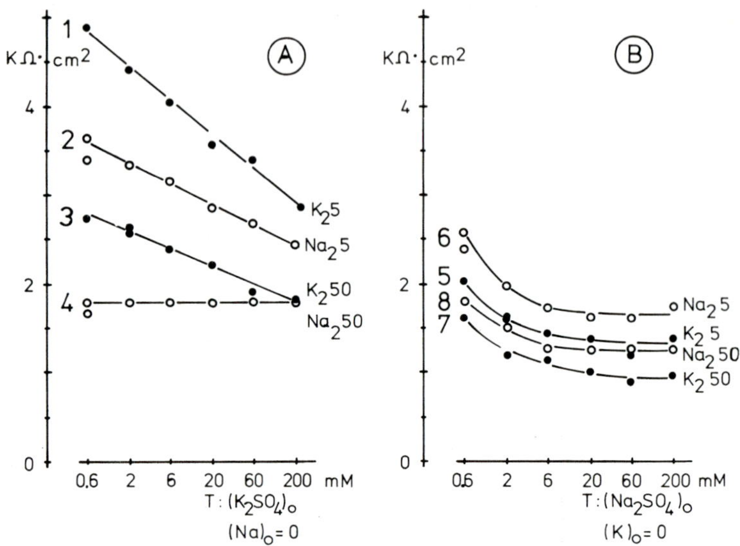

Fig. 3. Instantaneous R-c_0 curves obtained after preexposure to different P-solutions. The Na_2SO_4 or K_2SO_4 concentrations of P-solutions are written to the right of each curve. T-solutions (concentration on abscissa) were used in a sequence of increasing concentrations. In 5 cases, the last solution was a repeat with a low concentrated T-solution, to check for hysteresis.

While of the same general shape, the curves of fig. 3B appear shifted towards each other along the ordinate. This, however, need not be caused by resistance changes of the outer barrier: Different P-solutions – to each of which the skin is exposed for 30 min – might affect resistive structures in series to the Na-selective membrane, for instance via diffusion through shunt-pathways. For this reason, the slopes of instantaneous R-c functions reflect properties of the outer barrier more faithfully than do absolute R-values of the epithelium.

6. Interpretation

In fig. 3A, curve 1, total R is large and steeply dependent on $(K)_0$. In curve 4, where total R is small, there is no dependence on $(K)_0$. Curves 2 and 3 show intermediate features. This trend in resistance and slope values could be explained in terms of shunts: the $(K)_0$-dependent transport mechanism would be shunted by the conductance of electrolyte left in the vicinity of the "outer barrier" from the preexposure period. Let us recall, that the "outer barrier" is a parallel arrangement of two anatomically different transport pathways: apical cell membranes and extracellular shunts, including tight junctions of the zonula type. Therefore, conductances shunting the $(K)_0$-dependent conductance tested in fig. 3A could be

a) extracellular pathways for curve 3, where preexposure was to a Na-free medium, and
b) extracellular pathways plus Na-conductance g_m^{Na} of the Na-specific membrane for curves 2 and 4, where the preexposure medium contained Na.

Extracellular shunt. The shunt-conductance g_s will be determined by the physical dimensions of the extracellular shunt pathway and by the mean concentration c_s of electrolyte in the shunt. c_s can be expected to be changed by diffusion of electrolyte from and into the outer unstirred layers. Thus, in fig. 3A, c_s and g_s will have affected curves 3 and 4 more than curves 1 and 2.

Shunting by g_m^{Na} at zero $(Na)_0$. Comparison of curves 1 and 2 as well as 3 and 4 of fig. 3A shows, that the (presumable) shunting effect of Na-preexposure is larger than that of K preexposure. This might be due to additional shunting through the Na specific membrane, if Na is still available to this membrane after reducing $(Na)_0$ to zero. That Na-specific channels are involved in the shunting will be shown below by the effect of Amiloride on curves 2 and 4. If the Na-conductance g_m^{Na} can shunt the $(K)_0$-dependent conductance although $(Na)_0$ is zero, we have to look for a slowly exchanging Na-compartment in the vicinity of the Na-specific membrane. This compartment would be loaded with Na in the preexposure period and would supply the membrane with Na during exposure to Na-free test solutions. The possible localisation of this compartment will be discussed further below.

Conductance saturation. Fig. 3B gives information about the $(Na)_0$-dependent conductance. Comparing curves 4 and 8, we note that after preexposure to high $(Na)_0$-solutions, g_m^{Na} is large during brief exposure to zero $(Na)_0$ (curve 4), but can be increased further during brief exposures to increasing $(Na)_0$-values (curve 8). However, the dependence of g_m^{Na} on $(Na)_0$ disappears at large $(Na)_0$, i.e. the conductance saturates.

This behaviour would be expected when with increasing concentrations the resistance of the Na-selective membrane becomes so small, that series membranes in the depth of the tissue (which are unaffected from brief changes of $(Na)_0$) limit a further decrease of the epithelial resistance. However, microelectrode studies of frog skin resistance have shown, that at large $(Na)_0$ the larger part of the resistance is located in the "outer membrane" [3–5]. Therefore, the saturation must be due to properties of the outer epithelial surface, possibly of the Na-selective membrane channels themselves.

Generally speaking, the instantaneous R-c curve of a transport channel will have a large negative slope as long as quick changes of c_0 can vary the ion concentration in front of the rate limiting barrier of the channel. At large c_0-values another transport step in the membrane might become rate limiting, causing the steepness of the R-c function to decrease. (This form of saturation can occur at zero concentration at the inner end of the channel, but may begin with smaller c_0-values when the inner concentration is large. For a theoretical treatment see ref. [14], eq. 27–29.)

Our conclusion, that g_m^{Na} becomes independent of $(Na)_0$ at large $(Na)_0$ is supported further by flux – and impedance – studies performed in other laboratories: The short-circuit current [15, 16], as well as the Na flux from the outer solution into frog skin epithelium [17, 18] can saturate with increasing outside Na concentration. Similarly, the fractional Na conductance measured with low frequency a.c. was reported to saturate with increasing $(Na)_0$-values [20, 21].

7. Inhibitory effects of $(K)_0$

Fig. 4A and B show instantaneous R-c functions obtained from a different preparation, which was preexposed to Na-containing P-solutions and otherwise treated in the same way as the preparation of fig. 3. Similarly, fig. 5 shows curves obtained after preexposure to Na-free P-solutions. In contrast to fig. 3, all R-values belonging to the same P-solution have been collected in a separate diagram. The concentration scales of these diagrams mean brief increases of $(Na)_0$ (empty circles) or $(K)_0$ (filled circles). For ease of comparison, corresponding curves of different figures are labeled with the same numbers. Also shown is the effect of Amiloride (dashed curves), which will be discussed later.

Turning first to fig. 4, we find a striking difference to fig. 3 in responses after equilibration with solutions containing Na. With high $(Na)_0$ P-solutions, the instantaneous R-(K_2SO_4) function shows a positive slope (curve 4): *the resistance increases with $(K)_0$ while $(Na)_0$ of test solutions is zero, but can do so only when the skin was preexposed to Na-containing solutions.* We like to suggest that this represents an inhibitory decrease of the membrane Na-conductance g_m^{Na} caused by K, which can happen only if g_m^{Na} – as a result of preexposure with Na – is larger than zero to begin with.

Interestingly, the positive slope became larger when preincubation was done with P-solutions of lower Na-concentration (curve 2). This would be expected, if extracellular K and the Na left in the tissue from the preexposure

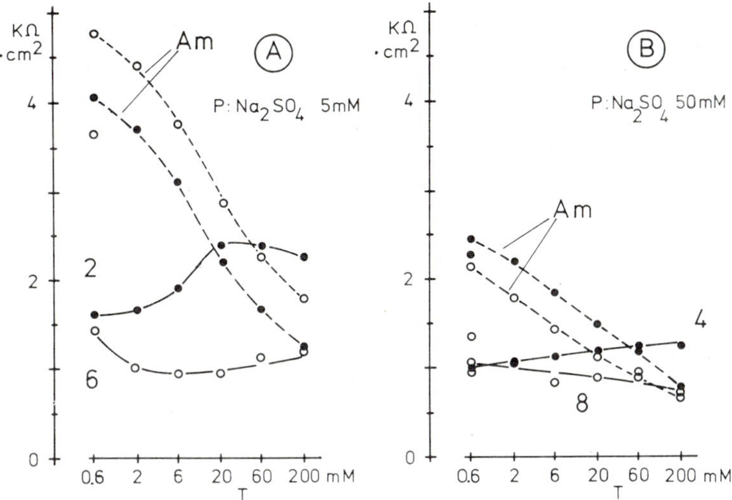

Fig. 4. Instantaneous R-c_0 curves. A: low, B: high $(Na)_0$ P-solution. Filled circles: K-T-solutions; empty circles: Na-T-solutions (concentrations on abscissa). Numbers point to corresponding curves of fig. 3. Amiloride concentration in P- and T-solutions: 8 mg/l (dashed curves). "Lonely" circles on the left side of each diagram are repeats with 0.6 mM Na_2SO_4 T-solution (containing Amiloride). These low values were recorded after using all Na-Amiloride T-solutions once, those of high concentration last. The repeats show that the highly concentrated solutions can increase c_s of the shunts within 2 s, an effect which becomes noticeable at very low outside concentrations (high R). It does not become noticeable at high concentrations (low R), were repeats are identical. Repeats with low concentrations agree better when taken a few minutes later (see fig. 5). The dashed curves of each diagram differ in amplitude by the same amount, by which the resistance during exposure to P-solution differed. When referred to this resistance (see fig. 2B), the curves become identical.

period would compete for the same transport channels. The same phenomenon was observed, when a constant Na-concentration was maintained in all test solutions [13]. At larger $(K)_0$-values of test solutions, the slope of curve 2 became zero, then turned negative. Apparently, after complete inhibition of g_m^{Na}, a $(K)_0$-dependent conductance remained. On the whole, the inhibition of g_m^{Na} by K was much more apparent in this preparation than in that of fig. 3.

The same difference in preparations is found when we turn to fig. 5. Here curves 5 and 7 show negative slopes (i.e. no saturation) in cases where P-solutions contained K rather than Na (in contrast to curves 5 and 7 of fig. 3). In the framework or our interpretation, the negative slope, i.e. the lack of satu-

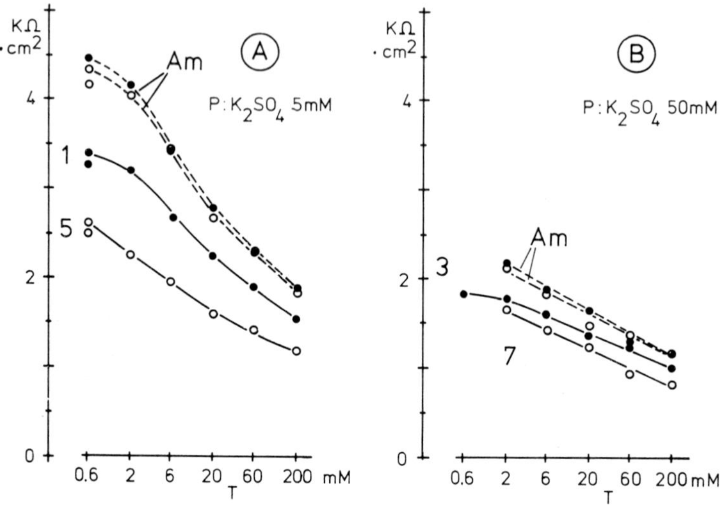

Fig. 5. Instantaneous R-c_0 curves. A: low, B: high $(K)_0$ P-solutions. See fig. 4.

ration would be due to inhibition of g_m^{Na} by K at low $(Na)_0$. We suggest that K remained in the vicinity of the Na-selective membrane during brief exposure to K-free solutions. Thus, at small $(Na)_0$ values, g_m^{Na} was small or zero, while at larger $(Na)_0$ the K-inhibition was gradually overcome by Na.

After preexposure to low $(K)_0$, slopes and absolute R-values in fig. 5 are larger than after preexposure to high $(K)_0$, as in fig. 3A.

Slopes and absolute R-values are found to be almost identical for brief $(K)_0$ and $(Na)_0$ exposures, if preexposure was to high $(K)_0$ (compare curves 3 and 7 of fig. 5). Apparently, in both cases only unspecific pathways (not involving g_m^{Na}) are used, possibly because g_m^{Na} is blocked by K. Accordingly, Amiloride is without effect after preexposure to high $(K)_0$ (see below).

Our results agree with reports from other laboratories, that in frog skin K can inhibit the uptake of Na from the outside solution [19]. A decrease of Na uptake with increasing $(Li)_0$ has also been described [18].

8. Inhibitory effect of Amiloride

Properties of the Na-selective membrane were investigated further by studying the effect of Amiloride on instantaneous R-c functions. This drug is known to block entry of Na from the outer bathing solution into the transport compartment [22, 23]. The effect of Amiloride can be described as follows:

1. R-values and slopes increase
2. saturation phenomena disappear
3. differences in slope between R-$(Na_2SO_4)_0$ curves and R-(K_2SO_4) curves disappear
4. the equilibration media retain some influence.

These results are readily explained by assuming that Amiloride reduces g_m^{Na} to zero. The remaining conductances are unspecific and can be increased by briefly raising $(Na)_0$ just as well as by briefly raising $(K)_0$. Short-time effects of $(Na)_0$ and $(K)_0$ are smaller after equilibration with high $(Na)_0$ and $(K)_0$ P-solutions because then the mean electrolyte concentration c_s in the shunt pathway is large even when external concentrations are briefly lowered.

Interestingly, slope and R-values are hardly changed by Amiloride after preexposure to a high $(K)_0$ P-solution (fig. 5). In agreement with the speculation offered at the end of section 7 we suggest, that in this case g_m^{Na} was already small before addition of Amiloride, possibly because of inhibitory effects of slowly exchanging K on g_m^{Na}.

9. Where is the slowly exchanging compartment?

In the experiment of fig. 3A, curve 4, g_m^{Na} was determined by the Na-concentration of the P-solution, with which the skin was preequilibrated: after preexposure to large outer Na-concentrations, g_m^{Na} could retain large values while $(Na)_0$ was temporarily reduced to zero.

What could be the nature of this "memory"? Apparently, in the situation described, Na is still available to the Na-specific transport channels of the membrane. It could be stored in many locations, like a) at the inner side of the membrane, b) in the channels themselves, c) at the outer surface of the membrane or in a space in front of and in contact with the outer surface of the membrane (like the extracellular space between *str. corneum* and *str. granulosum*), d) in the *str. corneum,* e) in the outer part of the extracellular shunt pathway, f) in gland ducts or g) even in special cells (like mitochondria-rich "flask"-cells). We do not know which of these choices or which combination of them is correct. However, since the matter may be important for studies of short time Na-isotope uptake, we want to report an observation which will at least exclude two choices from the above list.

As explained before, when the skin was first equilibrated with a solution of high $(Na)_0$, and then briefly exposed to a test solution of sufficiently low

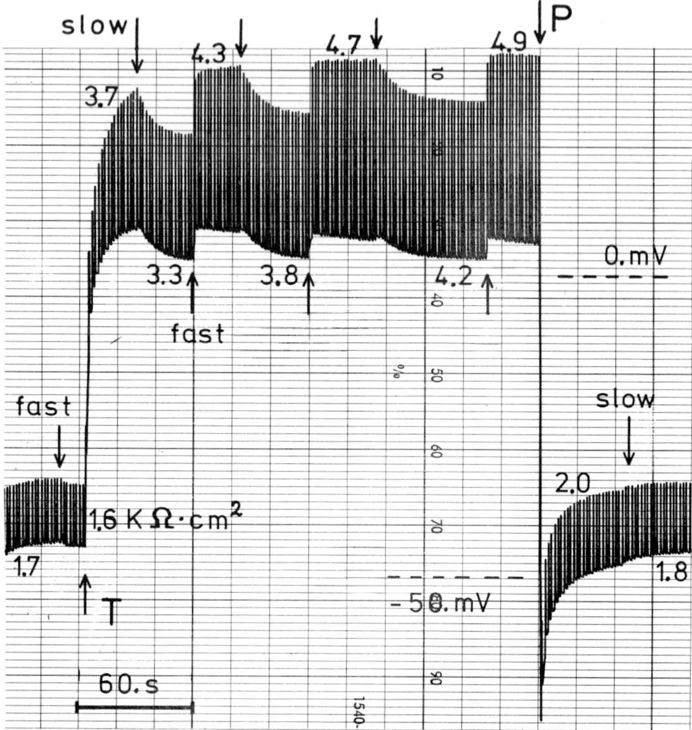

Fig. 6. Voltage and resistance response of abdominal skin (*Rana esculenta*) to a P-T-P-series. P-solution: Na_2SO_4 50 mM, T-solution: Na_2SO_4 0 mM. Both contained 1 mM $CaSO_4$, 5 mM Tris-SO_4 *p*H 6. The lower envelope is the change of transepithelial voltage (resting potential was −45 mV). This change is particularly slow when T-solution contains neither Na nor K. Brief upward deflections are responses to 10 $\mu A/cm^2$ current pulses. Thus the distance of both envelopes is proportional to skin resistance. Computed resistance values are marked on the record in KOhm · cm^2. Speed of flow was changed between 0.3 ml/s and 40 ml/s. Each decrease in flow-speed during exposure to T-solution caused a relatively slow hyper-polarisation and drop in resistance, each increase in flow caused a much faster opposite change.

$(Na)_0$, R increased (by a factor of 2 in fig. 3B). However, when during exposure to low $(Na)_0$ *the rate of flow* was decreased or the flow stopped, R decreased significantly. Slow flow, of course, means a thicker unstirred layer and thus less diffusional losses of Na towards the outside solution. The effect could be repeated several times *without changing* $(Na)_0$*: when the flow increased, R increased and vice versa* (fig. 6).

The repeatability of R-decreases in periods of slow flow is not explained

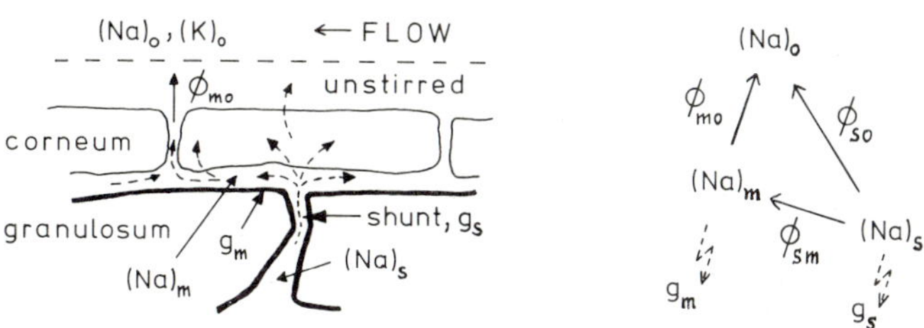

Fig. 7. Possible location of a slowly exchanging extracellular compartment, which can resupply the membrane with Na while $(Na)_0$ is briefly made zero. Resupplying is improved when the flow of the outer solution is made slower (see text and fig. 6).

by storage in Na-channels or by a slowly exchanging compartment at the outer surface of the membrane. When such a compartment would loose Na during a period of fast flow, its Na-concentration and thus g_m^{Na} could not regain its previous large value in the next period of slow flow. To explain our observation, *we need a Na-storing compartment which – in a period of slow flow – can resupply the membrane with Na lost in the previous period of fast flow.* This excludes choices b and c from our list (if they do not appear in combination with other choices).

A model which would work in the desired way is illustrated in fig. 7. The extracellular space between cells of the outer layer of the *str. granulosum* has, by way of example, been chosen as the Na-storing compartment. The store is loaded with Na during equilibration with P-solutions of large $(Na)_0$, and discharges when $(Na)_0$ is lowered. Distribution of discharged Na^+ over the whole outer membrane area could be facilitated by diffusion along fixed charges on the outer membrane surface and/or spacial restrictions (*vide* inner membrane of *str. corneum*). For the possible role of "surface conduction" in epithelia, see ref. [24, 25, 27]. The space in front of the membrane (concentration $(Na)_m$) can be thought to be functionally located between the Na-storing reservoir (concentration $(Na)_s$) and the bulk of unstirred layers as well as – further outside – the flowing solution (concentration $(Na)_0$). These three compactments may be connected by diffusional fluxes (ϕ) as shown in the right part of fig. 7. Constancy of $(Na)_m$) (and thus of R) means that ϕ_{sm} equals ϕ_{mo}. When ϕ_{mo} is increased by better stirring, $(Na)_m$ becomes smaller (R increases) such that ϕ_{mo} is lowered and ϕ_{sm} increased, until finally a new steady value of $(Na)_m$ (and R) is reached when ϕ_{sm} once again equals ϕ_{mo}.

We do not require a fast discharge ability or a large capacity of the Na-storing compartment, if the membrane can already respond to fairly small Na-concentrations. The membrane was in fact found to be quite sensitive. In one case, its resting voltage increased by 3.3 V in response to an increase of $(Na)_0$ from "zero" to 78 μM. At large $(Na)_0$, the mechanism outlined above might be overruled by direct equilibration between the outside solution and the space before the membrane, as well as between outside solution and extracellular shunt. The rate of flow had a much smaller effect while $(Na)_0$ was large (fig. 6).

10. Summary and conclusion

The outer, Na-selective membrane of frog skin epithelium was studied *in vitro* by observing its resistance changes in response to concentration steps in the outside bathing solution. Using a fast flow chamber with sufficient stirring capability, p.d. responses to concentration changes became complete within 0.3–1.5 s. Thus, exposure times to test solutions could be limited to 2 s.

Plotting the resistance (R) changes – which occurred in this time – against the Na or K concentration (c_0) of test solution yields "instantaneous R-c functions". Their slopes were taken to represent properties of the Na-selective membrane, its extracellular shunt-pathways and its protective layers. The slopes were found to depend strongly on the composition of the outer bathing solution, with which the skin was equilibrated before exposure to test solutions.

Instantaneous R-$(Na_2SO_4)_0$ curves were steep after preexposure to solutions that contained no Na, but much less steep after preexposure to Na. We speculate that this "remembering" of the composition of the preexposure medium was due to small amounts of Na leaking back from a slowly exchanging Na-compartment towards the outer membrane surface. The slow compartment could for instance be the outer part of the extracellular shunt-pathway. After blocking the Na-selective membrane with Amiloride, the memory effect became unnoticeable.

The instantaneous R-$(Na_2SO_4)_0$ curve usually showed a negative slope at small $(Na)_0$ and a horizontal branch at larger $(Na)_0$, regardless (in many cases) of which medium was used before for equilibration. We speculate that the horizontal course is due to conductance saturation of the Na-selective membrane (the concept is defined). The effect disappeared under exposure to Amiloride.

The R-$(K_2SO_4)_0$ curve could show a positive slope after preexposure to Na. We speculate that the membrane was still supplied with Na from the slowly exchanging compartment. When $(K)_0$ was transiently increased, K would block the Na-specific membrane channels, increasing R. The effect disappeared under exposure to Amiloride.

In this preparation, the membrane was quite sensitive to the blocking effect of K. After preexposure to high concentrations of K_2SO_4, the instantaneous R-$(Na_2SO_4)_0$ and R-$(K_2SO_4)_0$ curves were almost identical, the Na specificity had disappeared. We speculate that the membrane was still blocked by K, which was left in the slowly exchanging compartment during brief changes of the outside solution. Amiloride was almost without effect in this experimental situation.

In all cases studied, the slopes of instantaneous R-$(Na_2SO_4)_0$ and R-$(K_2SO_4)_0$ curves became almost identical under exposure to Amiloride (if they were not identical already, see above).

The interpretations developed in this paper are based on the conventional model of transcellular Na transport [7]. It includes the notion that Na enters an intracellular transport compartment by penetrating an apical cell membrane. This resistive "input" step could have a maximal transport rate (be saturable) but would prefer Na to K and be blocked by Amiloride [22, 23] but not by Ouabain [26]. For a different model of Na transport across epithelia, see ref. [24, 25].

Acknowledgement

It is a pleasure to thank Mr. H. Kilb for the excellent work he did in machining the fast-flow chamber. Our thanks are also due to Dr. N. Anna Nacimiento for performing preliminary experiments and to Mr. W. Ruffing for expert technical assistance.

References

[1] M.G. Farquhar and G.E. Palade, Functional organisation of amphibian skin. Proc. Nat. Acad. Sci. (Wash.) 51 (1964) 569–577.

[2] P.F. Parakkal and G.A. Matoltsy, A study of the fine structure of the epidermis of Rana pipiens. J. Cell Biol. 20 (1964) 85.

[3] T. Hoshiko, Electrogenesis in frog skin, in: Biophysics of Physiological and Pharmacological Actions. Amer. Ass. Advancem. Sci. (Wash.) 1961, p. 46.

[4] G. Whittembury, Electrical potential profile of the toad skin epithelium. J. Gen. Physiol. 47 (1964) 795–808.

[5] B. Lindemann and U. Thorns, Fast potential spike of frog skin generated at the outer surface of the epithelium. Science 158 (1967) 1473–1477.

[6] A. Martinez-Paloma, D. Erlij and H. Bracho, Localization of permeability barriers in the frog skin epithelium. J. Cell Biol. 50 (1971) 277–287.

[7] H.H. Ussing and E.E. Windhager, Nature of shunt path and active sodium transport

path through frog skin epithelium. Acta Physiol. Scand. 61 (1964) 484.

[8] C.L. Voute and H.H. Ussing, Some morphological aspects of active sodium transport in the epithelium of the frog skin. J. Cell Biol. 3 (1968) 625.

[9] R. Brennecke and B. Lindemann, A chopped-current clamp for current injection and recording of membrane polarization with single electrodes of changing resistance. T.I.T. J. Life Sci. 1 (1971) 53–58.

[10] J. Crank, The Mathematics of Diffusion (Oxford University Press, London, 1956), p. 45.

[11] W. Fuchs, U. Gebhardt and B. Lindemann, Delayed voltage responses to fast changes of $(Na)_0$ at the outer surface of frog skin epithelium, in: Passive Permeability of Cell Membranes, Series Biomembranes, Plenum Press, F. Kreuzer and J.F.G. Slegers, eds., in press.

[12] B. Lindemann, Experimentelle und theoretische Untersuchungen des Na- und Wassertransportes durch das Epithel der Froschhaut. Thesis, University of Saarland, Medical School Homburg (1968).

[13] B. Lindemann, Electrical excitation of the outer cell membrane in frog skin epithelium, in: Electrophysiology of Epithelia, ed. Giebisch, Symposia Medica Hoechst (Schattauer Publishing Company, 1971).

[14] K. Heckmann, B. Lindemann and J. Schnakenberg, Current-voltage curves of porous membranes in the presence of pore blocking ions. I: Narrow pores containing no more than one moving ion. Biophys. J. 12 (1972), in press.

[15] L.B. Kirschner, On the mechnism of active sodium transport across the frog skin. J. Cell. Comp. Physiol. 45 (1955) 61–87.

[16] A.C. Brown, Current and potential of frog skin in vivo and in vitro. J. Cell. Comp. Physiol. 60 (1962) 263–269.

[17] M. Cereijido, F.C. Herrera, W.J. Flanigan and P.E. Curran, The influence of Na concentration on Na transport across frog skin. J. Gen. Physiol. 47 (1964) 879–893.

[18] T.U.L. Biber and P.F. Curran, Direct measurement of uptake of sodium at the outer surface of the frog skin. J. Gen. Physiol. 56 (1970) 83–99.

[19] C.A. Rotunno, F.A. Vilallonga, M. Fernandez and M. Cereijido, The penetration of sodium into the epithelium of the frog skin. J. Gen. Physiol. 55 (1970) 716–735.

[20] A.C. Brown and K.G. Kastella, The AC impedance of frog skin and its relation to active transport. Biophys. J. 5 (1965) 591–606.

[21] P.G. Smith, The low-frequency electrical impedance of the isolated frog skin. Acta Physiol. Scand. 81 (1971) 355–366.

[22] J. Eigler and J. Crabbe, Wirkung von Diuretika auf den aktiven Natriumtransport an Amphibienmembranen, in: Renaler Transport und Diuretica, Intern. Sympos. Feldafing (1968) p. 195.

[23] W. Nagel and A. Dorge, Effect of Amiloride on sodium transport of isolated frog skin. Pflugers Arch. 316 (1970) R 47.

[24] M. Cereijido and C.A. Rotunno, Fluxes and distribution of sodium in frog skin. A new model. J. Gen. Physiol. 51 (1968) 280–289 S.

[25] M. Cereijido and N. Fraidenrach, Surface vs transcellular routes in the transport of sodium across epithelial membranes, in: Horizons in Surface Sciences, eds. Sears and Prince (Appleton-Century Crafts, N.Y., 1970).

[26] V. Koefoed-Johnsen, The effect of G-strophantin (Ouabain) on the active transport of sodium through the isolated frog skin. Acta Physiol. Scand. 42 (1958) 145 S.

[27] H.B. Steinbach, Movement of Na and K across and within frog skin. Am. J. Physiol. 212 (1967) 371–375.

STRUCTURE AND FUNCTION OF THE OUTER BORDER OF THE FROG SKIN

D. ERLIJ, T. MACHEN *, A. MARTINEZ–PALOMO
and
M.W. SMITH
Institute of Animal Physiology, A.R.C., Babraham, Cambridge, England and Centro de Investigación, Instituto Politécnico Nacional, México 14, D.F. México

1. Introduction

In this communication we will give a brief description of some studies in progress directed toward understanding the organization and function of the outer barrier of the frog skin.

Stimulus for the first two groups of experiments originated from observations made on frog skins bathed on the outside with Ringer's solutions containing small amounts of lanthanum. It was found that the membranes of the cells in the *s.corneum* allow the movement of lanthanum into the space separating the *s.corneum* and the *s.granulosum* [1, 2]. The penetration of lathanum from the outside solution is interrupted by the barrier formed by the outside membranes of the cells in the outer layer of the *s.granulosum* and the tight junctions that seal them together [1, 2].

The question raised by these observations that have been approached experimentally are: First, can the ionic lanthanum method show the pathway followed by large molecules across skins made leaky with outside hypertonic solutions? Second, what role does the external compartment, limited between the outer border of the *s.granulosum* and the outer anatomical border of the skin, play in the recent measurements [3, 4] of Na uptake from the outside solution into the skin? A third set of experiments was directed toward explor-

* Dr. Machen held a Bank of America–Giannini Foundation Post-doctoral Fellowship during the course of this study.

ing the possibility that the mechanism of Na uptake across the outer surface of the frog skin is linked in some manner to the excretion of H^+ ions [5–7].

2. The distribution of lanthanum in skins treated with hypertonic solutions

Ussing and Andersen [8] found that when the solution bathing the outside of the skin was made hypertonic by the addition of urea or other hydrophilic solutes the skin potential is reversibly and markedly reduced. Associated with this drop in potential there is a large increase in the passive permeability to anions and cations [9]. Furthermore the movement of substrates that normally do not penetrate cell membranes and whose rate of transfer across the skin is negligible, markedly increased during the action of the hypertonic solutions [10]. More recently Ussing [11] found that when skins were equilibrated with a Ba^{2+} containing solution on one side, and SO_4^{2-} solution on the other, no $BaSO_4$ precipitates were observed within the intercellular spaces of the control epithelia. However, when the outside solution was made hypertonic, $BaSO_4$ precipitates were localized with the electron microscope, within the intercellular spaces of the epithelium. On this basis Ussing [10, 11] has suggested that the hyperosmotic solution leads to an opening of tight junctions sealing together the cells of the outer layers of the skin.

As a first step in the use of lanthanum to test this hypothesis, we measured the movements of ^{140}La from the outside solution into and across the skins bathed with Ringer +400 mM Urea on the outside. We know that in control skins, after an initial period of rapid uptake that lasts for about 10 min, there is little or no further increase in the amount of ^{140}La contained in the skin. Furthermore no movement of ^{140}La into the inside solutions could be detected in these skins [2]. When 400 mM Urea was added to the outside solution after the constant level of ^{140}La had been reached a marked increase in the rate of uptake was observed. Furthermore ^{140}La now appeared at a measurable rat on the inside solution. A lanthanum permeability constant of 1×10^{-7} cm sec^{-1} was calculated.

When skins treated as outlined above, were fixed with glutaraldehyde and examined with the electron microscope, images like those in fig. 1 were found. In contrast with the observations in control skins, where La^{3+} penetrates from outside into the cells of the *s.corneum* and into the space between the *s.corneum* and the *s.granulosum,* in the skins treated with Ringer +400 mM Urea the La^{3+} penetrated beyond the outer border of the *s.granulosum* into the intercellular spaces of all the inner layers of the epidermis.

An interesting feature of the urea treated skins, illustrated in fig. 1, is that

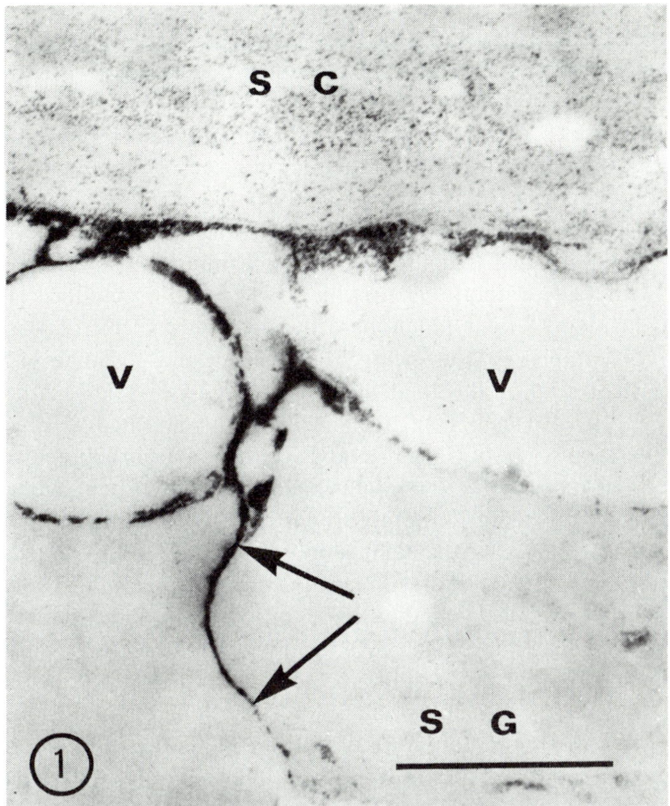

Fig. 1. The distribution of lanthanum in a skin treated with Ringer +400 mM urea for one hour. Lanthanum chloride appears as a dense irregular precipitate in the cytoplasm of the cells of the *s.corneum* (S.C.) and in the intercellular spaces separating the *s.corneum* from the *s.granulosum* (S.G.). In addition the electron dense tracer has penetrated into the lateral intercellular space between two cells of the *s.granulosum* at the level where tight junctions would normally be obliterating this space. The vacuoles formed during the treatment are indicated by *V*. The walls of the vacuoles are very thin, but examinations at high magnification shows that the electron dense tracer did not penetrate into them. Specimen fixed in glutaraldehyde and embedded in Epon. No counterstain was used. × 28 000 scale = 1 micron.

the extracellular spaces along the lateral surfaces of the outer layer of the *s.granulosum* are continuous, and allow the movements of La^{3+} throughout their extension, i.e., the tight junctions that normally prevent the inward movement of lanthanum and other tracers at this site are now open.

An additional feature of the epithelia treated with hypertonic solutions is illustrated in fig. 1. The cells in the *s.granulosum* develop numerous vacuoles. These vacuoles presumably arise also from the treatment with hypertonic solutions since they were not found in control skins.

Our observations are direct evidence indicating that the preferred route for the La^{3+} movements across the epithelium induced by the hypertonic treatment is through the extracellular route formed during the opening of tight junctions.

It has been suggested that the increases is permeability caused by treating epithelia with Ca-free solutions to which EDTA has been added also result from the opening of tight junctions. However, the available anatomical evidence is not completely convincing. Hays, Singer and Malamed [12] found that after treatment of the toad bladder with Ca-free solutions containing EDTA, the epithelial cells could be mechanically detached with great ease. Unfortunately no study of the appearance of the junctional complexes before mechanical separation was published. Sedar and Forte [13] working with frog gastric oxyntic gland found consistent changes only in the intermediate junctions. The changes in the tight junctions of the EDTA treated preparations were restricted to focal regions of membrane separation. These authors were unable to establish whether these openings in the *z.occludens* were continuous with the secretory surface. Cassidy and Tidball [14] working with EDTA treated rat intestine found only on one occasion an opening of the tight junctions. They concluded that this alteration was due probably to a distortion of the preparation and not a result of Ca depletion and also that the changes in epithelial permeability resulted from alterations in the cell membrane permeability and not from opening of intercellular junctions. Electron dense extracellular tracers have to be used when studying the permeability of tight junctions. However, lanthanum may not be the optimal tracer since, as a polyvalent cation, it may reverse the effects of the Ca free solutions.

3. Sodium uptake by the outside surface of the frog skin

The presence of a layer of cells interposed between the outer anatomical border of the skin and what is presumably the outer border of the transport compartment [2] raises the possibility that the movement of substances of different sizes may not be equally obstructed on their way to the outer sodium sensitive barrier. Such a situation may result in marked difficulties in obtaining accurate estimates of the Na uptake by the skin. Indeed, they may

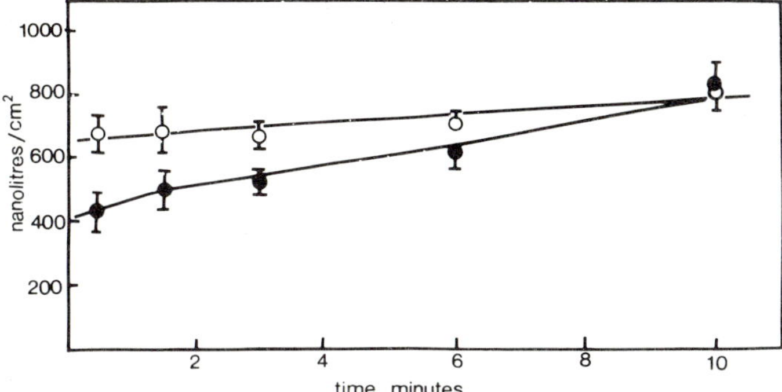

Fig. 2. The time course of the equilibration of mannitol-^{14}C and carboxy inulin-^{3}H spaces on the outside surface of the frog skin. Filled circles: inulin space. Empty circles mannitol space. Each point is the average of 24 determinations. Both tracers were measured in a single piece of skin in each determination. The vertical bars represent the standard error of the mean. The ordinates show the extracellular space calculated assuming that the tracers were present in the extracellular water adhering to the outside surface of the skin in the same concentration as in the outside bulk solution.

contribute importantly to the recent finding that sodium flux from the outside solution into the skin (J_{12}) is several times larger than sodium flux across the whole tissue (J_{13}) particularly when high outside sodium concentrations are used [3,4]. Therefore we decided to compare Na uptake using different extracellular space markers to determine the amount of extracellular sodium present [15].

Frog skins were mounted in chambers constructed as described for the intestine by Schultz, Curran, Chez and Fuisz [16]. The movements of ^{24}Na and ^{3}H- or ^{14}C-inulin was determined in all experiments; in addition the uptake of either ^{3}H-sucrose or ^{14}C-mannitol was also measured. The amount of mannitol that equilibrated with the skin remained constant between 30 sec and 10 min after its addition. On the other hand the inulin space, determined in the same experiment at 30 sec was only 0.65 times the mannitol space and it only reached the same value as the mannitol space between 6 and 10 min after the addition of the labelled solution (fig. 2). The sucrose space, which increased continuously with time, was always larger than inulin space.

The initial uptake of sodium, calculated by subtracting the inulin space, was plotted as a function of the outside sodium concentration, $[Na]_o$. The whole process could be described by the combination of a saturating and a linear component as described previously by Biber and Curran [4]. Thus sodium uptake (J_{12}) is given by

$$J_{12} = \frac{J_m [\mathrm{Na}]_o}{K_{\mathrm{Na}} + [\mathrm{Na}]_o} + \alpha [\mathrm{Na}]_o \,.$$

J_m, the maximal influx of the saturating component, was 1.27 μeq/hr/cm^2. K_{Na}, the apparent Michaelis constant was 15 mM and the coefficient α was 0.010 cm/hr. When sodium uptake was corrected for the mannitol space the relationship could still be described by the same process using identical values for K_{Na} and J_m but α now becomes equal to 0, i.e. the linear component of uptake has disappeared. An average value of J_{13} = 0.70 μeq/hr/cm^2 was determined in parallel experiments in which skins bathed with 115 mM Na Ringer's on both sides were mounted between Ussing type chambers.

The results in this section show that the estimations of the volume accessible to conventional extracellular tracers at the outer border of the skin depends on the tracer selected for the determinations. There are some findings that make it tempting to identify this space with the cells in the *s.corneum.*

Overton [17] found in 1904 that acidic aniline dyes penetrated into the cells of the *s.corneum* but did not move beyond the outer border of the *s.granulosum.* Thus the cell membranes of the *s.corneum,* in addition to being permeable to lanthanum [1, 2], allow the movement of organic molecules that do not penetrate into the cells of other layers. These observations and the finding that inulin and mannitol spaces reach an identical value when sufficient time for equilibrium is allowed, are most conveniently explained by proposing that these tracers move into the cells of the *s.corneum.*

Regardless of precise anatomical identifications our data indicate that the linear of the relationship between outside [Na] and J_{12} corresponds to movements of sodium into a space which is also accessible to mannitol and inulin. Inulin moves more slowly than mannitol into this space and for this reason the initial rates of sodium movement calculated using inulin as a space marker remain uncorrected for sodium present in the space.

Our conclusions are necessarily at variance from those implicit in determinations of J_{12} [3, 4] made before the role of the *s.corneum* was accounted for. The high values of J_{12} suggested that a great fraction of the sodium moving across the outer border of the epithelium when high $[\mathrm{Na}]_o$ were used had to flow back to the outside solution. It is interesting, however, that Biber and Curran [4] had already warned of the possibility that the linear component of J_{12} could result from the movement of sodium into a compartment unrelated to transepithelial transport.

4. The movements of H^+

We have studied the movements of H^+ from the skin into the outside solution because it has been suggested that the movement of sodium from the outside solution into the skin may result from the forced exchange of Na^+ by H^+ [5, 6] or that the outward movement of H^+ results from the potential difference generated by the inward movement of sodium [7].

In addition we have performed some experiments that could provide some clues on the origin of the H^+ secreted by the skin.

The experiments were carried out in isolated skins mounted in Ussing type chambers containing Ringer-Cl which in addition to the electrodes for measuring potential difference and passing current were designed to allow the measurement of hydrogen production on the outside solution by the *p*H-stat method. Also provisions were taken to use CO_2 free solution [18].

Our main observations can be summarized as follows:

a) Hydrogen moves from the skin into the outside solution at a rate of 0.040 μeq/hr/cm^2 while Na influx had a value of 0.38 μeq/hr/cm^2.

b) The rate of H^+ secretion is not significantly affected by substituting the Na in the outside solution either by K or Mg nor by inhibiting Na movements with amiloride.

c) The rate of H^+ production is not underestimated because it may have been neutralized by HCO_3 secreted into the outside solution in exchange for Cl^-. Substituting all the Cl^- by SO_4^{2-} in the outside solutions does not result in an increase in the rate of H^+ production.

d) The steady state rate of H^+ secretion is not affected by large changes in electrochemical potential gradient for H^+. Neither abolishing the potential difference across the skin nor a tenfold change in H^+ concentration in the outside solution affected significantly the steady state rate of H^+ secretion.

e) The hydrogen secretion was abolished by the metabolic inhibitors dinitrophenol (1×10^{-4} M) and Antimycin A (1.5×10^{-6} M which also markedly reduced the potential difference across the skin.

f) Acetazolamide (5×10^{-3} M) blocked H^+ secretion without altering the potential difference across the skin.

Observations (a), (b) and (f) clearly indicate that H^+ and Na^+ movements across the outer border of the isolated frog skin are not coupled. The ratio of Na^+ to H^+ movements is very different from unity and Na movements can be abolished without any effects on H^+ secretion and conversely H^+ movements can be abolished without interruption of Na^+ uptake.

A second conclusion suggested by these results is that the H^+ secretion does not result from movement of hydrogen following its electrochemical potential gradient since tha rate of secretion is not affected by marked changes in either potential or $[H^+]$. Furthermore the results using metabolic inhibitors suggest that the H^+ secretion is coupled in some manner with the energy producing reactions of the cell. The inhibition caused by Antimycin A could indicate that the acidification of the outer surface may be due to either the movement of respiratory CO_2 into the outside solution or to H^+ liberated during electron movement along the respiratory chain. However, the effects of dinitrophenol, an inhibitor that blocks oxidative phosphorilation without inhibiting respiration, suggest that the interruption of the respiratory chain caused by Antimycin A is not the direct cause of the inhibition of H^+ secretion. A more plausible interpretation for these observations is that both inhibitors act by reducing the levels of energy rich phosphate compounds. In addition the effects of acetazolamide support the idea that the acidification of the outside solution is not the result of respiratory CO_2 being liberated into the outside solution. This finding indicates that if respiratory CO_2 contributes to the production of secreted H^+ it has to be previously hydrated.

Although our findings clearly establish that the Na uptake through the amiloride sensitive path of isolated skins is not coupled to H^+ secretion, they do not offer an explanation for the findings of experiments in whole animals in which a coupling between Na and H^+ has been observed [5, 7].

References

[1] H. Bracho, D. Erlij and A. Martínez-Palomo, J. Physiol. 213 (1971) 50P.
[2] A. Martínez-Palomo, D. Erlij and H. Bracho, J. Cell Biol. 50 (1971) 277.
[3] C.A. Rotunno, F.A. Vilallonga, M. Fernández and M. Cereijdo, J. Gen. Physiol. 55 (1970) 716.
[4] T.U.L. Biber and P.F. Curran, J. Gen. Physiol. 56 (1970) 83.
[5] F. Garcia-Romeu, A. Salibian and S. Pezzani-Hernandez, J. Gen. Physiol. 53 (1969) 816.
[6] R.D. Keynes, Quart. Rev. Biophys. 2 (1969) 177.
[7] F. Garcia-Romeu, Phil. Trans. Roy. Soc. B 262 (1971) 163.
[8] H.H. Ussing and B. Andersen, Proc. 3rd Int. Congr. Biochem. Brussels Academic Press, (New York, 1955) p. 434.
[9] H.H. Ussing and E.E. Windhager, Acta Physiol. Scand. 61 (1964) 484.
[10] H.H. Ussing, Ann. N.Y. Acad. Sci. 137 (1966) 543.
[11] H.H. Ussing, Phil. Trans. Roy. Soc. B 262 (1971) 85.
[12] R.M. Hays, B. Singer and S. Malamed, J. Cell Biol. 25 (1965) 195.
[13] A.W. Sedar and J.G. Forte, J. Cell Biol. 22 (1964) 173.
[14] M.M. Cassidy and C.S. Tidball, J. Cell Biol. 32 (1967) 685.
[15] D. Erlij and M.W. Smith, J. Physiol. 218 (1971) 33 p.

[16] S.G. Schultz, P.F. Curran, R.A. Chez and R.E. Fuisz, J. Gen. Physiol. 50 (1967) 1241.
[17] E. Overton, Verh. phys.-med. Ges. Würzburg, 36 (1904) 277.
[18] M.G. Emilio, M.M. Machado and H.P. Menano, Biochim. Biophys. Acta 203 (1970) 394.

CELLULAR POOLS INVOLVED IN ACTIVE K-TRANSPORT ACROSS THE ISOLATED CECROPIA MIDGUT

William R. HARVEY and John L. WOOD
Department of Biology, Temple University, Philadelphia, Pennsylvania, and Department of Zoology, University of Cambridge, Cambridge, England

1. Introduction

A new analysis of the kinetics of potassium movements across the isolated silkworm midgut leads us to the conclusion that the transport route is through the epithelial cells. This conclusion is different from that reached by Harvey and Zerahn [1] who could not eliminate a pathway between the cells. The key observation forcing this reexamination of the route followed by the transported K was the finding by Wood [2] that the lag-time for ^{86}Rb-labeled K-influx across the *Antheraea pernyi* midgut is much longer than that reported by Harvey and Zerahn for the ^{42}K-influx across the *Hyalophora cecropia* midgut. We will now show that the lag-time for the ^{42}K-influx across the *H. cecropia* midgut is likewise longer than that reported by Harvey and Zerahn. The long lag-time corresponds to a large potassium pool in the midgut and the large pool in turn leads to the conclusion that the K-transport proceeds through the cells. This cellular pathway is supported by ^{42}K- and ^{22}Na-loading kinetics.

1.1. *General review of the midgut transport system*

Although Keynes [3] reviewed the midgut transport system recently, several of its characteristics are worth mentioning here as background for the kinetic analysis. When the midgut is isolated from *H. cecropia, A. pernyi* or any of several other Lepidopterous insects and is bathed with identical physiological solutions on both sides, it produces a spontaneous potential difference (PD). For example, in a solution containing 32 mM/1 K and other ions (Sl of

Table 1
Composition of solutions used to bathe the isolated midgut (mM/l).
All solutions are at *p*H of 8.3.

	S1 (32-S)	35(K + Rb)-S	2-S-Tris	35-S-Tris	50-S-Tris	100-S-Tris	K-32	Na-32
KCl	30	15	2	35	50	100	30	–
$KHCO_3$	2	5	–	–	–	–	2	–
NaCl	–	–	–	–	–	–	–	30
$NaHCO_3$	–	–	–	–	–	–	–	2
RbCl	–	15	–	–	–	–	–	–
$MgCl_2$	5	5	–	5	–	5	–	–
$CaCl_2$	5	5	5	5	5	5	–	(±5)
Tris base	–	–	5	5	5	5	–	–
Sucrose	166	166	250	150	250	30	166	166
Glucose	–	–	–	5	–	5	–	–

table 1), the lumen-side is about 140 mV positive to the blood-side. The short-circuit current, I_{sc}, produced by the midgut in this solution amounts to as much as 3 mA for a midgut sphere approximately 6 cm^2 in area. Folding of the midgut surface and of the plasma membranes of the epithelial cells increases the effective surface area some $10^2 - 10^3$ times, so that the I_{sc} expressed on the basis of the area of the plasma membrane would correspond to 5 – 50 peq./cm^2/sec.

The active transport of potassium from blood-side (hemolymph-side) to lumen-side in solutions containing 32 mM/l of K was found to account for the I_{sc} within an error of 17% by Harvey and Nedergaard [4], within an error of 10% by Harvey, Haskell, and Zerahn [5], and more recently, within an error of 1% by Wood [2]. However, in appropriate bathing solutions any one of the other alkali metal ions can be transported. Rb is transported even at high K concentrations and almost at the same rate as K [6], but Cs is transported only at low K concentrations [7]. Although K, Rb, and Cs are transported in the presence of Ca^{2+} and Mg^{2+}, Na and Li are transported only when these divalent cations are absent. Even with Ca^{2+} and Mg^{2+} absent, Na and Li are transported only when the K concentration is low, and then only at about half the rate at which K is transported [8]. Not only are all of the alkali metal ions actively transported, but in fact it is possible to prepare bathing solutions in which any one of them is transported exclusively. The net flux of the particular single alkali metal species being transported then accounts for all of the I_{sc} [7, 8]. In this sense the midgut can be said to possess a K-pump, a Rb-pump, a Cs-pump, a Na-pump, and a Li-pump, although it is likely that the same mechanism transports all of these ionic species. However, although the apparent affinity of the pump for the ions is in the order K>Rb>>Cs>Na>Li [9] (sequence IV of Diamond and Wright [10]) we will show that this apparent affinity is a complex function of (1) the access of the particular ion to the pump and of (2) the true affinity of the pump for each ionic species.

In the living insect the pump presumably transports just K because the Na concentration was found by Quatrale (11) to be no more than 2 mM/l in hemolymph, midgut tissue, and midgut contents. The K-pump in the midgut apparently plays a major role in regulating both hemolymph and cellular K concentrations. The larvae eat leaves which have a K concentration with an average value of 239 ± 10 (S.E.M.) mM/l [11]. Not surprisingly the K concentration in the midgut contents is similarly high and is variable at 230 ± 18 mM/l [11]. However, the K concentration in the blood is only about 1/10 of this value at 22.8 ± 1.0 mM/l [11]. Therefore, the total electrochemical gradient across the midgut *in vivo* would be approximately 180 mV (assuming a midgut PD of 120 mV plus the concentration ratio corresponding to 60 mV). The midgut K-pump has the capacity to operate against this steep gradient and presumably regulates the hemolymph K concentration at its low level. Finally, the K concentration in the midgut tissue is 90.2 ± 4.1 mM/l [11]. The hypothesis that the K-pump also regulates the cell K level is developed in § 2.5.

1.2. *Previous studies relating to pump location*

Several lines of evidence place the K-pump on the apical plasma membranes of the epithelial cells. Anderson and Harvey described the structure and ultra-structure of the midgut and were able to locate the pump in the one-cell-thick epithelium [12]. They showed that the epithelium is composed of large columnar cells along with somewhat smaller goblet cells whose apical surface is invaginated to form a goblet cavity. The plasma membrane lining the goblet cavity is folded outwardly to form microvillus-like projections each containing a large mitochondrion. On the inner leaflet of the plasma membrane of these projections are spike-like units. The close association of these units with the mitochondria in the projections led Anderson and Harvey to follow the suggestion of Gupta and Berridge [13] that the units may be associated with active K-transport. Accordingly, the pump would be located in the apical plasma membranes of the goblet cells. However, a basal location for the pump was not ruled out because the basal plasma membranes of the columnar cells possess numerous infoldings in intimate association with mitochondria resembling the arrangement found in many Na-transporting tissues. Therefore, Anderson and Harvey were unable to choose between an apical and a basal location for the pump on the evidence available to them. Meanwhile, Berridge, Gupta, Oschman, and Smith [13, 14, 19] reported several additional occurrences of spike-like units on membranes across which K-transport is believed to occur and concluded that the units play a key part in K-transport. Therefore, the location of the spike-like units associated with K-transport on the plasma membrane lining the goblet cavity would now lead us to choose the apical plasma membrane as the site of the K-pump on these structural grounds.

More direct evidence for an apical location is provided by Wood, Farrand, and Harvey [15]. By analyzing the potential profiles obtained from microelectrode impalements of the midgut they were able to conclude that the K-pump must be on the apical plasma membranes of the epithelial cells. They found a large positive step across the apical barrier and found that it is oxygen dependent. Since the pump is known to be both electrogenic [17, 15] and oxygen dependent [16, 5], the existance of the oxygen dependent, potential step across the lumen barrier is strong evidence that the pump is located in the apical plasma membranes of the epithelial cells. Moreover, Wood *et al.* could explain the prior finding that the K-pumping rate depends on the blood-side K concentration [4, 6] by their demonstration that a potassium diffusion potential exists across the basal barrier. Potassium could enter the cells passively down the electrochemical gradient across the blood-side barrier and be pumped to the lumen by the electrogenic pump on the apical barrier [15].

However, this attractive hypothesis seemed to be ruled out by some of the kinetic evidence presented by Harvey and Zerahn [1] which seemed to show that the transport pool was so small, and that the specific activity of K in the midgut tissue during the tracer steady-state was so low, that a pathway between the cells could not be excluded. We will presently show that the transport pool is not small and will account for the low specific activity in the transporting cells. First it seems advisable to review the theoretical effects of pool size and location on influx kinetics and to review the findings of Harvey and Zerahn.

1.3. *Size and location of pools through kinetic studies*

The kinetics of tracer influx is affected by the size and location of pools in the transport route (fig. 1). The simplest system is that of a pump on a barrier separating two compartments, i.e. no pool present (fig. 1A). It is assumed that the two compartments are well stirred, quasi-infinite reservoirs, that the

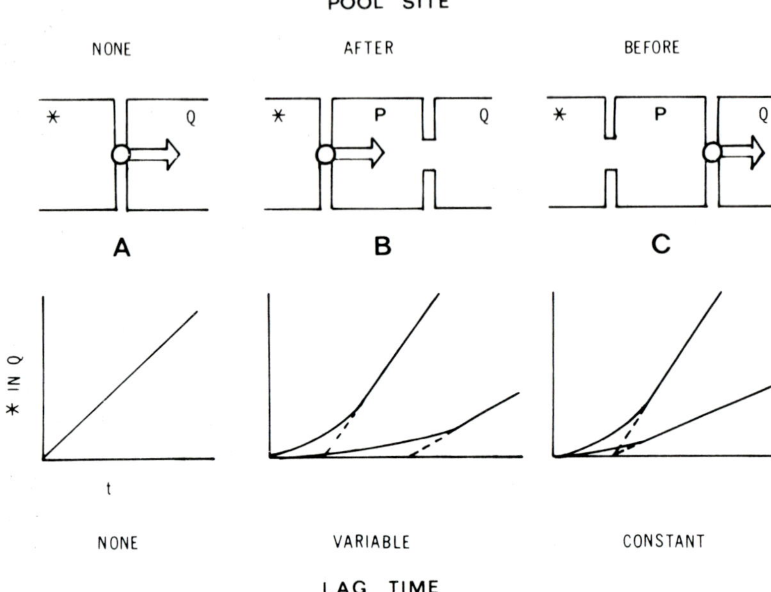

Fig. 1. Theoretical effects of pool size and location on kinetics. The effects of pool size on the tracer influx kinetics are dependent on the location of the pump (thick arrow) relative to the compartment to which the tracer is added (asterisk) and to the compartment from which samples are removed (Q). Below each model, the amount of tracer accumulated in (Q) is plotted as a function of the time elapsed from the addition of tracer to (*) (see §1.3.). Model A. In a two compartment system (i.e., no pool), there is no lag-time (i.e., the steady state influx intersects the abscissa at $x = 0$). Model B. In a three compartment system with the pool (P) after the pump, there is a variable lag-time which is inversely proportional to the pumping rate if the pool size (S_p) remains constant. Model C. In a three compartment system with the pool (P) before the pump, there is a constant lag-time if the pool becomes labeled at a rate independent of the pumping rate. (After Wood [2].)

tracer is uniformly and instantaneously mixed upon addition, and that the pump operates at a constant rate. Upon the addition of the tracer to compartment (*) of the two compartment system, the tracer will appear in compartment (Q) at a constant rate with no delay. The effect of introducing an intermediate compartment into the system (fig. 1B and 1C) is to cause a delay in the attainment of a constant influx by an amount of time proportional to the size of the compartment. This delay, which is called the lag-time, will be designated by the symbol (x). However, the exact kinetics of tracer

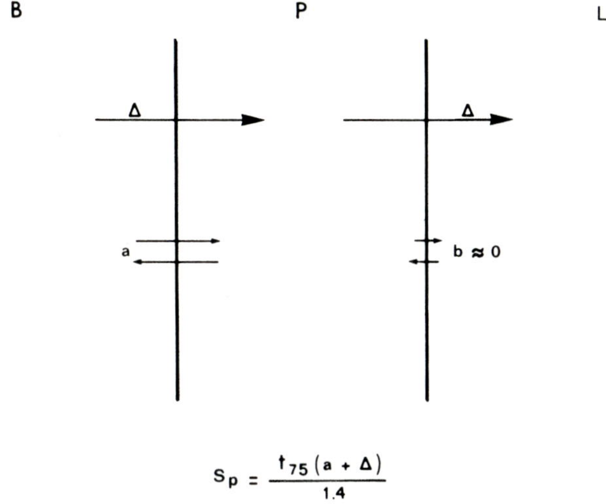

$$S_p = \frac{t_{75}(a + \Delta)}{1.4}$$

Fig. 2. A generalized model of the midgut for deriving equations. The midgut is represented as a two barrier, three compartment system; B = blood-side compartment, P = intermediate pool, and L = lumen-side compartment. No assumptions are made about the location of the pump. However, the passive exchange across the lumen-side barrier, b, is assumed to be zero. This assumption allows all of the remaining unidirectional fluxes to be written in terms of two variables, the passive exchange across the blood-side barrier, a, and the steady state influx into the lumen-side compartment, Δ. Harvey and Zerahn [1] used this model for the midgut to derive equations for the influx kinetics across the midgut. The amount of substance in the pool, P, is (S_p) and is related to the 75% mixing time of isotope in the pool (t_{75}) and the unidirectional fluxes (a + Δ) by eq. (1). (After Harvey and Zerahn [1].)

influx will be determined by whether the pool is located before or after the pump. If the pool is located after the pump (fig. 1B), and hence is labeled by the pump, the lag-time will also be determined by the pumping rate. If on the other hand, the pool is located before the pump (fig. 1C), and the pool is labeled by some process independent of the pump, then the lag-time will remain constant at all pumping rates.

This treatment is made formal by the following equation from Harvey and Zerahn [1]. As diagrammed in fig. 2, let (Δ) signify the (active) influx and (a) the passive exchange of K across the blood-side barrier. Then K enters the transport pool, P, from the blood-side at the rate (a + Δ) and leaves from this side at rate (a). K leaves to the lumen-side only by rate (Δ) because the exchange between the lumen-side and the tissue (b) was found by Harvey and Zerahn [1] to be negligible. We define S_p as the amount of K in the transport

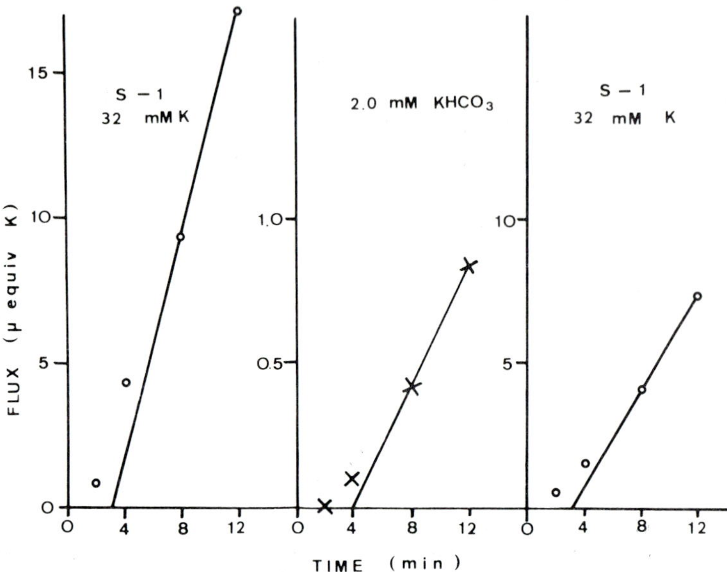

Fig. 3. Review of original study of influx kinetics across the midgut. A representative experiment from Harvey and Zerahn [1] in which the time-course of ^{42}K-measured K-influx (^{42}K-influx) from blood-side to lumen-side across a midgut isolated from *H. cecropia* was determined at different pumping rates achieved by varying the external K concentration. The curves show the lag-time in Sl (32 mM K), in low K (2 mM K) (note ordinate expanded 10 times), and again in Sl. These results show that the lag-time is short and constant at 3 to 4 min, although the influx varied from 119 to 6 and back to 49 μeq. K/hr. (Reprinted with permission from Journal of Experimental Biology.)

pool (assumed constant with time) and $S_{p*}(t)$ as the amount of ^{42}K in the pool at time, t. Taking the time for 75% mixing of ^{42}K in the pool (t_{75}), i.e., letting $S_{p*}(t) = 0.75\ S_p$, we obtain:

$$S_p = \frac{t_{75}(a + \Delta)}{1.4}. \qquad (1)$$

We will show that this equation correctly predicts the mixing-time: influx relationship for the midgut.

Let us now turn to the findings of Harvey and Zerahn [1]. The lag-time can be defined as the time required for all the intermediate pools in the transport route to become 100% labeled or, in other words, the time when the tracer exchange has reached a steady state. Harvey and Zerahn defined the steady-state in terms of the time for the tracer to reach a constant influx across the midgut. They projected the constant influx (Δ) back to the time axis in fig. 3 and designated this intercept as the lag-time (x). The relationship between the lag-time (x) and the 75% mixing-time (t_{75}) is discussed in §2.1. For our present purposes it is sufficient to point out that in fig. 3 (taken from Harvey and Zerahn, [1]) the "lag-time" is very short and remains constant at about three minutes while the K-influx varies 50 fold (note the change in scale of the ordinate). They obtained this same, small, constant lag-time whether the influx was altered by changing the potassium concentration as in this figure, by changing the potential difference across the midgut, or by altering the oxygen tension. The very short lag-time indicated that only a small fraction of the tissue K, at most, could be involved in a transport pool.

2. Results and discussion

2.1. *Mixing-time and pool size*

We will now show that the lag-time is appreciably longer than three minutes and that the corresponding pool size can amount to over half of the tissue K. Wood [2] developed a technique for mounting the midgut as a flat sheet and was able thereby to increase the accuracy of short-circuiting the midgut as well as to increase the length of time over which the midgut could be maintained in a steady-state. With these improvements he found that the net ^{86}Rb-labeled K-influx agreed with the I_{sc} within 1% (S.E.M.) error. Fig. 4 shows the time-course of ^{86}Rb-labeled influx across the *A. pernyi* midgut. The data is plotted according to the graphical method of Andersen and Zerahn [18] from which the lag-time can be seen to be 3 min. However, Wood's demonstration of agreement between influx and I_{sc} within 1% means that when the tracer steady-state is reached the time-course of the influx must be parallel to that of the I_{sc}. However, as can be seen from fig. 5 even when the ^{86}Rb-influx is constant with respect to time, it is still changing with respect to the I_{sc} and does not reach the same value as the I_{sc} until 40 min after the isotope is added. Accordingly, we must redefine the tracer steady-state in terms of the time required to reach a constant tracer influx *relative to the I_{sc} decay*. The dotted line in fig. 5 shows the tracer influx corrected for I_{sc} decay by the methods described by Wood [2]. The actual time for the current corrected influx to reach 75% of its final level is about ten minutes. It is now realized that the short time-lag shown in fig. 4 is an artifact created by the I_{sc} decay. The longer mixing-time resulting from the new definition of

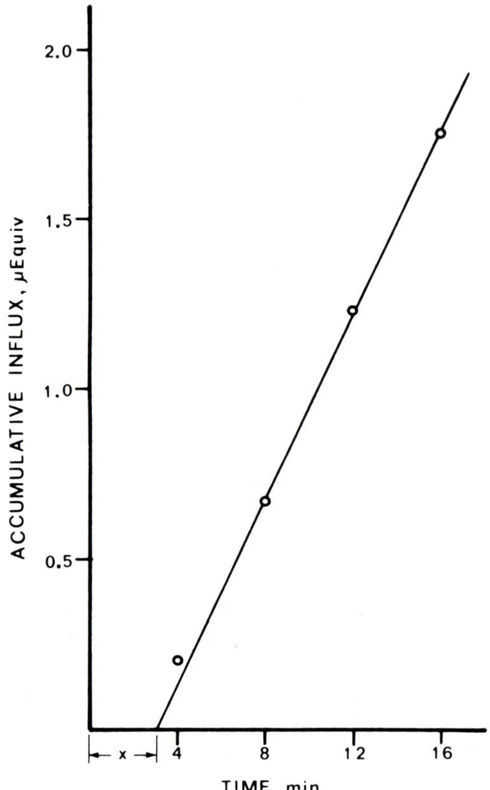

Fig. 4. The apparent lag-time of the raw ^{86}Rb-measured K-influx (^{86}Rb-influx) across the midgut of *A. pernyi*. The results of a typical ^{86}Rb-influx experiment are plotted to show the effects of I_{sc} decay on the apparent lag-time. The straight portion (8–12 min) of the influx curve (circles, not corrected for I_{sc} decay) is extrapolated back to the abscissa to yield an apparent lag-time intercept distance (x) of 3 min. This apparent lag-time for the ^{86}Rb measured K-influx across the *A. pernyi* midgut is identical with the lag-time for the ^{42}K-measured K-influx across the *H. cecropia* midgut found by Harvey and Zerahn [1] as shown in fig. 1. (After Wood [2].)

the steady-state indicates that a sizeable fraction of the tissue K is involved in the transport pool.

The longer mixing-time is not due to some unwarranted assumption in the correction for current decay nor is it simply a characteristic of ^{86}Rb-influx across the *A. pernyi* midgut. In fig. 6, data from a determination of ^{42}K-influx across the *H. cecropia* midgut are plotted. The midgut was equilibrated as a flat

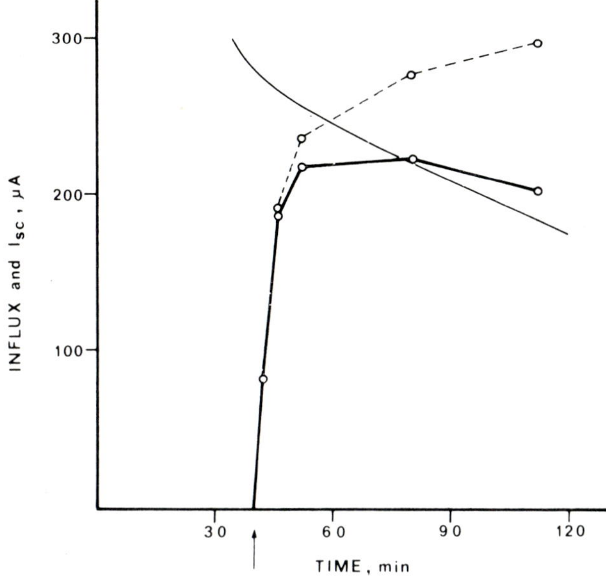

Fig. 5. The time course of ^{86}Rb-measured K-influx across the midgut of *A. pernyi*. The results of the same experiment (fig. 4) are replotted to show the time-course of the tracer measured influx (solid lines connecting circles) relative to the simultaneous time-course of the I_{sc} (thin line). The tracer is added (arrow) to the blood-side compartment of the chamber at 40 min (NB, numbers on abscissa refer to the time elapsed from mounting the midgut in the chamber). Although the raw influx levels off quickly in about 8 min, it is changing relative to the I_{sc} for at least 40 min after the tracer is added. The new criterion for the tracer steady state (§2.1.) is applied to the raw influxes to give the current corrected influx (dotted line connecting circles). The 75% mixing-time of the current corrected influx in this experiment is approximately 10 min. It is now realized that the short lag-time of 3 min obtained from the raw influxes (fig. 4) is an artifact created by the decay in the I_{sc}. (After Wood [2].)

sheet in 35-S-Tris solution for 120 min so that the decay in the I_{sc} had become very slight. *The longer mixing-time can now be seen directly from the raw influx data without any corrections for current decay* (fig. 6). The time to reach 75% mixing is 13 min. This same long mixing-time can be seen in the graphical plot in fig. 7. It should be noted that the 75% mixing-time is in fact about 1.4 times the *x*-intercept distance which Harvey and Zerahn referred to as the lag-time. Provided that (a) is small with respect to (Δ) this relationship follows directly from eq. (1) when (Δ) is replaced by its equivalent expression, S_p/x. The use of (x) rather than the 75% mixing-time by Harvey and

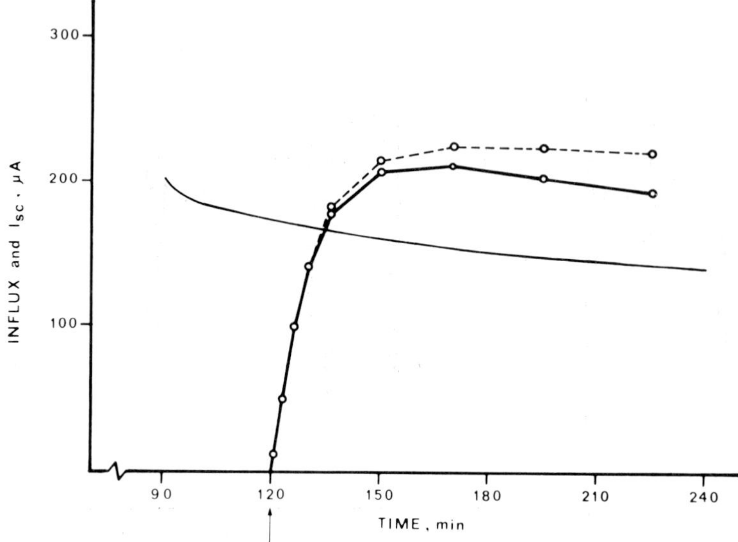

Fig. 6. The time-course of ^{42}K-measured influx across the midgut of *H. cecropia*. The results of a typical ^{42}K-influx experiment plotted as in fig. 5 to show the time-course of the raw influx (heavy solid line) relative to the simultaneous time-course of the I_{sc} (thin solid line). The addition of the tracer (arrow) was delayed until 120 min after the midgut was mounted in the chamber to allow the decay in the I_{sc} to become minimal. Now, the long delay before the tracer reaches a steady-state can be seen directly from the raw influx data without any corrections for I_{sc} decay. The dotted line shows the time-course of the influx corrected for the slight decay in the I_{sc} which occurred over the two hour sampling period (120–240 min). The 75% mixing time (t_{75}) for this experiment from the current corrected influx is approximately 13 min.

Zerahn [1] was a second factor leading them to report a short lag-time and a small pool size. The pool size corresponding to 13 min can be calculated from eq. (1) and amounts to 1.3 μeq. of K. The same pool size can also be read from the plot in fig. 7 as the vertical distance between the two parallel lines (after Andersen and Zerahn [18]). We now formally propose that this influx pool forms the transport pool and accordingly we designate the influx pool by *S*p. The wet weight of the midgut used in this determination was 29.1 mg and, assuming that the gut contains 65 μeq. K/g wet weight [1], this result means that 66% of the tissue K is involved in the transport pool.

2.2. *Pool and pump location*

The next question is whether this pool is located before or after the pump.

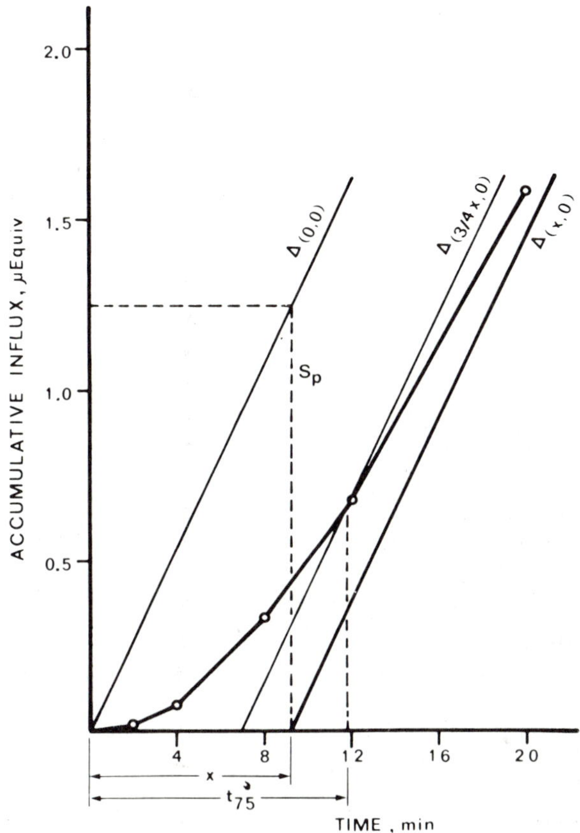

Fig. 7. The time-course of ^{42}K-measured influx plotted kinetically. The results of the same experiment (fig. 6) are replotted by the graphical method. The current corrected in fluxvalues are used to compute the accumulation of tracer in (Q) (solid line connecting open circles) and these points are plotted against elapsed time. The steady-state tracer influx ($\Delta_{x,o}$) is extrapolated back from the points at 60 and 90 min (not shown for reasons of scale factor) with a lag-time intercept distance, x, of 9.2 min. Now, it can be seen that when the current corrected influx values are plotted by the method used by Harvey and Zerahn [1], the lag-time intercept is much greater than when it is plotted from the raw data and that the mixing-time is greater still. The 75% mixing-time (t_{75}) for this experiment is 11.2 min. Since (t_{75}) does not equal a factor of 1.4 times (x), (a) is not zero, but is nearly so. The vertical distance between ($\Delta_{x,o}$) and ($\Delta_{o,o}$) is in fact equal to the size of the influx pool, (S_p).

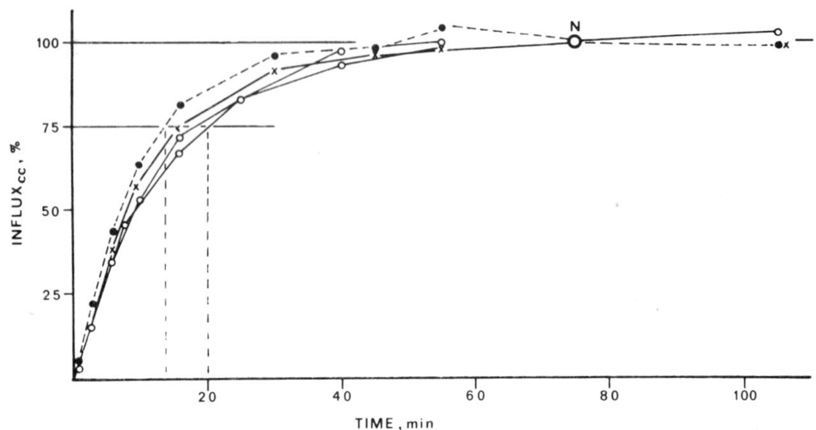

Fig. 8. The effects of varying the pumping rate on the influx mixing-time. The location of the pump relative to the transport pool was examined by altering the pumping rate, Δ, (see § 1.3.), through variations in the external K concentration and oxygen tension (K-35-Tris + 100% 0_2, ---o---; K-35-Tris + air, ---•---; and K-100-Tris + 100% 0_2, ---x---). The time-course of the corrected influxes from four different experiments are plotted, based on normalizing the 75 minute point (N; time elapsed from tracer addition). The 75% mixing-time (intercept of the influx curves with the 75% level) remains virtually constant for the different test conditions. These results confirm the constancy of the lag-time found by Harvey and Zerahn ([1]; see fig. 3). The constancy of mixing-time at different pumping rates seen here indicates that the pump is located after the transport pool (see § 2.2.).

This question can be answered by studying the dependence of the mixing-time on the pumping rate. In confirmation of the results of Harvey and Zerahn, we were unable to change the mixing-time. From fig. 8 we see that the mixing-time is long and constant for ^{42}K-influx across the *H. cecropia* midgut, whether it is in 35 mM K, 100 mM K, oxygen, or air. This constancy of mixing-time is most easily explained by placing the large pool before the pump. Moreover, now that we have shown that the pool is large we must place it in the epithelial cells because there is not enough volume elsewhere in the midgut to contain such a large amount of K at a reasonable concentration. Placing the pool in the cells allows us to support the conclusion that the pool is before the pump with an electrochemical argument as follows. Wood et al. [15] found that even in 2 mM/l K bathing solutions the cells are sufficiently negative that K could enter them passively up to a cellular K concentration of 37 mM/l. However, it is impossible for K to leave the cells passively toward the lumen-side. In the first place Harvey and Zerahn [1] found that there is virtually no passive exchange across the lumen-side barrier, a reasonable finding if the pump is located in that barrier. Moreover, the potential across the lumen-side barrier can be as high as 180 mV with the lumen positive to the cells [15]. Even in 2 mM K the K concentration in the cells would have to be 2000 mM/l for K to leave passively. K can be transported even when the external concentrations are 75 mM or higher which could require an internal concentration near the membrane of more than an order of magnitude greater than that allowed by the solubility product of any potassium salt. Therefore, if there is a large transport pool in the cells, then this pool must be before the pump and the pump must be in the apical plasma membranes of the epithelial cells. Recall that structural evidence and evidence from microelectrode impalements also placed the pump in the apical plasma membranes of the epithelial cells (§ 1.2).

2.3. ^{22}*Na-loading pools*

If other alkali metal ions such as Na are transported by the same mechanism as K and Rb, as is suspected, then there should be a similar pool awaiting transport for these ions. Since there is normally only 1 mM/l of Na in the tissue [11] it should be easy to demonstrate a sodium pool if it is present under conditions in which Na would be transported. The key to the demonstration of Na-transport was found by Harvey and Zerahn in 1971 [8] to be the omission of Ca^{2+} and Mg^{2+} from the bathing solutions. Recently, Weinberg, working in our laboratory, has shown that the critical ion is calcium. In the experiment shown in fig. 9, the midgut was first bathed in K-32 and then the solution was changed to Na-32. The decay was more rapid than in K-containing solutions. At the third arrow Ca^{2+}, sufficient in quantity to bring its concentration to 4 mM/l, was added to the sodium solution on the blood-side and the Na-current disappeared. This result shows that the Na-transport by the midgut is inhibited by the presence of calcium.

This information was used to study the ^{22}Na-loading of the tissue. The loading level was determined by cutting a midgut into eight pieces, incubating four in a ^{22}Na solution with calcium and four without calcium. Pieces of tissue were removed from the incubation solution at 15 minute intervals, rinsed, blotted, and counted. The lower curve in fig. 10 shows the ^{22}Na-loading in the presence of calcium. The fairly high level of sodium is probably due to extracellular sodium that was not removed in blotting the tissues. The upper curve shows the ^{22}Na-loading in calcium-free solution. The difference between the two curves after 60 min. of incubation indicates that 15 mM/l of Na more has entered the tissue in the absence of Ca^{2+}. Furthermore, the difference curve (not shown) yields a time-course of loading similar to the

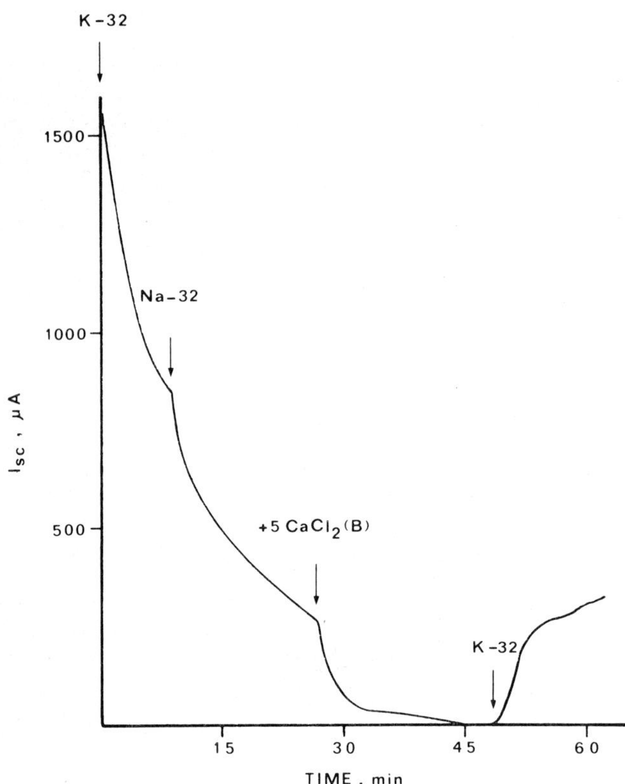

Fig. 9. The effects of Ca^{2+} on Na-transport by the midgut. The effects of Ca^{2+} on Na-transport were studied by examining the time-course of the I_{SC} produced by a midgut isolated from *H. cecropia* in different bathing solutions. The midgut was initially bathed (first arrow) in K-32 to establish the K-current. The decay in the I_{SC} seen here is rather rapid, but it is not unusual for Ca^{2+} free solutions. At the second arrow, the bathing solutions were changed to Na-32 to establish the Na-current in the absence of Ca^{2+}. At the third arrow, Ca^{2+} (5 mM/l) was added to the blood-side bathing solution, leading to the rapid loss of Na-current. Slower acting effects of Ca^{2+} on the Na-current have been found on the lumen-side. At the fourth arrow, the bathing solutions were changed back to K-32, leading to a return of a K-current. It has not been possible to reestablish a Na-current by adding Na-32 without Ca^{2+}, once Ca^{2+} has been added. These results show that Ca^{2+} inhibits Na-transport by the midgut.

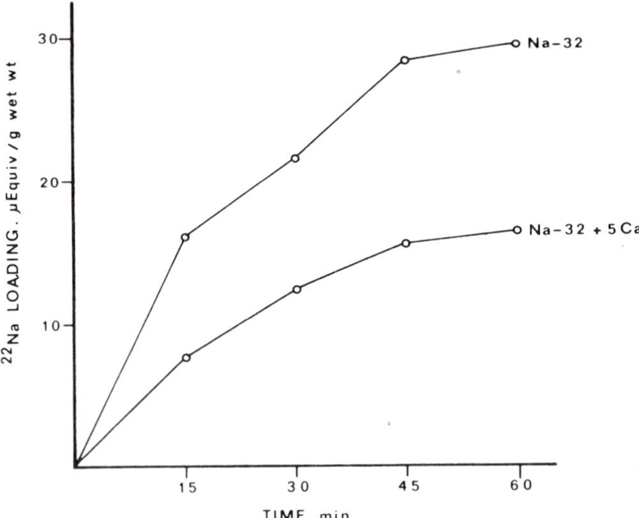

Fig. 10. The effects of Ca^{2+} on ^{22}Na-loading of the midgut. The ^{22}Na-loading of the midgut was studied by the methods given in § 2.3. The lower solid curve shows the time-course of ^{22}Na-loading in Na-32 plus 5 mM Ca^{2+}, and the upper solid curve shows the time-course of ^{22}Na-loading in Na-32 without Ca^{2+}. More ^{22}Na enters the tissue in the absence of Ca^{2+}. The difference in loading levels at 60 min indicates that about 15 mM/l of Na more has entered the tissue in the absence of Ca^{2+}. Furthermore, the difference curve (not shown) between the ^{22}Na-loading with and without Ca^{2+} has a 75% labeling (mixing) time of about 15 min. These results show that more Na is present in the midgut under conditions when Na would be transported and suggest the existence of a Na pool in the midgut. If there is a Na pool in the midgut, then Ca^{2+} probably inhibits Na-transport by interfering with the entrance of Na into the transport pool (§ 2.6.).

K-influx data with a 75% mixing-time of 15 min. In summary there is a large sodium pool in the tissue under conditions when the midgut is transporting sodium and a much smaller pool under conditions when it is not. These data suggest, moreover, that the reason why sodium is *not* transported when Ca^{2+} is present is that Ca^{2+} interferes with the entrance of Na into the pool awaiting transport. Demonstrating this large sodium loading pool in the tissue is necessary, but of course not sufficient, to demonstrate that a cellular pool is involved in Na-transport.

2.4. ^{42}K*–loading pool*

The evidence presented thus far supports the hypothesis that there is a

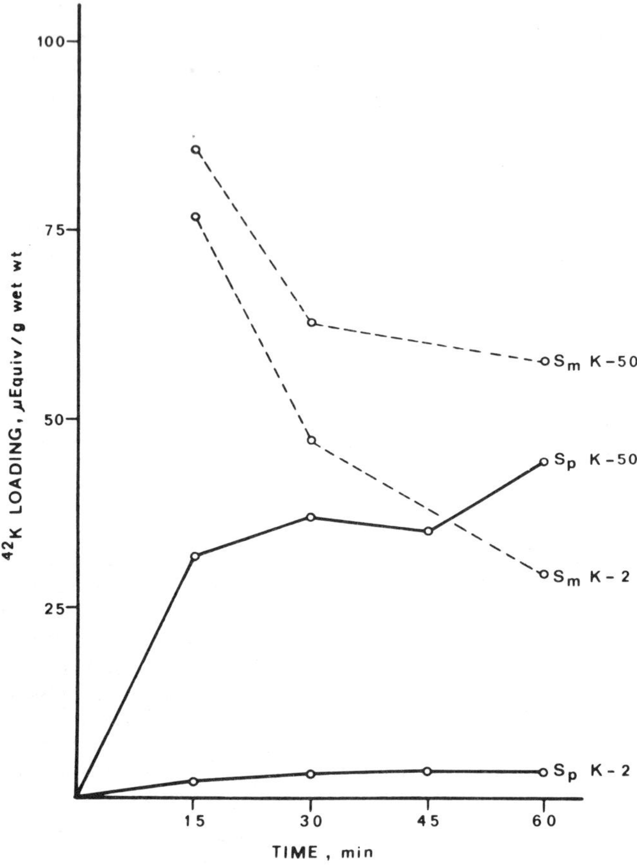

Fig. 11. The effects of external K on ^{42}K-loading and on total tissue K. The ^{42}K-loading and total tissue K in the midgut at different external K concentrations were determined by the methods given in §2.3. The upper solid curve shows the time-course of ^{42}K-loading (S_p) (computed on a wet weight basis) in high K (K-50-Tris), and the lower solid curve shows the time-course of ^{42}K-loading (S_p) in low K (K-2-Tris). The upper dotted line shows the time-course of the changes in the total tissue K (S_m) in high K (same samples), and the lower dotted line shows the time-course of the changes in the total tissue K (S_m) in low K (same samples). The external K concentration affects both the level of ^{42}K-loading and the level of total tissue K. The mean specific activity of the K in the midgut is given by the ratio of the loading level (S_p) to the total tissue K (S_m). Surprisingly, a large amount of the total tissue K does not become labeled ($S_m - S_p$) even after 60 min. Both of the ^{42}K-loading curves have the shape of saturation curves with 75% labeling (mixing) times of about 15 min. The ^{42}K-loading levels at 60 min will be equated with the transport pool (S_p) and are so identified in the figure. Further details of this experiment are given in the text.

large ion pool awaiting transport. However, before we can propose this hypothesis seriously we must deal with some contradictory evidence presented by Harvey and Zerahn [1]. Specifically, they found that in low K solutions the midgut was only 18% labeled after 10 min when the specific activity of the K arriving in the lumen was expected to be approaching 100%. This finding implies that less than 18% of the tissue K forms the transport pool in the low K solutions. Above we have just shown that 66% of the tissue forms the transport pool in high K solutions. Complicating the problem was the finding by Harvey and Zerahn that the tissue K concentration was virtually constant over this range of K concentrations in the bathing solutions. How can the apparent discrepancy in pool size implicit in these contradictory data be resolved?

To gain insight into this question we studied the time-course of ^{42}K-loading of the midgut using methods similar to those described in §2.3. for ^{22}Na-loading. Harvey and Zerahn determined that the exchange between midgut K and blood-side ^{42}K was approximately 2% per minute which implies that the tissue would be 100% loaded in about 50 min in 2 mM K. However, this result was not obtained as can be seen in fig. 11. The upper solid line shows that the ^{42}K-loading in 50 mM K is more or less as expected. However, the lower solid line shows that the ^{42}K-loading in 2 mM K levels off at a low value after about 30 min. This finding is contradictory to the result implied by Harvey and Zerahn and it indicates that a large fraction of the tissue K *does not exchange with the label.* To test this hypothesis directly we measured the total K from these same samples with a flame spectrophotometer. The time-course of the total tissue K in 50 mM K is shown in the upper dotted curve and that in 2 mM K is shown in the lower dotted curve. There is indeed a sizeable fraction of the total tissue K which has not exchanged after 60 min under both conditions. Moreover, we obtain a higher loading level in the 50 mM solution (42 μeq. K/g wet weight) than in the 2 mM K solution (4 μeq. K/g wet weight), just as we obtain a higher transport rate in the high K solution than in the low K solution. Therefore, the loading levels vary with the external K concentration in the same way as the transport rates vary.

We now propose that the loading level forms the pool awaiting transport. We define the loading level as the amount of ^{42}K which has accumulated in the tissue at 60 min. This loading level varies with the external K concentration in the same way that the pool size from the influx kinetics varies with the external K concentration. Moreover, the size of the pool estimated at high K concentration from the loading kinetics (42 μeq. K/g wet weight) is the same as that estimated from influx kinetics (43 μeq. K/g wet weight). Finally, the time to reach 75% loading is approximately the same as the 75% mixing-time from the influx data. Therefore, just as we designated the size of the influx pool as (S_p) in §2.1., we now designate the size of the loading pool also as (S_p). We propose that the influx pool and the loading pool are both synonymous with the actual pool of K awaiting transport across the midgut.

2.5. *Midgut model*

We now propose a model for the route followed by the transported K through the midgut (fig. 12). The blood-side K concentration, as modified by the electrical gradient across the blood-side plasma membranes, determines the rate at which K enters the midgut epithelial cells from the blood side, $a + \Delta$. The total K-influx into the pool ($a + \Delta$), determines the pool size, S_p. Assuming that the pool volume, V_p is constant, then S_p determines the value

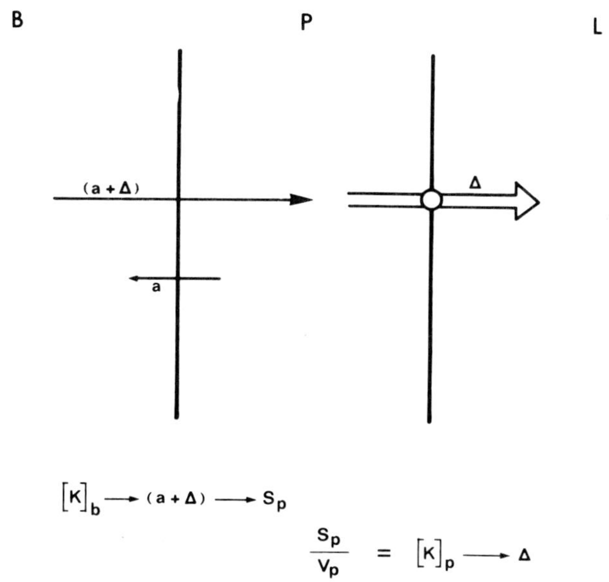

Fig. 12. A specific model for the midgut transport system. The model is based on fig. 2 but with the pump specifically located on the lumen-side barrier and the transport pool, P, specifically identified as the exchangeable K in the tissue. K is transported through the midgut by a cellular route. K enters the cellular pool, P, from the blood-side compartment across the basal plasma membranes at a rate, $a + \Delta$, which depends on the blood-side K concentration, $[K]_B$. K is then pumped out of the cells across the apical plasma membranes at a rate, Δ, which depends on the pool's K concentration, $[K]_p = S_p/V_p$. The detailed operation of the Model is described in the text.

of S_p/V_p, which is the K concentration in the pool, $[K]_p$. The pump is concentration-dependent (probably with an s-shaped, velocity concentration function). It is located in the apical plasma membranes of the epithelial cells where it pumps potassium out of the cells into the lumen-side compartment. The relationship between K entry rate $(a + \Delta)$, mixing-time (t_{75}), and pool size (S_p) is expressed mathematically by eq. (1).

According to this model, potassium diffuses from the blood-side compartment into the exchangeable pool at the rate $(a + \Delta)$, loading it to the level, S_p. K exits from the pool across the blood-side plasma membranes at the rate (a) and *K is pumped out of the cells* by a pump located in the lumen-side plasma membranes at the rate (Δ). According to the model, the rate at which K is pumped out of the cells into the lumen-side compartment (Δ) is strictly determined by the rate at which K enters from the blood-side $(a + \Delta)$, and not *vice versa*. In other words the model requires that the cells pump out to the lumen the K which enters the pool, rather than that K enters the pool to replace the K which has been pumped out. In this way the K concentration in the pool, $[K]_p$, determines the rate of pumping, Δ, just as the concentration of reactants (along with the rate constant) determines the initial rate of a scaler chemical reaction. Moreover, this obligatory dependence of (Δ) on (S_p) leads directly to the constancy of the mixing-time as can be seen more clearly by solving eq. (1) for (t_{75}).

$$t_{75} = \frac{1.4S_p}{(a+\Delta)}.$$

When (a) is small with respect to (Δ), then it can be seen that:

$$t_{75} \propto \frac{S_p}{\Delta}.$$

Therefore, if $S_p \propto \Delta$, then t_{75} is constant.

Finally, we suggest that *in vivo* the pump regulates both the cellular K concentration (by the dependency of Δ on $[K]_p$) and the hemolymph K concentration (by the dependency of $(a + \Delta)$ on $[K]_B$). Changes in the hemolymph K concentration would induce precise changes in the K concentration of the cellular pool which in turn would precisely alter the rate at which K is pumped out of the cells to the lumen. In this simple but elegant feed-back system the cellular K concentration is set primarily by the kinetic constants of the pump in the apical membrane while the hemolymph K concentration is set primarily by the contribution of the cellular K concentration to the electrochemical gradient across the blood-side membrane.

2.6. *Affinity of pump for alkali metal ions*

The rate of translocation of ions across the midgut epithelium is determined by two factors: (1) the passive entry of the ions into the pool and (2) the true affinity of the pump for the ion. The passive entry of ions into the pool would depend not only on the electrochemical gradient across the blood-side membrane as developed in the model but also on the permeability of the membrane to ions. Presumably Ca^{2+} reduces Na entry into the pool by decreasing the permeability of the blood-side membrane to Na and thereby inhibits Na transport (fig. 10). Now that we have demonstrated that a transport pool determines the pumping rate, the true affinity of the pump for the different ionic species would be given by the transport ratio divided not by the concentration ratio in the bathing solution but by the concentration ratio of the ionic species in the transport pool.

3. Summary

1. The time for ^{42}K-measured influx across the isolated midgut to reach 75% of its final rate is around 13 min.
2. This mixing-time corresponds to a pool which amounts to 66% of the total tissue K in 35 mMK.
3. The transport pool as measured by both the influx kinetics and the loading kinetics is found to vary with the external K concentration.
4. A large sodium pool is found when the tissue is transporting Na but not in the absence of Na-transport.
5. Sodium-transport is inhibited by calcium which interferes with the entrance of sodium into the transport pool.
6. A model for K-transport across the midgut is proposed as follows:
   a. The blood-side K concentration, modified by the electrical gradient across the blood-side membrane, determines the rate at which K enters the midgut epithelial cells ($a + \Delta$).
   b. The total K-flux into the pool ($a + \Delta$) determines the pool size (S_p).
   c. Assuming that the pool volume (V_p) is constant, then (S_p) determines (S_p/V_p) which is the K concentration in the pool, $[K]_p$.
   d. The pump is concentration dependent and is located in the apical plasma membranes of the cells.
   e. Potassium is pumped out of the cells.
7. The mixing-time is constant at varying pumping rates, because the flux (Δ) varies directly with the pool size (S_p).

8. We postulate that the rate of translocation of ions across the midgut is determined by (1) the passive entry of ions into the pool which depends on the electrochemical gradient and the permeability of the blood-side membrane to K and by (2) the affinity of the pump for the transported ion.

A more complete version of this paper with special emphasis on the midgut model and with a comprehensive section on Methods is in preparation for publication elsewhere. Part of this research was conducted while J.L. Wood was a Ph.D. candidate at the University of Cambridge under the supervision of Prof. Weis-Fogh, Prof. Ramsay and Dr. Moreton. This research was supported in part by a research grant (AI-09503) from the National Institute of Allergy and Infectious Diseases, U.S. Public Health Service.

Note added in proof

Harvey and Zerahn [20] have pointed out that the analysis of the data by this kinetic treatment does not allow an absolute distinction between a pool located before or after the pump. More direct evidence is required to make this distinction for the large pool demonstrated in the present paper (see Harvey and Wood [21]).

References

[1] W.R. Harvey and K. Zerahn, J. exp. Biol. 50 (1969) 297.
[2] J.L. Wood, Ph.D. Thesis, Univ. of Cambridge (1972).
[3] R.D. Keynes, Quart. Rev. Biophysics 2 (1969) 178.
[4] W.R. Harvey and S. Nedergaard, Proc. natn. Acad. Sci. (U.S.A.) 51 (1964) 757.
[5] W.R. Harvey, J.A. Haskell and K. Zerahn, J. exp. Biol. 46 (1967) 235.
[6] S. Nedergaard and W.R. Harvey, J. exp. Biol. 48 (1968) 13.
[7] K. Zerahn, J. exp. Biol. 53 (1970) 641.
[8] W.R. Harvey and K. Zerahn, J. exp. Biol. 54 (1971) 269.
[9] K. Zerahn, Proc. Int. Union physiol. Sci. 8 (1971).
[10] J.M. Diamond and E.M. Wright, Ann. Rev. Physiol. 31 (1969) 581.
[11] R.P. Quatrale, Ph.D. Thesis, Univ. of Massachusetts (1966).
[12] E. Anderson and W.R. Harvey, J. Cell Biol. 31 (1966) 107.
[13] B.L. Gupta and M.J. Berridge, J. Cell Biol. 29 (1966) 376.
[14] D.S. Smith, Tissue and Cell 1 (1969) 443.
[15] J.L. Wood, P.S. Farrand and W.R. Harvey, J. exp. Biol. 50 (1969) 169.
[16] J.A. Haskell, R.D. Clemons and W.R. Harvey, J. cell. comp. Physiol. 65 (1965) 45.
[17] W.R. Harvey, J.A. Haskell and S. Nedergaard, J. exp. Biol. 48 (1968) 1.
[18] B. Andersen and K. Zerahn, Acta physiol. Scand. 59 (1963) 319.
[19] M.J. Berridge and J.L. Oschman, Tissue and Cell 1 (1969) 247.
[20] W.R. Harvey and K. Zerahn, in: Current Topics in Membranes and Transport (F. Bronner and A. Kleinzeller, eds.) 3 (1972) in press.
[21] W.R. Harvey and J.L. Wood, in: Transport Mechanisms in Epithelia, Alfred Benzon Symposium V, Copenhagen, September, 1972 (Munksgaard, Copenhagen).

RELATIONSHIP BETWEEN PERMEABILITY AND ACTIVE TRANSPORT ACTIVITY OF THE ISOLATED RAT INTESTINE. POSSIBLE INVOLVED MECHANISMS*

G. ESPOSITO, A. FAELLI, G. GAROTTA,
R. PAROTELLI and V. CAPRARO

Istituto di Fisiologia Generale,
Università di Milano, Italy

1. Introduction

In previous papers [1–3] we have pointed out that the mobility coefficient of acetamide and thiourea of the rat jejunum perfused *in vitro* with a basic perfusion fluid (Krebs-Henseleit bicarbonate solution added with D-glucose 13.9 mM), decreases when the active transport across the intestinal epithelium was inhibited by replacing NaCl with Tris-Cl in the basic incubation fluid. The decrease of permeability seems not to be due to a shrinkage of epithelial cell taking place in the above reported modification of the perfusing fluid. In fact after adding phlorhizin 10^{-4} M to the basic perfusion fluid, the cells shrink as in the Tris-Cl solution, without statistical reduction of the acetamide mobility. The decrease of acetamide permeability in Tris-Cl perfusing fluid seems not to be related solely to the decrease of intestinal active transport. If we compare a group of intestinal tracts perfused with a low glucose content to a group of intestinal tracts perfused with the above Tris-Cl solution, we observe the same fluid and glucose transepithelial transport; nevertheless the mobility of acetamide is lower in the intestines perfused with the Tris-Cl solution (table 1).

Therefore the lowering of the Na concentration could affect directly the

* This work has been supported by a research grant of Consiglio Nazionale delle Ricerche, Roma.

Table 1
Glucose and fluid transepithelial transport, mobility of acetamide across the intestinal wall. Number of experiments in parentheses

Incubation fluid	Glucose transepithelial * transport μmoles g^{-1} hr^{-1}	Fluid transepithelial * transport ml g^{-1} hr^{-1}	ω * μmoles g^{-1} hr^{-1} atm^{-1}
Krebs-Henseleit bicarbonate + glucose 1.39–13.9 mM	35 ± 9 (10)	2.59 ± 0.25 (10)	439 ± 24 (10)
Krebs-Henseleit bicarbonate + glucose 13.9 mM – NaCl substituted isosmotically with Tris-HCl	46 ± 14 (6)	2.57 ± 0.74 (6)	315 ± 30 (6)

* Mean values (± S.E.M.) obtained when the range of glucose transport is between 0 and 100 μmoles g^{-1} hr^{-1}. All data are referred to one gram dry tissue weight and one hour.

physico-chemical properties of cell membrane due, for instance, to the varied Na:Ca ratio of the perfusing fluid. However an effect of the active transport on passive permeability of acetamide cannot be disregarded.

2. Methods

The experiments were performed by recirculating the perfusion fluid through an open and everted intestinal tract [2]. Sprague-Dawley albino male rats, initially weighing about 250 g, semistarved over a 15 day period (final percent weight decrease: 15–25%), were used. Under barbituric narcosis, a tract of small intestine 10–15 cm long, was removed from the animal, about 10 cm from the pylorus. Each tract was everted and perfused at 28°C. The basic perfusion fluid, as mentioned above, gassed with a mixture of 95% O_2 and 5% CO_2, was added with acetamide 10 mM; labelled ^{14}C-acetamide was present in the serosal space only. In a second set of experiments glucose was present at the lower concentration of 1.39 mM. At the end of each individual experiment the dry weight of the intestine was determined. Transepithelial mobility coefficient, disregarding the drag effect of the volume flow, and net glucose transport throughout a 1 hr perfusion period, were calculated [2].

In another set of experiments the electrical conductance of the intestinal barrier was investigated by perfusing the intestine with different solutions. The transepithelial electrical potential and the current crossing the barrier at different electrical fields were determined and the corresponding resistance (or conductance) calculated [4] (fig. 2 and 3).

3. Results and discussion

From the mobility experiments, it seems that there is a linear correlation between the mobility of acetamide and the net active fluid and glucose transport through the intestinal wall (fig. 1). As a matter of fact in normal perfused intestines there is a good correlation between glucose transport and fluid transport [5]. Nevertheless the absence of glucose transport in phlorhizin-poisoned intestines does not indicate a total absence of active transport activity because the fluid transport is still high.

The relationship between the mobility of acetamide and the net glucose transport has been confirmed by experiments in which the same intestinal tract was initially perfused for 20 min with a low glucose perfusion fluid (1.39 mM) and then for an equal period with a high glucose perfusion fluid (13.9 mM). In the second period of experiment, net glucose transport increases together with the mobility coefficient of acetamide (16 ± 3%, $n = 10$).

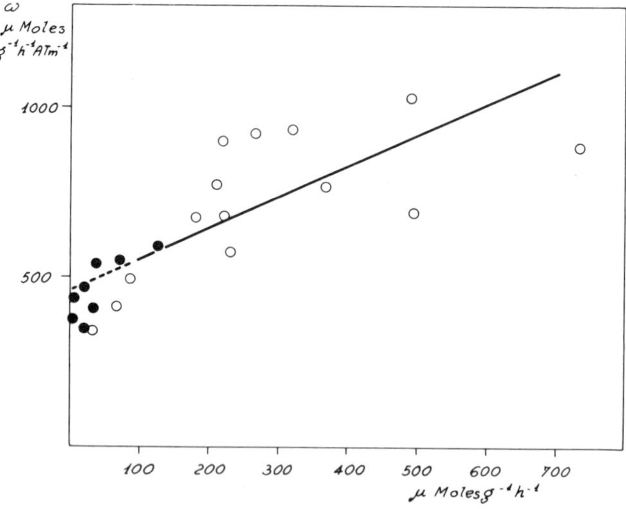

Fig. 1. Linear correlation between net glucose transport (abscissa) and mobility coefficient of acetamide (ordinate). Solid (•) and open (○) circles represent experiments in which glucose concentration of the incubating fluid was 1.39 and 13.9 mM respectively.

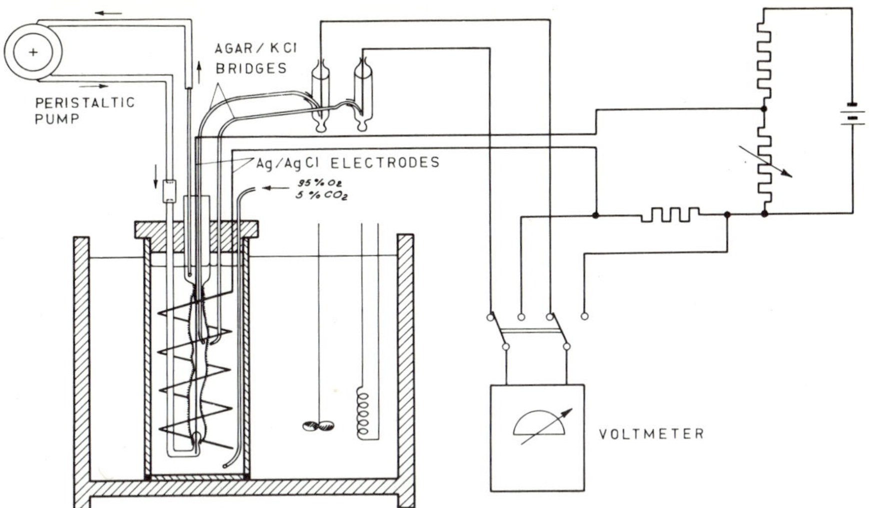

Fig. 2. Complete equipment used for the determination of electrical conductance of the intestine.

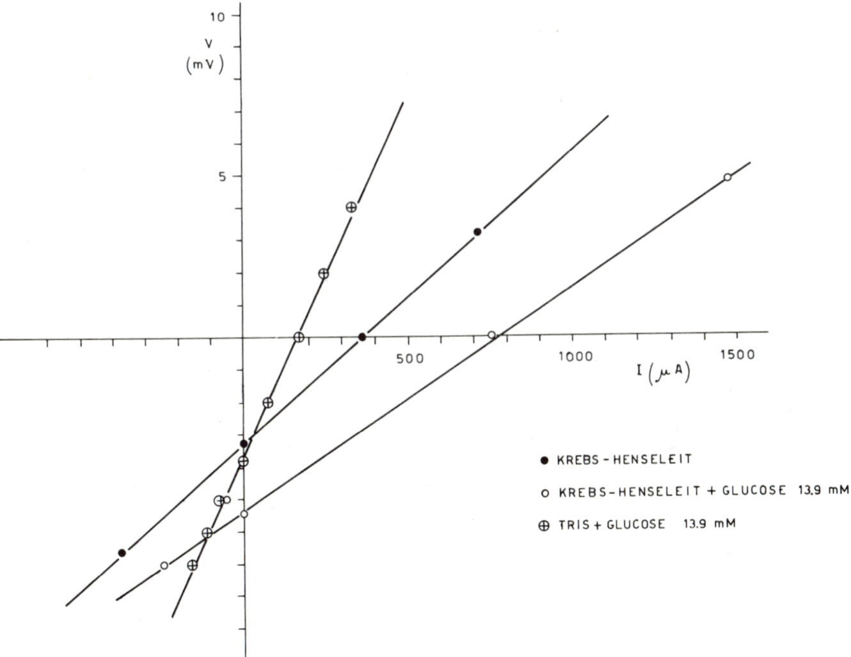

Fig. 3. Diagram showing linear relationship between the transintestinal potential difference and the correspondent current across the isolated intestine in different incubating media. These data have not been corrected for the resistance of the perfusing fluid (see ref. [4]).

Table 2
Conductance (Ω^{-1} g^{-1}) of the intestinal wall in different experimental conditions. Mean values ± S.E.M. referred to one gram dry tissue weight, are reported

Incubation fluid	Number of experiments	Conductance Ω^{-1} g^{-1}
Krebs-Henseleit bicarbonate	8	4.62 ± 0.61
Krebs-Henseleit bicarbonate + glucose 13.9 mM	9	7.17 ± 0.60
Krebs-Henseleit bicarbonate + glucose 13.9 mM – NaCl substituted isosmotically with Tris-HCl	5	1.42 ± 0.17

In control experiments (low glucose content) and during the same experimental periods there is no change of acetamide permeability. The results from the resistance experiments are reported in table 2 which shows that the presence of glucose increases the conductance of the intestinal barrier. Such an increase can also be obtained in the same intestinal tract by adding glucose 13.9 mM to the Krebs-Henseleit bicarbonate solution (percent increase 120 ± 36; number of experiments = 10). As to the mechanism of the relationship between the glucose transport and the mobility of the acetamide, the hypothesis may be put forward that the intracellular level of cyclic AMP is

Table 3
Glucose transepithelial transport and mobility of acetamide across the intestinal wall. Mean values ± S.E.M. referred to one gram dry tissue weight and one hour, are reported

Incubation fluid	Number of experiments	Glucose trans-epithelial transport μmoles g^{-1} hr^{-1}	ω μmoles g^{-1} hr^{-1} atm^{-1}
Krebs-Henseleit bicarbonate + glucose 1.39 mM	8	37 ± 14	465 ± 31
Krebs-Henseleit bicarbonate + glucose 13.9 mM	14	279 ± 51	721 ± 55
Krebs-Henseleit bicarbonate + glucose 1.39 mM + cyclic AMP 4 mM	8	141 ± 21	813 ± 24
Krebs-Henseleit bicarbonate + glucose 1.39 mM + Na-butyrate 4 mM	7	86 ± 14	554 ± 49

involved. It is known that cyclic AMP increases the permeability of the mucosal border of some epithelial cells [6]. In fact when cyclic AMP 4 mM (dibutyryl derivative) is added to the serosal perfusion fluid containing a low glucose concentration (1.39 mM) the mobility coefficient of acetamide becomes as high as that obtained by perfusing the intestine with a high glucose concentration (13.9 mM). Presumably also the increase of glucose transepithelial transport could be explained by an increase of permeability of the brush border. Control experiments in which Na-butyrate 4 mM was added to the serosal side, show no effect (table 3).

If we consider the whole intestinal barrier, besides the above explanation one has to take into account the widening of the intercellular spaces as a consequence of the increased glucose and water transport. As a matter of fact this condition increases the total serosal surface available for the diffusion of this molecule. The opening of the intercellular spaces has been observed in bullfrog intestine [7] when there is an osmotic flow from the mucosal to the serosal side of the intestine due to an osmotic gradient.

4. Conclusions

(i) there is a linear correlation between net active transport of glucose and the mobility of acetamide in rat jejunum.

(ii) the behavior of the electrical conductance does support the fact that active glucose transport is correlated to the electrolyte passive permeability of the intestine.

(iii) cyclic AMP added to the perfusion fluid has the same effect as the high glucose concentration and increases the mobility of the acetamide through the intestinal barrier.

References

[1] G. Esposito, A. Faelli and V. Capraro, Experientia, 25 (1969) 603.

[2] G. Esposito, A. Faelli and V. Capraro, Permeability and Function of Biological Membranes ed. L. Bolis et al. (North-Holland Publ. Co., Amsterdam, 1970) pag. 74.

[3] A. Faelli, G. Esposito, G. Garotta and V. Capraro, Experientia 27 (1971) 652.

[4] R.J.C. Barry, D.H. Smyth and E.M. Wright, J. Physiol. (London) 181 (1965) 410.

[5] V. Capraro, A. Bianchi and C. Lippe, Experientia, 19 (1963) 347.

[6] J. Orloff and J. Handler, Amer. J. Med. 42 (1967) 757.

[7] K. Loeschke, C.J. Bentzel and T.Z. Csaky, Am. J. Physiol. 218 (1970) 1723.

THE FUNCTIONING OF INSECT MALPIGHIAN TUBULES

S.H.P. MADDRELL
A.R.C. Unit of Invertebrate Chemistry and Physiology,
Department of Zoology, Downing Street, Cambridge, England

1. Introduction

When one reflects on the complexities revealed by the detailed analysis of transport of ions across a single membrane, one is somewhat reluctant to attempt the analysis of transport across an epithelium where ions probably move across two membranes in series. There is the further possibility that some movement may occur between the cells, that is to say in parallel with the trans-cell route. Further than this, Malpighian tubules of insects carry out not only ion transport but the movement of large volumes of water at the same time.

However, while the prospect of examining in detail the performance of pumps on each membrane of a Malpighian tubule cell may be daunting, it ought to be possible to decide what types of pump exist and where in the tubule cells they are located and so gain an understanding of how the tubules operate.

It will be the main aim of this paper to attempt just such an analysis for the fluid-secreting regions of typical Malpighian tubules. As will be shown, the discoveries so far made do not lead to an inescapable conclusion but do allow the erection of a reasonable hypothesis which further research could test.

There exist in some blood-sucking insects, Malpighian tubules which secrete fluid at a mugh higher rate than do more typical tubules. The later part of this paper describes the secretory performance of the tubules from one such insect, pointing out in what ways it has been specialised to achieve a higher rate of secretion.

2. The function of Malpighian tubules

Before looking in detail at the ion and water movements involved it will be appropriate to recall the position that Malpighian tubules hold in the anatomy and physiology of the insect.

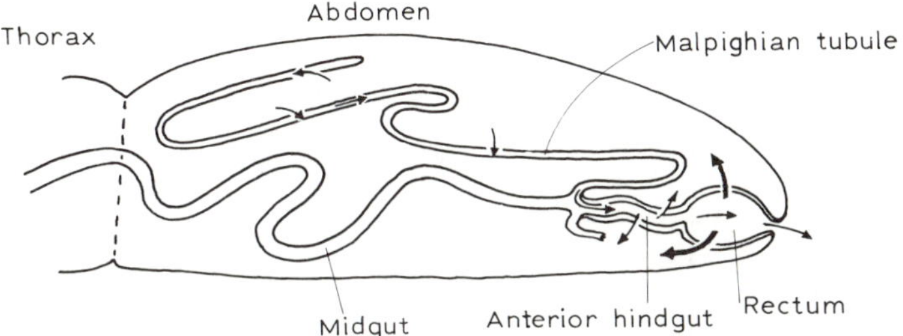

Fig. 1. The relationship of a single Malpighian tubule to the alimentary canal in an insect. The arrows indicate the direction and intensity of solute and water traffic.

Fig. 1 shows in diagrammatic form the relationship of a typical Malpighian tubule to the rest of the alimentary canal. Fluid is taken up by the Malpighian tubules from the insect's haemolymph (blood) which freely bathes them. The tubules discharge this secreted fluid into the alimentary canal at the junction of the mid- and hind-gut. The fluid is at this stage rich in many components of the haemolymph and the elimination of this fluid from the insect would rapidly drain it of many useful materials. The hindgut – particularly the rectum – therefore resorbs those substances that are useful. Substances that are not required are not resorbed and so are eliminated. As Ramsay [1] has pointed out, it is a neat feature of this system that it ensures automatic excretion of any new toxic molecules encountered by the insect, without providing specific transport mechanisms for their removal. As with vertebrate kidneys, the flow of fluid through the system is quite brisk, the whole blood volume being filtered every few hours.

3. Malpighian tubules as experimental material for the study of fluid transport

Not only are Malpighian tubules of interest because of their important function for the organisms possessing them but they make convenient material for the study of transepithelial transport of fluid. All Malpighian tubules

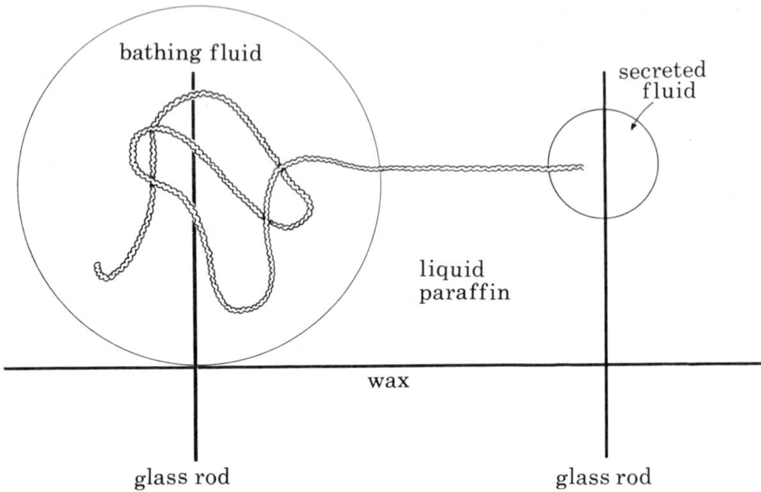

Fig. 2. The arrangement used to observe the secretory behaviour of an isolated Malpighian tubule.

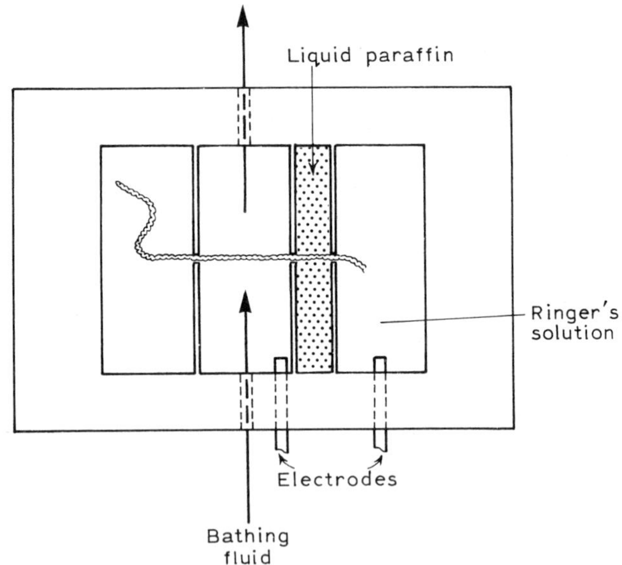

Fig. 3. Plan view of a flow cell which allows rapid changes of the fluid bathing an isolated Malpighian tubule.

consist of an epithelial layer in the form of a tube whose wall is only one cell thick. In some tubules there is only one cell type, and even in the tubules with two cell types, there are only a small minority of the second cell type. Because the epithelium is in the form of a tube it is very easy to collect, analyse and measure the rate of secretion of fluid produced *in vitro*. The drawbacks to the system are that it is not easy to short circuit the epithelium or to change the fluid bathing the luminal surfaces of the cells.

Two main methods are useful in studying isolated Malpighian tubules. One is to isolate the tubule very simply into a drop of saline solution under liquid paraffin (mineral oil) and to pull the cut end of the tubule out onto a glass rod where the secreted fluid accumulates (fig. 2). By positioning an electrode in each of the two drops the trans-wall potential difference can be followed. This method is particularly suitable for experiments where only small amounts of bathing solution are available as for example samples of haemolymph suspected to contain acceleratory hormones, or for tracer experiments.

A method which allows the bathing solution to be changed rapidly is to set a tubule up in a flow cell of the type shown in fig. 3.

4. Ion movements in Malpighian tubules

Which ions are important in secretion is revealed by the composition of the fluid secreted by Malpighian tubules. Here it must be emphasized that the analyses to be given are based on work on the tubules of only five insects *. There may well be more than a million different species of insects so that any generalization based on these figures should be accepted with some reservation! Nonetheless it is also true that although the five tubules for which reasonable complete figures are available are from different groups of insects, the figures do agree to a great extent.

The composition of the fluid produced by typical Malpighian tubules bathed in a saline solution similar in composition to the haemolymph is shown in table 1.

From table 1 it is clear that in comparison with the bathing fluid,

1) the concentrations of potassium ions and phosphate ions in the secreted fluid are elevated,

* *Carausius morosus* [2–4]; *Calliphora erythrocephala* [5, 6]; *Tipula paludosa* [7]; *Calpodes ethlius* [8]; *Schistocerca gregaria* (unpublished observations of Klunsuwan and Maddrell).

Table 1
The ionic composition of the fluid secreted by Malpighian tubules

	Concentration in the bathing medium (mM)	Concentration in the secreted fluid (mM)
K^+	10	125
Na^+	125	20
Ca^{2+}	2	<1
Mg^{2+}	5	<1
Cl^-	120	95
phosphate	20	50

2) other ions are at a lower concentration in the secreted fluid, though substantial amounts of chloride ion are present.

5. The movements of cations

It seems that the high concentration of potassium ions in the secreted fluid may be particularly significant – not only is the tubule lumen usually about

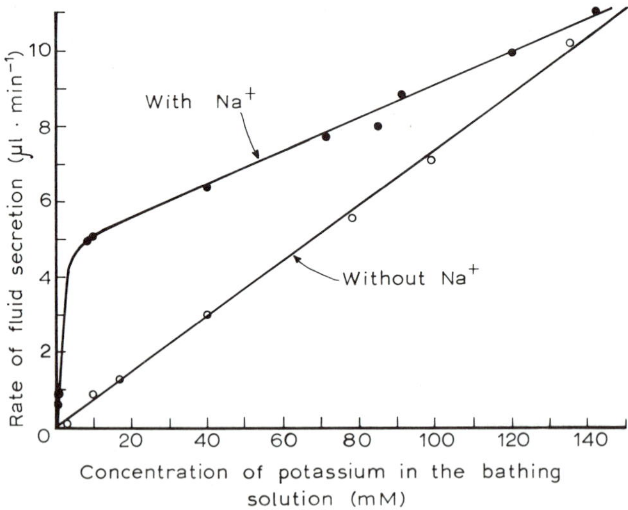

Fig. 4. The dependence of the rate of secretion of fluid by isolated Malpighian tubules of *Calliphora* on the potassium concentration of the bathing medium. (From Berridge [5].)

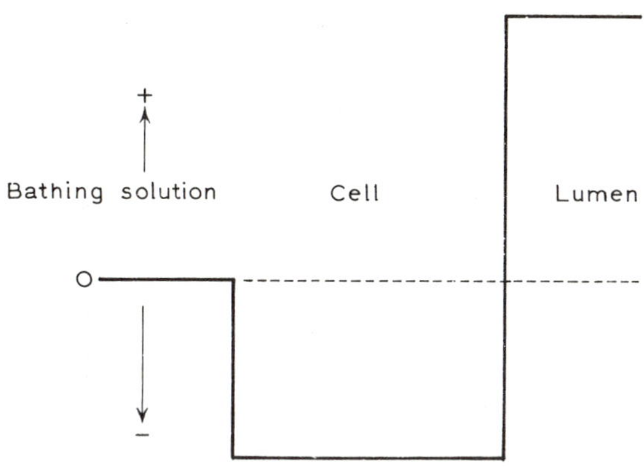

Fig. 5. The potential profile across a Malpighian tubule cell.

30 mV positive with respect to the bathing solution, but the rate of fluid secretion depends very much on the concentration of potassium ions in the bathing fluid (fig. 4). Secretion is much less affected by changes in the phosphate concentration. Given these facts it seems very likely that potassium movements are active and provide the driving force for fluid secretion. Only very few potential measurements have so far been made with intracellular microelectrodes. Those that have been made agree in showing the sort of profile shown in fig. 5. It seems quite probable then that the potassium pump is electrogenic and sited on the apical cell membrane.

Potassium entry into the cell from the basal side may well be passive because of the potential drop on this side of the cell. What one does not know is the intracellular concentration of potassium so that one can only guess at the direction of the electrochemical potential gradient across the basal cell wall. There is the further point that at low potassium concentrations, the rate of secretion of potassium becomes sensitive to the sodium concentration in the bathing fluid (fig. 4). Because the basal cell membrane is thought to be relatively impermeable to sodium ions [9, 10], it follows that this effect is likely to take place at the basal cell wall and not deeper within the cell – especially as 1–3 mM of sodium is sufficient to produce the effect. Quite probably then, potassium entry into the cell is sodium-dependent.

The low rate of secretion by Malpighian tubules in potassium-free sodium-rich solutions (fig. 4) may well be due to a rather restricted access of sodium ions to the cell interior from which however they can be pumped either by the apical potassium pump or by an apical sodium pump. Thus sodium ions may be transported by a process with some similarities to the transport of sodium by the cecropia midgut described by Professor Harvey on p. 310. It is worth pointing out here that secretion by Malpighian tubules is as unaffected by treatment with ouabain as is secretion by the cecropia midgut.

6. The movements of anions

It seems likely that anions are transported passively across the wall of Malpighian tubules. The transepithelial potential generated by the potassium pump would provide the driving force for this movement. In general, anions with the smallest hydrated radius support the highest rates of secretion (fig. 6). This, as Diamond and Wright [11] have shown, probably means that the cell membranes are characterized by sites of low electric field strength. However, it is clear from fig. 6 that an anomalous position is held by phosphate ions. These ions appear to cross the wall of the tubule in a facilitated fashion. Thus copper ions, which prevent secretion by tubules bathed in a solution containing anions other than phosphate, do not affect secretion by tubules in a phosphate-based solution. Similarly arsenate ions, which block phosphorylation, greatly slow secretion by tubules bathed in a phosphate solution but do not affect secretion by tubules bathed in a solution based on other anions.

Whether these facilitated phosphate movements are active or not is un-

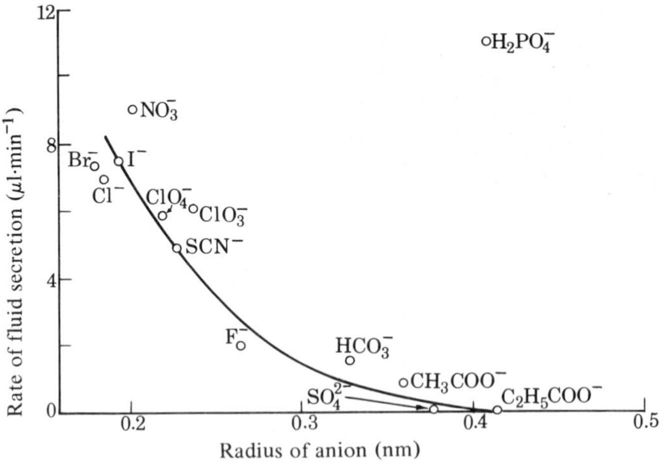

Fig. 6. The dependence of the rate of secretion of fluid by isolated Malpighian tubules of *Calliphora* on the hydrated size of the anion in the bathing medium. (Redrawn from Berridge [6].)

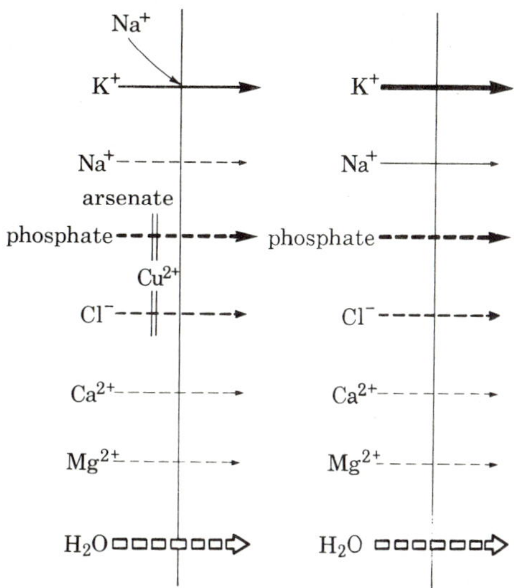

Fig. 7. Hypothetical schem for the operation of "typical" Malpighian tubules. Active movements are indicated by unbroken lines and passive movements by broken lines.

clear. Since these ions are not much concentrated in the secreted fluid and the lumen is at a positive potential they do not move up an electrochemical gradient. One may suppose, therefore, as a first hypothesis that their transport is passive. As far as ion movements are concerned the operation of Malpighian tubules can be summarized in the scheme shown in fig. 7. It should clearly be understood that this scheme is no more than a hypothesis consistent with the results; its value is in the experiments it suggests.

7. Movements of water

Fig. 7 suggests that water movements across the walls of Malpighian tubules are passive. What is the evidence for this and how might water movements be linked to the ion movements described?

Two pieces of evidence are relevant. One is that the fluid secreted by Malpighian tubules is very slightly hyperosmotic to the bathing fluid over a wide range of osmotic concentrations (fig. 8). The second is that the rate of fluid secretion is inversely related to the osmotic concentration of the bathing fluid (fig. 9). These two facts taken together mean that the rate of solute transport is unaffected by the osmotic concentration – all that alters is the

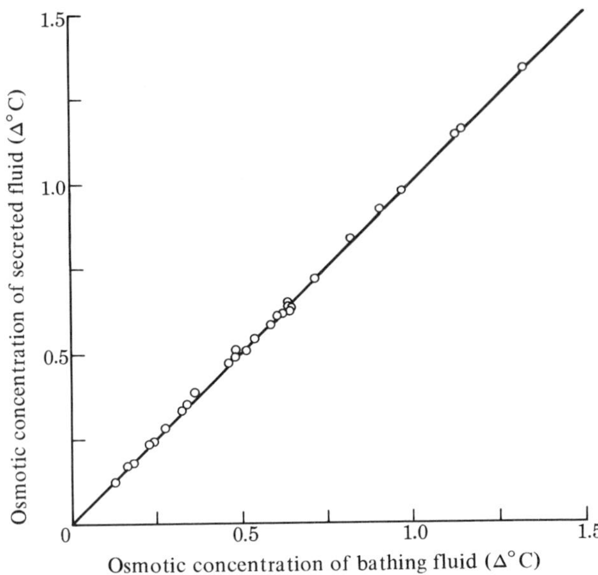

Fig. 8. The dependence of the osmotic concentration of the fluid secreted by isolated Malpighian tubules of *Rhodnius* on the osmotic concentration of the bathing fluid. (From Maddrell [12]).

rate at which water crosses the tubule wall and this changes precisely to make the fluid secreted very slightly hyperosmotic. While it is not impossible that there is an alternative explanation it seems very likely that ion movements are the driving force and water movements are a passive consequence of them.

How this linking of water and solute fluxes is brought about is not yet certain. However, when one looks at the ultrastructure of Malpighian tubules it is very noticeable that the cell membranes are very much folded on both sides of the cells (fig. 10). These folds provide just the sort of structures which would be expected to maintain standing osmotic gradients if ions are pumped across the membranes – in the manner described by Diamond and Bossert [13].

The trouble with this suggestion is that the basal infoldings and microvilli are not very long. As a result, the maintained gradients will be small. If a flow of iso-osmotic fluid is to result, the osmotic permeability of the cell walls will have to be high. This is perhaps to be expected because other fluid transporting systems have high osmotic permeabilities [13]. However, a further feature of the infoldings and microvilli may contribute to promoting fluid flow. This is that they are arranged so as to produce a parallel array of channels opening alternately to the cytoplasm and to the extracellular fluid (fig. 11). As a

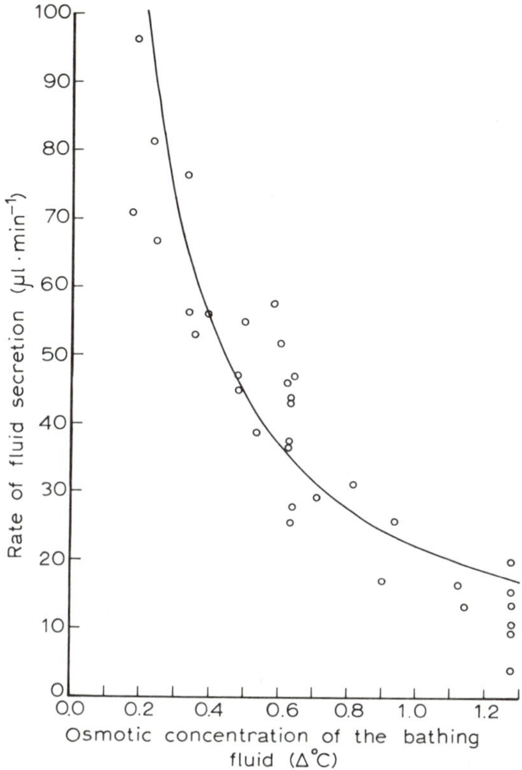

Fig. 9. The dependence of the rate of secretion of fluid by isolated Malpighian tubules of *Rhodnius* on the osmotic concentration of the bathing fluid. (From Maddrell [12].)

result, ion movements from one side of the membrane will tend not only to establish a hyper-osmotic fluid on the other side but a hypo-osmotic fluid on the side from which transport is occurring because neither fluid can equilibrate as quickly with the cytoplasm or extracellular fluid as would otherwise be the case. Thus, the effective gradient across the cell wall is made steeper and so solute/solvent coupling is made more efficient. Not only will this be the case but ion transport in the system will tend to be concentrated on the side from which ions are moved. This is because ions will be reduced in concentration in the fluid flowing in the channels on that side and so fewer ions will be available for transport deeper within the system. This arrangement, as Diamond and Bossert [13] have shown, will produce a more nearly iso-osmotic secretion.

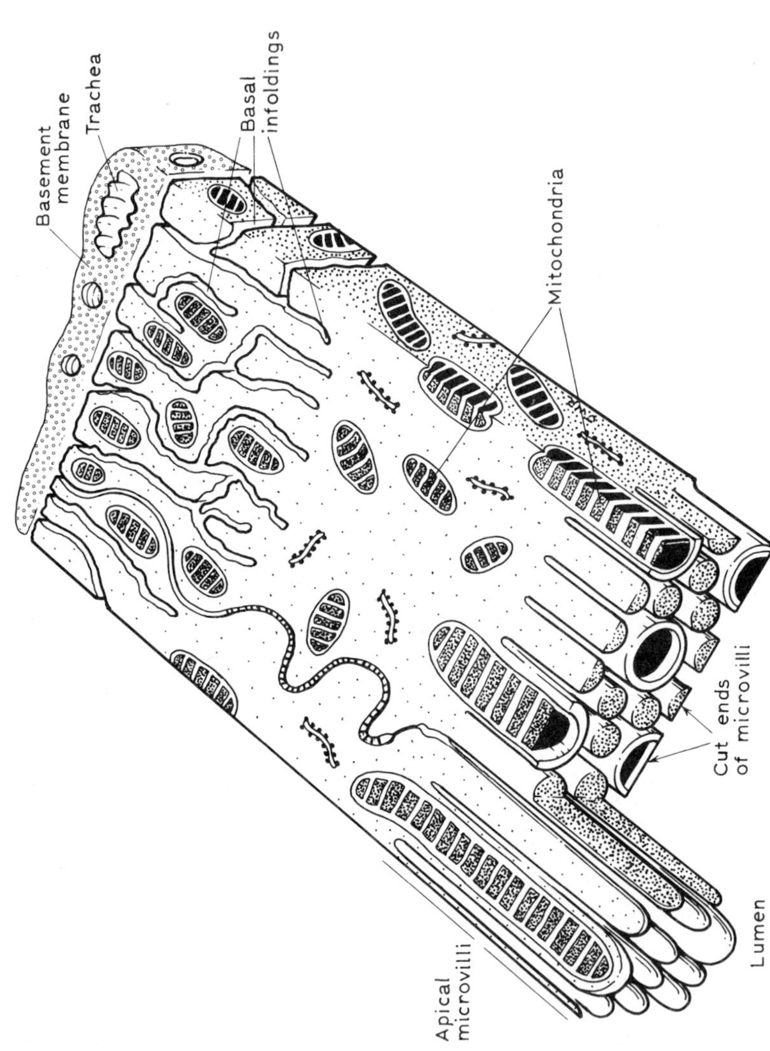

Fig. 10. Diagrammatic representation of part of a transverse section of the wall of a typical Malpighian tubule. × 10 000.

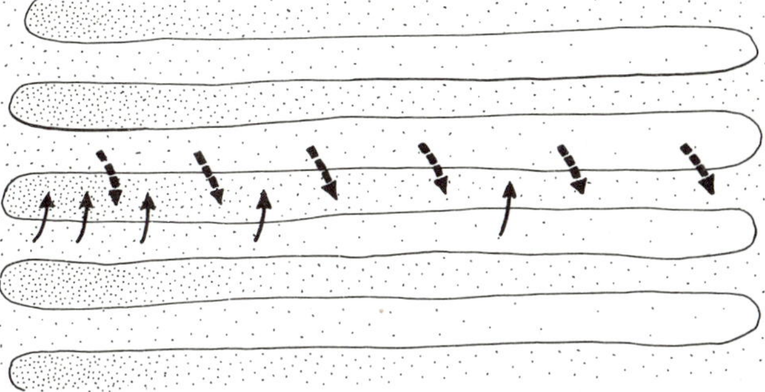

Fig. 11. Coupling of ion and water movements at a cell membrane thrown into a series of close-packed folds. Active solute movements are indicated by the smaller continuous arrows and passive water movements by the larger broken arrows.

While the folding of the cell walls will tend to promote a flow of fluid along the lines set out above, it nonetheless remains possible that this mechanism plays only a minor role, because the osmotic pressure differences that can be maintained by such a system are small. Taylor [14] suggests that the folding of the cell walls is primarily a device to increase the surface areas on both sides of the cell and that the driving force for water movements is in fact the overall osmotic pressure difference between the lumen and bathing solution. As Taylor points out there is in several Malpighian tubules an osmotic pressure difference of the right sense and about the correct size to promote water flow into the lumen. He calculates that an osmotic pressure difference of, say, 10 mOsm and an effective osmotic permeability of 10^{-4} cm sec^{-1}. Osm^{-1} (or 1 mOsm and 10^{-3} cm sec^{-1} Osm^{-1}) would be sufficient to account for the measured rates of transport.

Assuming the osmotic pressure measurements so far made are accurate, then it follows that the fluid in the lumen of most Malpighian tubules is sufficiently hyper-osmotic to the bathing fluid to provide a driving force for water movement a good deal larger than that which could be established by standing gradients associated with the basal infoldings and apical microvilli. As Taylor [14] points out, cryoscopic osmotic pressure determinations may be subject to small systematic errors such as the possible precipitation of some solutes on freezing or the fact that the colligative properties of solutions may not be the same slightly below 0°C as they are at room temperature. Nonetheless, it does at the moment seem likely that simple osmotic movements across the tubule wall maintained by an osmotic pressure difference between lumen and bathing solution could make a very important contribution to the secretion of fluid by Malpighian tubules.

8. Fluid secretion by Malpighian tubules of blood-sucking insects

This paper has so far described the way in which "conventional" Malpighian tubules operate. However, the Malpighian tubules of recently fed blood-sucking insects operate very much faster. For example, the Malpighian tubules of tsetse flies and of kissing bugs can secrete a volume of fluid equivalent to that of the whole insect in about 30 min. In absolute terms the Malpighian tubules of *Rhodnius* (the blood-sucking insect whose Malpighian tubules have been investigated) can secrete fluid at rates of about 3.3 μl min^{-1} cm^{-2} [12]; this is several times faster than, say, the rate of fluid movements in the mammalian gall-bladder. As another measure of how fast is this rate of secretion, it can be calculated that a volume of fluid equal to that of the cells in the wall flows through them every 16 sec. The ion fluxes involved are equally impressive; chloride ions cross the wall at a rate of 10^4 pmol sec^{-1} cm^{-2}.

For the purposes of this paper it will be appropriate merely to describe the ways in which the Malpighian tubules of *Rhodnius* differ from the more slowly secreting tubules of other insects. But before going on to do this it should be pointed out that *Rhodnius* Malpighian tubules in *in vitro* experiments require the presence of a stimulant in the bathing solution. Without such a stimulant the rate of secretion is extremely low. The stimulant used can be any one of the following: (i) the naturally occurring diuretic hormone extractable from the central nervous system [15]; (ii) 5-hydroxytryptamine, which, at concentrations greater than 3×10^{-8} M, is a potent mimic of the natural hormone [16] or (iii) cyclic AMP at concentrations greater than 3×10^{-5} M [16]. In life, the Malpighian tubules of *Rhodnius* normally only secrete at a high rate immediately after a blood meal; they then secrete at a much reduced rate until the next meal which may not occur for some months.

The dramatic acceleration on feeding is achieved by the release into the haemolymph of the potent diuretic hormone mentioned above.

9. Transport of cations by *Rhodnius* Malpighian tubules

The most obviously different feature in the operation of *Rhodnius* Malpighian tubules is that sodium transport plays an important role in secretion. So fast is this transport that potassium can be entirely omitted from the bathing medium without secretion being slowed. This is very different from the behaviour of other Malpighian tubules (fig. 12).

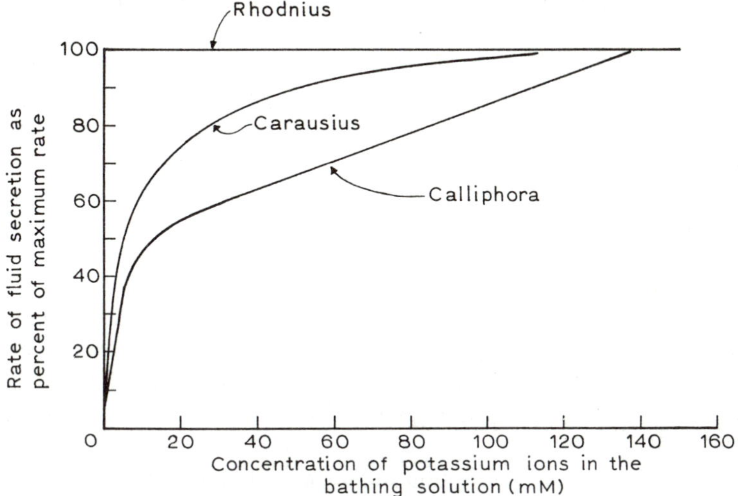

Fig. 12. The dependence of the rate of secretion of fluid by isolated Malpighian tubules on the potassium concentration in the bathing solution.

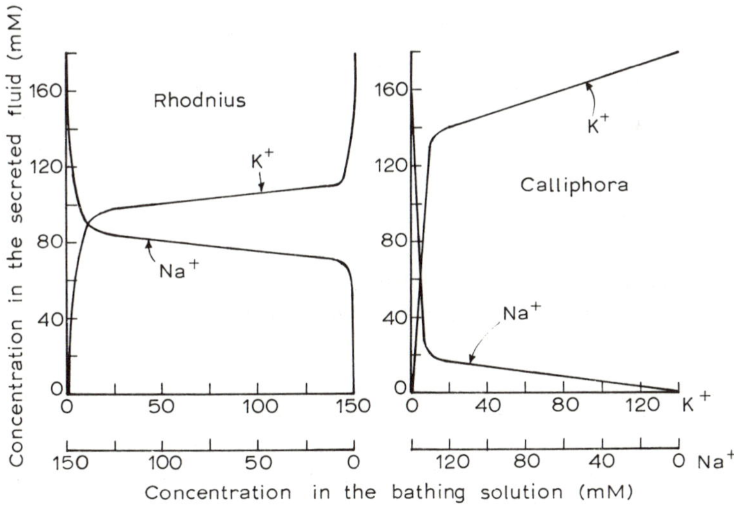

Fig. 13. The dependence of the ionic composition of the fluid secreted by isolated Malpighian tubules of *Rhodnius* and *Calliphora* on the ionic composition of the bathing solution. (From Maddrell [12].)

This ability to transport sodium is also shown by the sodium-rich fluid which is secreted by tubules bathed in solutions poor in sodium ions (fig. 13). In a solution containing equal concentrations of sodium and potassium, sodium is transported at a rate nearly equal to that at which potassium is transported. Again this is very different from the behaviour of other Malpighian tubules (fig. 13).

10. Transport of anions by *Rhodnius* Malpighian tubules

As far as anions are concerned the tubules of *Rhodnius* agree with other Malpighian tubules in secreting fastest in solutions of smaller anions. However they will not secrete in a phosphate-based solution and so do not seem to have the same facilitation of phosphate movements that other tubules have.

Although *Rhodnius* tubules secrete so fast, they are not slowed by a decrease in chloride concentration of the bathing solution until this drops to below 30 mM (fig. 14). This compares with other Malpighian tubules (fig. 14) which are affected much sooner. This argues either that the Malpighian tubule wall in *Rhodnius* is more permeable to chloride ions or that chloride ions are actively transported.

Electrical measurements show that the lumen of the tubule in *Rhodnius* is at a negative potential with respect to the bathing solution, so that chloride movements are thermodynamically uphill. Their active transport is a distinct

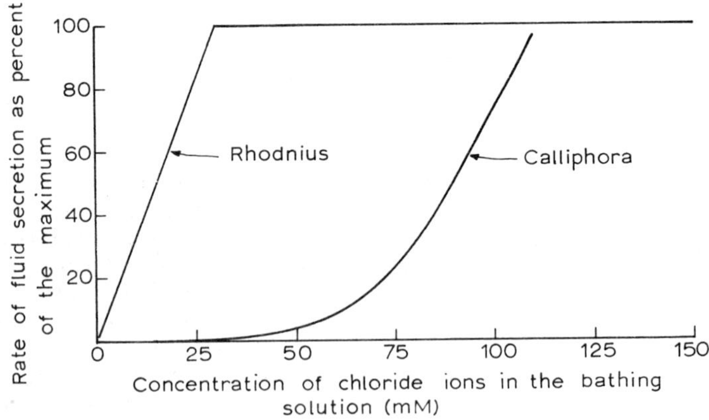

Fig. 14. The dependence of the rate of secretion of fluid by isolated Malpighian tubules of *Rhodnius* and *Calliphora* on the chloride concentration of the bathing solution.

Table 2
The greatest electrochemical gradients against which movements of each major ion have been recorded in Malpighian tubules of *Rhodnius*

		Bathing solution (mM)			Secreted fluid (mM)			Trans-wall potential difference (lumen with respect to bathing solution) (mV)
		Na	K	Cl	Na	K	Cl	
Sodium		9	141	155	85	100	180	+ 45
Potassium	i	0	150	155	0	185	180	+158
	ii	141	9	26	95	90	170	+ 30
Chloride	i	141	9	155	95	90	180	−105
	ii	150	0	30	185	0	170	− 20

possibility therefore, but one must all the time bear in mind the strong likelihood of both solute and solvent drag effects in such a fast fluid transporting system.

It does seem possible that, since both major cations sodium and potassium are capable of being highly concentrated in the lumen as are chloride ions, all

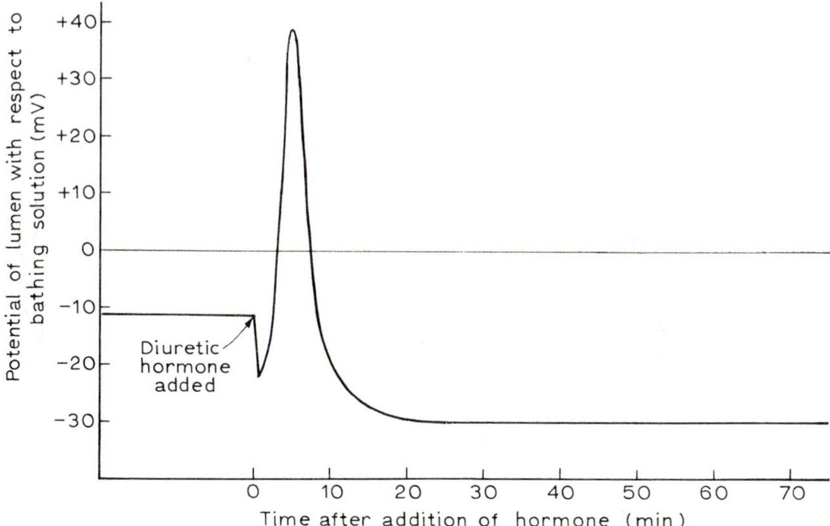

Fig. 15. Changes in the trans-wall potential difference of an isolated Malpighian tubule of *Rhodnius* induced by treatment with the diuretic hormone from this insect.

three ions may be actively transported. In table 2 are recorded the maximum electrochemical potential differences against which these ions have been shown to move. From table 2 it seems very probable that all three ions are actively transported, even bearing in mind the probability of drag effects.

A further indication of this possibility comes from measurements of the transwall potential difference during stimulation by the diuretic hormone (fig. 15). After a small initial fall, the luminal potential rapidly goes positive – perhaps cations are being moved towards the lumen. Then the potential goes negative and stays there while secretion persists.

If sulphate ions are substituted for chloride ions the potential rises to a high level and stays there. This is reminiscent of the similar situation in frog skin and probably is indicative of active cation transport in this case also.

One other indication of the possibility of the active transport of both cations and anions is that the trans-wall potential is much affected by temperature – so much so that it can even change in sign. Such a change might be

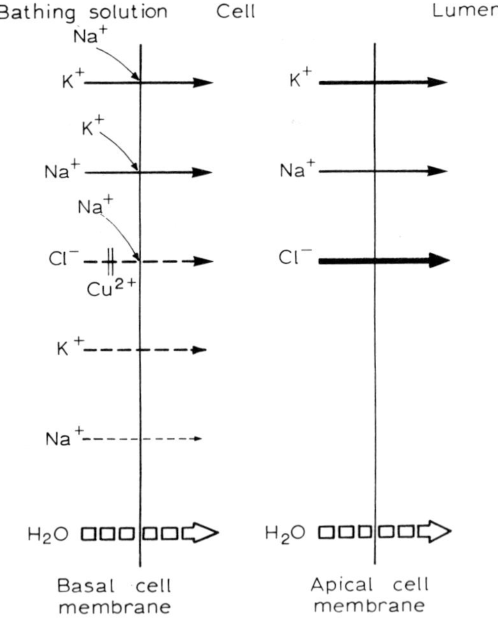

Fig. 16. Hypothetical scheme for the operation of Malpighian tubules of *Rhodnius*. Active movements are indicated by unbroken lines and passive movements by broken lines.

expected if the transport mechanisms for cations and anions have different temperature sensitivities.

Based on these sorts of observations it is possible to construct a hypothetical scheme for the operation of Malpighian tubules of *Rhodnius* (fig. 16). Basically it differs from that of the other Malpighian tubules in postulating active electrogenic pumps for sodium, potassium and chloride ions on the luminal side.

It may be that the existence of three pumps in parallel for the major ions provides an explanation for the astonishingly high rates of secretion which these tubules can reach. It will be most interesting to see if the tubules of other blood-sucking insects also have such a system. For this reason it will be valuable to investigate the Malpighian tubules of the tsetse fly, a blood-sucking insect not at all closely related to *Rhodnius*.

11. Conclusion and discussion

The picture that emerges from this examination of the operation of Malpighian tubules, is that typically they are driven by an active electrogenic transport of potassium across the apical cell wall into the lumen. Anions passively follow this cation transport and these ionic movements are osmotically linked to water movements. Secretion is apparently speeded up in the specialised tubules of *Rhodnius* by their having pumps for all three major ions involved.

It is worth ending by reflecting on the similarity of the transport systems of the cecropia midgut and typical Malpighian tubules. Each has as its central action electrogenic potassium pumps on the luminal side of the cells.

The midgut receives during feeding a fluid extremely rich in potassium ions [17], so that most of the ionic influx into the cell *in vivo* is across the luminal face. The basal cell wall is practically impermeable to sodium ions, so although the pump is capable of handling them, no transport of sodium occurs. As the ion traffic *in vivo* mostly involves movements into and out of the cells across the luminal face, rather little trans-epithelial transport occurs and so no large net water flux across the wall is achieved. By contrast, Malpighian tubules have a more permeable basal cell wall so that potassium, chloride and to a lesser extent, sodium ions gain access to the cell from the bathing solution. Their transport into the lumen directly or indirectly by the potassium pump results in a large transepithelial flux which entrains a large net flux of water across the wall. It is interesting that two insect organs with such similar potassium pumps on their luminal membranes have achieved such different transport properties, merely by controlling the access of ions to the pump.

References

[1] J.A. Ramsay, Excretion by the Malpighian tubules of the stick insect, *Dixippus morosus* (Orthoptera; Phasmidae): amino acids, sugars and urea. J. exp. Biol. 35 (1958) 871–891.

[2] J.A. Ramsay, Active transport of water by the Malpighian tubules of the stick insect, *Dixippus morosus* (Orthoptera; Phasmidae). J. exp. Biol. 31 (1954) 104–113.

[3] J.A. Ramsey, The excretion of sodium, potassium and water by the Malpighian tubules of the stick insect, *Dixippus morosus* (Orthoptera; Phasmidae). J. exp. Biol. 32 (1955) 200–216.

[4] J.A. Ramsay, Excretion by the Malpighian tubules of the stick insect, *Dixippus morosus* (Orthoptera; Phasmidae): calcium, magnesium, chloride, phosphate and hydrogen ions. J. exp. Biol. 33 (1956) 697–708.

[5] M.J. Berridge, Urine formation by the Malpighian tubules of *Calliphora*. I. Cations. J. exp. Biol. 48 (1968) 159–174.

[6] M.J. Berridge, Urine formation by the Malpighian tubules of *Calliphora*. II. Anions. J. exp. Biol. 50 (1969) 15–28.

[7] G.M. Coast, Formation of urinary fluid by Malpighian tubules of an insect. J. Physiol. 202 (1969) 102–103.

[8] H.B. Irvine, Sodium and potassium secretion by isolated insect Malpighian tubules. Am. J. Physiol. 217 (1969) 1520–1527.

[9] D.E.M. Pilcher, The influence of the diuretic hormone on the process of urine secretion by the Malpighian tubules of *Carausius morosus*. J. exp. Biol. 53 (1970) 465–484.

[10] S.H.P. Maddrell, Fluid secretion by the Malpighian tubules of insects. Phil. Trans. Roy. Soc. Lond. B. 262 (1971) 197–207.

[11] J.M. Diamond and E.M. Wright, Biological membranes: the physical basis of ion and nonelectrolyte selectivity. A. Rev. Physiol. 31 (1969) 581–646.

[12] S.H.P. Maddrell, Secretion by the Malpighian tubules of *Rhodnius*. The movements of ions and water. J. exp. Biol. 51 (1969) 71–97.

[13] J.M. Diamond and W.H. Bossert, Standing-gradient osmotic flow. A mechanism for coupling of water and solute transport in epithelia. J. Gen. Physiol. 50 (1967) 2061–2083.

[14] H.H. Taylor, Water and solute transport by the Malpighian tubules of the stick insect, *Carausius morosus*. The normal ultra-structure of the type 1 cells. Z. Zellforsch. mikrosk. Anat. 118 (1971) 333–368.

[15] S.H.P. Maddrell, Excretion in the blood-sucking bug, *Rhodnius prolixus* Stål. I. The control of diuresis. J. exp. Biol. 40 (1963) 247–256.

[16] S.H.P. Maddrell, D.E.M. Pilcher and B.O.C. Gardiner, Pharmacology of the Malpighian tubules of *Rhodnius* and *Carausius*: the structure-activity relationship of tryptamine analogues and the role of cyclic AMP. J. exp. Biol. 54 (1971) 779–804.

[17] W.R. Harvey and S. Nedergaard, Sodium-independent active transport of potassium in the isolated midgut of the cecropia silkworm. Proc. natn. Acad. Sci. (U.S.A.) 51 (1964) 757–765.

ION TRANSPORT ACROSS THE PROXIMAL CONVOLUTION OF THE MAMMALIAN KIDNEY

K.J. ULLRICH
Max-Planck-Institut für Biophysik, Frankfurt/Main, Germany

One principle feature of renal function is the isotonic reabsorption in the proximal convolution of approximately 60% of the filtered sodium, 50% of the filtered chloride and 90% of the filtered bicarbonate. The work of our group during the last four years was designed (1) to determine the driving forces for the transport of the different ions across the proximal tubular wall [1]; (2) to clarify the interrelationship between the two active transport processes which participate in the proximal isotonic reabsorption, namely the Na^+ reabsorption and the HCO_3^- reabsorption [2, 3].

To get some information about the driving forces we determined the transport coefficients of the principle ions Na^+, Cl^-, and HCO_3^- according to an equation derived by Dr. Sauer [1, 24]

$$J_i^{(V)} = 0 = -\sigma_i \bar{c}_i J_v - P_i^{(V)} \left[\Delta c_i + \frac{z_i F \Delta\varphi}{RT} \bar{c}_i \right] + J_{i\,\text{active}}^{(V)} . \tag{1}$$

This equation requires that all measurements are made in a stationary state where the concentration change with time dc/dt is zero, because in this situation $J_i^{(V)}$, the ion transport relative to the volume transport equals zero.

Three transport terms will then cancel, the concentration change due to reflection of the substance at the tubular wall $\sigma_i \bar{c}_i J_v$, the concentration change due to diffusion $[\Delta c_i + P^{(V)}(z_i F \Delta\varphi / RT)\bar{c}_i]$ and the concentration change due to active transport $J_{i_\text{active}}^{(V_i)}$ [1]. (σ_i = reflection coefficient, $\bar{c}_i$ the mean concentration across the epithelial wall, i.e. $\frac{1}{2}(c_\text{lumen} + c_\text{interstitium})$, J_v is the transtubular water flux, $P_i^{(V)}$ is the permeability coefficient measured at $J_v = 0$, Δc_i and $\Delta\varphi$ are the transtubular concentration difference and electrical potential difference measured with KCl bridges respectively, and $z_i F$ and RT have their usual meaning).

A useful form of eq (1) is the following:

$$\left[\Delta c_i + \frac{zF}{RT}\Delta\varphi\,\bar{c}_i\right] - \frac{\sigma_i}{P_i^{(V)}}\bar{c}_i J_v + \frac{J_{i\,\mathrm{active}}^{(V)}}{P_i^{(V)}}. \tag{2}$$

As will be shown below the transtubular water fluxes, J_v, have been varied and the concentration and electrical potential differences have been measured, the latter by Dr. Frömter [1]. A plot of $[\Delta c_i + (z_i F\Delta\varphi/RT)c_i]$ against $-c_i J_v$ gives $\sigma_i/P_i^{(V)}$ as slope and $J_{i\,\mathrm{active}}^{(V)}/P_i^{(V)}$ as intercept on the ordinate. Since P_i can be determined by an independent procedure, σ_i and the active transport term $J_{\mathrm{active}}^{(V)}$ can be evaluated.

The method used to measure the transtubular water flux in a stationary situation is as follows (fig. 1): The convoluted tubule was punctured by a double barrelled glass capillary. A droplet of colored castor oil was then injected which blocks the free flow. Thereafter a droplet of isotonic Ringer solution which had approximately the stationary Na^+, Cl^- and HCO_3^- concentration was injected so as to split the oil droplet into two. When the injected fluid is reabsorbed the split oil columns reapproach one another. This process is photographed in 5 sec intervals. J_V was determined using these photographs in a manner published by Györy [4]. The transtubular water flux was calculated according to the equation:

$$J_V = \ln 2\,\frac{\pi r^2}{t_{1/2}}.$$

J_V was varied by adding raffinose to the capillary perfusate, whereas addition of raffinose to the luminal perfusate brought J_V to zero.

The stationary concentration difference Δc_i was measured in a second set

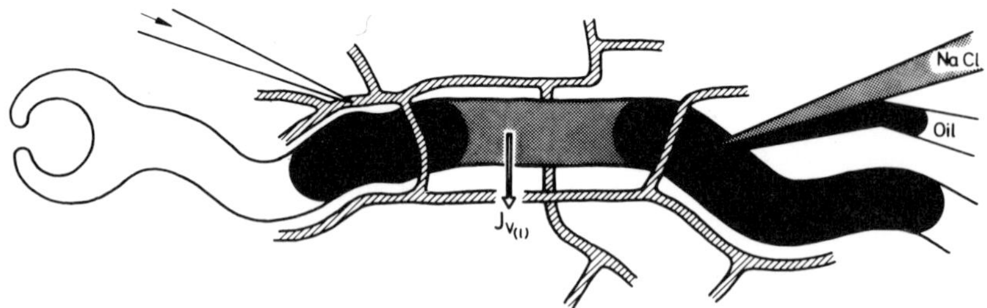

Fig. 1. Scheme of the shrinking droplet technique in the proximal convolution with a double barrelled capillary and simultaneous perfusion of the renal blood capillaries.

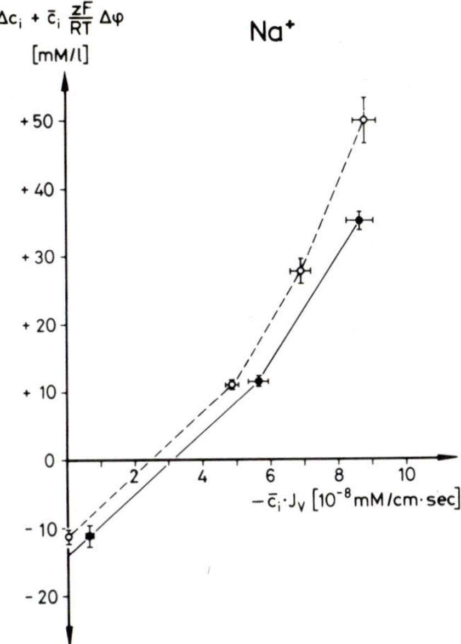

Fig. 2. Electrochemical potential difference ($\Delta c_i + \bar{c}_i zF\Delta\varphi/RT$) for sodium across the proximal convolution plotted against the transtubular water flux J_V multiplied by the mean transepithelial concentration ($\bar{c}$). ●– – –● with and ○– – –○ without acetate in the capillary perfusate. The abscissa to ordinate ratio equal the reflection coefficient: permeability ratio $\sigma_i/P_i^{(V)}$. The ordinate intersect equals the active transport rate divided by permeability coefficient ($J_{i_{active}}^{(V)} : P_i^{(V)}$) (Ullrich et al. [1]).

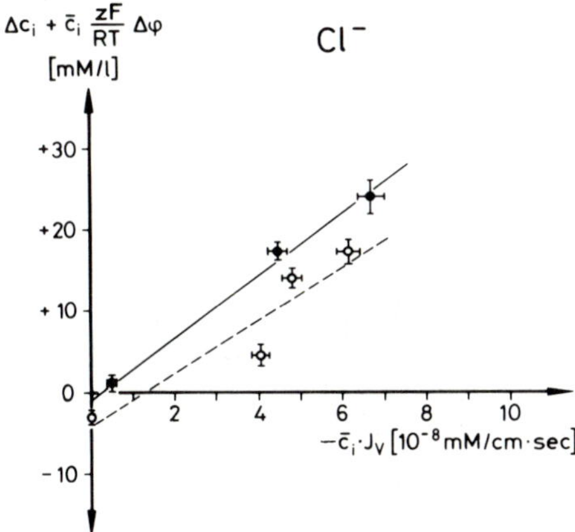

Fig. 3. Electrochemical potential difference ($\Delta c_i + c_i zF\Delta\varphi/RT$) for chloride across the proximal convolution plotted against the transtubular water flux J_V multiplied by the mean transepithelial concentration (c). ●– – –● with and ○– – –○ without acetate in the capillary perfusate. Otherwise as in fig. 2.

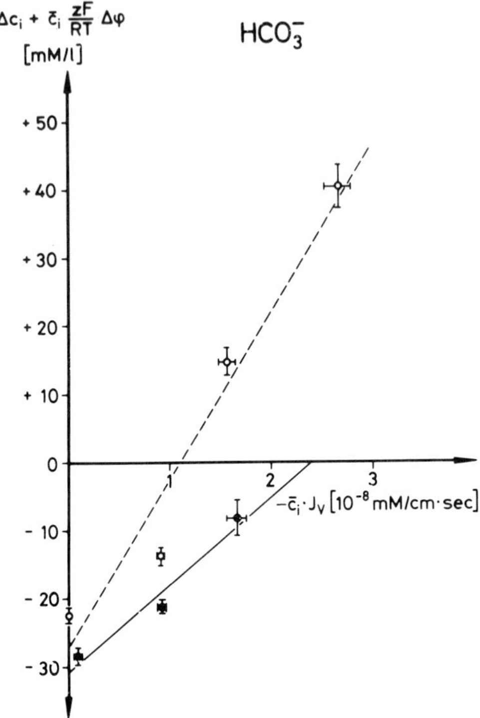

Fig. 4. Electrochemical potential difference ($\Delta c_i + \bar{c}_i zF\Delta\varphi/RT$) for bicarbonate across the proximal convolution plotted against the transtubular water flux J_V multiplied by the mean transepithelial concentration ($\bar{c}$). •–––• with and ○–––○ without acetate in the capillary perfusate. Otherwise as fig. 2.

of experiments with the various capillary perfusates by withdrawing the shrinking droplet after it had been in contact with the tubule wall for more than 15 sec: the probe was analysed for Na^+ and Cl^-. $\Delta\varphi$ was measured in the same situation. The bicarbonate concentrations of the corresponding stationary solutions were calculated as anion deficit. Data gained for Na^+ and Cl^- are shown in figs. 2 and 3. Each figure contains data from experiments with acetate as metabolic fuel in the capillary perfusate and data from experiments where acetate was omitted. The slope being as already mentioned: $\sigma_i/P_i^{(V)}$ and the ordinate intersection: $J_{i\,active}^{(V)}/P_i^{(V)}$. Note that the chloride line starts near the zero point which means that Cl^- behaves passively.

Before we can calculate the coefficients σ_i and $J_{i\,active}^{(V)}$ from the slope $\sigma_i/P_i^{(V)}$ and the ordinate intersection $J_{i\,active}^{(V)}/P_i^{(V)}$, we need to know the

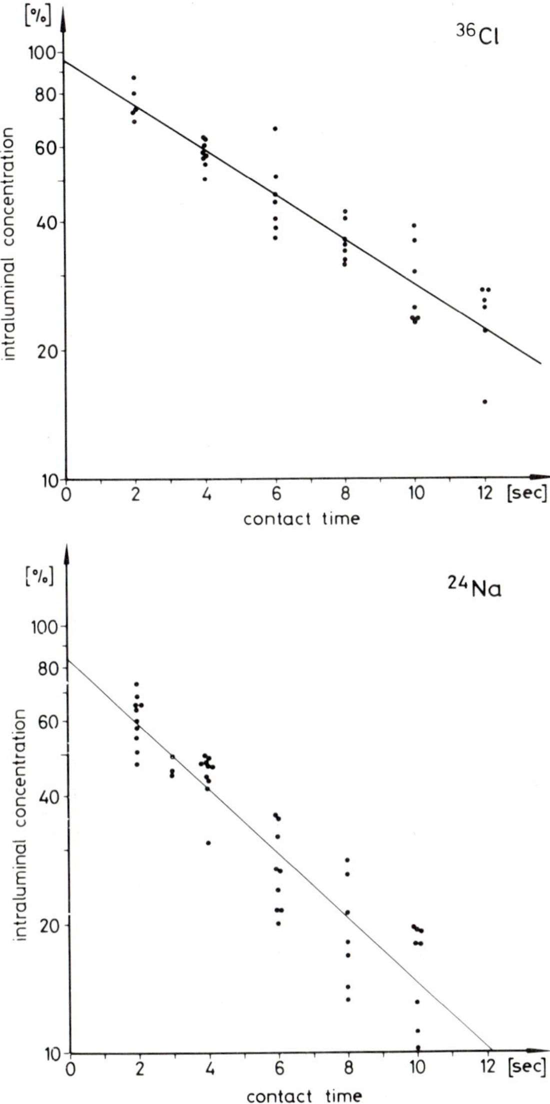

Fig. 5. Evaluation of the tracer Na^+ and Cl^- permeability of the proximal convolution with the stop flow microperfusion method. The luminal tracer concentration is plotted against the contact time. The ^{24}Na outflux was measured with artificial capillary perfusion (110 NaCl, 35 $NaHCO_3$,10 Na acetate, 3 $CaCl_2$), while the ^{36}Cl efflux was estimated with blood in the capillaries. (Ullrich et al. [1].)

permeability coefficient $P_i^{(V)}$ for the ions in question. They were measured in several ways: 1. With a stop flow microperfusion method where the tubule was filled with a solution at which J_V was zero. The solution contained either ^{24}Na or ^{36}Cl. After 2–12 sec, the injected droplet was reaspired. By pooling 2–7 time samples collected from the same tubule, enough fluid was obtained so that duplicate determinations of the tracer concentrations could be performed. In fig. 5, the tracer concentrations plotted against the contact time are shown for two series of experiments. The permeability was calculated according the equation

$$P_{i\,app.}^{(V)} = \frac{r^2 \pi \ln 2}{\tau_{1/2}}$$

where r is the luminal radius, and $\tau_{1/2}$ is the time at which the tracer concentration decreases to half of its original value. In order to obtain the true tracer permeability from these experiments $P_{i\,app.}^{(V)}$ has to be divided by a correction term $[1 - \frac{1}{2}(\Delta c_i/\bar{c}_i)]$. This term arises from the fact that the concentrations of the unlabelled ion species are not identical on both sides of the membrane. An uncertainty however remains from the fact that the mean concentration for the labelled species cannot be clearly defined, since tracer is added only to one side [1].

Furthermore, a microperfusion method has been used to measure $P_i^{(V)}$ [5, 6]: The free flow is diverted by creating a leak in the tubule near the glomerulus. Downstream from an oil blockade the tubule is perfused with a solution at which $J_V = 0$ using a high precision pump. When, possibly, the same tubule was punctured repeatedly and the permeability was calculated as the average of all individual slopes. Otherwise a regression line was drawn through all values except those within 200 μ of the infusion site (fig. 6). The equation used to calculate P was as follows: $P = \dot{V} \ln 2/l_{1/2}$; $\dot{V}$ is the tubular perfusion rate, $l_{1/2}$ is the tubular length at which the tracer concentration decreases to half of its original value. The data where the luminal concentration of inactive Na^+ was different from the interstitial Na^+ concentration were corrected by the factor $[1 - \frac{1}{2}(\Delta c_i/\bar{c}_i)]$. Data gained with and without acetate in the perfusate are shown in fig. 6.

Table 1 summarizes the tracer data gained during the last 6 years with the different methods used. The only values in the literature which could be compared with ours are those of Morel and Murayama [7]. To calculate σ_i and the active transport rate $J_{i\,active}^{(V)}$ we used the stop flow microperfusion data 15.5×10^{-7} cm^2/sec for ^{24}Na and 13.3×10^{-7} cm^2/sec for ^{36}Cl and

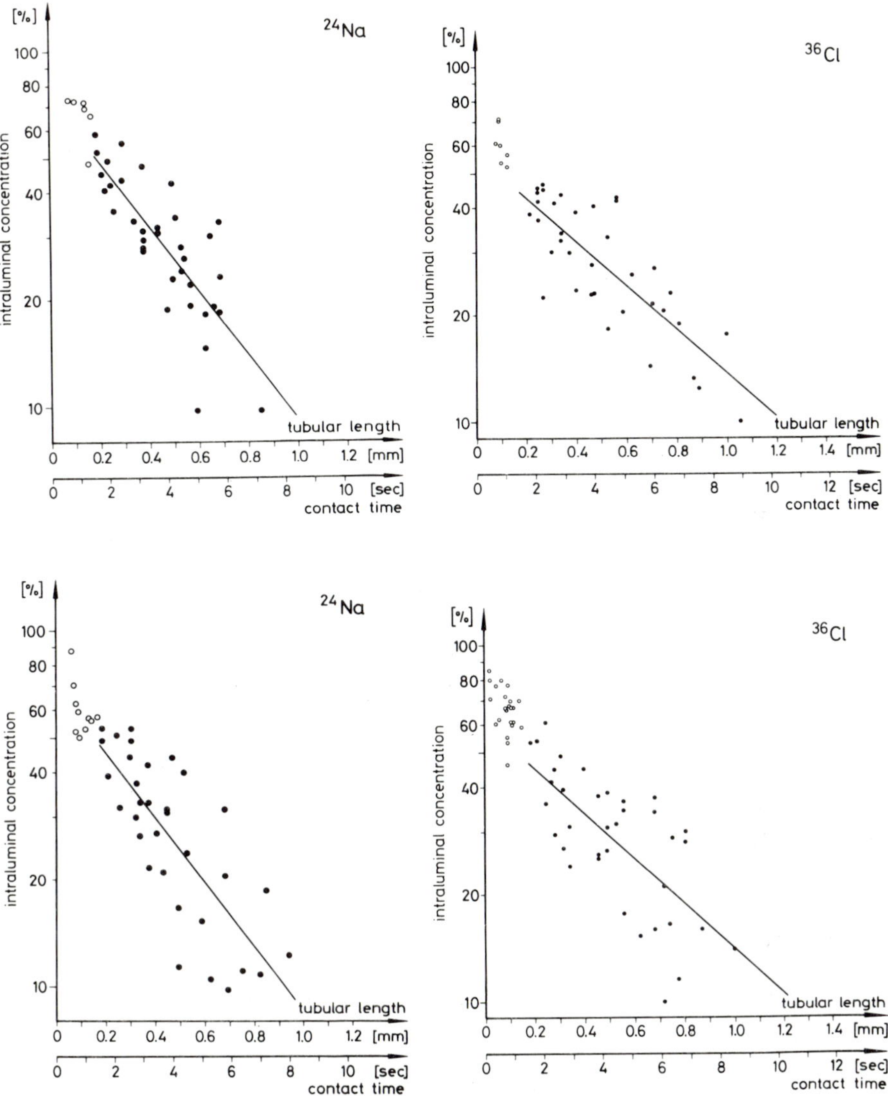

Fig. 6. Decrease of ^{24}Na and ^{36}Cl concentration within the tubular fluid plotted against conctact time and tubular length respectively. The method was microperfusion with artificial capillary perfusion. The permeability coefficient is evaluated by the regression line drawn through the filled points only. The open circles, which are points near the infusion site, are omitted. a) Capillary perfusate with 10 mmol/1 acetate; b) Capillary perfusate without acetate. (Ullrich et al. [1]).

Table 1
Sodium and chloride permeability $\times 10^{-7}$ cm^2/sec of the proximal convolution gained with the stop flow microperfusion and microperfusion method and different capillary perfusates. + = values of Morel and Murayama [7]. The sodium values are corrected by $1 - \frac{1}{2}\Delta c_i/\bar{c}_i$ to eliminate the effect of active transport. (Ullrich et al. [1].)

	Capillary perfusate	^{24}Na	^{36}Cl
Stop flow micro-perfusion	blood		13.3 ± 0.8 (SE)
	Ringer with 10 mM acetate	15.5 ± 0.8 (SE)	
Micro-perfusion	blood	11.9 17.7 15.3+	9.1 – 15.5
	Ringer with acetate	16.3 ± 1.9 (SE)	12.3 ± 1.5 (SE)
	Ringer without acetate	16.4 ± 2.1 (SE)	12.2 ± 2.1 (SE)

got the values shown in table 2; σ_{Na} = 0.70 with and without acetate, σ_{Cl} = 0.50 and 0.43 respectively. The rate of active sodium transport was 2.17 $\times 10^{-8}$M/cm · sec and 1.72 respectively which amounts to 35 and 39% of the net sodium transport measured in the experiments comparable to free flow conditions in the normal rat. This means that more than 60% of the net sodium transport in the proximal convolution is passive, due to diffusion and solvent drag. The chloride transport however seems to be passive almost entirely. Since the bicarbonate permeability is not known, neither the reflection coefficient nor the active transport rate can be calculated. However, since the present results suggest that the reflection coefficient for bicarbonate is considerably higher than for chloride, active transport rates of 2.4 $\times 10^{-8}$ mM/cm sec (in the experiments with acetate) and 1.1 $\times 10^{-8}$ mM/cm sec (in the experiments without acetate) can be estimated, if σ_{HCO_3} is taken to be approximately 1.0. A comparison of these figures with the respective net transport rates which are 0.92 $\times 10^{-8}$ mM/cm sec in both groups of experiments suggests that in the case of bicarbonate the active transport is probably equal to or greater than the net transport.

From these data one has to conclude that two active transport processes Na^+ and H^+ (HCO_3^-) transport participate in the isotonic absorption in the proximal convolution. In connection with the finding that the Na^+K^+ ATPase is located at the basal infoldings [22] and that the hydrogen ion transport

Table 2
The reflection coefficients (σ_i) and the rates of active transport $J^{(V)}_{i_{active}}$ for Na^+ and Cl^- (Ullrich et al. [1]).

			without acetate	with acetate
Na	$\frac{\sigma_i}{P_i^{(V)}}$	10^5 cm^{-2} sec	−4.51	−4.54
	$P_i^{(V)}$	10^{-7} cm^2 sec^{-1}	15.5	15.5
	σ_i	–	0.70	0.70
	$\frac{J^{(V)}_{i_{active}}}{P_i^{(V)}}$	mM/l	−11.1	−14.0
	$J^{(V)}_{i_{active}}$	10^{-8} mM/cm sec	−1.72	−2.17
Cl	$\frac{\sigma_i}{P_i^{(V)}}$	10^5 cm^{-2} sec	−3.28	−3.80
	$P_i^{(V)}$	10^{-7} cm^2 sec^{-1}	13.3	13.3
	σ_i	–	0.43	0.50
	$\frac{J^{(V)}_{i_{active}}}{P_i^{(V)}}$	mM/l	−4.2	−0.55
	$J^{(V)}_{i_{active}}$	10^{-8} mM/cm sec	−0.55	−0.07

$\sigma_i/P_i^{(V)}$ was taken from the slope of the regression line in case of chloride (fig.3) and from the slope of the lines connecting the two points near the origin in case of sodium (fig.2). Like all other values those for $P_i^{(V)}$ were taken from the stopflow microperfusion data (table 1).

occurs across the luminal brush border, one has furthermore to conclude that the two active transport processes are located on opposite cell sites.

The permeability coefficient measured with isotopes can be used to solve eq. (1), only when there is no isotope interaction and when ion flows are not coupled. One test for isotope interaction is the comparison of the partial conductances of the different ions with the corresponding tracer permeabilities. When partial conductances evaluated by Frömter [8, 9] are brought to suitable dimensions (X 10^{-7} cm^2/sec), we get a P_{Na} : P_{Cl} : of 17.8 : 14.5 (or relatively 1 : 86). The corresponding tracer values are 15.5 : 13.3 (or relatively 1 : 0.86). The absolute and relative values agree fairly well. Therefore, isotope interaction, at least in the transtubular flux, if present at all must be small.

The next question which arises is: Does the active H^+ secretion (or bicarbonate absorption) depend upon the active sodium absorption? Three years ago when we performed our first capillary perfusion experiment, we made an observation which helps to answer this question. As shown by fig. 7, we observed that the isotonic absorption from the proximal convolutions depends on the bicarbonate concentration in the capillary perfusate [2, 10]. Furthermore we found that with increasing capillary bicarbonate concentration the luminal bicarbonate concentration also rises (fig. 8a). Therefore one gets the following correlation: With increasing luminal HCO_3 concentration, the bicarbonate transport increases. Sodium transport is simultaneously increased but in a fashion showing saturation (fig. 8b).

We next asked whether the effect on the isotonic absorption was specific for bicarbonate or not? Table 3 shows that a series of chemically very different substances could partially replace HCO_3^-. The two sulfonamides, glyco-

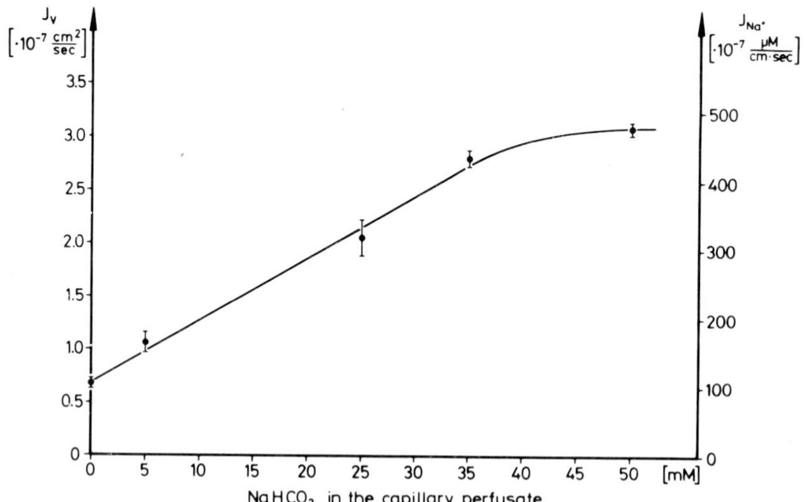

Fig. 7. Relation of the isotonic fluid (J_V) and sodium absorption (J_{Na^+}) on the bicarbonate concentration in the capillary perfusate. (Ullrich et al. [2]).

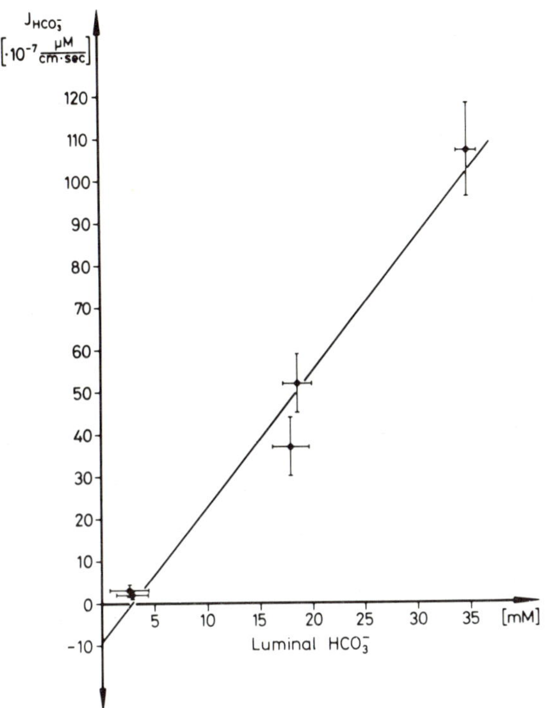

Fig. 8a. Relation between luminal bicarbonate concentration and transtubular bicarbonate reabsorption from the proximal convolution. (Ullrich et al. [2].)

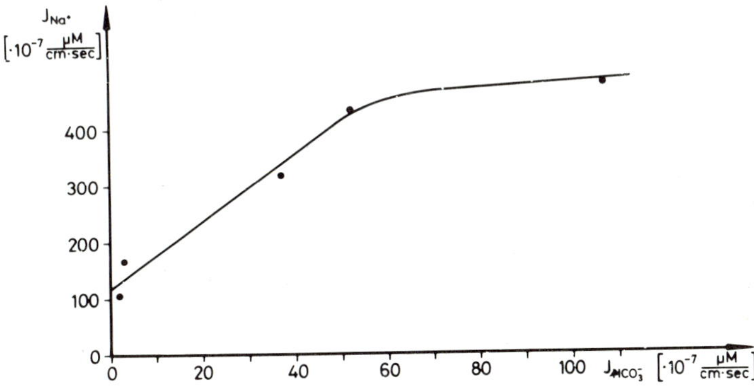

Fig. 8b. Relation between bicarbonate- and sodium-reabsorption from the proximal convolution. (Ullrich et al. [2].)

Table 3
Effect of different buffers on the isotonic reabsorption in the proximal convolution. Mean values ± SE, n = number of measurements; in brackets number of animals. (Ullrich et al. [2])

30 mM buffer (Na-salt)	pK	Reabsorption rate: J_V ($\times 10^{-7}$ cm^3 cm^{-1} sec^{-1})	
Bicarbonate	6.3	2.75 ± 0.07	n = 82 (15)
Glycodiazine	5.7	2.74 ± 0.07	n = 72 (9)
Sulfamerazine	7.1	2.74 ± 0.07	n = 38 (3)
Butyrate	4.8	2.48 ± 0.16	n = 34 (2)
Propionate	4.9	2.03 ± 0.11	n = 23 (2)
Acetate	4.8	1.85 ± 0.08	n = 32 (3)
α-Aminoisobutyrate	2.6 (pK$_1$), 4.4 (pK$_2$)	1.10 ± 0.07	n = 14 (2)

diazine and sulfamerazine, substitute totally for bicarbonate. Butyrate, propionate, acetate, and to a smaller extent aminobutyric acid, are also effective. Paraminohippuric acid, which is not reabsorbed from the lumen, does not promote isotonic absorption at all. From this one may suggest that the active reabsorption of the buffer anions are essential for proximal isotonic transport and that a nonionic diffusion may be involved. The data could be explained with the scheme of fig. 9 originally proposed by Pitts [23]. The buffer anion concentration (A^-) within the tubular lumen varies proportionally with the buffer anion concentration within the capillary perfusate and may be limiting for the rate of H^+ transport. H^+ and A^- combine within the lumen to form AH which easily crosses the luminal cell border. Within the cell, AH again dissociates into H^+ and A^-. The H^+ is actively transported into the lumen and A^- diffuses across the basal site into the interstitium.

The sodium-potassium pump located at the basal border of the cell may be influenced by the luminal H^+ pump in several ways. Possibility 1: The H^+ secretion could be coupled with a Na^+ antiport. Possibility 2: The intracellular A^- concentration would influence the passive permeability of the luminal barrier. More Na^+ would be delivered into the cell and to the sodium pumping sites with both possibilities. A third consideration is that the intracellular pH might be affected and influence the Na-K pump secondarily. With a low H^+ pumping rate the cell pH would shift into the acid range and would become alkaline at a high pumping rate. This possibility could be easily ruled out. A variation of the CO_2 pressure from 0 to 20% and also a considerable change in the undissociated form of glycodiazine (which certainly influences the intracellular pH) does not affect the isotonic reabsorptive rate [2]. Thus,

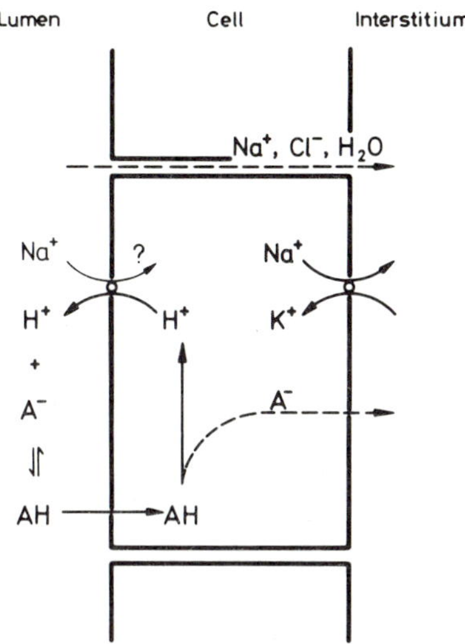

Fig. 9. Tentative scheme of the connection between H^+ secretion and buffer anion (A^-) reabsorption via the undissociated acid (AH) located at the luminal cell side, and the active sodium transport mechanism located at the basal cell side.

increased intracellular sodium pool as cause of an increased Na^+ transport possibilities 1 and 2 seems more likely.

Our finding of a buffer effect on transmembranal transport has two parallels in the literature. Singer, Sharp, and Civan [11] observed an increased short circuit current and Na^+ transport in the toad bladder when propionate, acetate, butyrate or pyruvate was added to the serosal surface. To explain this effect, the authors discussed a lyotropic action on membrane macromolecules of the mucosal surface thereby increasing the permeability for sodium. A similar increase of ion permeability was observed by Flemström [12] when he gave acetate, propionate, and lactate to the mucosal side of the isolated gastric mucosa. Simultaneously, the H^+ secretory rate increased. It seems therefore to be a common biological phenomenon that transepithelial Na^+ transport rises when some weak acids are transported into the cell. The HCO_3^- dependence of the transtubular Na^+ transport may also belong in this category. I should mention that a HCO_3^- effect on sodium transport was observed in quite a number of epithelial membranes, namely gall bladder [13], pancreas [14], proximal convolution [2, 10], parotid duct [15] nasal salt gland [16], and turtle bladder [17].

We further focused our interest at the H^+ transport mechanism located at

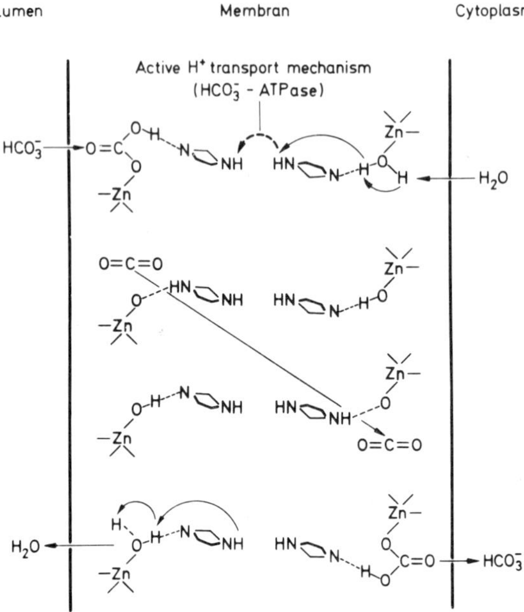

Fig. 10. Tentative scheme of the connection between Zn containing carbonic anhydrase and the active H^+ transport mechanism in the luminal cell membrane (modified according to Wang [21]).

the luminal brush border. Our working hypothesis for the HCO_3^-/H^+ transport is depicted in the following scheme (fig. 10), which shows successive carbonic anhydrase catalysed steps of HCO_3^- dehydration at the luminal site and successive carbonic anhydrase catalysed hydration steps of CO_2 at the cytoplasmatic membrane site. In between, an active proton translocation step must be located, possibly a HCO_3^- stimulated ATPase. Acetazolamide is supposed to block the HCO_3^- buffered system via the carbonic anhydrase. If this would be the only site of attack, acetazolamide should not inhibit the HCO_3^- free system which is buffered with the other buffers mentioned above. Fig. 11 (upper part) shows that with rising acetazolamide concentrations in the capillary perfusate, the isotonic absorption in a HCO_3^- buffered system is heavily blocked. The glycodiazine buffered system is however also blocked especially with higher concentrations. Since we could not detect an effect of acetazolamide on the Na^+K^+ stimulated ATPase in proximal tubular homogenate, we may conclude that acetazolamide in higher doses blocks the H^+ transporting system itself. As shown in fig. 11 (lower part) qualitatively

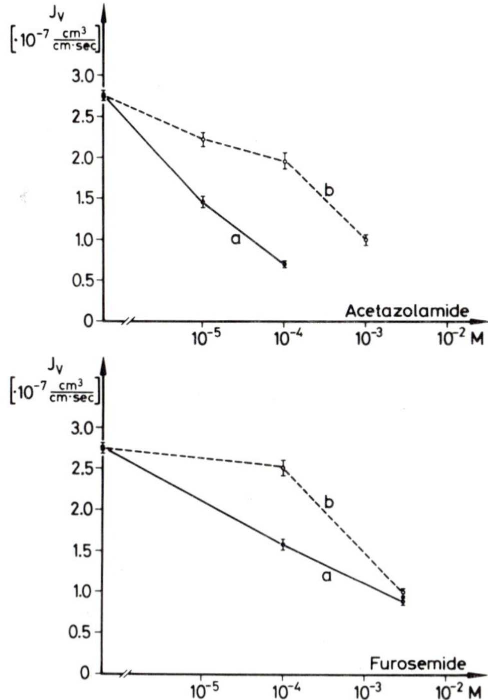

Fig. 11. Effect of acetazolamide, upper part, and effect of furosemide, lower part, on the proximal isotonic reabsorption $\times$ 10^{-7} cm^2 sec^{-1} a) when 30 mM bicarbonate buffer b) when 30 mM glycodiazine buffer was used as capillary perfusate (Radtke et al. [3]).

similar results were gained with furosemide. Because of the known dual action of thiocyanate on the H^+ transport mechanism [18] and on carbonic anhydrase, we tested this substance and found that 10^{-2}M inhibits the isotonic transport of a bicarbonate buffered system by 34% and the glycodiazine buffered system by 14%. It is tempting to consider some parallelism between the dual action of acetazolamide on the one hand and furosemide and thiocyanate on the other. Furosemide however may exert an effect on the energy producing apparatus in higher doses [19] and thiocyanate in 10^{-2}M concentrations may exert lyotropic effects on quite a number of macromolecules. Therefore only the dual action of acetazolamide on the carbonic anhydrase as well as on the H^+ transporting system seems to be proven.

Schwartz, Rosen and Steinmetz [20] have also reached the conclusion that acetazolamide had a dual action. They observed that acetazolamide diminished the H^+ secretion in the turtle urinay bladder. CO_2 could restore the secretory rate to normal but only at an acetazolamide dose of lower than 5×10^{-5}M. At the higher dose (5×10^{-4}M) which was tested, this was not the case. The dual action of acetazolamide can be explained in the following way. If the histidine molecule in the carbonic anhydrase [21] and a histidine molecule supposed in the translocation mechanism are the same, or if the supposed histidine of the translocation mechanism is very near the inhibitor binding Zn [21], then it is conceivable that a sulfonamide molecule bound to the carbonic anhydrase-Zn inhibits the enzyme and influences the proton translocation as well.

In conclusion one can say: About 60% of the sodium absorption from the proximal tubule of the rat kidney could be accounted for by passive forces: solvent drag and diffusion. The other half of the isotonic sodium absorption is by definition active. Beside an active Na^+ transport mechanism located at the basal cell site, an active H^+ transport mechanism located in the luminal brush border participates in proximal isotonic absorption. The H^+ secretory mechanism is responsible for the reabsorption of bicarbonate and similar buffers apparently via nonionic diffusion. The buffer reabsorption is essential for isotonic absorption, because only 20% of the normal isotonic absorption persists when the buffer reabsorption is suppressed either by omission of suitable buffers or by carbonic anhydrase inhibitors and H^+ transport blocking agents respectively. The H^+ ion transport seems to be linked to the sodium transport through an H^+-NA^+ exchange mechanism across the luminal brush border thereby determining the amount of sodium ions which are offered to the sodium transport mechanism located at the opposite cell side.

References

[1] K.J. Ullrich, C.A. Baldamus, E. Frömter, K. Luer, H.W. Radtke, G. Rumrich and F. Sauer, J. Membrane Biology, in press.

[2] K.J. Ullrich, H.W. Radtke and G. Rumrich, Pflügers Arch. 330 (1971) 149.

[3] H.W. Radtke, G. Rumrich, Eva Kinne-Saffran and K.J. Ullrich, Kidney, submitted.

[4] A.Z. Györy, Pflügers Arch. 324 (1971) 328.

[5] K.J. Ullrich, E. Frömter and K. Baumann, in: Laboratory Techniques in Membrane Biophysics, eds. H. Passow and R. Stämpfli (Berlin – Heidelberg – New York, Springer, 1969) p. 106.

[6] H.W. Radtke, G. Rumrich, S. Klöss and K.J. Ullrich, Pflügers Arch. 324 (1971) 288.

[7] F. Morel and Y. Murayama, Pflügers Arch. 320 (1970) 1.

[8] E. Frömter and C.W. Müller, IV. International Congress of Nephrology, Stockholm (1969).

[9] E. Frömter, C.W. Müller and T. Wick, in: Electrophysiology of Epithelial Cells, ed. G. Giebisch (F.K. Schattauer, Stuttgart–New York, 1971) p. 119.
[10] G. Rumrich and K.J. Ullrich, J. Physiol. 197 (1968) 69–70 P.
[11] I. Singer, G.W.G. Sharp and M.M. Civan, Biochim. Biophys. Acta 193 (1969) 430.
[12] G. Flemström, Acta Physiol. Scand. 82 (1971) 1.
[13] J.M. Diamond, J. Gen. Physiol. 48 (1964) 1.
[14] I. Schulz, Pflügers Arch. 329 (1971) 283.
[15] H. Knauf, Personal communication.
[16] M.R. Hokin, Life Sci. 5 (1966) 1829.
[17] C.F. Gonzales, Y.E. Schamoo and W.A. Brodsky, Biochim. Biophys. Acta 193 (1969) 403.
[18] R.P. Durbin and D.K. Kasbekar, Fed. Proc. 24 (1965) 1377.
[19] V.D. Jones, Fed. Proc. 30 (1971) 811.
[20] J.H. Schwartz, S. Rosen and P.R. Steinmetz, Fed. Proc. 30 (1971) 421 Abs.
[21] J.H. Wang, Proc. Nat. Acad. Sci. 66 (1970) 874.
[22] R. Kinne, J.E. Schmitz, Eva Kinne-Saffran, Pflügers Arch. 329 (1971) 191.
[23] R.F. Pitts, Progress Cardiovasc. Dis. 3 (1961) 537.
[24] C.A. Baldamus, H.W. Radtke, G. Rumrich, F. Sauer and K.J. Ullrich, J. Membrane Biol. 7 (1972) 377.

TRANSCELLULAR AND PARACELLULAR PATHWAYS FOR ION FLOW ACROSS EPITHELIA

G. WHITTEMBURY, L. MATEU and F.A. RAWLINS

Centro de Biofísica y Bioquímica, Instituto Venezolano de Investigaciones Científicas, IVIC, P.O. Box 1827, Caracas, Venezuela

I would like to make a few comments about the relative role of transcellular and paracellular pathways for ion and water flow across some epithelia. Fig. 1 illustrates these two pathways in the kidney tubule. The transcellular pathway was supported by measurements of ion concentrations in tubular lumen, cell interior and blood compartment (which bathes the interstitial or peritubular spaces) and by measurements of electrical potentials and of ion fluxes between these compartments. These measurements led to the suggestion, based in the model of Koefoed Johnsen and Ussing [1] that Na transport across the kidney tubule would take place in two steps [2, 3] : (a) Na ions would enter the cell from the tubular lumen, down its electrochemical potential difference. (b) This entry of Na ions would follow Na extrusion across the peritubular cell membrane (from cell to interstitial space) by a Na for K exchange pump located at or near this cell membrane. A further requirement was the existence of different permeability characteristics of the luminal and of the peritubular cell membranes [2, 3] . Evidence is large for the existence of a Na for K exchange pump in the kidney (cf. [4–7] . Lately, evidence has accumulated indicating the existence of another Na pump which extrudes Na with Cl [6, 7] . The permeability characteristics of luminal and peritubular cell membrane have been found to be different and as required by the model [8, 10] . Changes in cell volume and composition when transport is impaired [6] and the demonstration that Na transport is electrogenic [4] further support this transcellular route.

In addition, the existence of a pathway has been proposed [8–11] based mainly on the observation that in the *Necturus* proximal tubule the measured transtubular resistance is, at least, 10 times smaller than the sum of

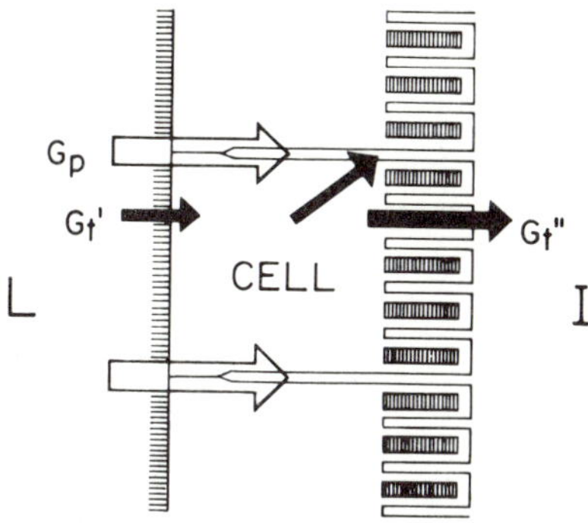

Fig. 1. Schematic representation of part of a kidney tubule, showing tubular lumen (L), the cell proper and interstitial compartment (I). The morphological asymmetry brush border at the luminal aspect and mytochondria and infoldings at the interstitial or peritubular aspect of the cell membrane are illustrated. Na ions would enter the cell (from the tubular lumen) down its electrochemical potential gradient, because the sodium conductance of the luminal membrane is larger than that of the peritubular membrane [10]. In a second step Na ions would be expelled at the peritubular membrane by active Na transporting mechanisms [6, 7]. The dark arrows illustrate these transcellular pathways for ion flow. Their different size indicates that the luminal cell membrane conductance, G'_t has different characteristics from that of the peritubular membrane conductance, G''_t. The paracellular pathway is illustrated by the big empty arrow. Its size indicates that the paracellular conductance G_p is larger than the transcellular conductance, in the proximal tubule. In the distal tubule G_p has a lower value. The value of G_p may vary according to the experimental condition. For example it may be increased in the proximal tubule by saline loading [20] or in amphibian skins by application of hydrostatic pressure [13] or of hypertonic urea solutions [18].

the luminal cell membrane resistance plus the peritubular cell membrane resistance. The value of this paracellular conductance increases during saline loading [20]. These measurements have been qualitatively confirmed in the rat and dog proximal tubule [9]; A slightly different situation prevails in the distal tubule, where the paracellular pathway has a resistance much lower than that of the proximal tubule [10].

We have attempted to localize such paracellular pathways, and have used lanthanum salts in a soluble form (not in a colloidal form [12]) in the perfused toad kidney, a viable preparation that permits ample manipulation of the perfusion fluids [6]. Fig. 2 is an electronmicrograph showing that lanthanum, a cation not much bigger than sodium can cross the tight junction of the toad kidneys proximal tubule. These experiments confirm the existence of a pathway in parallel to the cellular route and locate it at the level of the tight junction.

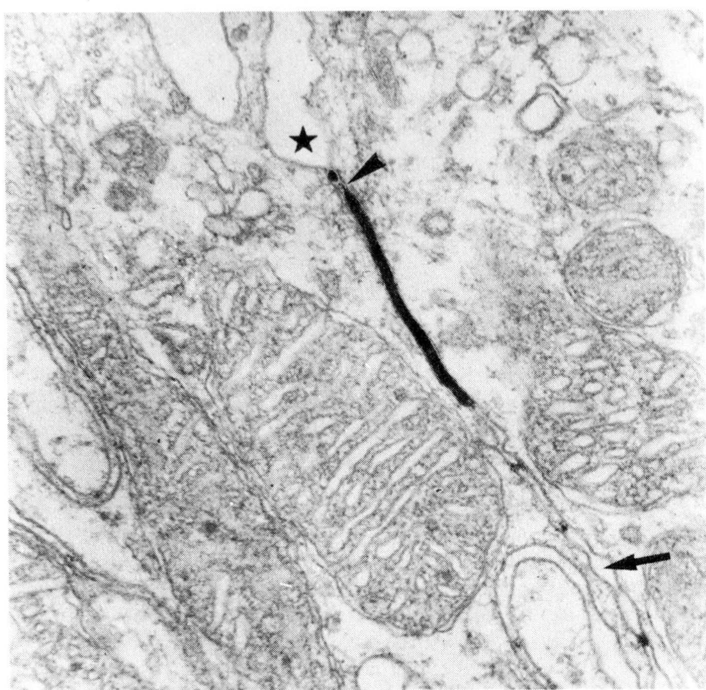

Fig. 2. Electromicrograph of a proximal tubular cell of a doubly perfused toad kidney. Lanthanum salt was added via the aortic circulation in a soluble form. After filtering through the glomerulus it entered the lumen (star) and crossed the tight junctions (arrow head) filling the intercellular space (arrow). Then it precipitated. Thus the paracellular pathway is located at the tight junction as has been previously suggested [8–11]. Kidneys were fixed with 5% glutaraldehide, dehydrated and embedded in Epon 812. Fine sections were examined with a Siemens Elmiskop 1A Electronmicroscope (from F.A. Rawlins and G. Whittembury, unpublished observations).

In the abdominal amphibian skins measurements of electrical potentials with microelectrodes located within cells of the different layers of the epithelium [13] indicated that below the corneal layer (which has practically no resistance) there exists a layer with a conductance much higher for Na than for K, and that its Na conductance increases further by the addition of antidiuretic hormone [13]. Since these measurements were obtained with microelectrodes localized within the cells, they constitute evidence that ions pass through the cells from outside in rather than between the cells as Cereijido's interesting proposition suggests [14]. As proposed by Koefoed

Johnsen and Ussing model [1] the cell membrane beyond the transporting layer is mainly K permeable [13]. However at variance with it all cells located in strata deeper than the transporting layer are highly K permeable. Voute and Ussing [15] have also located this transporting layer to a place immediately, below the corneal layer, in agreement with the electrical measurements [13]. Furthermore, the volume changes they observed secondary to given variations in the rate of transport also underscore the transcellular ion route. An important question is whether ions after crossing the first transporting layer then go to the next cell layer, or whether they have been already transported into the intracellular space. Recent evidence [16] indicates that electrical communication is preferentially horizontal in the toad skin epithelium and that vertical communication (that is communication in the outside inwards direction, from one cell layer to the next) is almost negligible. If these measurements are taken to represent ion flow they would indicate that after crossing the first transporting layer a large fraction of the transported ions probably have already passed into the intercellular spaces.

Normally in these abdominal amphibian skins a paracellular route seems practically closed, as can be gathered from electrical measurements [17] and from morphological evidence using tracers of extracellular space [18, 19]. However, application of hydrostatic pressure, from inside out [13] or use of hypertonic urea in the outside bathing solution [18], opens the tight junction which then becomes permeable to extracellular tracers, which may be visualized with the aid of the electron microscope [13, 18-]. Yet, despite these opening of the paracellular pathways, the epithelium continues transporting salt inwards.

One could then envisage transport across epithelia as occurring transcellularly, with the paracellular pathway acting as a parallel shunt (or leak). In view of its variable magnitude among epithelia, the paracellular pathway could be represented by a variable conductance. This parallel conductance would be highest in the proximal tubule of the mammalian kidney under usual conditions; it would be not so high in the proximal tubule of *Necturus*, it would be smaller in the kidney distal tubule. Finally the abdominal skins of amphibia should have the smallest conductance.

As mentioned above, the absolute value of this parallel conductance also varies in the same epithelia according to the experimental condition (saline loading, hydrostatic pressure, hypertonic solution). Thus it is interesting to speculate that changes in the conductance of this paracellular pathway may constitute a mechanism that may modulate the efficiency of the pumping mechanism.

References

[1] V. Koefoed Johnsen and H.H. Ussing, Acta Physiol. Scand. 42 (1958) 298.
[2] G. Whittembury, J. Gen. Physiol. 43 No 5, supp. (1960) 43.
[3] G. Giebisch, Circulation 21 (1960) 879.
[4] G. Whittembury, in: Electrophysiology of Epithelial Cells, ed. G. Giebisch, Symposia Medica Hoechst (F.K. Schattauer Verlag, Stuttgart-New York, 1971) p. 154.
[5] G. Giebisch, E. Boulpaep and G. Whittembury, Phil. Trans. Roy. Soc. Lond. B. 262 (1971) 175.
[6] G. Whittembury and J. Fishman, Arch. ges. Physiol. 307 (1969) 138.
[7] G. Whittembury and F. Proverbio, Arch. ges. Physiol. 316 (1970) 1.
[8] E.L. Boulpaep, in: Transport und Funktion intracellularer Elektrolyte, ed. F. Kruck (Urban and Schwarzenberg, Munchen, 1967) p. 98.
[9] E.L. Boupaep and J.F. Seely, Am. J. Physiol. 221 (1971) 1084.
[10] G. Giebisch, J. Gen. Physiol. 51 (1968) 315s.
[11] E.E. Windhager, E.L. Boulpaep and G. Giebisch, Proc. 3rd Intern. Congr. Nephrol. Washington DC, Vol. 1 (Karger, Basel, 1967), p. 35.
[12] J.P. Revel and M.J. Karnovsky, J. Cell Biol. 33 (1967) C7.
[13] F.A. Rawlins, L. Mateu, F. Fragachan and G. Whittembury, Arch. ges. Physiol. 316 (1970) 64.
[14] M. Cereijido, J. Gen. Physiol. 51 (1968) 280s.
[15] C.L. Voute and H.H. Ussing, J. Cell Biol. 36 (1968) 625.
[16] L. Mateu, M. Gimenez, R. Callarotti and G. Whittembury, Biophysical J. 10 (1970) 34a.
[17] H.H. Ussing and E.E. Windhager, Acta Physiol. 61 (1964) 484.
[18] H.H. Ussing, in: Electrophysiology of Epithelial Cells, ed. G. Giebisch, Symposia Medica Hoechst (F.K. Schattauer Verlag, Stuttgart-New York, 1971) p. 3.
[19] A. Martínez-Palomo, D. Erlij and H. Bracho, J. Cell. Biol. 50 (1971) 277.
[20] E.L. Boupaep, Am. J. Physiol., in the press.

THE INFLUENCE OF HORMONES ON ACTIVE SODIUM TRANSPORT BY AMPHIBIAN EPITHELIA INCUBATED IN HYPOTONIC FLUID *

J. CRABBÉ

WITH THE TECHNICAL ASSISTANCE OF A. DECOENE

Endocrine Unit, Department of Physiology U.C.L., Louvain, Belgium

Ten years ago, MacRobbie and Ussing [2] had reported that when the inside of frog skin is exposed to solutions the tonicity of which was decreased with respect to Ringer's fluid, electrical potential difference across the preparation increased – and reversibly so. Therefore hypotonicity was recently used in an attempt to further our knowledge concerning the mechanism of action of hormones stimulating transepithelial sodium transport.

Exposure of the inner surface of sodium-transporting epithelia of the toad, *Bufo marinus,* to a hypotonic solution, results rapidly in a sizable, lasting increase in transmembrane potential and in short-circuit current. This is illustrated in fig. 1, in the case of the urinary bladder. Short-circuit current still largely reflects sodium influx under such circumstances (table 1) and, as could be expected, bladder tissue is also more hydrated; furthermore, sodium pool size increases in proportion to the sodium-transporting activity of the tissue. Incidentally, the latter observation is in keeping with the conclusion formulated by MacRobbie and Ussing according to whom swelling of the epithelial cells induced by hypotonicity results in increased permeability of their apical cell border to sodium.

There is evidence that amphotericin B acts at the apical border of toad bladder epithelium, disrupting at that site the saturable permeability barrier that sodium ions have to cross [5]. Although this antibiotic is a trickier tool

* An abstract of the studies dealt with here, has been published [1].

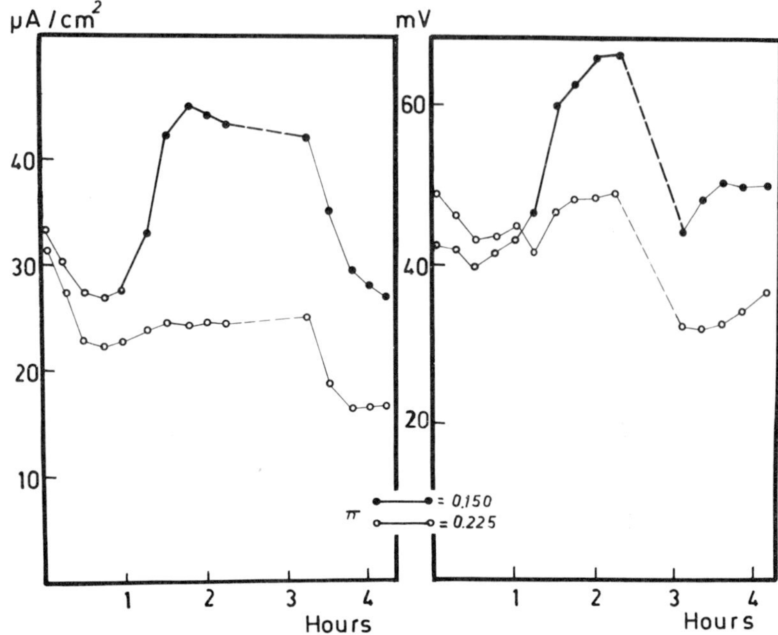

Fig. 1. Stimulation of sodium transport by toad bladder incubated in hypotonic fluid. Both hemibladders of male *Bufo marinus* were incubated simultaneously as devised by Ussing and Zerahn [3]; one served as a control, *i.e.* its inner surface was exposed throughout to Ringer's solution of the following composition: NaCl, 115 mM; $KHCO_3$, 2.5 mM; $CaCl_2$, 1.0 mM. The matched tissue was treated likewise for the initial hour, whereupon a solution identical with respect to $KHCO_3$ and $CaCl_2$, but containing only 78 mM of NaCl, was substituted, so as to decrease by a third the tonicity of the solution (black dots). On the outside, the solution made of NaCl, 78 mM; $KHCO_3$, 2.5 mM; and $CaCl_2$, 1.0 mM, was present from the outset; sucrose was added to bring the osmolality back to 0.225 when such was the value for the solution used on the inside, *i.e.* for the membrane incubated in standard conditions (open dots) and during the initial hour in the case of the matched, treated one. Readings were interrupted during the penultimate hour of incubation. N = 6 paired experiments.

than we would like when it comes to functional "dissection" of the mechanism of transepithelial sodium transport, it might be mentioned that the stimulating effect of hypotonicity on short-circuit current is wiped out by exposure to amphotericin B, in the case both of the bladder and of the colon (fig. 2). These observations thus provide an additional argument for hypotonicity "weakening", as it were, that apical barrier.

In view of the likelihood that hypotonicity results in changes of the prop-

Table 1
Stimulation of sodium transport by toad bladder incubated in hypotonic solution (mean values ± S.E. of 10 paired experiments)

		Ringer's $\pi = 0.225$	Dilute Ringer's $\pi = 0.115$
Final short-circuit current ($\mu A/cm^2$)		38.5 ± 2.7	54.7 ± 4.4
$\frac{\text{Sodium influx}}{\text{S.C.C.}} \times 100$		99.6 ± 5.4	92.5 ± 3.0
Sodium tissue pool size (μeq)		0.166 ± 0.021	0.240 ± 0.027
Tissue	dry weight (mg)	12.5 ± 1.1	12.1 ± 0.9
	hydration (μl H_2O/mg D.W.)	403 ± 9	472 ± 9

For these experiments, hemibladders of male *Bufo marinus* were incubated simultaneously according to Ussing and Zerahn [3]; one was exposed to solution of the usual tonicity while the latter was reduced to one half in the case of the matched preparation; this was achieved by dilution of NaCl from 115 mM to 57.5 mM. In order to obtain identical conditions with respect to sodium transport, NaCl concentration on the outside was uniformly 57.5 mM; tonicity of the final solution was brought back to 0.225 by means of sucrose in the case of the control.

Sodium influx and tissue pool were determined as detailed elsewhere [4]. S.C.C. denotes mean short-circuit current (recorded at 5 min intervals) during the final half-hour of incubation that lasted for less than one hour altogether.

The small discrepancy between sodium influx and S.C.C. during exposure to hypotonicity, might be accounted for by a loss of tissue potassium [2]; more recent experiments indeed indicated a drop of tissue potassium contents from 152 ± 6 to 122 ± 5 μEq/mg dry weight.

erties of the apical border of epithelial cells of amphibian bladder, colon and skin, at least with respect to sodium movement, it was thought that additional information could be obtained by means of this manipulation, concerning the site of action of hormones such as aldosterone [6, 7, 8], insulin [9, 10] and vasopressin [11] which are known to stimulate transepithelial sodium transport.

When the effect of vasopressin on the sodium-transporting activity of the toad bladder was examined, it appeared that hypotonicity failed to modify the amplitude of the surge in short-circuit current attending introduction of the hormone in the incubation fluid (fig. 3).

That hypotonicity does not amplify the influence of vasopressin on short-circuit current is further illustrated by the data summarized in table 2, from experiments performed on 6 pairs of hemibladders. Their inner surface was

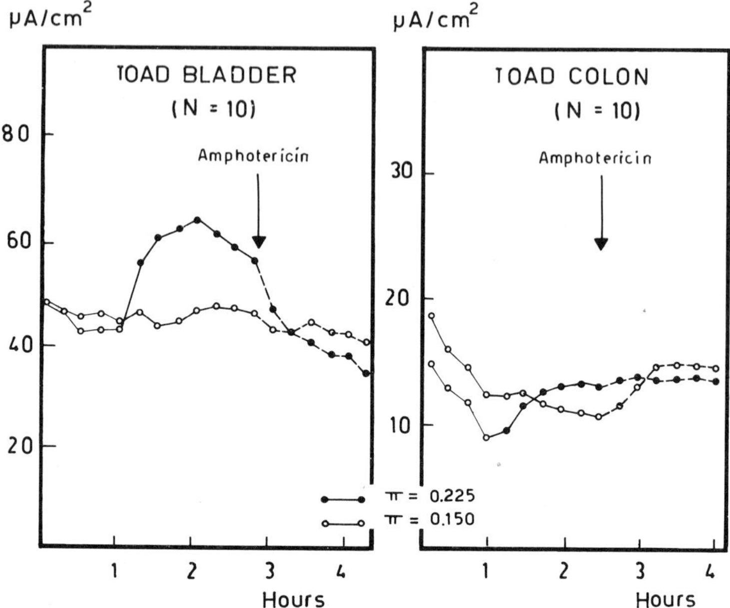

Fig. 2. Effect of amphotericin B on short-circuit current recorded across toad bladder and colon incubated in solution of different tonicity. The inner surface of paired bladder halves and paired pieces of colon mucosa of male *Bufo marinus* was first incubated in Ringer's solution (NaCl: 115 mM; $KHCO_3$: 2.5 mM; $CaCl_2$: 1.0 mM) according to Ussing and Zerahn [3]. After 1 hr, one preparation of each pair (black dots) was exposed to dilute Ringer's (NaCl: 78 mM; $KHCO_3$: 2.5 mM; $CaCl_2$: 1.0 mM). From the outset, fluid on the outside had the same ionic composition as that of dilute Ringer's, and sucrose was added in the case of the control to eliminate an osmotic pressure gradient across the tissue. After 2½ hr, amphotericin B was introduced in the outer solution for both preparations of each pair; the final concentration was 2.5 µg/ml in the case of toad bladder tissue, and 12.5 µg/ml when colon mucosa was treated. The lack of stimulating effect of amphotericin B on short-circuit current across the control bladder preparations probably results from the high baseline activity of the tissue: it has been our experience that the amplitude of the response to amphotericin B is inversely related to the rate of sodium transport prior to its use. Furthermore, dilution of sodium in the external solution results in a decreased effect of amphotericin B on short-circuit current [6].

uniformly exposed to a solution in which magnesium had replaced sodium (composition: $MgCl_2$, 57.5 mM; $KHCO_3$, 2.5 mM; $CaCl_2$, 1.0 mM); on the outside, the solution contained NaCl, 78 mM; $KHCO_3$, 2.5 mM and $CaCl_2$, 1.0 mM. For one preparation of each pair, incubation was carried out in these solutions to which sucrose had been added to raise osmotic pressure from 0.150 to 0.225.

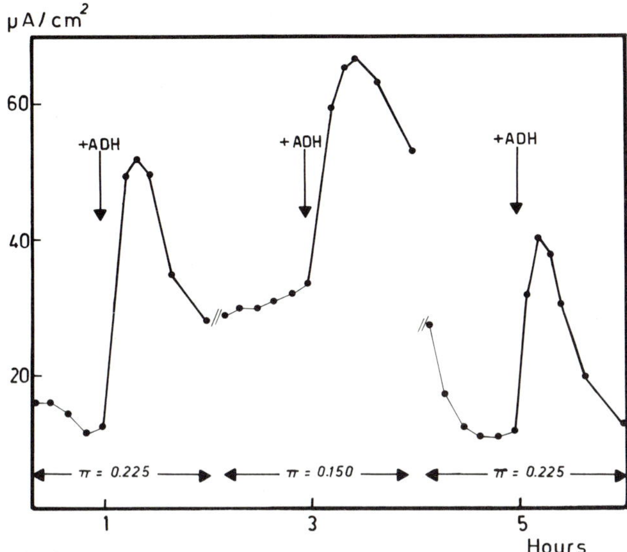

Fig. 3. Influence of tonicity of incubation fluid on the "natriferic" response of toad bladder to antidiuretic hormone. Four hemibladders (from 2 male *Bufo marinus*) were set up for incubation according to Ussing and Zerahn [3]; for the initial and final 2 hr, Ringer's solution (NaCl: 115 mM; $KHCO_3$: 2.5 mM; $CaCl_2$: 1.0 mM) was used on the inside. During the intermediate 2-hour period, a solution containing only 2/3 that NaCl concentration was substituted. On the outside, the latter, dilute solution (NaCl, 78 mM; $KHCO_3$, 2.5 mM; $CaCl_2$, 1.0 mM) was used throughout, with sucrose added during the first and last thirds of the incubation period to eliminate osmotic pressure gradients. In the middle of each of these 3 phases, lysine-vasopressin, 125 mU/ml, was introduced in the solution on the inside. The amplitude of the response to the hormone averaged, for the 20-min period that followed its use (in $\mu A/cm^2$): 31.3 before and 22.8 after exposure to dilute Ringer's ; during exposure to the latter, the corresponding value was 25.4.

Of interest is the fact that the D.C. conductance of the tissue was increased by hypotonicity, as was the case with vasopressin also.

Such data are in keeping with the hypothesis of an action of hypotonicity at the apical border of the epithelial cells involved in active transmembrane sodium transport, as this is the accepted site of action of vasopressin as well [12].

This observation prompted an appraisal of the repercussion of hypotonicity on the effect of insulin – a hormone also capable of stimulation of sodium transport across the amphibian tissues under consideration. Toad colon mucosa appears especially suitable in this respect, as it reacts both to insulin

Table 2
Effect of tonicity of incubation solution on stimulation of sodium-transporting activity of toad bladder by anti-diuretic hormone (means of 6 paired experiments)

Time (min)	Short-circuit current ($\mu A/cm^2$)		Transepithelial potential (mV)	
	π = 0.150	π = 0.225	π = 0.150	π = 0.225
−15	41.2	21.5	35.0	24.7
0	40.5	21.6	35.8	25.4
+ 5	46.7	23.4	–	–
10	51.3	30.3	–	–
15	56.2	41.6	35.9	35.7
30	57.1	47.3	34.8	39.5
45	53.1	47.2	31.8	38.3
60	47.2	45.1	25.8	36.0
75	44.3	42.6	23.6	33.1
90	42.3	39.0	21.0	30.4

Incubation of paired hemibladders of male *Bufo marinus* was carried out according to Ussing and Zerahn [3] in solutions as detailed in the text. After 2 h lysine-vasopressin, 125 mU/ml, was added on the inside: the moment of this addition is taken as "zero".

and to hypotonicity *in vitro*. From fig. 4, the effect of the hormone showed a tendency toward a decrease after less than one hour of exposure; switching to hypotonicity had as a result that the insulin effect was "revived" and sustained.

If it is accepted that insulin, unlike hypotonicity, acts close to, or at, the "pump" site *i.e.* the cell borders *other* than the apical one, an amplification of the effect of insulin upon resorting to hypotonic fluid is amenable to interpretation in view of a disposition in series of these 2 factors.

In the case of aldosterone, the situation is more intricate, as appears from fig. 5. In order to evaluate the consequences arising from changes in incubation fluid tonicity, matched bladders were preincubated overnight, one membrane of each pair being exposed from the outset to aldosterone. In doing so, stimulation of sodium transport by this steroid hormone is quite apparent the following morning and, more important yet, the treated preparation is stable. Glucose exerts at that stage an enhancing influence on the aldosterone effect.

Relevant to the issue being considered is that hypotonicity magnified the aldosterone effect only after addition of glucose. This is interpreted as indicating that, in the presence of glucose after prolonged incubation, aldosterone-treated preparations react to hypotonicity in proportion to the activity

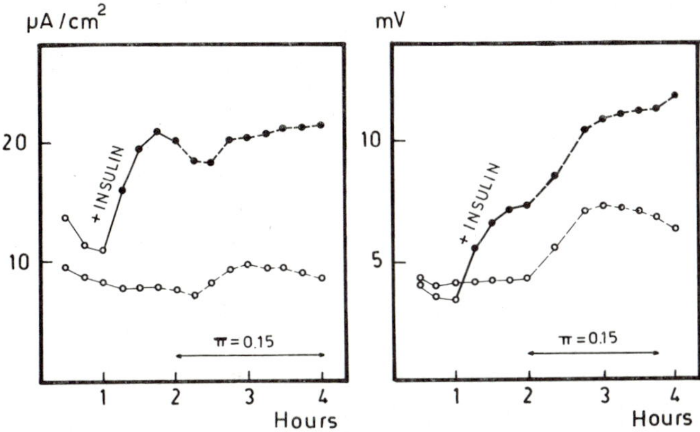

Fig. 4. Stimulation by insulin of active sodium transport by toad colon mucosa as a function of the osmotic pressure of the incubation fluid. In 8 instances, the colon mucosa of male *Bufo marinus* was separated mechanically from the underlying tissue layers and divided in 2 pieces that were incubated according to Ussing and Zerahn [3]. Ringer's solution (NaCl, 115 mM; $KHCO_3$, 2.5 mM; $CaCl_2$, 1.0 mM) was used on the inside for the first 2 hr, after which incibation went on with a solution containing 78 mM NaCl, 2.5 mM $KHCO_3$ abd 1.0 mM $CaCl_2$. This solution was present from the outset on the side corresponding to the mucosal (i.e. outward-facing) surface of the preparation, but, during the initial 2-hr phase, enough sucrose was added to eliminate osmotic pressure gradients across the preparation. After 1 hr of incubation, insulin, 125 mU/ml, was introduced into the solution on the inside (black dots).

recorded when the fluid is isotonic – which suggests an arrangement in series of the mechanisms involved in the overall response of the tissue. This stands to reason as it would be very difficult to ascribe the glucose effect to anything else than improved "pumping" of sodium. On the other hand, the fact that this substrate effect is also demonstrable for untreated preparations in hypotonic media, raises the possibility of this substrate effect being a *secondary* phenomenon in the case of aldosterone treatment.

In conclusion, incubation of amphibian epithelia such as toad bladder, colon and skin in hypotonic media, brings about a lasting, reversible stimulation of their sodium-transporting activity. There are reasons to ascribe this stimulation to an increased permeability toward sodium at the apical border of the cells involved in this process.

Hypotonicity fails to modify the amplitude of this type of response to antidiuretic hormone; the effect of insulin on the other hand, appears magnified by this manipulation. With aldosterone, the same magnification phenomenon is clearly demonstrable only in the presence of glucose to which,

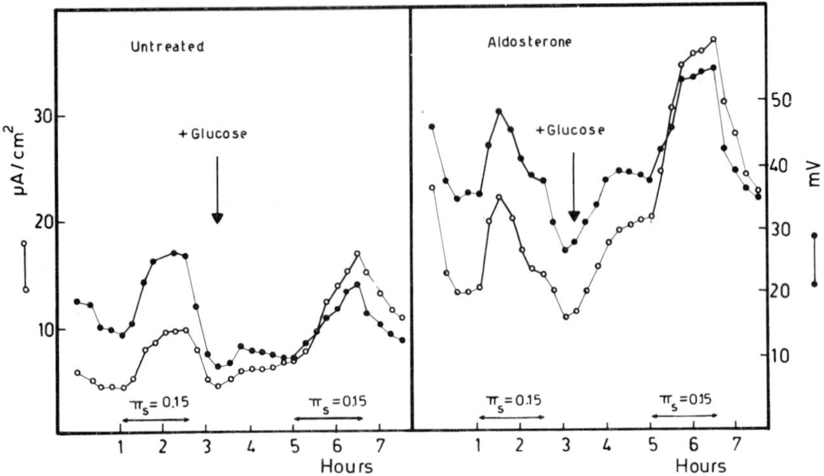

Fig. 5. Influence of hypotonicity on the behaviour of toad bladder stimulated by aldosterone. In 8 instances, paired hemibladders were set up for incubation according to Ussing and Zerahn [3] in the evening, in Ringer's fluid (NaCl, 115 mM; $KHCO_3$, 2.5 mM; $CaCl_2$, 1.0 mM) to which antibiotics had been added (Penicillin sodium, 100 U/ml; Streptomycin sulphate, 1 mg/ml). One of the preparations was exposed in addition to aldosterone, 5×10^{-8} M, through the night. The following morning, the solutions were replaced by antibiotic-free Ringer's and, on the outside, the sodium chloride concentration was reduced to 78 mM; short-circuit current and transmembrane electrical potential were measured thereafter at 15 min intervals. After 1 hr, Ringer's on the inside was replaced with the dilute solution (NaCl 78 mM; $KHCO_3$, 2.5 mM; $CaCl_2$, 1.0 mM) and incubation went on for 90 min. Whereupon standard Ringer's (π = 0.225) was reintroduced on the inside. Approximately one hour later, glucose, 10 mM, was added everywhere, which brought about a further large increase in sodium-transporting activity of the aldosterone-treated tissues. This was enhanced reversibly when hypotonic fluid was used anew.

incidentally, untreated preparations also react when incubated in hypotonic media – this raises the question of the significance of the substrate effect that had been considered specific for aldosterone-treated preparations.

Acknowledgments

Dr. F.C. Herrera kindly provided assistance in shipments of *Bufo marinus* from Venezuela. Aldosterone (Aldocorten), insulin (pork, 10 times recrystallized), vasopressin (synthetic lysine-8-vasopressin) and amphotericin (Fungizone) were generous gifts of CIBA, Novo, Sandoz and Squibb, respectively.

This work has been supported financially by the Fonds National de la Recherche Scientifique (Belgium).

References

[1] J. Crabbé and A. Decoene, Hormonal influences on sodium transport by amphibian epithelia incubated in hypotonic media. Proc. I.U.P.S.: XXV int. Congress, München, vol. 9 (1971) p. 123.

[2] E.A.C. MacRobbie and H.H. Ussing, Osmotic behaviour of the epithelial cells of frog skin. Acta Physiol. Scand. 53 (1961) 348.

[3] H.H. Ussing and K. Zerahn, Active transport of sodium as the source of electric current in the short-circuited isolated frog skin. Acta Physiol. Scand. 23 (1951) 110.

[4] J. Crabbé and P. de Weer, Relevance of sodium transport pool measurements in toad bladder tissue for the elucidation of the mechanism whereby hormones stimulate active sodium transport. Pflügers Arch. 313 (1969) 197.

[5] N.S. Lichtenstein and A. Leaf, Effect of amphotericin B on the permeability of the toad bladder. J. clin. Invest. 44 (1965) 1328.

[6] J. Crabbé, Suppression, by amphotericin B, of the effect exerted by aldosterone on active sodium transport. Arch. int. Physiol. 75 (1967) 342.

[7] G.A. Porter and I.S. Edelman, The action of aldosterone and related corticosteroids on sodium transport across the toad bladder. J. clin. Invest. 43 (1964) 611.

[8] G.W.G. Sharp and A. Leaf, Biological action of aldosterone *in vitro*. Nature 202 (1964) 1185.

[9] R. André and J. Crabbé, Stimulation by insulin of active sodium transport by toad skin: influence of aldosterone and vasopressin. Arch. int. Physiol. Biochim. 73 (1966) 538.

[10] F.C. Herrera, G. Whittembury and A. Planchart, Effect of insulin on short-circuit current across isolated frog skin in the presence of calcium and magnesium. Biochim. Biophys. Acta 66 (1963) 170.

[11] A. Leaf, J. Anderson and L.B. Page, Active sodium transport by the isolated toad bladder. J. Gen. Physiol. 41 (1958) 657.

[12] A.D.C. Macknight, A. Leaf and M. Civan, Effects of vasopressin on the water and ionic composition of toad bladder epithelial cells. J. Membr. Biol. 6 (1971) 127.

ACTION OF POSTHYPOPHYSEAL HORMONES, ADENOSINE 3′5′ MONOPHOSPHATE AND THEOPHYLLINE ON CHOLESTEROL CONTENT OF THE EPITHELIAL CELLS IN TOAD URINARY BLADDER

D. CREMASCHI, S. HENIN and M. CALVI
Istituto di Fisiologia Generale, Università di Milano, Italy

Recently the role of cholesterol in affecting the structure and the permeability of the thin lipid membranes to water and non-electrolytes was pointed out [1–3]. Experiments, carried out on the plasma membranes of the erytrocytes suggest the same conclusions [4]. As emphasized by Finkelstein [1], in the toad urinary bladder as well the low water permeability could be related to the high amount of cholesterol found in the luminal membranes of the epithelial cells. If so, posthypophyseal hormones might reduce the cholesterol level in the membrane, when they increase water, urea and Na^+ permeability.

To test this hypothesis, we have labelled the cellular cholesterol by mucosal incubation of the tissue in 1,2 T cholesterol-Ringer solution. Modification in the rate of cholesterol washing out after hormonal treatment have been investigated.

Methods

Female *Bufo bufo* were used and urinary bladders were excised from the killed and pithed animal. Of the two halves of the bladder, one was employed as control and the other for the experiment. In the controls the net water transport across the tissue or the 1,2 tritiated cholesterol washing out were measured.

a) in the first set of controls, the water transport was gravimetrically determined under an imposed 1:10 osmotic gradient, obtained by the 1:10 dilution of the luminal Ringer solution. After three 10 min periods, Pitruitrin (Parke and Davies) (50 mU/ml) or Adenosine 3′5′ monophosphate (2 mM) or Theophylline (5 mM) were added during the 4 following periods. The transport was measured as $\mu l\ cm^{-2}\ hr^{-1}$, considering the bladder as a sphere.

b) in the second set of controls the lobe was opened and put between to lucite chambers. The luminal Ringer solution contained 1,2 T cholesterol (30–50 Ci/mM, circa 3 μCi/ml) over 4 hr. After this incubation period the mucosal medium was replaced by cholesterol free Ringer solution, washed three times and the washing out rate of labelled substances from the tissue measured every 20 min, renewing the mucosal and serosal solutions. Thus the washing out kinetics without any treatment was determined.

The hemibladders used for the experiment were prepared and incubated with labelled cholesterol, like in b), but after three 20 min experimental periods, in the serosal medium Pituitrin (50 mU/ml) or Adenosine 3′5′ monophosphate (2 mM) or Theophylline (5 mM) were present over 4 periods. In the subsequent 2–3 periods fresh Ringer solution was used again as a serosal fluid.

The radioactivity was measured with a liquid scintillation counter (Packard, Spectrometer 3003). The radioactivity measured in the first period was taken as a reference value (100%) to have comparable data. In some of these experiments the electric potential difference across the tissue was measured in every experimental period by a microvoltmeter (Keithley 155). Calomel electrodes and saturated KCl 3% agar bridges were employed. The Ringer solution used had the following composition: NaCl 110 mM, KCl 2 mM, $CaCl_2$ 2.5 mM, Na_2HPO_4 3.7 mM, NaH_2PO_4 1.7 mM, $pH = 7$.

It was bubbled with air, at room temperature (circa 23°C). The experiments were carried out in spring and autumn. The washed out labelled material was chromatographied on silica gel G thin layers (0.5 mm thick). The eluent used was benzene-ethylacetate (9:1) solution (22°C). 8 mm strips were scraped and mixed with a scintillation fluid (Toluene 1000 ml; 2,5 Diphenyloxazole 5 g; Dimenthyl 1,4 bis-[2-(4 methyl-5-phenyloxazolyl)] -benzene 300 mg).

Results and discussion

In the urinary bladder of *Bufo marinus* posthypophyseal hormones are known to selectively increase water, urea and Na^+ permeability of the muco-

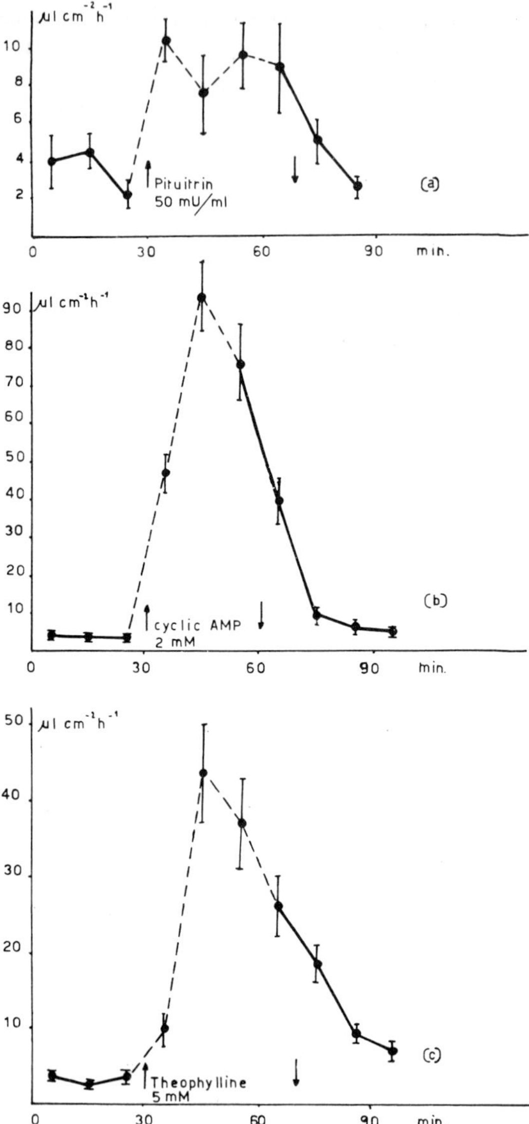

Fig. 1. Net osmotic transport of water across urinary bladder of *Bufo bufo* vs time: treatment with Pituitrin (50 mU/ml), cyclic AMP (2 mM), Theophylline (5 mM). The luminal medium is 1:10 Ringer solution. The arrows stand for the beginning and the end of the treatment; the dashed line for the treatment period. Number of experiments: (a) = 5, (b) = 4, (c) = 7.

sal plasma membrane [5–7]. The adenosine 3′5′ monophosphate (cyclic AMP) is the intracellular hormonal messenger in the epithelial cells. The same hormonal effect on water and Na^+ seems to be present in *Bufo bufo*. Cyclic AMP and Theophylline (which increases the cyclic AMP intracellular level)

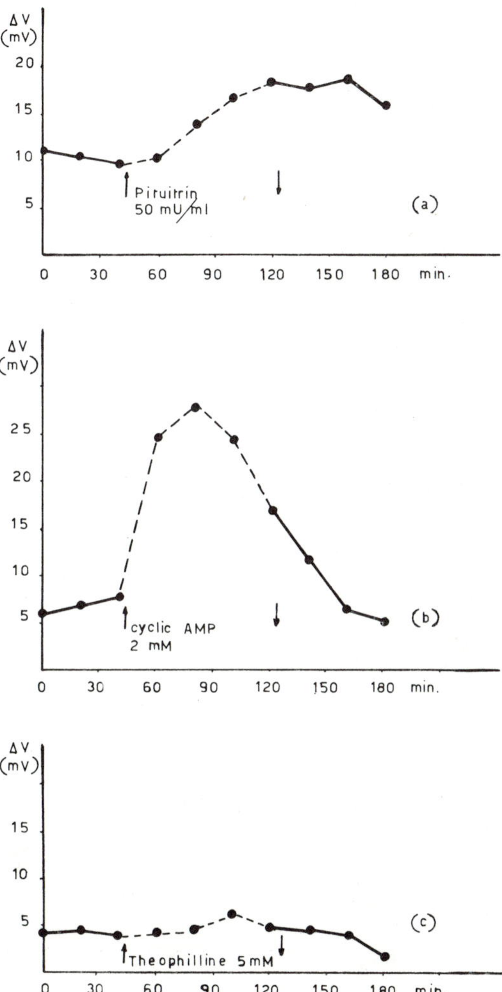

Fig. 2. Transepithelial potential difference across urinary bladder of *Bufo bufo* vs time: treatment with Pituitrin (50 mU/ml), cyclic AMP (2 mM), Theophylline (5 mM). The arrows stand for the beginning and the end of the treatment, the dashed line for the treatment period.

mimic this action (figs. 1 and 2). Moreover, previous papers have given evidence to the fact that also urea permeability is enhanced and that a modification of the mucosal barrier is responsible for such an increase [10–12]. Therefore the two substrates seem to be completely comparable.

If the hormone, by its messenger, reduces the cholesterol content in the mucosal plasma membrane, a removal of labelled material from the membrane and an enhanced washing out rate may be expected. On the basis of the figs. 3, 4 and 5, the suggestion seems to be true. These figures report the mean amount of luminal washing out in every experimental period vs time. It is seen that the control kinetics is described by an exponential function. The washing out rate of the experiment equals the control only when the hormone (or cyclic AMP, or Theophylline) is absent in the serosal perfusion fluid, whereas it is significantly higher during the treatment period. The statistical probability values (*P*) for each set of mean data are reported.

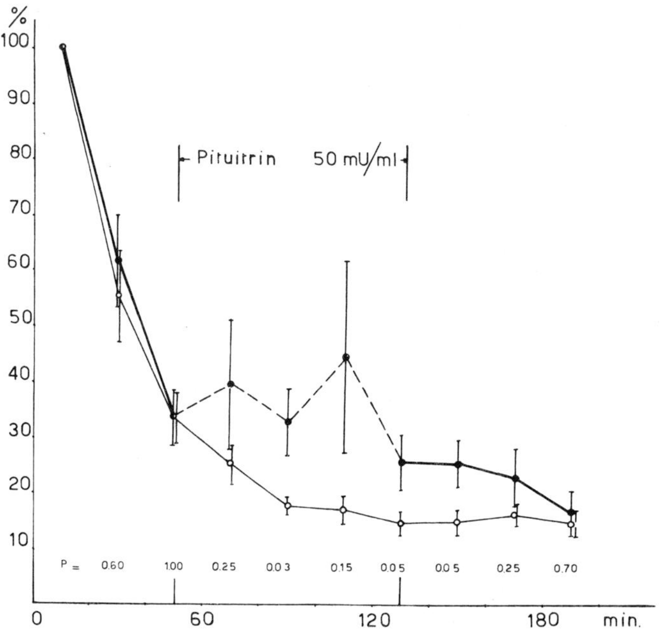

Fig. 3. Luminal washing out of labelled material in urinary bladder of *Bufo bufo* vs time: treatment with Pituitrin (50 mU/ml). The washing out of the first period is taken as a reference value (100%). Periods: 20 min. —○—: controls (N = 7). —●—: experiments (N = 10). The dashed line stands for the treatment period. The statistical probability values (*P*) for each set of mean data are reported in every period.

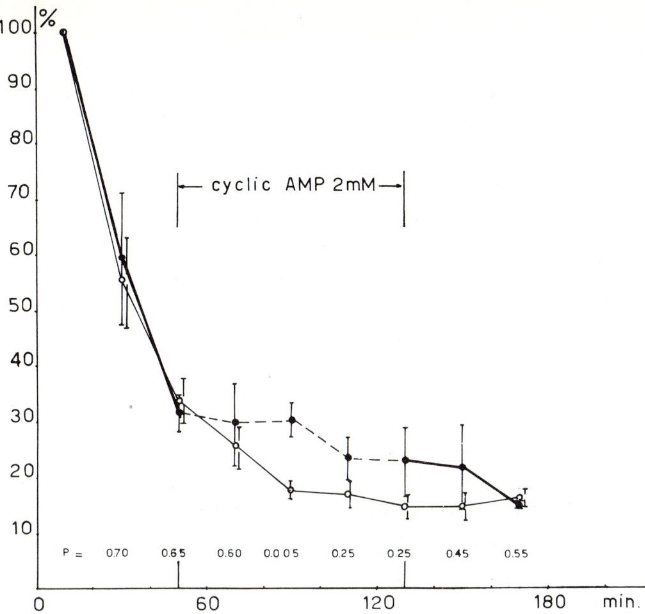

Fig. 4. Luminal washing out of labelled material in urinary bladder of *Bufo bufo* vs time: treatment with cyclic AMP (2 mM). Controls: N = 7. Experiments: N = 4. For other explanations see fig. 3.

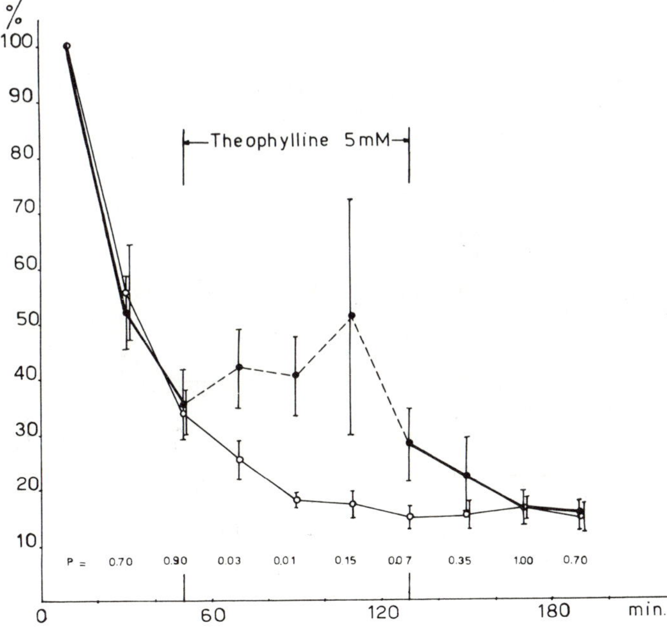

Fig. 5. Luminal washing out of labelled material in urinary bladder of *Bufo bufo* vs time: treatment with Theophylline (5 mM). Controls: N = 7. Experiments: N = 8. For other explanations, see fig. 3.

These results have been obtained with preparations which were in good functional condition, and the effect is likely to be directly related to the hormonal action. It should be noted that:

a) The labelled cholesterol was purchased as a benzene solution. Therefore a small amount of benzene (10–30 μl/ml) was added to the Ringer solution. In some experiments measures of the transepithelial p.d. were carried out to ascertain that the benzene had not damaged the preparation. Fig. 2 shows the results of these measures: a p.d. of several mV is present under basal conditions, even after a 4 hr incubation in cholesterol-benzene Ringer solution and it increases after the treatment.
b) Since a cellular swelling was observed after hormonal treatment [12], a displacement of cholesterol from the tissue could be possible, as a consequence of the swelling. Therefore experiments were carried out with a 1:4 diluted serosal Ringer solution instead of hormonal treatment, to cause a swelling of the cells. Fig. 6 shows the results of such experiments: no significant difference from the controls is observed.

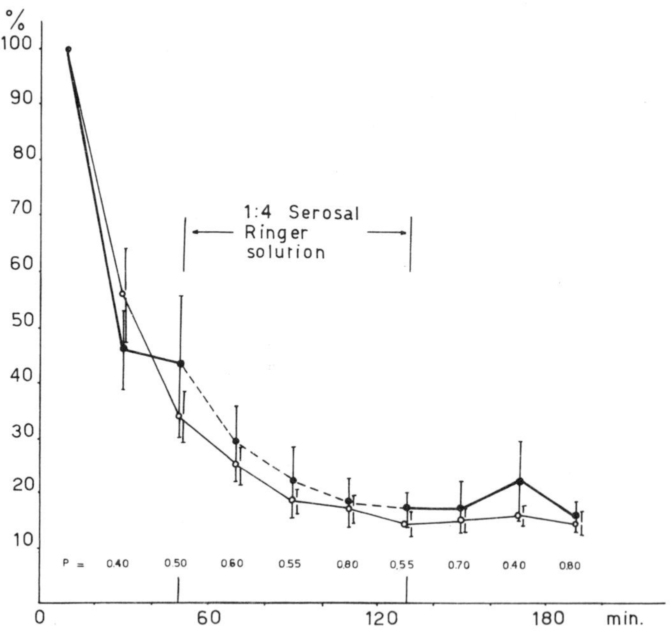

Fig. 6. Luminal washing out of labelled material in urinary bladder of *Bufo bufo* vs time: cellular swelling is caused by the 1:4 dilution of the serosal medium. Controls: N = 7. Experiments: N = 3. For other explanations, see fig. 3.

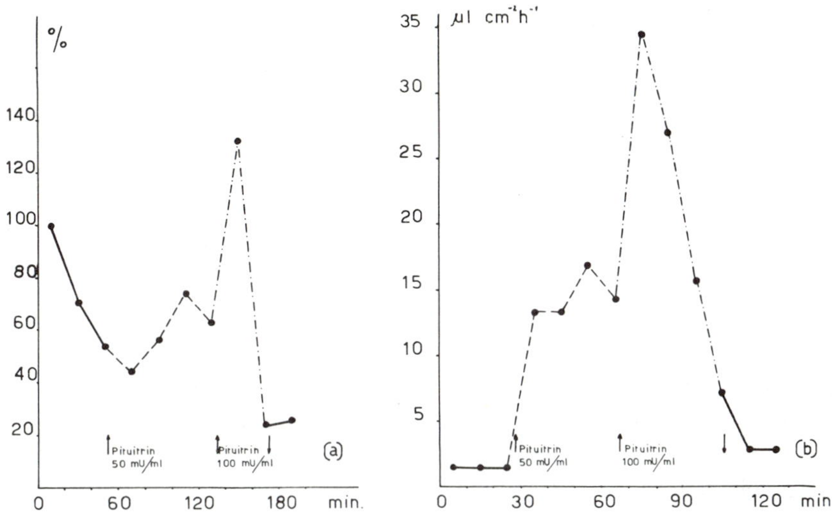

Fig. 7. Luminal washing out of labelled material (a) and net osmotic transport of water (b) in the urinary bladder of *Bufo bufo* vs time: treatment with Pituitrin (50 mU/ml, 100 mU/ml). The washing out of the first period is taken as a reference value (100%). The arrows stand for the beginning and the end of the treatment, the dashed line for the treatment period.

c) Finally in the same tissue an increased hormonal dose seems to increase the effect on cholesterol washing out as well as on osmotic water transport (fig. 7).

The kinetics of the serosal washing out generally does not seem to change after treatment; but no final conclusion can be drawn because of the low detectable radioactivity. The excess cholesterol washing out is likely to come from the mucosal plasma membrane. In fact the enhanced rate of washing out would not be due to an increased outflux from the intracellular medium as posthypophyseal hormones selectively increase Na^+, water and urea permeabilities only [5–7]. Under our experimental conditions transepithelial fluxes of labelled cholesterol were measured and they were not found to increase when the hormone was added (0.22 pM cm^{-2} hr^{-1} before treatment, 0.21 pM cm^{-2} hr^{-1} after treatment, with a mucosal concentration of 67 pM/ml). The low serosal washing out suggests also that the labelled material comes from the mucosal plasma membrane.

The next problem is to determine if the hormone merely displaces or metabolized the cholesterol. The cholesterol washed out during treatment with Pituitrin or cyclic AMP was chromatographied on silica gel G thin layers, to identify the labelled material. One peak only corresponding to the cholesterol (R_f = 0.33), was found in both cases using benzene-ethylacetate (9:1) as an elution solvent. Therefore a displacement seems likely to occur, within the limits of the used method. We are carrying out further chromatographical experiments with different eluents to elucidate this important point.

Summary

The importance of cholesterol in the mechanism of action of posthypophyseal hormones on the permeability of luminal plasma membranes in toad urinary bladder is pointed out. The effect of Pitruitrin, Adenosine 3′5′ monophosphate, and Theophylline on the washing out kinetics after luminal incubation of the tissue in labelled cholesterol Ringer solution is studied. A significant increase in the washing out rate is obtained after hormone treatment.

References

[1] A. Finkelstein and A. Cass, Nature 216 (1967) 717.
[2] C. Lippe, J. Mol. Biol. 39 (1969) 669.
[3] J.S. O'Brien, J. Theoret. Biol. 15 (1967) 307.
[4] K.R. Bruckdorfer, R.A. Demel, J. de Gier and L.L.M. van Deenen, Biochim. Biophys. Acta 183 (1969) 334.
[5] A. Leaf, J. Gen. Physiol. 43 (1960) 175.
[6] A. Leaf and R.M. Hays, J. Gen. Physiol. 45 (1962) 921.
[7] R.H. Maffly, R.M. Hays, E. Lamdin and A. Leaf, J. Clin. Invest. 39 (1960) 630.
[8] J.S. Handler, R.W. Butcher, E.W. Sutherland and J. Orloff, J. Biol. Chem. 240 (1965) 4524.
[9] J. Orloff and J.S. Handler, Am. J. Med. 42 (1967) 757.
[10] C. Lippe, D. Cremaschi, V. Nicolella and V. Capraro, Boll. Soc. It. Biol. Sper. 43 (1966) 413.
[11] C. Lippe, D. Cremaschi, V. Nicolella and V. Capraro, Boll. Soc. It. Biol. Sper. 43 (1966) 417.
[12] C. Lippe, D. Cremaschi and V. Capraro, Comp. Biochem. Physiol. 19 (1966) 179.
[13] L.D. Peachey and M. Rasmussen, J. Biophys. Biochem. Cytol. 10 (1961) 529.

NON-ELECTROLYTE PERMEABILITY ACROSS LIPID BILAYER MEMBRANES

E. GALLUCCI, S. MICELLI and C.LIPPE

Istituto di Fisiologia Generale, Università di Bari, Italy

After the discovery of thin lipid membranes [1] made with lipids of biological origin, several researches have been performed on the electrical properties of these bilayers, but only a few works have been made on non-electrolyte permeability [2–5] and their mechanism of transport has not been clarified yet.

The present work is the development of previous papers [6–10] on the permeability of non-electrolytes through artificial thin lipid membranes of different composition.

Methods

Thin lipid membranes were formed with egg-lecithin or phosphatidyl-inositol dissolved in decane or with oxidized cholesterol dissolved in mixture of equal volumes of *n*-octane-dodecane.

The membranes were formed at 28°C by brushing the solution on a small circular aperture (0.07 cm^2) of a teflon cell. The cell was similar to that described by Läuger et al. [11]. The bathing fluid was a 10 mM-KCl solution. The concentration of the tested molecule was 0.1 or 1 mM. The non-electrolyte permeability was measured as follows. When membrane became "black" about 10 μl of water, containing about 10 μC of a labelled molecule were added to one side, and the same quantity of unlabelled molecule was added to the opposite one. After 40′ a small amount of fluid from both compartments was collected. Stirring of the solutions was achieved by convection currents provided by small temperature differences between the bottom and the top of the chambers. The labelled molecules we used were: [^{14}C]urea,

[^{14}C] thiourea, [^{35}S] N-methyl-thiourea, [^{35}S] N-N′-dimethylthiourea (from Radiochemical Centre, Amersham); [^{14}C] dimethyl-sulfoxide, [^{14}C] acetamide, [^{14}C] N-methyl-urea (from N.E.N., Frankfurt/M); [^{14}C] formamide, (from Philips-Duphar, Petten, Holland); [^{14}C] N-N-dimethyl-formamide (from Mallinckrodt Nuclear, Orlando, Fa.).

Results and discussion

It can be seen (table 1) that there are several discrepancies between the oil-water partition coefficients and the permeability coefficients:

a) Urea and thiourea show quite similar permeability coefficients, but the oil-water partition coefficient of thiourea is about ten times larger.

b) In lecithin membranes the permeability of acetamide and formamide is 30–40 times larger than that of thiourea, in spite of similar oil-water partition coefficients.

These discrepancies may be explained on the basis that partition coefficients between two bulk phases may be different from the partition coefficients between water and a thin, organized membrane. These discrepancies could be due also to the very low surface charge of the fatty acids used for the partition coefficient measurements, as compared to that of the lipids used

Table 1

Permeability coefficients (P) of non-electrolytes across lipid bilayer membranes

Molecule	Membrane of phosphatidyl-choline in n-decane P cm sec^{-1} × 10^{-6} mean values ± S.E.	Membrane of oxidized cholesterol in n-octane-dodecane P cm sec^{-1} × 10^{-6} mean values ± S.E.	Oil-water partition coefficients
urea	3.65 ± 0.61 (6)	0.89 ± 0.22 (5)	0.15×10^{-3}
N-methyl-urea	39.20 ± 7.56 (6)	2.89 ± 1.06 (4)	0.44×10^{-3}
acetamide	143.2 ± 49.1 (6)	17.5 ± 0.4 (4)	0.83×10^{-3}
formamide	160.0 ± 32.8 (6)	6.21 ± 2.34 (6)	0.76×10^{-3}
thiourea	4.64 ± 0.55 (8)	0.89 ± 0.23 (5)	1.20×10^{-3}
N-methyl-thiourea	36.73 ± 12.14 (6)	5.87 ± 1.47 (5)	–
N,N′-dimethyl-thiourea	107.7 ± 34.2 (8)	80.3 ± 17.7 (7)	–
N,N′-dimethyl-formamide	113.4 ± 17.2 (7)	147.38 ± 31.4 (6)	4.9×10^{-3}
dimethyl-sulfoxide	89.9 ± 25.9 (7)	48.4 ± 19.2 (7)	–

The figures in parentheses represent the number of experiments.

Table 2
Urea and thiourea permeability through phosphatidyl-inositol membranes

Molecule	P cm sec^{-1} × 10^{-6} mean value ± S.E.
urea	12.85 ± 2.74 (6)
thiourea	49.4 ± 11.46 (7)

The figures in parentheses represent the number of experiments.

in the present work. This latter interpretation is supported, even though indirectly, by the finding that oxidized cholesterol membranes, possessing a surface charge lower than that of lecithin ones, show a lower discrimination power between the more polar amides.

This fact suggests that the interface interaction of some molecules with the polar heads of lecithin plays an important role in determining the overall permeability across the lipid film. In support to this theory we have observed that the urea and thiourea permeabilities are larger across phosphatidyl-inositol membranes than across lecithin ones (table 2). These two phospholipids differ from each other because lecithin has no net charge and phosphatidylinositol has a net negative charge.

It can be seen that oxidized cholesterol membranes are less permeable than the lecithin ones. Since phospholipids are absent in our cholesterol membranes, the decrease in permeability is due to the presence of sterols *per se*, and not to their interaction with lecithins. In other words, the effect of sterols on the permeability cannot be related to the well known condensing effect of sterols on lecithin monolayers [12].

The decrease of the permeability of cholesterol membranes as compared to that of the lecithin ones is particularly evident for the more polar molecules such as urea, thiourea and formamide. Such effect can also be seen for molecules possessing one methyl group such as acetamide, *N*-methyl-urea and *N*-methyl-thiourea, while it is virtually absent for molecules possessing two methyl-groups.

We can attempt to interpret this phenomenon by considering a mechanism involving at least two steps: the boundary phenomena at the membrane-water interface, and the passage across the apolar core of the membrane. The interactions of the more polar molecules with the polar heads of lecithins could facilitate their penetration in the membrane. This effect should be reduced in membranes possessing a lower surface charge such as membranes of oxidized cholesterol. The presence of two methyl groups would determine a steric hindrance that could reduce the possibility of interface interaction between the polar groups of these molecules and the phospholipids polar heads. For these reasons the permeability of these molecules across the two types of membranes should be virtually equal.

The permeability coefficients to several non-electrolytes are of the same order of magnitude as those found in biological membranes [2, 6].

This observation demonstrates that water-soluble molecules can cross also a non-porous structure, as the lipid bilayer. This phenomenon should be taken into account especially when equivalent pore radius is measured in biological membranes.

Acknowledgement

This work has been supported by a research grant from C.N.R., Roma.

References

[1] P. Mueller, D.O. Rudin, H.T. Tien and W.C. Wescott Nature 194 (1962) 979.
[2] H.I. Vreeman, Koninkl. Ned. Acad. Wet. Ser. B., 69 (1966) 564.
[3] R.C. Bean, W.C. Shepherd and H. Chan, J. Gen. Physiol. 52 (1968) 495.
[4] T.E. Andreoli, V.W. Dennis and A.M. Weigl, J. Gen. Physiol. 52 (1969) 133.
[5] R.E. Wood, F.P. Wirth and H.E. Morgan, Biochim. Biophys. Acta 163 (1968) 171.
[6] C. Lippe, B. Giordana and D. Cremaschi, Boll. Soc. It. Biol. Sper. 44 (1968) 1052.
[7] C. Lippe, J. Mol. Biol. 39 (1969) 669.
[8] E. Gallucci, C. Storelli and C. Lippe, Boll. Soc. It. Biol. Sper. 46 (1970) 613.
[9] C. Lippe, E. Gallucci and C. Storelli, Arch. Internat. Physiol. Bioch. 79 (1971) 315.
[10] E. Gallucci, S. Micelli and C. Lippe, Arch. Internat. Physiol. Bioch. 79 (1971) 881.
[11] P. Läuger, W. Lesslauer, E. Marti and J. Richter, Biochim. Biophys. Acta 135 (1967) 20.
[12] L. de Bernard, Bull. Soc. Chim. Biol. 40 (1958) 11.

PART V

COUPLING MECHANISMS

INTRODUCTION

E. HEINZ

Institute of Vegetative Physiology, Johann Wolfgang Goethe University, Frankfurt am Main, Germany

Introduction

In the last few days many systems have been discussed here, in which ions and molecules are transported with the expenditure of energy against concentration gradients into, out of, or across cells. It seems more than appropriate that this last session will deal with the question where this energy comes from and how it is fed into the proper transport system. We therefore shall now discuss the various possible ways of coupling as a means that energy from exogenous processes is made available to endogenous transport. I have been asked to give a more general introduction to this session, in order to help those who are less familiar with this subject to understand and integrate the material presented. I shall therefore try to give a short outline of the field to define some crucial concepts and to formulate the most acute problems and questions which are still open and to which we hopefully expect answers from the next speakers. Let me first start with the definition and characterization of active transport and its possible coupling to other processes.

We speak of active transport if a solute (*i*) is driven across a membrane, or barrier, by a force other than its conjugate driving force. The conjugate driving force is the (negative) electrochemical potential gradient (ECP) of *i*:

$$-\frac{\mathrm{d}\bar{\mu}_i}{\mathrm{d}x} = r_{ii}\, J_i\,.$$

J_i is the flow of *i* and r_{ii}, a resistance coefficient. In the steady state, J_i can be considered constant throughout the thickness (δ) of the transport region, so that we can integrate the above gradient, and we obtain the difference of the

ECP of the solute i between the two solutions on either side of the region:

$$-\Delta\bar{\mu}_i = \int_0^{\delta} r_{ii}\, \mathrm{d}x\, J_i$$

or, following the usual notations of irreversible thermodynamics:

$$X_i = J_i R_{ii}\,; \qquad X_i = -\Delta\bar{\mu}_i \qquad \text{and} \qquad R_{ii} = \int_0^{\delta} r_{ii}\, \mathrm{d}x\,.$$

In order that other than conjugate forces become effective, as in active transport, the flow of i has to be "coupled", either to the flow of other substances (solutes or the solvent), J_j, across the same barrier, or to the "advancement" of a chemical reaction, J_r. In the first case we speak of osmo-osmotic, in the second case, of chemi-osmotic coupling. Accordingly, we can expand the above equation to include the other flows and the reaction rate

$$X_i = J_i R_{ii} + \sum_{j=2}^{n} R_{ij} J_j + R_{ir} J_r\,.$$

It is obvious that in osmo-osmotic coupling J_i depends also on the ECP gradients of other substances, including the solvent, X_j, and in chemi-osmotic coupling, on the affinity of a chemical reaction, A_r. This dependence, in which we are mainly interested, can be better visualized by using the L-notation.

$$J_i = L_{ii} X_i + \sum_{j=2}^{n} L_{ij} X_j + L_{ir} A_r\,.$$

It has become customary now to call the transport of i, to the extent that it is driven by A_r via chemi-osmotic coupling, "primary" active, and to the extent that it is driven by X_j via osmo-osmotic coupling, "secondary" active. The above simple notation holds only if the system is in the steady and near equilibrium. It does, moreover, give no information on the underlying molecular mechanism.

In the first part of our session we shall consider chemi-osmotic coupling, i.e. *primary active transport*. In the last decade the cases of truly primary active transport seem to have dwindled, as an increasing amount of transport systems has turned out to depend on electrolyte ions, such as Na-, K-, and

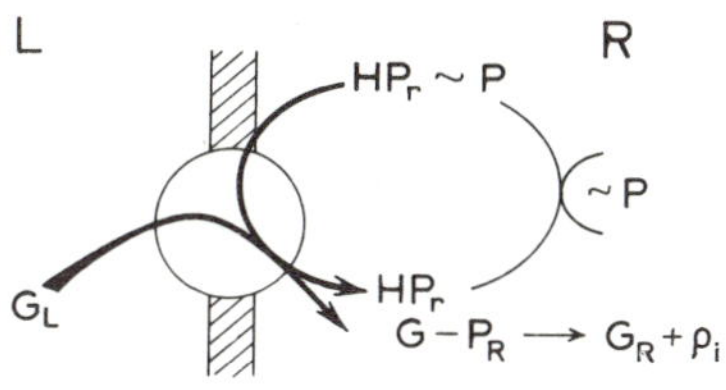

$$1.\quad HP_r + {\sim}P \longleftrightarrow HP_r{\sim}P$$

$$2.\quad G_L + HP_r{\sim}P_R \;\text{—○→}\; Hp_r + G_R + p_i$$

Fig. 1. Model of phosphotransferase system. For explanation see text.

H-ions. Among the few remaining systems in which a chemi-osmotic coupling is undebated are the phosphotransferase systems for sugar transport, the Na- and K-, the Ca-, and the HCO_3-activated ATPases for electrolyte transport, and perhaps some redox systems for proton transport, such as are assumed to operate in the mitochondrial membrane.

The phosphotransferase systems have been presented in one of the previous sessions. The main features are summarized in fig. 1: The coupling takes place between the accumulative transport of a sugar and the cleavage of an energy-rich phosphate. It requires a "heterophasic" reaction, i.e. a reaction between reactants which come from different sides of the barrier. Here, for instance, the energy-rich phosphate comes from inside the cell, and the sugar from outside. The reaction takes place within the barrier, and the transphosphorylation step is to be identical with the translocation step for the sugar. Thus the affinity of the overall dephosphorylation reaction becomes the immediate driving force to pull the sugar through the barrier, which it cannot penetrate otherwise. It should be recalled that such a coupling has already been postulated by P. Mitchell in 1960 for his model of "Group Translocation". We see on the following figure, in which I tried to visualize Mitchell's idea, that after renaming some of the symbols we obtain basically the same mechanism as that supposed to operate in the phosphotransferase system (fig. 2).

The phosphotransferase mechanism is at present possibly the best elucidataed case of chemi-osmotic coupling. It seems, however, to occur only in a limited number of microbiological systems. Dr. Kepes will show us some of his recent findings concerning that system which may tell that the mechanism is not quite as simple as I tried to make it appear in this introduction.

Far less well elucidated is the molecular mechanism of the ATPase sys-

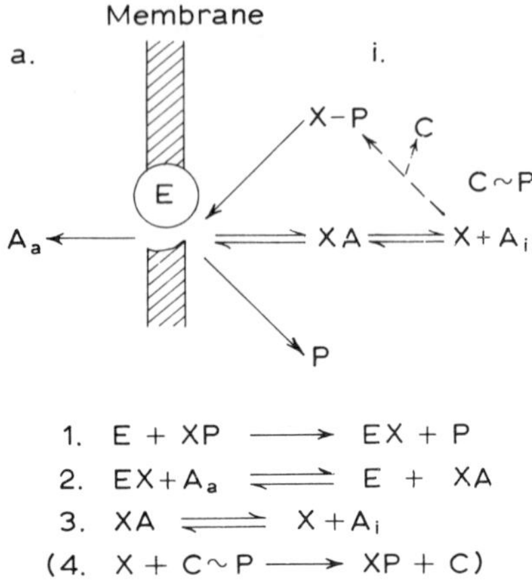

Fig. 2. Translocase model after P. Mitchell. E is a spacially oriented enzyme within the membrane, catalyzing reaction 1 and 2. A_a is an outside substrate, X-P is the phosphorylated group, X is the group combining with the substrate to form XA, C~P is a high energy phosphate of C. The model differs from the phosphotransferase system only in that not the phosphate group but the X-group combines with the transported substrate. (From: E. Heinz, In: Biochemie des aktiven Transportes, 12. Kolloquim Gesellschaft für Biol. Chemie 1961 (Springer, Berlin, 1961) p. 173).

tems. The main difficulty here may be that in this case the transported substrates, electrolyte ions, cannot directly participate in a true chemical reaction involving covalent bonds. We may again postulate a "heterophasic" reaction to be essential for this mechanism. We could, for instance, assume that the enzyme catalizing the fundamental reaction, the hydrolytic cleavage of an energy-rich phosphate, is so oriented in the barrier that it accepts the energy-rich phosphate only from inside the cell, and the water molecule, at least part of it, the OH-ion, only from outside the cell. Such a reaction will forcibly transfer a negative charge across the barrier, which in turn may be electrically coupled to the movement of a K^+. The H-ion left behind on the outside of the cell could by a specific cation-exchange lead to the extrusion of a Na-ion. This model, albeit energetically plausible, is a pure speculation since at present nothing is known whether it occurs in any system (fig. 3).

As to the systems of proton transport, even less is known about them than about the previously discussed ones. Formerly it was believed that the H^+/O_2 ratio, i.e. the number of protons transported per oxygen molecule consumed, is restricted to 4, the number of electrons a molecule of oxygen can accept. More recently, however, Mitchell has devised a model, which clearly does

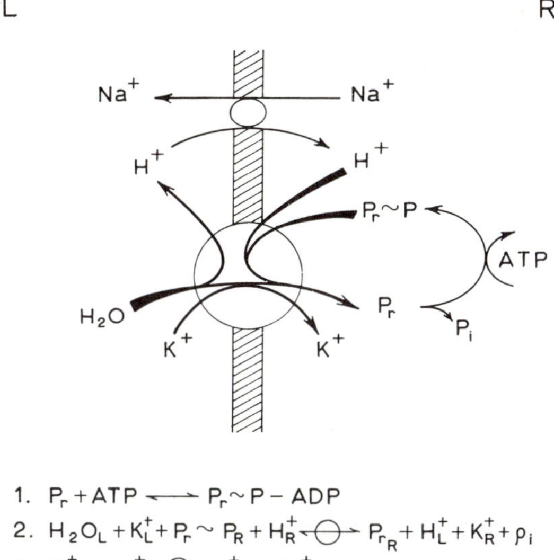

$$1.\quad P_r + ATP \rightleftharpoons P_r\sim P - ADP$$

$$2.\quad H_2O_L + K_L^+ + P_r \sim P_R + H_R^+ \rightleftharpoons P_{r_R} + H_L^+ + K_R^+ + p_i$$

$$3.\quad H_L^+ + Na_R^+ \rightleftharpoons H_R^+ + Na_L^+$$

Fig. 3. Hypothetical model of ATPase system for electrolyte transport. The lower circle within the membrane represents a spacially oriented enzyme which catalyzes the (heterolytic) cleavage of an energy-rich phosphoryl compound (P_r-P). The upper circle represents a carrier which specifically catalyzes the exchange between H- and Na-ions by counter-transport (antiport). For further details see text.

without this limitation. I have tried to illustrate it in fig. 4. We see that the number of protons transported depends on the number of loops of the electron transfer across the membrane, and can, within the limits of energy available from the overall redox potential, very well exceed four. The resulting proton gradient across the membrane is supposed to drive oxydative phosphorylation by a chemi-osmotic coupling which is the other way round as the chemi-osmotic coupling mentioned before, though the underlying mechanism may be similar in both cases. This similarity is also suggested by the fact that other supposedly chemi-osmotic coupled transport systems may under certain conditions be reversible. I recall the observations of Glynn on the cation transport in red cells and the already mentioned paper of Hasselbach who could show that, by a reversion of the calcium uptake in sarcoplasmic reticuli, resynthesis of ATP can be enforced.

The second part of this session will deal with *secondary active transport*. The active transport of certain amino acids, of sugars, and of other organic solutes, mostly, but not only, in animal cells has been shown to depend on

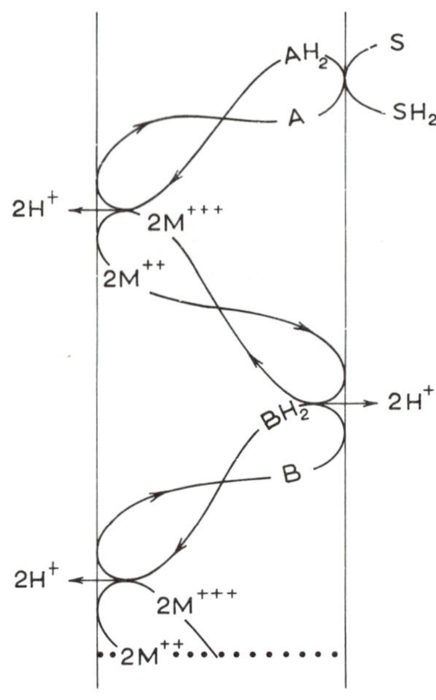

Fig. 4. Redox pump for H^+. The reaction pathways of the redox steps are represented by loops. The mechanism can be extended by further loops at the lower end. Essential is that a hydrogen transporting molecule alternates with an electron carrier. The figure is a modified representation of Mitchell's hydrogenation/dehydrogenation mechanism.

the presence of electrolyte gradients. It is assumed that the entrance of amino acids and sugars into the cell is coupled to the (downhill) movement of Na-ions in the same direction (co-transport or symport).

The coupling in "co-transport" is usually attributed to the formation of a ternary complex between the carrier, the transported substrate, and the co-substrate. It should be added that the mere formation of such a complex would not yet warrant energetic coupling: One has rather to assume that the attachment of the co-substrate to the carrier increases either the affinity for the substrate, or the velocity of the carrier, or both. In the pure affinity type model only K_m, and in the pure velocity type model, only V_{max} is supposedly affected by the co-substrate. In many cases, however, both, K_m and J_{max} are changed at the same time. Besides coupling by co-transport, also coupling by counter-transport (antiport) has been suggested, in which for example the inward transport of amino acids is energetically linked to the (downhill) movement of K-ions of the cell. The interactions between carrier and co-substrates in counter-transport are supposedly analogous but inverse to those in co-transport. Later in the session we will have the opportunity to listen to two experts in this field, Dr. Christensen who was the first to suggest this kind of coupling, and Dr. Eddy who has carried out the most extended studies on this kind of coupling.

The great number of investigations on coupling by co-transport and counter-transport leave no doubt that it exists. The serious question, however, remains and can hardly be answered at the present time, namely whether this coupling is energetically adequate. It seems that in many cases so far investigated the combined gradients of sodium ions and potassium ions, inwardly and outwardly directed, respectively, do not fully account for the energy required to accumulate the amino acids or sugars to the extent actually observed. So, even if we take an energetic coupling between electrolyte fluxes and organic solute fluxes for granted, we still are in doubt whether nevertheless primary active transport takes place, i.e. whether not a part, or even the major part of the required energy stems from a chemical reaction via chemi-osmotic coupling.

A special type of coupling, in some respect different from the above characterized ones, is that between solute flow and the bulk flow of the solvent, formerly called "solvent drag". It does not matter here whether the solvent flow is driven by a hydrostatic gradient or by an osmotic one. This kind of coupling may also lead to a thermodynamically uphill movement of solute. It differs from the previous ones with electrolyte ions in two ways: 1) it is not stroichiometric, i.e. there is no fixed relation between the number of solute molecules and water molecules moving together, and 2) it depends on a direct interaction between the two co-substrates, solute and water, whereas in the previous cases a carrier is considered essential for the coupling. Undoubtedly, this solvent-solute coupling is of wide biological importance, especially for the reabsorption of water in the intestine and the kidneys. I can be short because Dr. P.F. Curran, a well-known expert in this field, will now report his experiments on this subject.

SOLUTE-SOLVENT INTERACTIONS AND WATER TRANSPORT

Peter F. CURRAN

Department of Physiology, Yale University School of Medicine, New Haven, Connecticut, U.S.A.

The general category of coupling mechanisms includes a rather wide range of phenomena involving both coupling of flows to metabolism and coupling of one flow to another. Included in the latter category are the solute-solvent interactions that appear to be involved in water transport across epithelial membranes. Phenomena of this type occur in secretory epithelia such as gastric mucosa and Malpighian tubules and in absorptive structures such as intestine and kidney tubules. In all these cases, net transport of solute and solvent takes place across intact layers of cells from blood to lumen or in the reverse direction. There is now a fairly long history of interest in the interrelations between solute and solvent transport in these systems and I will review the major features in a general way. My discussion is necessarily in the nature of a review because the conceptual developments that have lead to working hypotheses for the coupling between solute and solvent are now several years old. However, some ideas I shall try to develop are, in certain respects at least, fundamental to any consideration of coupling so that a review seems appropriate. In addition, the considerations will illustrate that structural features of a complex membrane system can play a major role in its functional properties, particular with respect to coupling.

The fundamental observation that calls attention to the general problem of solvent transport in epithelial systems is the fact that water transport against an osmotic pressure difference can be demonstrated quite easily. I shall begin by illustrating two examples of this behavior. Fig. 1 shows the results of some studies of Whitlock and Wheeler [1] on rabbit gall bladder *in vivo*. The lumen of the gall bladder contained Ringer's solution made progressively hypertonic by the addition of sucrose. Net water absorption is shown as a function of the

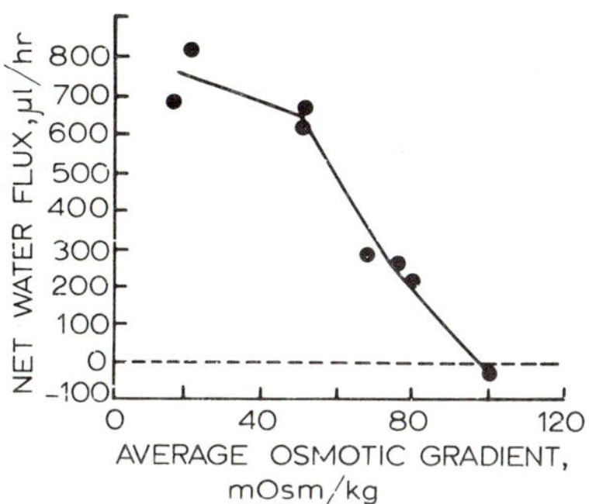

Fig. 1. Water absorption from rabbit gall bladder *in vivo*. Reproduced from Whitlock and Wheeler [1].

osmotic pressure difference between lumen and plasma. It is apparent that water absorption continues until the osmotic pressure in the lumen is 100 mOsmoles per liter greater than the plasma. Fig. 2 shows a similar behavior in an *in vitro* preparation of rat small intestine [2]. In this case, the fluid in the lumen was made hypertonic by addition of NaCl, and net volume flow across the tissue is shown as a function of the osmotic pressure difference

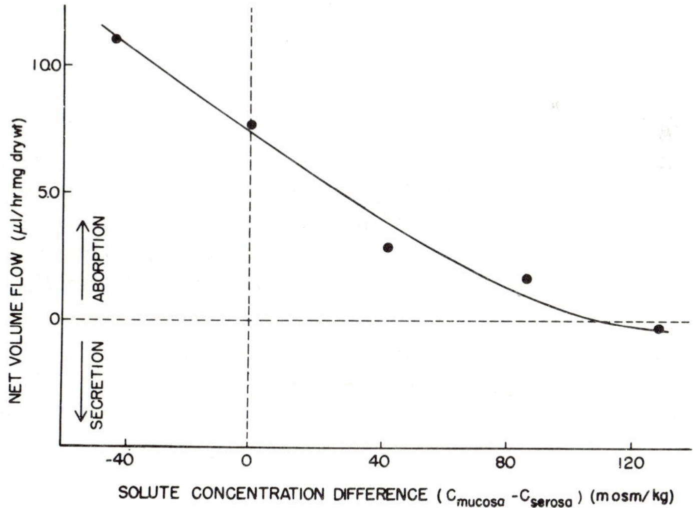

Fig. 2. Water absorption by rat intestine *in vitro*. Reproduced from Curran [3]. Data from Parsons and Wingate [2].

between the fluid on the mucosal and serosal sides of the tissue. Net water flow from mucosa to serosa (net absorption) occurs against a substantial osmotic pressure difference. Observations similar to these two have been made on a variety of epithelial tissues using several different techniques so there is little doubt that it is a real phenomenon.

One of the simple definitions of an active transport process is that it is one that can bring about net movement of a substance against a chemical (or electrochemical) potential difference. In these terms, the experiments illustrated in figs. 1 and 2 would be interpreted as indicating that these tissues possess an active transport system for water. Observations that the ability of these tissues to cause solvent transport against an osmotic pressure difference required intact metabolism [2] seemed perfectly consistent with this view. However, for some of us at least, the concept of a specific active mechanism for water transport was not particularly attractive and alternative explorations were sought.

I think it is worthwhile to consider this problem in terms of certain general concepts suggested some years ago by Kedem [4] even though this represents something of a departure from historical sequence. Kedem used the formalism of non-equilibrium thermodynamics to obtain a general expression for the flow of a substance across a membrane. It is suitable for either a simple membrane or for a complex membrane system. For the case of water flow (or volume flow, which is the quantity usually determined) the expression has this form:

$$J_\mathrm{w} = \frac{\Delta\mu_\mathrm{w}}{R_\mathrm{ww}} - \sum_{j=1}^{n} \frac{R_\mathrm{wj}}{R_\mathrm{ww}} J_\mathrm{j} - \frac{R_\mathrm{wr}}{R_\mathrm{ww}} J_\mathrm{r} \,. \tag{1}$$

J_w is net water flow (which can be taken as approximately equal to volume flow). $\Delta\mu_\mathrm{w}$ is the difference across the membrane of the chemical potential for water and since the experiments are usually carried out in absence of a hydrostatic pressure difference $\Delta\mu_\mathrm{w}$ can be replaced by $-\bar{V}_\mathrm{w}\Delta\pi$. R_ww is the resistance of the membrane system to water flow. The summed terms express the possibility that water flow may be coupled to the flow of other substances and the relation is determined by a coupling coefficient, R_wj, which in principle can have any value including zero. The final term is taken to represent active transport and expresses a direct coupling between water flow and a metabolic reaction that proceeds at a rate J_r. A non-zero value of the coefficient R_wr indicates the presence of active transport. This expression is quite useful. It immediately indicates that transport against a chemical potential difference can arise because of coupling as well as from a metabolically driven active transport system. This is a particularly important concept since a flow such as J_w might be coupled to another flow which is in turn metabolically driven. Thus, J_w could depend on metabolism indirectly even if it is not itself directly coupled to metabolism.

Let us now return and consider water transport in terms of this rather general concept. Quite a long time ago now we began to obtain some evidence in rat intestine that water movements were fairly closely related to the rate of net sodium transport across the tissue [5, 6]. We found that perfusion of the intestinal lumen with isotonic solutions containing various concentrations of sodium led to striking changes in water movement. Net water absorption from isotonic solutions decreased as sodium concentration decreased, and at luminal sodium concentrations below about 50 mM, fluid was secreted rather than absorbed.

Fig. 3 shows the results of a series of experiments in which net sodium flux and volume flow were measured in an *in vitro* preparation of rat ileum [6]. Variations in net sodium movement were caused by varying the sodium chloride concentration in the mucosal solution while the osmotic pressure of the solution was kept constant by addition of mannitol. Volume flow varied approximately linearly with sodium flow and there was no net movement of volume in the absence of net sodium flux. It is instructive to consider these results in terms of eq. (1). Since in these studies $\Delta\pi$=0 the first term may be

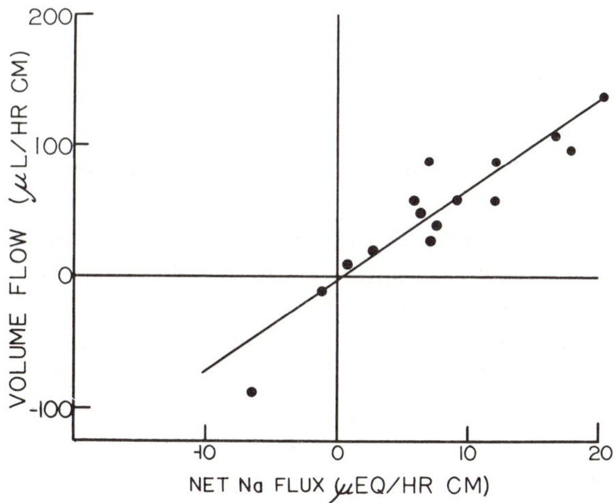

Fig. 3. Relationship between net sodium flux and volume flow across rat ileum *in vitro*. Reproduced from Curran [5].

neglected. The linear relation between J_w and J_{Na} indicates that there is a coupling phenomenon so that at least one of the summed terms is non-zero. Finally, the zero intercept in fig. 3 shows that $J_w=0$ when $J_{Na}=0$ and indicates that the last term in eq. (1) does not play a role in volume or water movement across intestine. Thus on the basis of this type of observation, we would conclude that water is not subject to direct active transport in intestine. Its movement can be ascribed entirely to differences in osmotic pressure and to coupling to fluxes of solute, particularly sodium. As discussed by Schultz and Curran [7] observations similar to those shown in fig. 3 have been reported by a number of investigators. This analysis tells us nothing, of course, about possible mechanisms of coupling.

The slopes of lines such as the one shown in fig. 3 can be interpreted as a measure of the tonicity of the transported fluid. In general the rates of solute and water transport are such that the fluid in the lumen remains essentially isotonic although the general tendency is toward a slight hypertonicity of transport; that is, solute is transported slightly more rapidly than solvent. Careful studies of Powell and Malawer [8] on rat intestine *in vivo* have illustrated this quite clearly and also shown that all tissues are not identical in this respect. They found a linear relation between solute and water flow in the upper and lower small intestine and in the colon. However, the slopes of the lines differed markedly. In jejunum, absorption was essentially isotonic (325 mOsm/l) while it became significantly hypertonic in the ileum (405 mOsm/l) and colon (515 mOsm/l).

Given the idea of some kind of coupling between solute and solvent flow, one of course asks what type of mechanism could be involved. In 1960, Durbin [9] and I [6] suggested independently and in a somewhat different manner a plausible mechanism that still appears reasonable although with certain fairly significant modifications. The hypothesis depends on two fundamental ideas: first that active solute transport creates a gradient of osmotic pressure within the epithelial tissue, and second that certain structural features of the tissue lead to a directed solvent transport as a result of this gradient.

The idea regarding structural features initially centered around the concept of a reflection coefficient first introduced by Staverman [10]. This coefficient is a measure of the effectiveness of a solute in causing osmotic flow across a particular membrane. In formal terms, in the absence of a hydrostatic pressure difference, volume flow caused by a concentration difference is given by:

$$J_v = L_p \sigma_s \Delta\pi_s \,, \qquad (2)$$

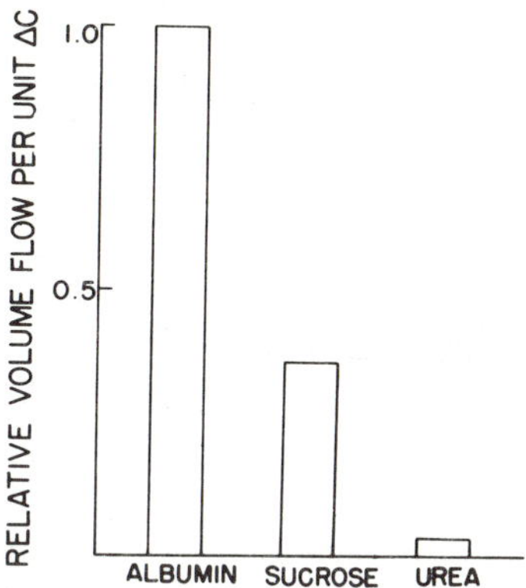

Fig. 4. Relative volume flow across dialysis membrane caused by concentration differences of three solutes. The flow caused by albumin is equal to that caused by a hydrostatic pressure difference equivalent to $RT\Delta C$. Data from Durbin [9].

where L_p is the filtration coefficient of the membrane and σ_s is the reflection coefficient. To a first approximation, σ_s varies between zero and one for different solutes and a given membrane. It is unity for a solute to which the membrane is impermeable and is zero for a solute that cannot be distinguished from water by the membrane. Thus as $\sigma_s \to 0$, a concentration difference for that solute will cause no volume flow; the effective osmotic pressure approaches zero. Let me illustrate this important point with some results obtained by Durbin [9] in studies of volume flow across a dialysis membrane caused by concentration differences of various solutes. Fig. 4 shows relative volume flow per unit concentration difference across this membrane for three solutes. Albumin causes a maximal flow, sucrose a much smaller flow and urea almost no flow. Thus $\sigma = 1$ for albumin and $\sigma \to 0$ for urea.

This concept can be utilized in a relatively simple manner to suggest a mechanism to account for coupling between solute transport and water movement in epithelia such as intestine. The idea is based on the realization that epithelia are complex structures having two or more barriers in a series array. Fig. 5 shows a simple scheme that could explain the ability of active solute transport to cause solvent flow in a system involving two membranes arranged in series and separated by a closed compartment. For simplicity, we will assume that only a single solute is present, initially at equal concentrations in all compartments. The solute is actively transported across membrane α into the central compartment, causing an increase in concentration in the central region relative to the two external solutions. The net effect of the concentra-

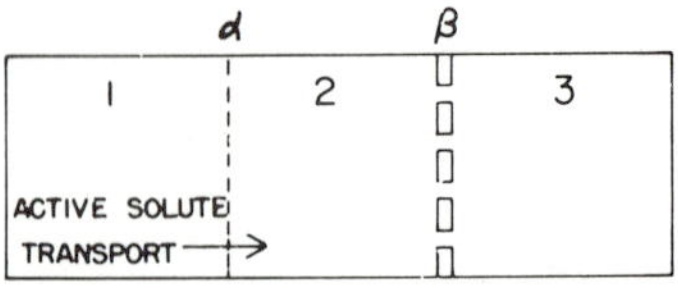

Fig. 5. A scheme for coupling active solute transport to volume flow. Reproduced from Curran [3].

tion difference generated by the active transport depends on the properties of the two membranes, particularly their reflection coefficients for the solute. Assume that membrane α is relatively tight and has $\sigma \to 1.0$ for the solute. The effective osmotic pressure across this membrane will be relatively high and will cause fluid to flow into the central compartment from compartment 1. If membrane β is rather loose with $\sigma \to 0$, the solute concentration difference will not cause significant flow from compartment 3 to compartment 2. However, since the central compartment is closed and cannot expand, the entrance of fluid will generate a hydrostatic pressure in that compartment and this pressure will cause a flow of fluid mainly across the loose membrane, β. The net result of the operation of the system is thus a net flow of fluid from compartment 1 to compartment 3 in the absence of an overall driving force for volume flow (an osmotic pressure difference). The flow is caused by, or coupled to, the active solute transport and the coupling is determined by a particular set of structural features. An oversimplified analysis indicates that in this system, net volume flow would be given by

$$J_{\mathrm{v}} = \frac{L_{\mathrm{p}\alpha} L_{\mathrm{p}\beta}}{L_{\mathrm{p}\alpha} + L_{\mathrm{p}\beta}} (\sigma_\alpha - \sigma_\beta)\, \Delta\pi_\alpha \,. \tag{3}$$

Thus, the fundamental feature needed to give coupling is that $\sigma_\alpha \neq \sigma_\beta$.

In order to see whether such a system would really work as predicted, we constructed a simple artificial model using dialysis tubing for one membrane and a sintered glass disc for the other [11, 12]. Active transport could be simulated by simply introducing a relative high solute concentration into the central compartment. Fig. 6 shows one simple set of results obtained with this system. The solute was sucrose; it was present at 0.5 M in compartment 2 and quite low in compartment 3. We have plotted quasi-steady state volume flow from 1 to 3 as a function of sucrose concentration in 1. Clearly, volume flow can be demonstrated against substantial differences in osmotic pressure and the behavior is really quite similar to that shown earlier for biological systems (figs. 1 and 2).

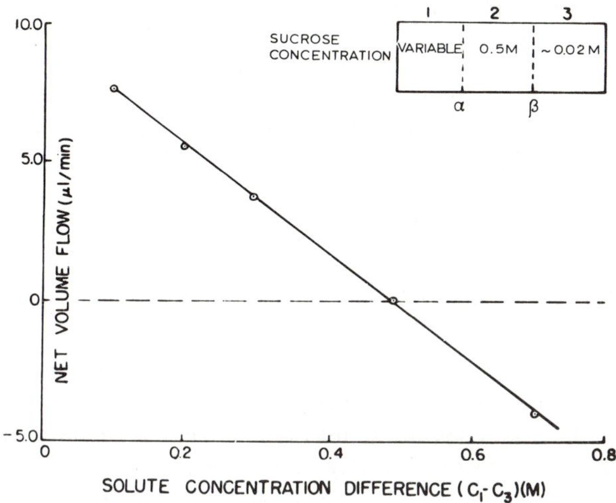

Fig. 6. Behavior of model system for solute-solvent coupling. Reproduced from Curran [3].

We might now ask how the structural features of this hypothetical model could be related to the actual biological situation in intestine or other epithelia. Fig. 7 is an attempt to illustrate possible relations. The lateral and/or serosal membrane of the cell would be equivalent to membrane α since this membrane is usually thought to contain the active Na transport system for

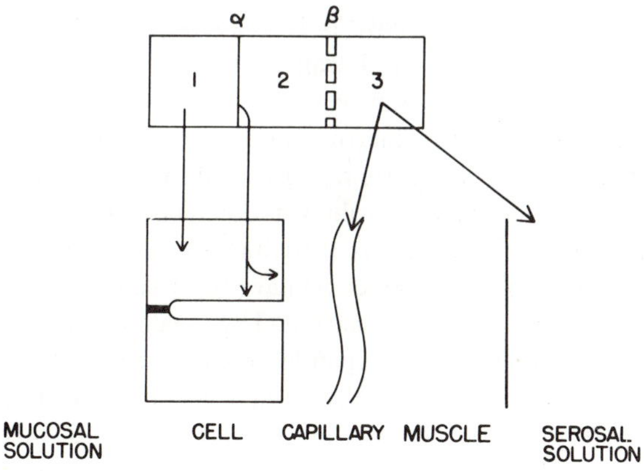

Fig. 7. Possible relationship between aspects of the model system and structural features of the intestine.

epithelia such as intestine [7]. This transport system extrudes solute into the lateral and subcellular spaces which could, in fact, play the dual role of the central compartment and membrane β. Note that the second membrane is not really necessary; the requirement is for a structure that prevents rapid diffusion of the actively transported solute. Alternatively the capillary membrane or perhaps the basement membrane could serve as membrane β. In this scheme, compartment 1 would be the cell interior and compartment 3 the plasma *in vivo* or the serosal solution *in vitro*.

An interesting and important aspect of this view is the suggestion that the lateral intercellular spaces may be involved in the coupling phenomena. The first clear evidence in favor of this possibility came from the electron microscopic studies of Kaye et al. [13] and of Tormey and Diamond [14] on the gall bladder. They examined these spaces carefully in gall bladders that were transporting fluid and solute at maximal rates and in ones in which fluid transport was inhibited by a variety of techniques. In non-transporting bladders, the lateral spaces were nearly obliterated and adjacent cellular membranes were close to one another. The average width of the lateral spaces was 0.02 micron. In bladders transporting fluid at high rates, the lateral spaces were markedly distended and had an average width of about 1 micron. These observations were interpreted as indicating that the major route of fluid movement is through the lateral intercellular spaces and led to the suggestion that these spaces play a specific role in the phenomenon of solute-solvent coupling, at least in this particular epithelial system. Tormey et al. [15] have recently obtained further evidence indicating that these spaces are indeed a major route for transfer of various substances across the gall bladder. Similar but less detailed observations have been made in other epithelia but distension of lateral spaces during fluid transport does not seem to be a uniform observation in all epithelia. However, the ability of the spaces to distend could be influenced by many factors and failure to observe distension does not necessarily mean the spaces are not involved in fluid transport.

Starting from the idea of a role for the lateral spaces in the coupling phenomenon, Diamond and Bossert [16] have developed a more detailed and specific version of the simple model we have been discussing. They called this the standing osmotic gradient mechanism. The concept is illustrated schematically in fig. 8, which shows an expanded view of the lateral space between two adjacent epithelial cells. The idea is that active solute transport elevates the solute concentration in the space above that in the cell causing an osmotic pressure difference. Solvent then flows into the space and since the volume of the space is constrained, the fluid is forced out the open end of the channel. Again the requirement is for a structural arrangement that prevents a rapid

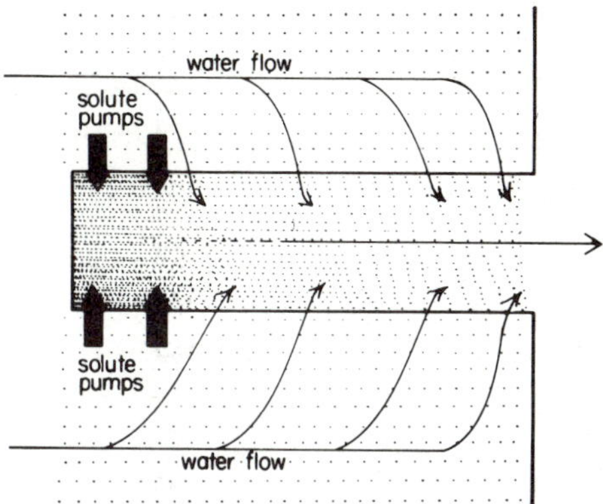

Fig. 8. The standing osmotic gradient system. Reproduced from Diamond and Tormey [17].

dissipation of the solute concentration difference via diffusion; that is, solute must be transported into a region from which it cannot diffuse away rapidly. As indicated by the stipling in this figure, solute concentration in the channel would decrease with distance, primarily due to continued entrance of fluid and under proper conditions, an essentially isotonic fluid would emerge from the open end of the channel. Diamond and Bossert [16] have carried out a detailed mathematical analysis of the behavior of this type of system. They have shown how the tonicity of the emergent fluid depends on various parameters of the system such as channel length and the region of solute input. The situation is relatively complex and I will not attempt to describe their results in any detail. (See Diamond [18] for a summary and also Segel [19] for further discussion.) In most cases, the calculated results are in accord with one's intuition about the behavior of such a system. For example, if other factors are constant, decreasing the length of the channel will make the emergent fluid more hypertonic.

Diamond and Bossert [20] have also shown that this type of system can work in the opposite direction. If solute is actively transported from the space into the cell, a standing osmotic gradient can be generated with an orientation opposite to that shown here and a coupled solvent movement in the opposite direction would occur. Fig. 9, taken from Oschman and Berridge [21], illustrates this type of system and shows how one could use it to account for secretion of fluid from the blood stream. Note that they have also suggested here a standing gradient system in the microvillus structure, a feature about which there is no direct evidence at present. There are also

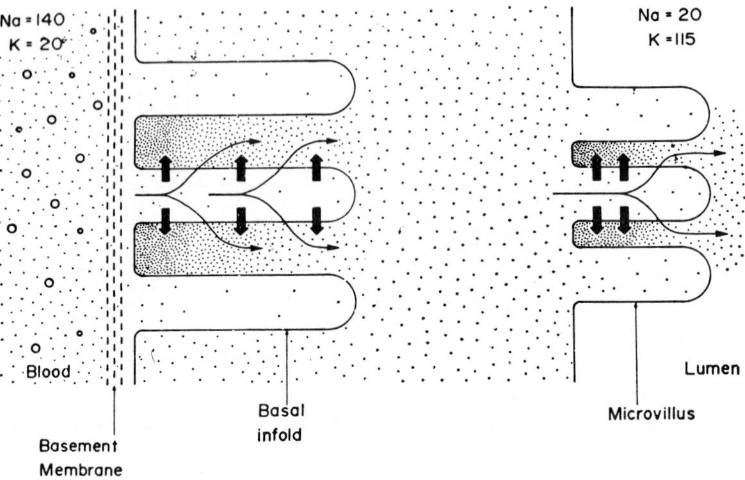

Fig. 9. Additional aspects of a standing osmotic gradient that might be involved in fluid secretion from blood to lumen. Reproduced from Oschman and Berridge [20].

alternative explanations for coupling during secretion. Secretory systems often involve structures such as the tubules in gastric mucosa or intracellular canniculi in parietal cells and other secretory cells. Transport of solute into the lumen of a tubule or a canniculus could also generate an appropriate osmotic gradient and give rise to a coupled secretion of fluid [22]. It is important to realize that in all these systems fluid transport takes place across the whole cell layer and that entrance of solvent into the cell must be considered as well as its exit at the opposite side. This problem has received little specific attention.

We might now ask whether there is any additional evidence to support these kinds of hypotheses to account for solute-solvent coupling. To the best of my knowledge, the only direct information available comes from the study of Wall et al. [23] on insect rectal pads. Using micro techniques, they were able to obtain fluid from intercellular spaces in this tissue during the process of fluid absorption. The fluid collected from the spaces was consistently hyperosmotic to the fluid in the lumen by an average of about 130 mOsmoles/liter, as predicted by the hypothesis we have been discussing.

There are a variety of other observations and theoretical considerations relating to this type of phenomena that could be discussed. However, my main aim has been to present a general outline of the problem rather than a detailed discussion and I think the above considerations, although by no means complete, are probably sufficient for that purpose. I have tried to provide some idea of the development of the main concept that we now use to explain coupling of solute transport to water movement. I should stress two things in conclusion. First, the coupling is thought to be determined by certain structural arrangements in relatively complex tissues. Second, the whole concept is a hypothesis; it should not be considered entirely proven even though it is attractive and appears to be capable of explaining a considerable variety of experimental observations.

References

[1] R.T. Whitlock and H.O. Wheeler, J. Clin Invest. 43 (1964) 2249.
[2] D.S. Parsons and D.L. Wingate, Biochim. Biophys. Acta, 66 (1961) 441.
[3] P.F. Curran, Federation Proc. 24 (1965) 993.
[4] O. Kedem, in: Membrane Transport and Metabolism, ed. A. Kleinzeller and A. Kotyk (Prague, Czech. Acad. Sci., 1961) p. 87.
[5] P.F. Curran and A.K. Solomon, J. Gen. Physiol. 41 (1957) 143.
[6] P.F. Curran, J. Gen. Physiol. 43 (1960) 1137.
[7] S.G. Schultz and P.F. Curran, Handbook of Physiology, Sect. on Alimentary Canal (Washington, D.C., Am. Physiol. Soc., 1968) Vol. 3, p. 1245.
[8] D.W. Powel and S.J. Malawer, Am. J. Physiol. 215 (1968) 49.
[9] R.P. Durbin, J. Gen. Physiol. 44 (1960) 315.
[10] A.J. Staverman, Rec. Trav. Chim. 70 (1951) 344.
[11] P.F. Curran and J.R. McIntosh, Nature, 193 (1962) 347.
[12] J.T. Ogilive, J.R. McIntosh and P.F. Curran, Biochim. Biophys. Acta 66 (1963) 441.
[13] G.I. Kaye, H.O. Wheeler, R.T. Whitlock and N. Lane, J. Cell. Biol. 30 (1966) 237.
[14] J.M. Tormey and J.M. Diamond, J. Gen. Physiol. 50 (1967) 2031.
[15] J.M. Tormey, A.P. Smulders and E.M. Wright, Federation Proc. 30 (1971) 422, Abs.
[16] J.M. Diamond and W.H. Bossert, J. Gen. Physiol. 50 (1967) 2061.
[17] J.M. Diamond and J.M. Tormey, Federation Proc. 25 (1966) 1458.
[18] J.M. Diamond, Federation Proc. 30 (1971) 6.
[19] L.A. Segel, J. Theoret. Biol. 29 (1970) 233.
[20] J.M. Diamond and W.H. Bossert, J. Cell. Biol. 37 (1968) 694.
[21] J.L. Oschman and M.J. Berridge, Federation Proc. 30 (1971) 49.
[22] S. Ito, J. Biophys. Biochem. Cytol. 11 (1961) 333.
[23] B.J. Wall, J.L. Oschman and B. Schmidt-Nielsen, Science, 167 (1968) 3924.

DISCUSSION REMARK TO THE PAPER OF CURRAN

E. HEINZ

A Dr. Hogben pointed out previously, the workers on gastric acid formation have already delt with many of the problems now under discussion, long before these drew the interest of other specialists. This also applies to the present topic: solute-solvent interaction, since about ten or more years ago we have pondered about the mechanism by which water moves into the stomach during acid secretion. We then came up with a model which, in my opinion, is in principle very similar to those put forward and experimentally tested much later. This may be illustrated by the following slide (fig. 1) taken from a publication in 1960. We assume two compartments, R and C, which are connected by a long thin channel (B). We further assume that a solute, in this case HCl, is actively transported from compartment R to compartment A. The resulting increase of osmotic pressure in A then causes the movement of water into the same compartment. If the walls of this medium compartment are sufficiently rigid, the hydrostatic pressure in it will rise. If, furthermore, the combining channel is narrow enough to prevent appreciate diffusion of HCl towards C and, on the other hand, wide enough to prevent an

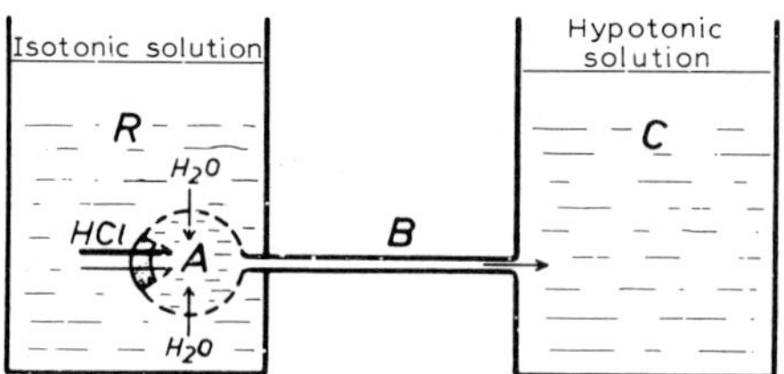

Fig. 1. Model of the movement of water against the gradients of its vapour pressure. (From: E. Heinz, Klinische Physiologie I (1960) 184). For details see text.

osmotic pressure difference between A, the hydrostatic pressure gradient will continuously drive solution from A to C. The net result is: water moves from a place with higher osmolarity (R) to a place with lower osmolarity (C), without a special H_2O-transport system. I believe that the principles underlying this old model account for the coupling between solvent flow and solute transport just as well as do the more recent models.

INTESTINAL TRANSPORT OF SUGARS. THE FUNCTION OF THE SUGAR CARRIER *

T.Z. CSÁKY

Department of Pharmacology, University of Kentucky, College of Medicine, Lexington, Kentucky, U.S.A.

The first structural model of the plasma membrane, the Danielli-Davson structure [1] consisting of a bimolecular layer of lipid and protein, was developed from studying its functional proporties. This model was proposed years before it could be verified in essence with the proper histological techniques using the electron-microscope. In spite of some elegant and highly sophisticated work which necessitated slight modifications and variations, the Danielli-Davson model still explains adequately the basic permeability properties of the plasma membrane. [2]

The most prominent function of the plasma membrane is to mediate the communication between the cell and its environment. The communication does not take place indiscriminately over the entire surface of the membrane, rather through highly specialized, most likely proteinaceous patches, receptors, with which the reacting substrates temporarily combine. The reacting substrate may be a nutrient molecule to be transported across the lipid membrane or a drug molecule which relais a chemical message from the environment to stimulate the cell to react.

In the followings I should like to offer some experimental facts and speculations concerning one of these receptors namely the sugar carrier. Over the years extensive work was conducted in my laboratory on the transport of sugars in epithelial cells primarily those of the intestine.

As mentioned above, from a functional point of vies, the plasma membrane can be considered essentially as a lipid barrier. Consequently, in order

* The research work reported in this paper was supported by grants from the U.S. Public Health Service.

to penetrate across the membrane, a substance has to be able to interact with the lipid i.e. it has to be fat soluble. This was suggested already at the turn of the century by Overton [3] and fully demonstrated experimentally by Collander [4, 5] and his co-workers. Thus a substance has to diffuse first through a watery layer and then, in order to reach the inside of the cell, across a lipid layer. Accordingly, Fick's law of diffusion has to be written by two separate equations:

Equation (1) expresses the diffusion of the substance in the unstirred watery extracellular space

$$\frac{\mathrm{d}n}{\mathrm{d}t} = DA\,\frac{\mathrm{d}c}{\mathrm{d}x} \tag{1}$$

where, dn is the number of molecules crossing the area A in the time dt when the concentration difference is dc over a distance dx, D being the diffusion constant.

Equation (2) describes the rate of passage across the lipid layer

$$\frac{\mathrm{d}n}{\mathrm{d}t} = D'Aq\,\frac{\mathrm{d}c}{\mathrm{d}x} \tag{2}$$

where q is the lipid-water distribution coefficient expressing the relative fat solubility of the substrate.

The overall permeation of a given solute across the plasma membrane can be characterized by the permeability equation (3)

$$\frac{\mathrm{d}n}{\mathrm{d}t} = PA\,\mathrm{d}c \tag{3}$$

where the permeability coefficient P includes not only the thickness of the membrane but other physical properties, such as the lipid/water distribution coefficient q of the solute. Consequently, the higher the q the higher the value of the permeability coefficient P. In case of very polar solutes, such as glucose, the value of q is infinitesimally small; consequently the value of P is infinitesimally small; i.e. such substances do not cross membranes. Yet by combining with the appropriate proteinaceous receptor (the carrier) on the cell surface, the sugar can penetrate the cell, thus permeate across the lipid layer. Clearly because of the carrier-mediation the value of P for glucose increases, consequently the value of q has to increase simultaneously. Thus it is logical to assume that the function of the carrier consists of rendering a water soluble substrate, such as glucose or other sugars, fat soluble. Physicochemically speaking the carrier enables the highly water-soluble sugar molecule to break its hydrogen bondings with the water and enter the lipid layer.

The assumption that the primary function of the carrier is to increase q is a logical deduction without solid experimental proof. The experiments described below were undertaken in order to test the validity of this assumption. We were looking for a substance which forms hydrogen bonds with sugar more avidly than water and at the same time is also fat soluble, assuming that such a substance may act as a model carrier for glucose. Dimethylsulfoxide (DMSO)

$$O{=}S\begin{matrix}\diagup CH_3\\ \diagdown CH_3\end{matrix} \rightleftharpoons O^{-}{-}S^{+}\begin{matrix}\diagup CH_3\\ \diagdown CH_3\end{matrix}$$

is such a substance. [6]

Experiments were carried out in rats using a procedure developed in our

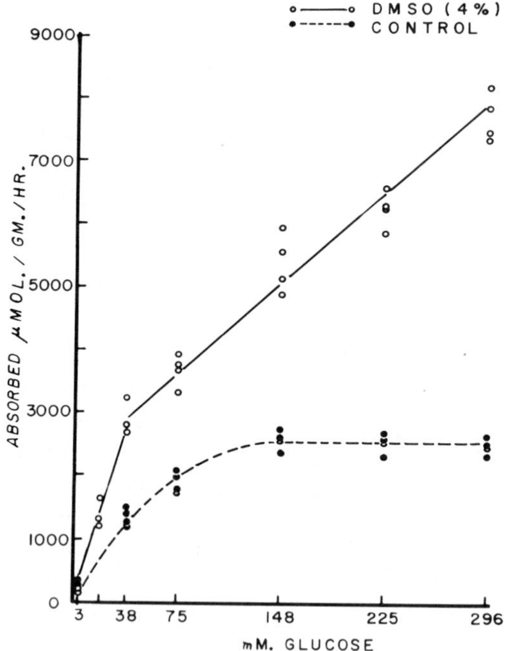

Fig. 1. Intestinal absorption kinetics of glucose with and without dimethylsulfoxide. (DMSO).

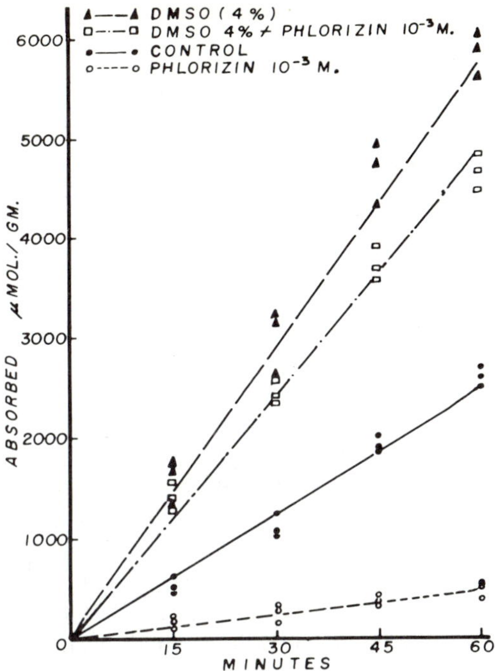

Fig. 2. Phlorizin-inhibited rate of absorption of glucose from the intestine of rat. No appreciable inhibition by phlorizin in the presence of DMSO.

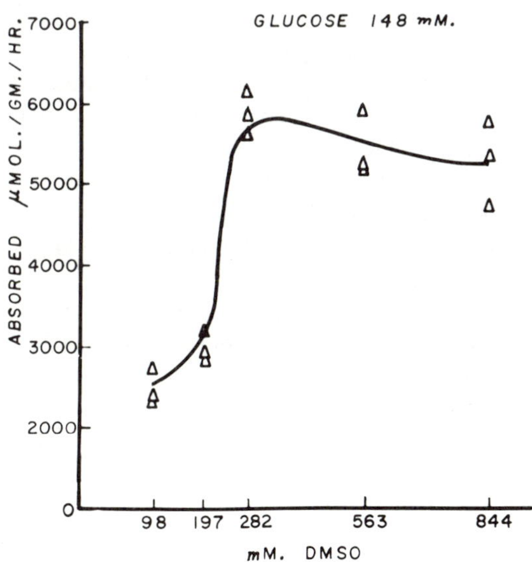

Fig. 3. Effect of relative DMSO concentration on the intestinal glucose transport. Maximum stimulation is achieved at a glucose:DMSO ratio of 1:2.

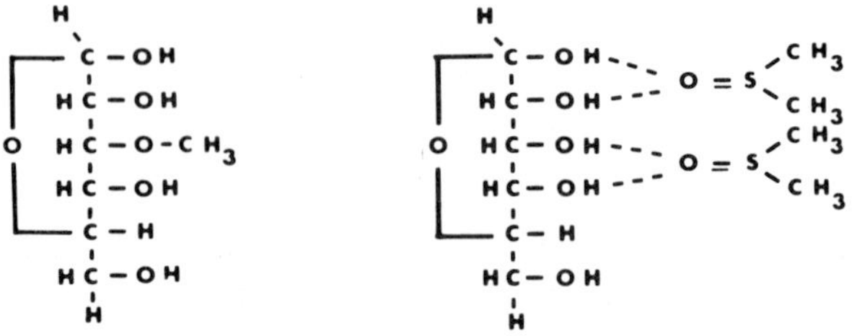

```
     H                          H
      \                          \
  ┌───C - OH               ┌─────C - OH- - -          ,CH3
  │   |                    │     |          O = S
  │  HC - OH               │    HC - OH- - -        `CH3
  │   |                    │     |                  ,CH3
  O  HC - O - CH3          O    HC - OH- - -  O = S
  │   |                    │     |                  `CH3
  │  HC - OH               │    HC - OH- - -
  │   |                    │     |
  └───C - H                └─────C - H
      |                          |
     HC - OH                    HC - OH
      |                          |
      H                          H
```

Fig. 4. Hypothetical glucose – DMSO complex (assumed to be hydrogen bonded).

laboratory [7]. In this a loop of intestine of an anesthetized animal is perfused *in situ* with an isosmotic glucose-Na_2SO_4 solution and the rate of disappearance of glucose from the lumen measured. When DMSO was added to the perfusate the rate of disappearance of sugar increased and at the same time the saturation type kinetics of glucose absorption, typical for a carrier-mediated transport, was completely eliminated and converted into a straight diffusion type kinetics. (fig. 1). [8] A further indication of the absorption of sugar without carrier-mediation in the presence of DMSO came from experiments with phlorizin. Phlorizin is considered a specific inhibitor of aldose transport by competing with the sugar molecule for the carrier. No inhibition of the sugar transport was observed when DMSO was present (fig. 2).

Fig. 3 shows that quantitatively two mols of DMSO have to be present for each mol of glucose for the optimal effect.

These results indicate to us that in the presence of DMSO the function of the intestinal sugar carrier was by-passed, or in other words DMSO itself acted as a model intestinal sugar carrier. It was assumed, that DMSO forms a complex with the sugar molecule. In such a hypothetical complex four polar hydroxyl groups in the glucose molecule would be masked (fig. 4) The assumption of such a hypothetical complex formation was strongly supported by the observation that DMSO affects the transport of glucose and galactose but not that of 3-0-methylglucose. In the case of the latter only one pair of adjacent OH groups is free for complexing.

The hypothetical complex between DMSO and glucose is a loose one, probably hydrogen-bonded. In order to test the hypothesis that DMSO acts as a model sugar carrier fat-soluble glucose complexes which are stable, covalent bonded compounds were selected and the kinetics of their intestinal transport examined. Acetone can react under proper conditions with the glucose molecule forming covalently bonded isopropylidene (Ip) compounds of the following basic structure [9]

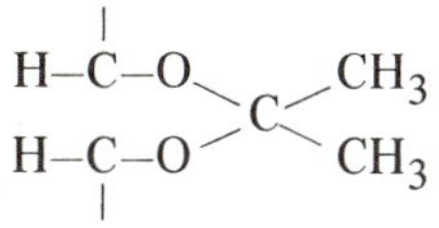

$$\begin{array}{l} \quad\;| \\ H{-}C{-}O\diagdown\;\;\diagup CH_3 \\ \qquad\quad\; C \\ H{-}C{-}O\diagup\;\;\diagdown CH_3 \\ \quad\;| \end{array}$$

Depending whether one or two molecules of acetone react with each sugar molecule 1–2, mono-Ip-glucose or 1–2,5–6, di-Ip-glucose is formed. The former posseses three, the latter one free hydroxyl groups. Preliminary experiments in the above described *in vivo* rat preparation indicated that the di-Ip-glucose is transported in the intestine considerably faster than glucose (about the same rate as glucose in the presence of DMSO), while the intestinal transport rate of the mono-Ip compound was below that of glucose.

The results of such model experiments allow speculation as to what the carrier accomplishes with glucose. It appears that, by combination with the carrier, the lipid solubility of the glucose increases but not quite to the same extent as with DMSO or as in the case of the di-Ip compound. Consequently, one may hypothesize that, after glucose has combined with the carrier, still more than one hydroxyls are left free to form hydrogen bonds with the water. However, the number of free hydroxyls must be less than three because the mono-Ip compound was transported at a slower rate than glucose. Moreover, these experiments justify the assumption that the essential function of the carrier is to increase the q (the fat-water distribution coefficient).

If the basic function of the carrier is to increase the value of P by increasing q (see eq. (3)) then the permeability eq. holds as long as free carrier sites are available. Consequently, the rate of transport is directly proportional to the concentration difference and as the latter approaches zero, the rate of the carrier mediated net transport also approaches zero. This is the case of a simple carrier-mediate equilibrating transport. The function of the carrier however is also involved in the uphill, concentrative transport which does not obey Fick's law.

Most, if not all, concentrative transport mechanisms are sodium-dependent or coupled to the transport of sodium. [10] The discovery of the ionic dependence of nonelectrolyte transports has raised the logical question: is every carrier-mediated transport sodium dependent, in other words: is the carrier inherently sodium dependent? To answer this question it is practical to turn to a simple model of carrier-mediated equilibrating transport such as in the red blood cell. We found no difference in the rate of carrier-mediated sugar entry into human or rabbit red blood cells suspended in sodium containing medium and in media without sodium [11] (figs. 5 and 6).

Subsequently it was found in other laboratories that the sugar entry in the

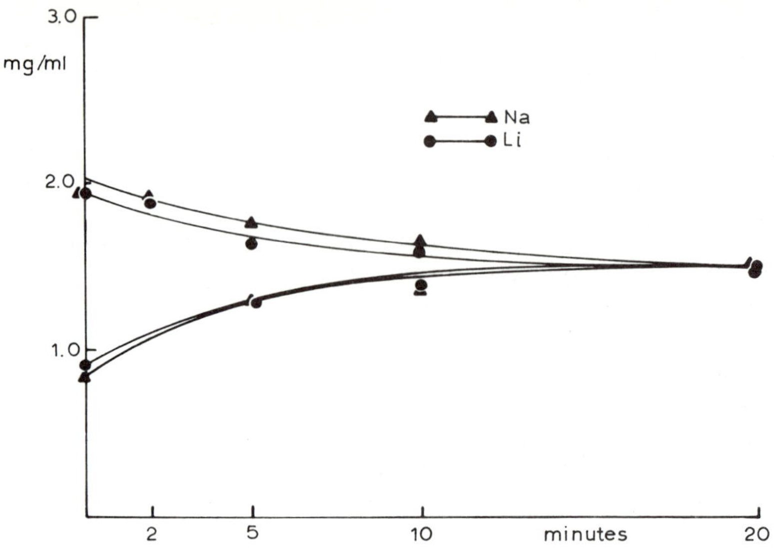

Fig. 5. Entry of 3-0-methylglucose into human red blood cells at 10°C suspended in a medium containing sodium or lithium, respectively. Top curves: decrease of the sugar concentration in the medium. Bottom curves: increase of the concentration of sugar in the red blood cells.

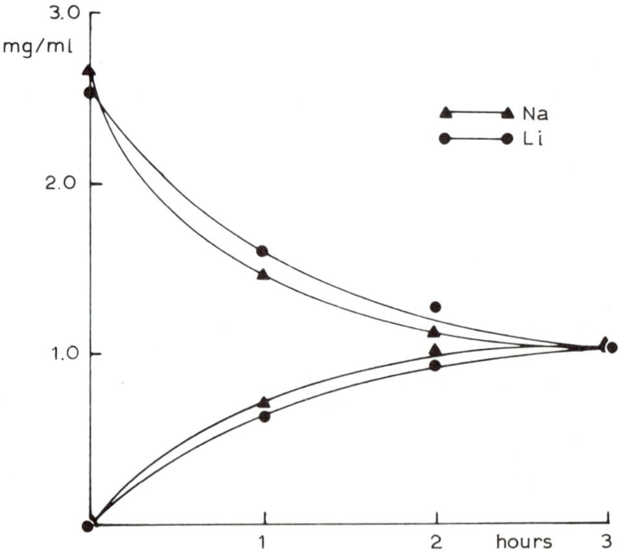

Fig. 6. Outflux of 3-0-methylglucose from the red blood cells (at 37°C) of the rabbit suspended in medium with or without sodium. Top curves: concentration of sugar in the red blood cells. Bottom curves: concentration of sugar in the plasma.

muscle, which is a carrier mediated equilibrating process, is also independent of sodium [12].

What is the situation then in a carrier-mediated system which is also capable of uphill transport? In such a system one has to distinguish between the simple fact of entry and the rate of entry, i.e. the time needed to reach a tissue-to-medium (*T*/*M*) value of 1.0. In the accumulating transport *T*/*M* values considerably higher than unity are rapidly obtained; clearly in such transport a *T*/*M* equilibrium will be achieved more rapidly than in a system which obeys Fick's law.

Studying polar cells such as the epithelium of the intestine or kidney particularly differentiated for uphill transport one has to observe the significant methodological point that these cells have two membranes, the brush border and the basal membrane, distinctly different both in their morphological and permeability properties. By simply immersing these cells or tissue slices in the medium which surrounds both membranes simultaneously one is likely to obtain permeability values which are too complex for a meaningful analysis and distorted as far as the physiological picture is concerned. With this in mind we examined the question: is sodium necessary for the carrier-mediated entry of sugars across the brush border into the intestinal epithelial cell when accumulation is not involved. The experiments were carried out in the isolated small intestine of the bullfrog which was filled, as a sausage, with a solution containing labelled 3-0-methylglucose in isosmotic media of varying ionic composition, then suspended in a solution of similar composition but without the sugar, After three hours at room temperature the intestinal sausage was opened, the mucosa gently blotted with filter paper, scraped off and analyzed. The *T*/*M* was calculated. By definition a *T*/*M* value of 1.0 means complete equilibrium between the intra- and extra-cellular space while values of 0.3 to 0.4 mean equilibrium between the medium and extracellular tissue space only. The latter was determined with inulin. Fig. 7 shows the results. It can be seen that without sodium in the medium the sugar penetrated the intracellular space. That the entry was mediated by the carrier is shown by the inhibition with phlorizin or large excess of glucose.

Thus, the carrier-mediated equilibrating transport in the intestinal epithelium is not dependent on the presence of sodium in the medium. Is it necessary then to assume the existence of two separate carriers: one for the equilibrating and another for the accumulating transport, the latter only being sodium dependent? This question is difficult to answer. No doubt the overall rate of transport in a system capable of accumulation is strongly stimulated by sodium ions, but whether the site of the stimulation is the carrier or another component of the transport system is still an open question. One can

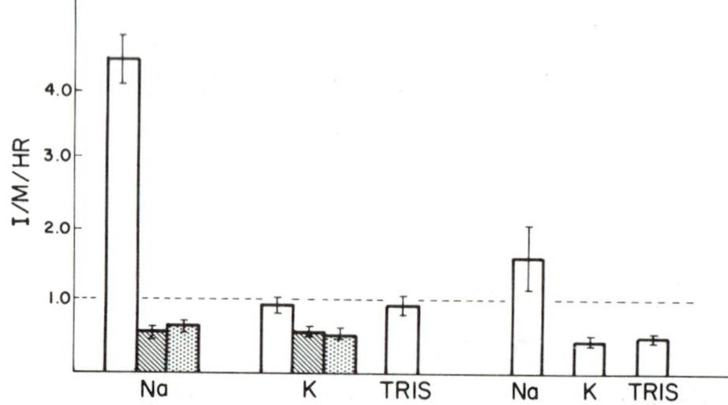

Fig. 7. Entry of labelled 3-0-methylglucose into the epithelial layer of the small intestine of the bullfrog. The concentration ratio between the intracellular space (I) and the medium (M) is plotted on the ordinate. Open columns: entry without inhibitors: cross hatch columns: entry in the presence of 10^{-3} M phlorizin; dotted columns: entry in the presence of large excess of glucose. Experiments shown in the first three sets of columns were performed in such a way that the polarity of the epithelial cells was observed. The last three sets of columns depict experiments in which rings of tissue were suspended in a medium which surrounded both the serosal and mucosal area. In the presence of sodium intracellular sugar-accumulation occurs, however when phlorizin or large excess of glucose is present only the apparent extracellular space is filled. In the absence of sodium the intracellular space reaches equilibrium with the extracellular through a transport mechanism mediated by a carrier as indicated by the inhibition by phlorizin or large excess of glucose. If the polarity of the cell is not observed the entry rates are much distorted.

readily measure the overall transport kinetics and from such measurements phenomenological constants can be calculated; but whether these constants assess accurately the molecular reaction between the substrate and the carrier is still a matter of speculation.

It is a popular belief that sodium activates the transporting carrier which then acts as a sugar pump. Without being dogmatic, I should like to call attention to a few experimental findings which would favor a "unitarian view", namely that there is only one type of carrier for both equilibrating and accumulating transport:

a) both carriers display similar if not identical substrate specificities

b) both are strongly inhibited and quantitatively to the same extent by phlorizin.

c) Until recently all carbohydrates were strictly divided into two major groups: those which were actively transported in the intestine and those which were not. Only sugars which had a restricted specific configuration could qualify for pumping although it was realized that sugars with different

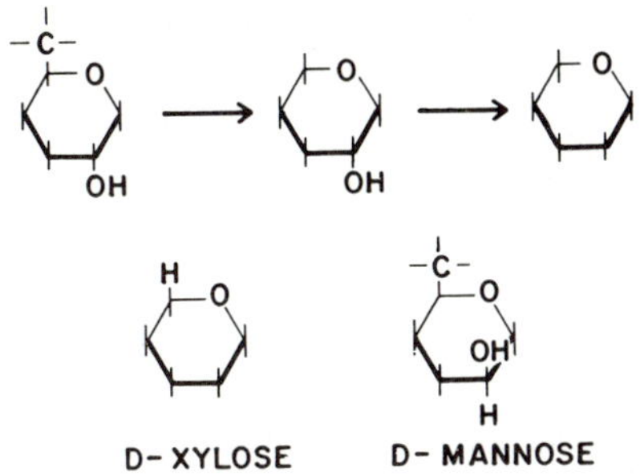

D-XYLOSE D-MANNOSE

Fig. 8. Modification of the configuration of sugar thought to be essential for active transport. The configuration of D-xylose and D-mannose are shown in the lower line.

structures could still be substrates for carrier mediated, non-active transport. [13, 14] The postulated specific configuration is shown on fig. 8. This configuration had to be modified when we discovered that D-xylose, which was listed among the non-actively transported species, is also actively transported. [15, 16] Another adjustment was necessary when it was demonstrated that D-mannose is also actively transported. [17] Thus at the present it is safe to assume that a sugar to be actively transported has to display the basic structure of a pyranose and it is not unreasonable to assume that eventually, as the experimental tools improve, accumulative transport may be demonstrated for every sugar which can combine with the specific carrier.

Going a step further the question may be raised: is the carrier primarily responsible for the accumulative sugar transport or is there another mechanism, the pump, or the transducer, which converts chemical into osmotic energy. After careful examination of the available experimental evidence I have recently proposed the theory that, at least in the intestine, there exists only one pump, namely the sodium pump and all other uphill transports are energized through this single mechanism. [10] This hypothesis would assign to the carrier a function analogous to that of ionophoric cyclyc peptides which render polar ions fat soluble enabling them to cross the lipid barrier of the membranes. Experiments presented in this paper favor such an analogy.

References

[1] J.F. Danielli and H. Davson, A contribution to the theory of permeability of thin films. J. Cell. Physiol. 5 (1935) 495.

[2] R.W. Hendler, Biological membrane ultrastructure. Physiol. Rev. 51 (1971) 66.

[3] E. Overton, Ueber die allgemeinen osmotischen Eigenschaften der Zelle, ihre Bedeutung für die Physiologie. Vierteljahrschrift der Naturforschenden Gesellschaft in Zürich, 44 (1899) 88.

[4] R. Collander and H. Bärlund, Permeabilitäts-studien an *Chara ceratophylla.* Acta Botan. Fennica 11 (1933) 1.

[5] R. Collander, The permeability of plant protoplasts to small molecules. Physiol. Plantarum 2 (1949) 300.

[6] P.G. Sears, W.D. Siegfried and D.E. Sands, Viscosities, densities and related properties of solutions of some sugars in dimethylsulfoxide. J. Chem. Eng. Data 9 (1964) 261.

[7] T.Z. Csáky and P.M. Ho, Intestinal of D-xylose. Proc. Soc. Exp. Biol. Med. 120 (1965) 403.

[8] T.Z. Csáky and P.M. Ho, Effect of dimethylsulfoxide on the intestinal sugar transport. Proc. Soc. Exp. Biol. Med. 122 (1966) 860.

[9] E.F. Armstrong and K.F. Armstrong, The Carbrohydrates (London, Longmans Green and Company, 1934) 5th ed. p. 85–87.

[10] T.Z. Csáky, Physiological considerations of the relationship between intestinal absorption and electrolytes and nonelectrolytes. in: Intestinal Transport of Electrolytes, Amino Acids and Sugars, eds. W. McD. Armstrong and A.S. Nunn, Jr. (Charles C. Thomas, Springfield, 1971).

[11] T.Z. Csáky, A possible link between the active transport of electrolytes and nonelectrolytes. Fed. Proc. 22 (1963) 3.

[12] D.M. Kipnis, The role of Na^+ and K^+ on -amino-isobutyric acid transport in striated muscle, in: eds. B. Chance, R.W. Estabrook, and J.R. Williamson (New York, Academic Press, 1965) Control of Energy Metabolism.

[13] R.K. Crane, Intestinal absorption of sugars Physiol. Rev. 40 (1960) 789.

[14] T.H. Wilson, Intestinal Absorption (Philadelphia, W.B. Saunders, 1962).

[15] T.Z. Csáky and U.V. Lassen, Active intestinal transport of D-xylose. Biochim. Biophys. Acta 82 (1964) 215.

[16] U.V. Lassen and T.Z. Csáky, Active transport of D-xylose in the isolated small intestine of the bullfrog, J. Gen. Physiol. 49 (1966) 1029.

[17] T.Z. Csáky and P.M. Ho, Active transport of D-Mannose in the small intestine. Life Science 5 (1966) 1025.

SOME OBSERVATIONS FROM AMINO ACID TRANSPORT*

Halvor N. CHRISTENSEN

The Department of Biological Chemistry, The University of Michigan, Ann Arbor, Michigan, U.S.A.

1. Juxtaposition and chemical linkage between the cosubstrates of linked transport

In this presentation I want to discuss the geometric relations assumed at transport receptor sites between the amino acid and Na^+ cosubstrates. The sketch of fig. 1 shows the recognition sites for the two side by side, although we had no basis for such a placement of the two cosubstrates during the years 1952 to 1958 when Riggs and I fashioned the ion-gradient hypothesis for the energization of amino acid transport [1–3]. Now we believe we are able to show within an Angstrom unit where the alkali-metal ion binds, in relation to the amino acid molecule.

This ability began with the recognition of a certain degree of equivalence at the binding stage between Na^+ and a cationic group on the sidechain of the amino acid. This equivalence was first seen for a transport system for basic amino acids, which system can in general be inhibited by certain neutral amino acids plus Na^+, provided that the neutral amino acid has a suitable structure [4–6]. Subsequently we noted, for a Na^+-requiring neutral amino acid system we call *ASC*, that arginine and the like will produce in converse a Na^+-independent inhibition [7–10]. Table I shows the relations between these two situations. It should be emphasized that the preference for the defining substrates of each of these systems is fifty-fold, and that we have rather overwhelming evidence for their separate operation.

* The experiments desribed here that derive from the author's laboratory have been supported in part by Grant HD-01233 from the National Institute for Child Health and Human Development, The National Institutes of Health, U.S.P.H.S.

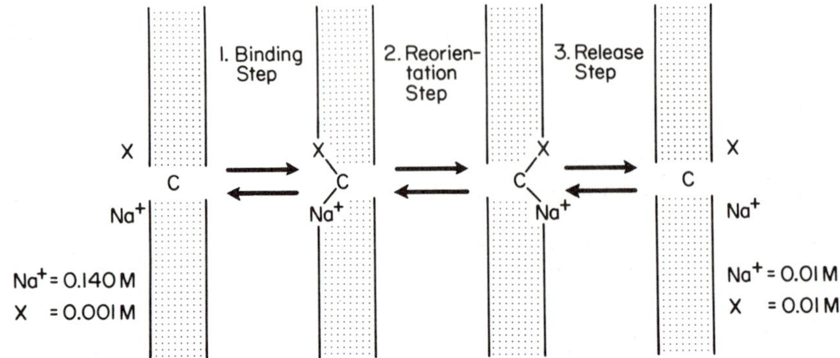

Fig. 1. A representation of linked transport between Na^+ and an organic metabolite, X, corresponding to the apparent operation of System A of the Ehrlich ascites tumor cell. The energy inherent in the steady-state electrochemical gradient of Na^+ is in this case visualized as having concentrated X into the cell (right) by ten-fold. The representation of the linkage of X and Na^+ in juxtaposition at the carrier site here, because it anticipated the evidence summarized in this report, was arbitrary. Reproduced with permission from H.N. Christensen and R.A. Cellarius, Introduction to Bioenergetics (a learning program), (W.B. Saunders Co., Philadelphia, 1972).

Table 1

Interactions between basic amino acids and (a neutral amino acid plus Na^+) occur in two different transport systems

System	Preferred substrate	Inferior* reaction with	Leads to exchange between these?	Sobriquet
$\overset{+}{Ly}$	basic amino acids	neutral amino acid + Na^+	yes	Na^+-dependent inhibition of a Na^+-independent system
ASC	certain neutral amino acids + Na^+	basic amino acids	no	Na^+-independent inhibition of a Na^+-dependent system

* The K_i values for the inferior reactant are characteristically about fifty times those for the preferred substrate. Furthermore, the interaction with System Ly can be observed in the adult rabbit red cell, in which no manifestation of System ASC survives. Hence the two systems presumably cannot be manifestations of a single entity.

Table 2
Evidence for nature of interaction between Na^+ and neutral amino acids in systems *ASC* and $\overset{+}{Ly}$

1. Competition between (neutral amino acid + Na^+) and (cationic amino acid without Na^+).
2. Guanidine and methylguanidine cations are more reactive than NH_4^+ in association with neutral amino acid, in correspondence to the greater reactivity of arginine than lysine.
3. Hydroxyl group on a specific carbon atom causes mutual enhancement of transport reactivity of the neutral amino acid and Na^+.
4. When hydroxyl group is on the right carbon atom selectivity among alkali-metal ions is sharply modified (System $\overset{+}{Ly}$).
5. Flux ratio, Na^+ to neutral amino acid, is sharply increased in System *ASC* of pigeon red blood cell when the hydroxyl group is on the right carbon atom.
6. The hydroxyl group of hydroxyproline must be *trans* to the carboxyl group.

The nature of the evidence for the geometric position of Na^+, accumulated first in one transport system and then in another, is summarized in table 2. I will illustrate the documentation for the last four of the points listed. First let me point out that a hydroxyl group on a certain carbon atom of the amino acid sidechain enormously enhances the interaction between the neutral amino acid and Na^+. This effect will be illustrated for System *ASC* of the pigeon red blood cell. The results were virtually identical in the rabbit reticulocyte, although the transport system in question is rapidly lost on maturation of the reticulocyte to the adult mammalian red blood cell. Results are similar for the Ehrlich cell, although the rates are higher in that cell.

Table 3 summarizes the characteristics of System *ASC*. In the nucleated

Table 3
Summary of characteristics of transport system *ASC*

Formally defined for: The Ehrlich ascites-tumor cell. Probably ubiquitous in higher animal cells.

Substrates: 3-, 4- and 5-carbon linear-chained hydroxyaliphatic and mercaptoaliphatic amino acids; the prolines, asparagine, glutamine.

Characterizing features: Na^+-dependency (first-order); does not tolerate an *N*-methyl group (distinction from System *A*).

V_{max}: Characteristically variable; for methionine and norleucine, indetectibly low.

pH *sensitivity*: Intermediate between Systems *A* and *L*.

Stereospecificity: Exceptional.

Exchanging properties: Weak in the Ehrlich cell; obligatory, in the rabbit reticulocyte and the pigeon red blood cell.

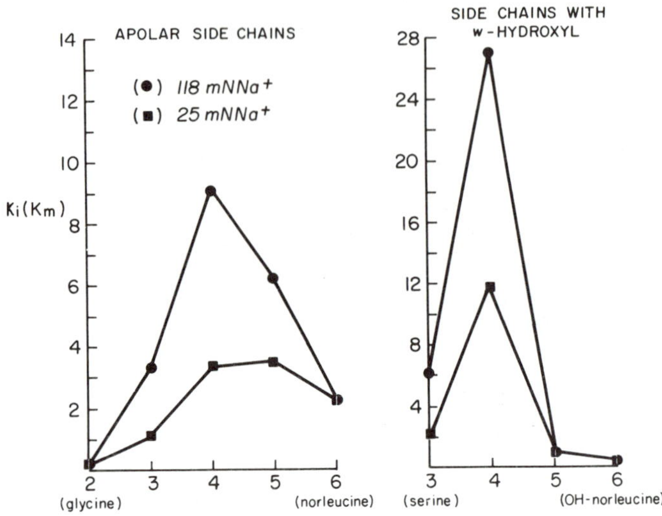

Fig.2. Effect of chain length on the apparent affinity of neutral amino acids for System *ASC* of the pigeon red blood cell, determined at 25 and 118 mN Na^+. The values are reciprocals of K_i or K_m values. Unsubstituted aliphatic amino acids are shown at the left, the corresponding ω-hydroxy amino acids at the right. Reproduced with permission from E.L. Thomas and H.N. Christensen, J. Biol. Chem. 246 (1971) 1682.

and the reticulated red cell, the system cannot function for net transport, being "locked into exchange"; but that is not true for the Ehrlich cell. I will develop later this peculiar feature of System *ASC* in the various red cells.

Reactivity of amino acids with System *ASC* is plotted in the form $1/K_i$ in fig. 2 as a function of chain length. If there is no omega hydroxyl group (graph at left), a mild preference is shown for sidechains of intermediate length, and total loss of transport occurs at a length of six carbons. Enhancement of Na^+ reactivity runs in parallel, as may be seen by comparison of the results at 25 and at 118 m*N* Na^+. But when a terminal hydroxyl group is present the peak is enormously higher (note difference in the two ordinate scales). The same effect is seen with a sulfhydryl or a carboxamide group. We can by such analyses recognize three preferred positions:

On carbon 4 (No. 3 nearly as good) in *ASC* system in all observed occurrences,

On carbon 4 (No. 5 somewhat inferior) in $L\overset{+}{y}$ system in red blood cells,

On carbon 5 (No. 4 somewhat inferior) in $L\overset{+}{y}$ system in Ehrlich cell.

We take these optimal positions to be indications of the position taken at

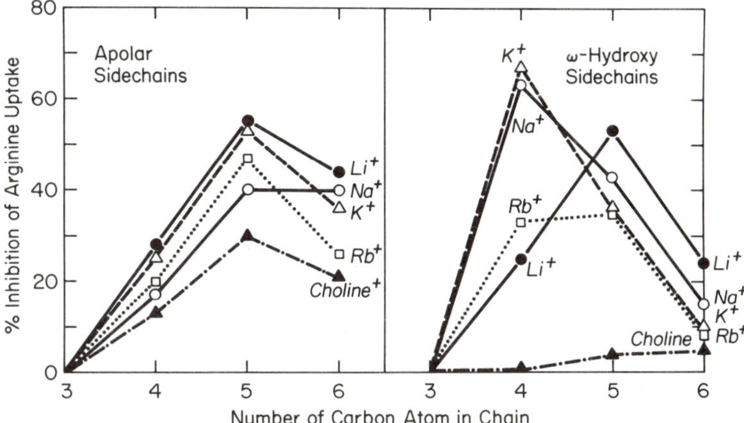

Fig. 3. Effect of chain length of neutral amino acids on the selectivity of the alkali-metal ion dependence of their inhibition of arginine uptake by the rabbit reticulocyte. Uptake of 5×10^{-5} *M* arginine was observed during 3 min at 37° in the presence or absence of the indicated amino acid at 50 mM. The indicated alkali-metal ions were at 72 m*N*, choline + neutral amino replacing isoosmotically the Na^+ and K^+ of Krebs-Ringer phosphate medium. Reproduced with permission from E.L. Thomas, T.-C. Shao and H.N. Christensen, J. Biol. Chem. 246 (1971) 1677.

the receptor site in each case, either by the sodium ion or by the distal cationic group on the amino sidechain.

Let me turn then to the changing selectivity for the alkali metal ion as cosubstrate, an effect seen largely in System $L\overset{+}{y}$. (In the *ASC* system, in contrast, Na^+ is the obligatory cation for almost all substrates. In System *A*, Na^+ is greatly preferred to Li^+, but important effects of amino acid structure on that preference are now being studied.) If there is no hydroxyl or other polar group on the sidechain, the several alkali-metal ions show no great differences in effectiveness in permitting neutral amino acids to react with the cationic transport System $L\overset{+}{y}$ (fig. 3). The sequence in general is $Li^+ > Na^+ > K^+ > Rb^+ > Cs^+ >$ choline, but with rather small differences among these cations as to activity. That sequence of the metal ions is the one expected of electrovalent bonding when there is no special reaction to upset that order. But as soon as there is a hydroxyl group on carbon 4 for the rabbit reticulocyte, Na^+ and K^+ become far more effective than the other ions. The same basic sequence is seen for System $L\overset{+}{y}$ in the Ehrlich cell (fig. 4). I have shown here the results only for Li^+ and Na^+; the other three alkali-metal ions were relatively ineffective, in the usual order. Note that there is a sharp

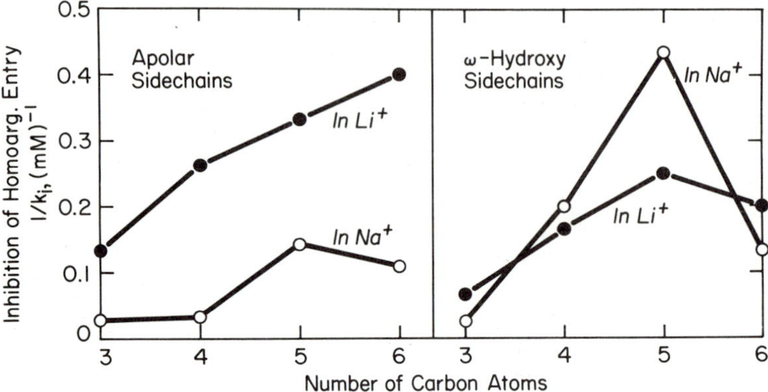

Fig. 4. Effect of chain-length of amino acids on their inhibition of homoarginine uptake by the Ehrlich cell in the presence of Na^+ or Li^+. Left, apolar, linear sidechains; right, the corresponding ω-hydroxy amino acids. Uptake of ^{14}C-homoarginine observed during 1 min at *p*H 7.4 and 37° in presence of 0, 2.5, 10 and 25 m*M* of the indicated amino acid. Na^+ or Li^+ at 105 m*N*, choline at 25 to 38 m*N*. The plot is of the reciprocal of the K_i values. K^+ and Rb^+, in that order, were very much less effective than Na^+. Reproduced with permission from the same source as fig. 3.

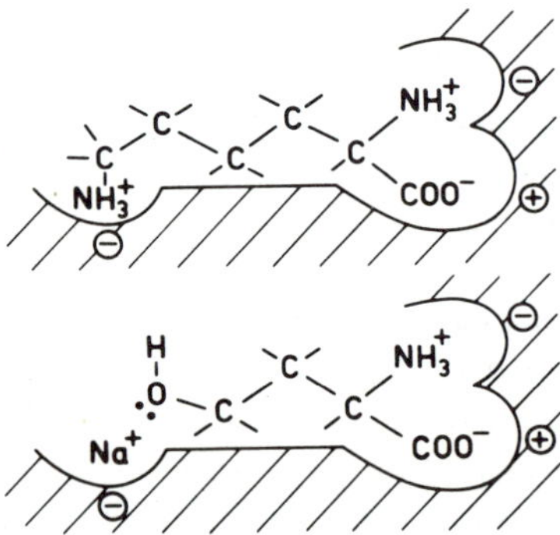

Fig. 5. A two-dimensional representation of the receptor site of System $L\overset{+}{y}$, occupied, above, by a normal substrate, lysine, and below, by homoserine and Na^+. One possible way in which the hydroxyl group helps prepare the site for Na^+ is represented by the shared electrons. Reproduced with permission from H.N. Christensen, Fed. Eur. Biochem. Soc. Symp. 20 (1970) 81.

Table 4
Comparison of reactivities of allo and threo amino acids with system ASC. The cells were brought to apparent internal levels of ^{14}C-alanine of 0.7 to 1.4 mM per kg cell water by 30-min incubation. Decline of ^{14}C content was then observed during 5 (for threonine) or 10 min at 37° at 20 mM external levels of the test amino acids as listed. Inhibition of uptake of ^{14}C-serine (0.02–0.8 mM) from the suspending medium was observed during 3-min intervals. The K_i values for DL amino acids were halved to obtain approximate values for the L isomers

Amino or imino acid	Pigeon red blood cells				Rabbit reticulocyte
	Stimulation of alanine exodus $[Na^+]_o$ = 118 mN (%)	K_i at $[Na^+]_o$ = 25 mN (mM)	K_i at $[Na^+]_o$ = 118 mN (mM)	$\frac{K_{i_{25}}}{K_{i_{118}}}$	Stimulation of alanine exodus $[Na^+]_o$ = 118 mN (%)
L-proline	15.7	2.0	1.8	1.1	8.2
DL-3-*cis* hydroxyproline	0.6	11	11	1.0	*
DL-3-*trans* hydroxyproline	14.9	0.08	0.07	1.1	*
L-4-*cis* hydroxyproline	3.8	2.0	2.0	1.0	2.0
L-4-*trans* hydroxyproline	30.7	0.27	0.16	1.7	13.3
DL-*allo*-free threonine	22.6	0.27	0.10	2.7	*
DL-*allo* threonine	22.1	0.24	0.10	2.4	*

* Not determined

inversion of the preference when the hydroxyl group falls on carbon 5 of the amino acid sidechain. Again the sulfhydryl group shares this effect. Fig. 5 shows the two-dimensional representation we reached two years ago for the spatial relations at Site $L\overset{+}{y}$ [12].

The prolines provided us the opportunity for testing the effect of not only the *position* but also of the *orientation* of the hydroxyl group. When the ring is closed, it is possible to differentiate between the hydroxyl group pointing in the same (*cis*) or the opposite (*trans*) direction as the carboxyl group. For this demonstration we return to System *ASC* using the pigeon red blood cell (table 4). Proline is only a fair substrate, and by comparing its K_i at two Na^+ levels, we see that this amino acid has only a weak interaction with Na^+. Introducing a *cis* hydroxyl group on carbon 3 has strongly unfavorable consequences for reactivity. We have evidence that the hydroxyl group in that orientation interferes with favorable apolar bonding occurring with the *cis*

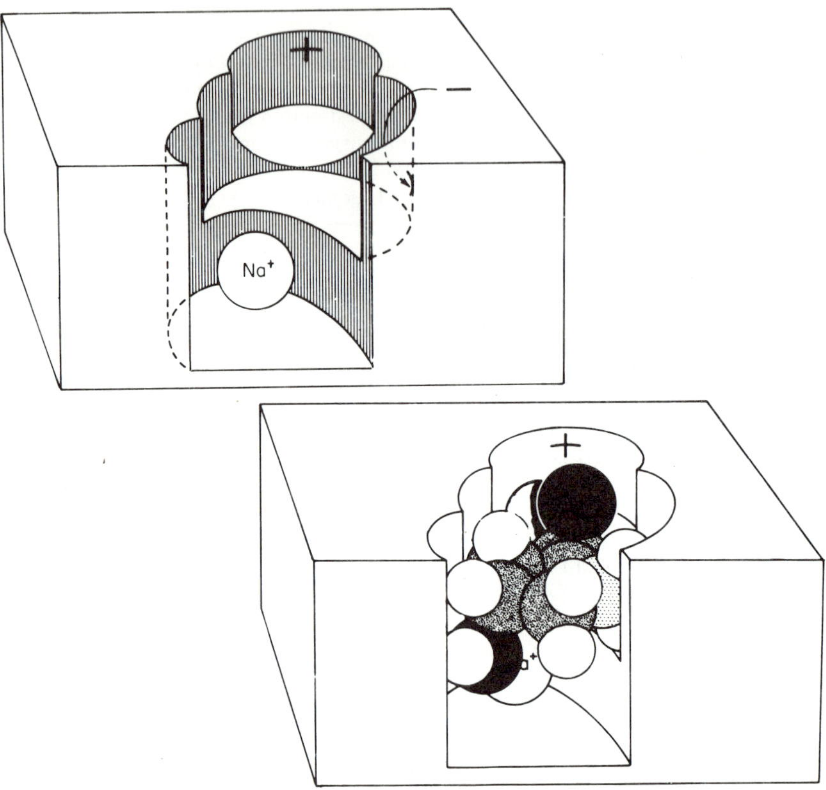

Fig. 6. A diagrammatic representation of the geometric relations proposed for transport site *ASC*, showing Na^+ (above) and both Na^+ and hydroxyproline (below) in suggested relative positions. No implication is intended that Na^+ must enter the site first, nor is the site conceived as static during its occupation. Reproduced with permission from E.L. Thomas and H.N. Christensen, Biochem. Biophys. Res. Commun. 40 (1970) 277.

aspect of the 3-methyl or -methylene group. Similarly, a *cis* 4-hydroxyl group does not enhance the reactivity of proline, nor its cooperativity with Na^+. But if the hydroxyl group is in the *trans* orientation, sharp increases take place in reactivity and in cooperativity with Na^+.

As control substances we may use allothreonine and threonine. With these open-chain compounds, the hydroxyl group is free to take either orientation, and the transport system makes no discrimination between the allo and threo forms (table 4).

Fig. 6 reproduces diagrammatically the orientation we visualize. Note the following components proposed for the recognition site: A point recognizing the carboxyl group; one recognizing the α-amino or -imino group; apolar areas bonding in a hydrophobic way with almost all methyl or methylene groups of the substrate; and finally a structure adjoining carbon 4, lying in a *trans* orientation with respect to the carboxyl group. This structure is the one that permits a properly orientated hydroxyl group to enhance so strongly the interaction with Na^+. We believe it is in fact the cosubstrate sodium ion, the presence of which is sensed and reported by the hydroxyl or sulfhydryl group, if it can reach it.

The pictures of figs. 5 and 6 show that we believe that Na^+ assumes a bridging function from the macromolecular site to the suitable placed O or S atom on the sidechain. We propose that for Na^+-dependent sugar transport a similar bridging function of Na^+ occurs to the oxygen on C2, whereby the C-2 hydroxyl group is introverted and the apolar aspect of the carbon chain is turned outward. Although substrate action can apparently occur with the C-2 hydroxyl group in the opposite orientation as in *D*-mannose or *L*-glucose, reactivity is exceedingly weak.

Table 5
Effect of a hydroxyl group on the coupling ratio, $\Delta\nu_{Na^+}/\Delta\nu_{a.a.}$, for entry into pigeon red blood cell by system of *ASC*. From B. Koser and H.N. Christensen, Biochim. Biophys. Acta 241 (1971) 9.

Unhydroxylated Amino Acid		Hydroxylated Amino Acid
proline	0.2	4-hydroxyproline, 3.2
alanine	2.5	serine, 4.0 (cysteine 5.0)
α-amino-*n*-butyric	1.8	threonine, 4.5

Table V shows an additional point of evidence. When one introduces a hydroxyl group in the correct orientation for System *ASC* of the pigeon red blood cell a sharp increase occurs in the flux augmentation ratio, i.e., the augmentation of the rate of Na^+ entry due to the presence of the amino acid, divided by the augmentation of the rate of amino acid entry due to the presence of Na^+ [13]. In this respect System *ASC* in reticulated and nucleated red blood cells is quite unusual. In complete contrast to the System *A* studied by Riggs and myself 1952 to 1958 and summarized here by Dr. Eddy, the present system apparently cannot be energized by the Na^+ gradient. Instead the system can serve only to redistribute that energy among its amino acid substrates. Indeed, it is what we may call a tertiary active transport system. In the exchange that takes place among the amino acid substrates, an apparently

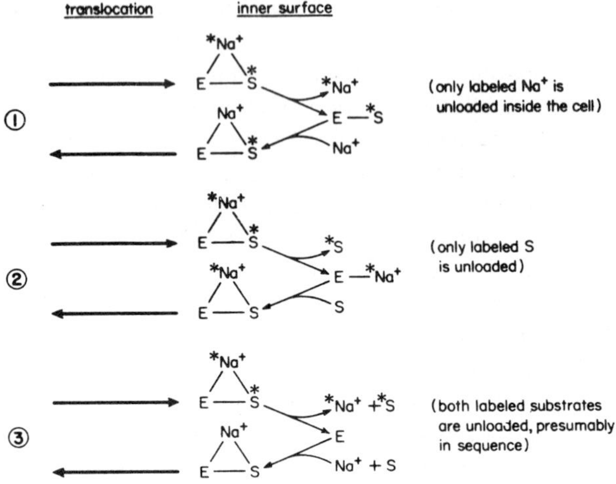

Fig. 7. Scheme for the translocation and dissociation steps of System *ASC* of the nucleated red blood cell of the pigeon. Recall that the hydroxylation of proline changes the coupling constant from 0.2 to 3.2. According to this model, even though process 2 is stimulated by that hydroxylation, process 1 is stimulated so much more that the contribution of process 2 falls from about 80% to one of less than 25% of the total recordable translocations, whereas process 1 rises from a contribution of less than 17% to one of 67% or more. Process 3, by which both substrates are unloaded consequent to the same translocation, still makes only a minor contribution to the total number of translocations. Reproduced with permission from B.H. Koser and H.N. Christensen, Biochim. Biophys. Acta 241 (1971) 9.

meaningless exchange of internal Na^+ for external Na^+ takes place. This exchange neither harnesses nor dissipates the energy of the Na^+ gradient. Furthermore, in this case the flux augmentation ratio has no relation to the composition of the ternary complex, which by all other kinetic evidence has a composition, one Na^+ to one amino acid molecule to one receptor molecule. The flux augmentation ratio does not depend on the time of observation; it does not depend on the concentration of either Na^+ or the amino acid substrate. It is determined by the structure of the two amino acids, external and internal, participating in the exchange. The most remarkable of the changes in the flux augmentation ratio shown in table 5 is the one that occurs on introducing the 4-*trans* hydroxyl group into proline, i.e., from a value of four or five amino acid molecules moving for each sodium ion, to almost the reciprocal value, 3.2 sodium ions transferred for each amino acid molecule.

Careful evaluation of this system has indicated a random order for formation and dissociation of the ternary complex. It is clear that translocation cannot occur unless both amino acid and alkali-metal ion have the correct structures and unless both are present on the two sides of the membrane [10]. We may conclude that the translocation event cannot in this case be rate-limiting to transport, since the same translocation event occurs for both cosubstrates, although their rates of transport are different. The rate limitation must instead be set by the dissociation-and-replacement event, as shown in fig. 7. Indeed it is quite likely that it is a displacement event. We must visualize three courses that can be followed, as in fig. 7:

1. Na^+ may be replaced from the ternary complex by internal Na^+, without replacement of the amino acid.
2. The amino acid can be replaced from the ternary complex by an internal amino acid molecule, without replacement of Na^+.
3. Both substrates may be replaced, in a random order.

If we reconsider the magnitude of the change in flux augmentation ratio caused by insertion of the 4-*trans* hydroxyl group, we see that it can be produced only by an extensive increase in the proportion of process I, which has to be largely at the expense of the proportion of process II.But we have already seen that the 4-*trans* hydroxyl group greatly increases the rate at which proline itself is transferred across the membrane. For the flux augmentation ratio to rise 16-fold, we must have a relatively much greater acceleration of the replacement of Na^+ from the ternary complex. Hence the hydroxyl group increases the rate of the limiting step for amino acid transport, the replacement of the amino acid; but it increases to a much greater extent the rate of replacement of Na^+, which is rate-limiting to Na^+ transfer [13].

This is indeed a strange transport system, as we see it in reticulated and nucleated red blood cells. By means of it, certain suitable amino acids stimulate an intense Na^+-for-Na^+ exchange diffusion, a behavior that our colleagues in alkali-metal ion transport should take into account. It is healthful for us to realize that alkali-metal dependency can have as different a significance as it does in System *A* of the Ehrlich cell, where the stoichiometric ratio appears to be one-for-one and energy is transferred, and for the *ASC* system of these red blood cells, on the other hand, where energy of the amino acid gradients is merely redistributed among various amino acids. Biologically, we may suspect that the cosubstrate role of Na^+ has lost its energizing function by an evolutionary differentiation (a retrogression, if you like) by which the carrier site loses its ability to reorientate unless both correct substrates are in place.

The other lesson of amino acid transport for the student of cation transport is how simply a suitable amino acid serves on one hand to generate a Na^+-recognizing site where none was present before, and for the principal secondary transport System A, to convert an apparently K^+-preferring site into a Na^+-preferring site.

2. Possibility of designing substrates showing trans inhibition on a direct mass-action basis

On theoretical grounds Heinz and Durbin predicted in 1957 the consequence of a much higher rate of movement of the loaded transport carrier than of the empty carrier, namely *trans* acceleration [14], a prediction confirmed experimentally the next year by Heinz and Walsh [15]. But at the same time Heinz and Durbin also considered the consequences of the converse situation, namely a much higher rate of movement of the empty carrier than of the loaded carrier. By neglecting the terms in the kinetic equation that became negligible under this condition, it appeared that a *trans* inhibition should result.

This proposal seems to have received little if any attention. It becomes especially important in evaluating the approach to a steady state of uptake of amino acids by yeast and *Neurospora*. The reversal of uptake by these cells is so slow as to have escaped observation, with one exception [16]. Instead influx gradually declines to a very small value. Eddy proposed in a brief communication in 1963 that either the accumulated amino acids or a metabolic product derived from them might immobilize the carrier at the inner surface of the membrane [16]. The first form of this proposal resembles the prediction of Heinz and Durbin, and it appears that Eddy probably made calculations similar to those of Heinz and Durbin.

In contrast, Ring, Gross and Heinz [17] and Grenson, Crabeel, Wiame and Bechet [18] have invoked special feedback loops whereby the amino acid accumulated through a transport system comes to occupy a modifier site, distinct from the transport receptor site, to inhibit further uptake. Both Grenson et al. and Pall have called attention to the similarity of the structural requirements for uptake and for *trans* inhibition in *Neurospora*, and Pall proposed that two sites might indeed be identical, so that *cis* inhibition would arise from a mass-action effect intrinsic to the transport system [19]. In contrast, these structural requirements for the two effects were strikingly different in the experiments of Ring *et al.* in *Streptomyces hydrogenans* [17]. Accordingly, *trans* inhibition by the mechanism discussed by Heinz and Durbin appears to be excluded for that species.

Pall has made a new proposal that an important condition for the phenomenon of *trans* inhibition as seen in *Neurospora* is a large difference in two

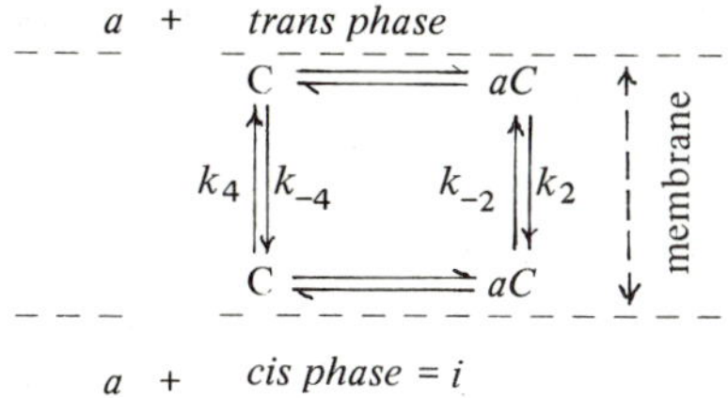

Fig. 8. Diagram for presentation of model. For simplicity the process is taken to be a facilitated diffusion. A perturbation is produced by placing a large concentration of *a* in phase *i*, which may be either the phase inside or that outside the cell. The following results are predicted:

Condition	Asymmetry generated for total carrier, C_t	Flux asymmetry generated
I. $k_2, k_{-2} \simeq k_4, k_{-4}$	$C_t^i = C_t^o$	none
II. $k_2, k_{-2} \ll k_4, k_{-4}$	$C_t^o > C_t^{i*}$	*trans* acceleration
III. $k_2, k_{-2} \gg k_4, k_{-4}$	$C_t^o < C_t^{i**}$	*trans* inhibition

(Similar effects may be expected for uphill transports.)
*The asymmetry across the membrane being due chiefly to the uneven distribution of the free carrier, *C*.
**The asymmetry in C_t distribution across the membrane is, however, chiefly due to the asymmetry in *aC* distribution, $[C^o]$ and $[C^i]$ being kept more nearly equal by their more rapid interconversion.

rate constants, corresponding to $k_2 >> k_{-2}$ in fig. 8 [19]. Such an asymmetry, as I have indicated above, is inherent in the uptake of amino acids in *Neurospora* and yeast. In examining that proposal, Dr. D.L. Oxender and I came to the view that a more general condition for *trans* inhibition is the much slower migration of the loaded than of the empty carrier, a conclusion that brought our attention to the same, insufficiently recognized proposal of Heinz and Durbin.

Fig. 8 illustrates the proposal as applied to a facilitated diffusion. If k_2, $k_{-2} << k_4, k_{-4}$, a high concentration of *A* in phase *i* can lead to the sequestering of the carrier-substrate complex *AC* in its orientation toward phase *i*, depleting to a greater or lesser degree the entire membrane of all other forms of the carrier. It is acceptable but not necessary that process k_4, k_{-4} be as fast as the association-dissociation steps. The consequence is that the steady state of an uptake process initiated by supplying substrate to a cell is not approached in the usual way for molecular processes, namely by an acceleration of back-flux until it equals the forward flux. Instead the forward flux slows as uptake proceeds, and the backward flux increases much less than it otherwise would need to do to equal the forward flux. The proposal applies also to uphill transports without substantial change.

Certain amino acids appeared to show a weak *trans* inhibition in the experiments described for the Ehrlich cell by Oxender and myself in 1963 [20, 21] and indeed also in the paper of Heinz and Walsh [15]. The case of α-aminoisobutyric acid is notable (fig. 3 in ref. [21]) because this amino acid appears to have its transport limited to one system. In the present experiments I have considered the possibility that amino acid analogs might be selected or designed to produce *trans* inhibition by looking for atypically low values for V_{max}. Such an amino acid is already known in α-(methylamino)-

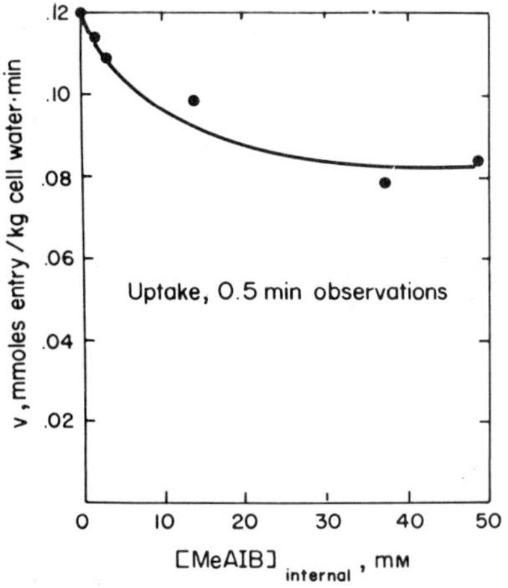

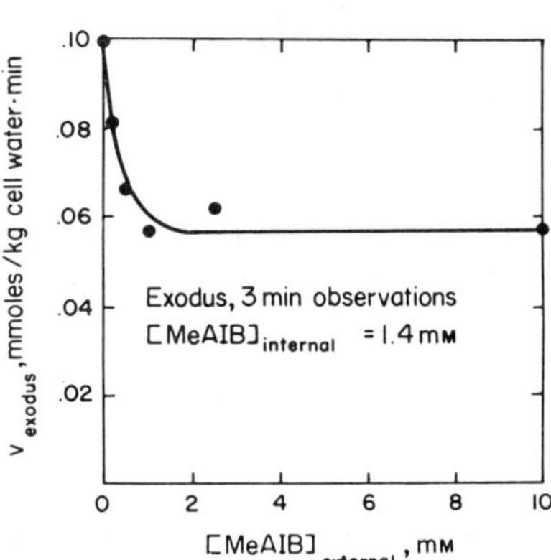

Fig. 9. *Trans* inhibition of the uptake (left) and exodus (right) of α-(methylamino)-isobutyric acid for the Ehrlich cell. (Left) The cells were loaded to several calculated internal concentrations by incubation for 30 min in MeAIB concentrations ranging from zero to 100 mM in Krebs-Ringer bicarbonate medium as usual. (Simultaneous parallel tests, using ^{14}C-MeAIB, served to estimate the cellular levels reached.) Uptake of ^{14}C-MeAIB by these washed loaded cells was then observed during 30 sec at 37°. (Right) The cells were brought to a calculated internal level of 1.4 mM in ^{14}C-MeAIB by incubation 15 min in a medium containing the amino acid at 0.1 mM. Exodus from the washed cells was then observed during 3 min at 37° into 600 volumes of Krebs-Ringer bicarbonate medium containing unlabeled MeAIB at the indicated concentrations (25).

isobutyric acid (see fig. 13 in ref. [22]), which has its transport limited to one system in the Ehrlich cell. The lowness of V_{max} should increase the probability that steps k_2 and k_{-2} might be significantly slower than steps k_4 and k_{-4}. Fig. 9 shows that this criterion has led to the observation for this amino acid of a fairly strong *trans* inhibition, both in the inward and the outward direction. For AIB, however, we observe only *trans* stimulation.

Conceivably modification or substitution of the N-methyl group could produce further slowing of migration and even stronger *trans* inhibition.

References

[1] H.N. Christensen and T.R. Riggs, J. Biol. Chem. 194 (1952) 57.
[2] H.N. Christensen, T.R. Riggs, H. Fischer and I.M. Palatine, J. Biol. Chem. 198 (1952) 1.
[3] T.R. Riggs, L.M. Walker and H.N. Christensen, J. Biol. Chem. 233 (1958) 1479.
[4] H.N. Christensen and J.A. Antonioli, J. Biol. Chem. 244 (1969) 1497.
[5] H.N. Christensen and M.E. Handlogten, Fed. Eur. Biochem. Soc. Lett. 3 (1969) 14.
[6] H.N. Christensen, M.E. Handlogten and E.L. Thomas, Proc. Nat. Acad. Sci. (U.S.A.) 63 (1969) 948.
[7] H.N. Christensen, E.L. Thomas and M.E. Handlogten, Biochim. Biophys. Acta 193 (1969) 228.
[8] E.L. Thomas and H.N. Christensen, Biochem. Biophys. Res. Commun. 40 (1970) 277.
[9] E.L. Thomas, T.-C. Shao and H.N. Christensen, J. Biol. Chem. 246 (1971) 1677.
[10] E.L. Thomas and H.N. Christensen, J. Biol. Chem. 246 (1971) 1682.
[11] H.N. Christensen, Adv. in Enzymol. 32 (1969) 1.
[12] H.N. Christensen, Fed. Eur. Biochem. Soc. Symp. 20 (1970) 81.
[13] B.H. Koser and H.N. Christensen, Biochim. Biophys. Acta 241 (1971) 9.
[14] E. Heinz and R.P. Durbin, J. Gen. Physiol. 41 (1957) 101.
[15] E. Heinz and P.M. Walsh, J. Biol. Chem. 233 (1958) 1488.
[16] A.A. Eddy, Biochem. J. 88 (1963) 34p.
[17] K. Ring, W. Gross and E. Heinz, Arch. Biochem. Biophys. 137 (1970) 243.
[18] M. Grenson, M. Crabeel, J.M. Wiame and J. Bechet, Biochem. Biophys. Res. Comm. 30 (1968) 414.
[19] M. Pall, Biochim. Biophys. Acta 233 (1971) 201.
[20] D.L. Oxender and H.N. Christensen, Nature 197 (1963) 765.
[21] D.L. Oxender and H.N. Christensen, J. Biol. Chem. 238 (1963) 3686.
[22] Y. Inui and H.N. Christensen, J. Gen. Physiol. 50 (1966) 203.
[23] H.N. Christensen and M.E. Handlogten, J. Biol. Chem. 243 (1968) 5428.

PROTON MOVEMENTS ACCOMPANYING THE ABSORPTION OF AMINO ACIDS BY YEAST: A POSSIBLE ANALOGUE OF THE MAMMALIAN SYSTEMS?

A.A. EDDY

Department of Biochemistry, University of Manchester Institute of Science and Technology, Manchester, England

The mammalian systems

It is now widely believed that Na^+ ions and K^+ ions are intimately involved in the concentration of various neutral amino acids and simple carbohydrates by certain mammalian tissues. Such preparations may accumulate *in vitro* a solute concentration that is twenty times the concentration in the Ringer solution bathing the tissue. While the precise function of Na^+ ions and K^+ ions in these systems is a controversial matter, many of the observations are consistent with the so-called cosubstrate hypothesis [1, 2], the elements of which can be illustrated by considering the effects of Na^+ ions and K^+ ions on the absorption of glycine by mouse ascites tumour cells [3–5].

(I) The influx of glycine into the tumour cells is markedly stimulated by the presence of Na^+ ions in the extracellular phase, their function being competitively inhibited by extracellular K^+ ions. (2) The absorption of glycine stimulates both the influx of Na^+ ions and, in certain circumstances at least, the efflux of K^+ ions. These effects occur in the presence of metabolic inhibitors, such as sodium cyanide, that appear virtually to stop energy metabolism and thereby prevent the normal functioning of the sodium pump. We have shown that the preparations of the tumour cells depleted of ATP both absorb an additional 0.9 equivalent of Na^+ ions with each equivalent of glycine and expel an additional 0.6 equivalent of K^+ ions. (3) The close link which exists in this system between a component of the influx of Na^+ ions and the influx of glycine itself is illustrated by the observations depicted in fig. 1. The tumour cells were equilibrated with a Ringer containing 33 meq.

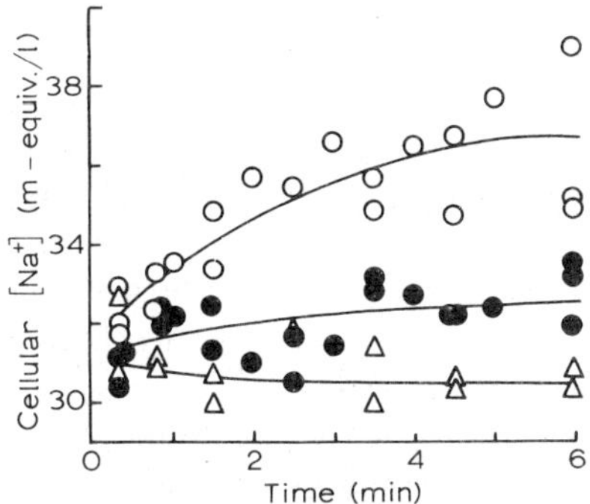

Fig. 1. The increase in cellular $[Na^+]$ brought about by the addition of 12 mM-glycine to the starved mouse ascites tumour cells. The effect was not reproduced with L-arabinose. The Ringer solution contained 33 meq. of Na^+/1, 131 meq. of Li^+/1, 1 meq. of K^+/1, 0.1 mM-ouabain and 2 mM-sodium cyanide, Glycine (○) was added to one portion of the suspension, 12 mM-L-arabinose (△) to a second portion and the third portion (●) served as a control. Eddy [6] gives a more detailed account.

of Na^+/1, under conditions where the cellular ATP content was relatively small. Cellular $[Na^+]$ and extracellular $[Na^+]$ were then almost equal and had reached steady values. After the addition of 12 mM-glycine, cellular $[Na^+]$ increased up to about 36 meq/1 during the next 3–4 min, in which the cellular concentration of glycine reached about 14 mM. The system thus behaved as though the spontaneous flow of glycine into the tumour cells carried with it a stream of Na^+ ions, the downhill movement of glycine creating a concentration gradient with respect to $[Na^+]$ between the cellular and extracellular phases. The reciprocal effect of the ionic gradients on the amino acid gradient have also been studied. We have shown that when the tumour cells are manipulated so that a concentration gradient exists across the cell membrane with respect to either $[Na^+]$ or $[K^+]$, then a gradient of glycine concentration tends to be set up. Neither ATP nor exchange reactions with the endogenous complement of amino acids appear to be involved necessarily in these phenomena. Even though the molecular basis of the kinetic observations is largely unknown, they indicate rather clearly that the spontaneous diffusion of the Na^+ ions and the K^+ ions across the plasmalemma is coupled by means of an appropriate carrier system to the parallel flow of the amino acid molecules. Such a concept, due originally to Christensen, is conveniently referred to as the ion gradient hypothesis.

Amino acid transport in yeast

There is now some reason for thinking that these ideas may be relevant to the problem of the mechanism of the absorption of amino acids by yeast (*Saccharomyces cerevisiae* and *Saccharomyces carlsbergensis*). In that system the ratio of the cellular to the extracellular concentrations of amino acid may exceed the factor 10^4. Appropriate strains of brewers and bakers yeast absorb at least 1 μmole of glycine, for instance, per mg dry weight of yeast, from a 0.2 mM solution of glycine at 30°C near *p*H 5. Energy metabolism is required for that purpose and is maintained either by the utilization of endogenous substrates, or by supplying a compound such as glucose. We found that, in contrast to the mouse cells, the yeast system could function when Na^+ ions were virtually absent from the extracellular phase. Instead H^+ ions appeared to be involved, especially when energy metabolism was restricted. Thus the influx of glycine and of various other amino acids was then at least six times faster at *p*H 4 than at *p*H 7. It was also significantly retarded by raising the concentration of K^+ ions in the extracellular phase. Hence, it seemed possible that in the yeast system H^+ ions and K^+ ions served functions analogous to those of Na^+ ions and K^+ ions, respectively, in the mouse system [7]. It will be recalled that Mitchell [8] had earlier argued, also on rather indirect evidence, that H^+ ions might be a cosubstrate of the lactose transport system of *Escherichia coli*.

To develop further the analogy with the mouse cells, one may next ask how the presence of the amino acids affects the transport of H^+ ions and K^+ ions into the yeast. Bakers yeast has long been known to absorb K^+ ions from the environment, apparently by exchanging them with cellular H^+ ions, in a reaction that depends on energy metabolism and leads to protons accumulating in the extracellular phase. When both 2-deoxyglucose and antimycin A are present energy metabolism virtually stops. A flow of K^+ ions out of the yeast then occurs and is accompanied by a net uptake of H^+ ions. We have shown that the preparations containing these metabolic inhibitors nevertheless absorbed, relatively rapidly, amounts of various amino acids up to about 20 nmole/mg dry weight of yeast.

At *p*H 4.5 the phase of rapid absorption of the amino acid was complete within about three minutes at 30°C. It appears not to involve, necessarily, either (1) the chemical transformation of the absorbed amino acid or (2) a process of exchange diffusion with amino acids already present in the yeast cells. The finding that glycine, for instance, was concentrated at least 100-fold when it was absorbed under these conditions is specially significant in relation to the ion gradient hypothesis, because energy metabolism ap-

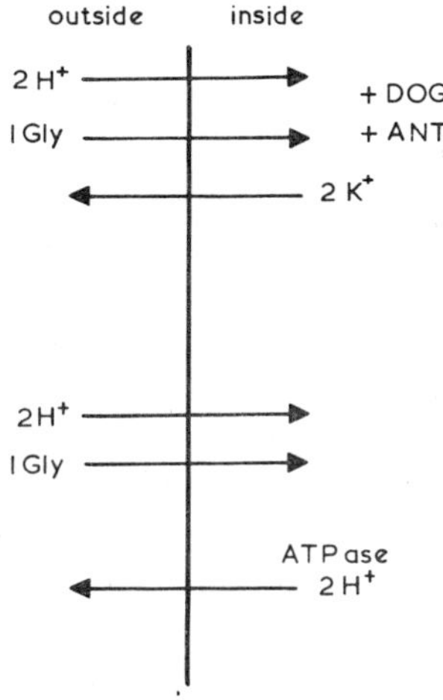

Fig. 2. Selected models of the relations between the ionic fluxes and the amino acid movements. Eddy and Nowacki [10] give a more detailed account. A stoichiometry of 2 H^+ per glycine equivalent is assumed. The upper diagram refers to the system depleted of ATP in the presence of antimycin and deoxyglucose (ANT + DOG). The lower diagram shows one way of coupling the system to a hypothetical proton pump driven by ATP (ATPase system).

peared to be too slow to provide the driving force [9]. According to that hypothesis, the driving force might be partly derived from the spontaneous flow of protons, from the extracellular phase, into the cellular phase at about *p*H 6.5, and partly derived from the flow of K^+ ions out of the yeast into the extracellular phase. Strong support for that interpretation came from the observations [10] showing that the absorption of various amino acids, in the presence of the two metabolic inhibitors that depleted the yeast of ATP, accelerated both the influx of H^+ ions and the efflux of K^+ ions (fig. 2).

Some preliminary experiments, providing a further example of the induced flux of protons, with other strains of yeast than those used in our original work, may be mentioned in this connexion. These are illustrated in fig. 3 which shows how the *p*H changed when small amounts (0.5 μmole) of glycine or methionine were added to preparations of two strains of yeast*. One strain B (fig. 3, lower line) exhibited no induced proton movements, absorbed the amino acids relatively slowly (0.7 nmole of glycine/min per mg dry wt) and is believed to lack the functions of the so-called general amino acid permease

* I am indebted to Dr. M. Grenson for generously making these cultures available.

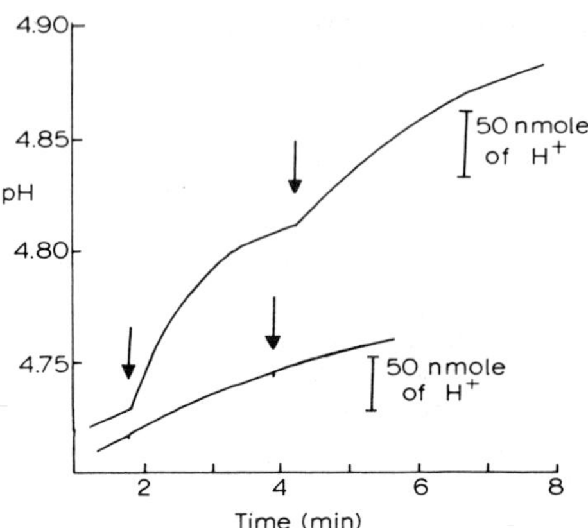

Fig. 3. Effect of glycine and of L-methionine on the change of *p*H with time in a yeast suspension depleted of ATP. The yeast was harvested from a mineral salts medium and kept at 25°C for 2 hr in the presence of 2% (w/v) glucose to deplete the cellular amino acid pool. The cells (50 mg dry wt) were then suspended in 5 mM-tris buffer (5 ml) brought to *p*H 4.5 with citric acid and containing 5 mM-deoxyglucose and antimycin A (0.2 μg/mg dry wt). The additions of amino acid (0.5 μmole) were made at the times indicated by the arrows. The upper line refers to yeast strain A (code MG 276) which absorbs amino acids relatively rapidly; glycine was added first, then methionine. The lower line refers to strain B (code 2512 C) carrying the *gap* mutation and with defective amino acid absorption. Methionine was added first, then glycine. The interpretation of such observations is discussed by Eddy and Nowacki [10].

[11]. The other strain A (fig. 3, upper line) exhibited proton displacements. It carries the general permease and absorbed both amino acids relatively rapidly (6 nmole of glycine/min per mg dry wt). These observations associate the proton movements with the functioning of the general permease. Subsequent work has shown that the various specific amino acid permeases carried by these strains also function with H^+. The use of such mutant yeast strains may help to unravel some of the complex problems presented by the study of the stoichiometry of the flows of H^+ ions and K^+ ions induced during the absorption of the amino acids [10].

It is already clear that during energy metabolism there are circumstances where (1) the influx of K^+ ions is not affected by the presence of the amino acids and (2) the yeast appears to secrete the same number of H^+ ions as are apparently absorbed with the amino acid (fig. 2). As in the model of the lactose transport system described by Mitchell [8], it may be valid under these conditions to regard the electrochemical gradient of H^+ ions across the cell membrane as supplying the force driving the accumulation of the amino acid. Much further work seems to be needed, however, before firm conclusions about the functions of the ions can be drawn.

References

[1] H.N. Christensen, in: Membranes and Ion Transport, ed. E.E. Bittar (Wiley-Interscience, London, 1970) vol. 1, p. 365.
[2] S.G. Schultz and P.F. Curran, Physiol. Rev. 50 (1970) 637.
[3] A.A. Eddy, Biochem. J. 108 (1968) 195.
[4] A.A. Eddy, Biochem. J. 108 (1968) 489.
[5] A.A. Eddy and M.C. Hogg, Biochem. J. 114 (1969) 807.
[6] A.A. Eddy, Biochem. J. 115 (1969) 505.
[7] A.A. Eddy, Proc. 2nd Symp. Yeast (Bratislava, Slovak Academy of Sciences, 1966) p. 457.
[8] P. Mitchell, Biochem. Soc. Symp. no. 22 (1963) p. 142.
[9] A.A. Eddy, K. Backen and G. Watson, Biochem. J. 120 (1970) 853.
[10] A.A. Eddy and J.A. Nowacki, Biochem. J. 122 (1971) 701.
[11] M. Grenson, C. Hou and M. Crabeel, J. Bact. 103 (1970) 770.

DISCUSSION REMARK TO THE PAPERS OF H. CHRISTENSEN AND A.A. EDDY

E. HEINZ

The data of Christensen, Eddy and others leave no doubt that the flow of electrolyte ions is energetically coupled to the transport of amino acids in Ehrlich cells. This has been confirmed by many other investigators, and also by us, as is shown on the next slide (fig. 1). Here the net flow of amino-iso-butyric acid (AIB) into Ehrlich cells is plotted versus the "vectorial" driving force derived from all known electrochemical potential gradients between cells and medium, including the (conjugate) gradient of AIB itself. We assume an 1 : 1 co-transport with Na- and an 1 : 1 counter-transport with K-ions, and complete coupling (fig. 2). It is seen from the two regression lines on fig. 1 that the transport rate rises with increasing "vectorial" driving force. It is noteworthy that the regression lines do not pass through the origin. In other words, even at zero vectorial driving force there is a substantial AIB net transport into the cells, whereas this net transport becomes zero, or reverse, only at an opposing driving force of about 4 000 Joules or more. If the gradients are correctly plotted, the data indicate the presence of an extra driving force of at least 4 000 Joules which is not accounted for by the gradients. This extra driving force must be covered by another energy source, for instance, via direct chemi-osmotic coupling, by metabolism. Before drawing this conclusion, however, we must make sure that the gradients of Na^+ (and K^+) are correctly inserted. It could be, as has been suggested by Eddy, that owing to compartmentalization the actual Na^+-concentration near the inner face of the cell membrane is lower than the overall cell concentration of Na^+, and the driving gradients accordingly higher. It is known for all cells that the nuclei contain more Na^+ than the cytosol. This we could confirm for Ehrlich ascites tumor cell nuclei, which we started to isolate using non-aquaeous extractants. Preliminary results seem to indicate, however, that the difference in Na^+ contents between nuclei and cytoplasm is far to small to approximately account for the above mentioned deficit.

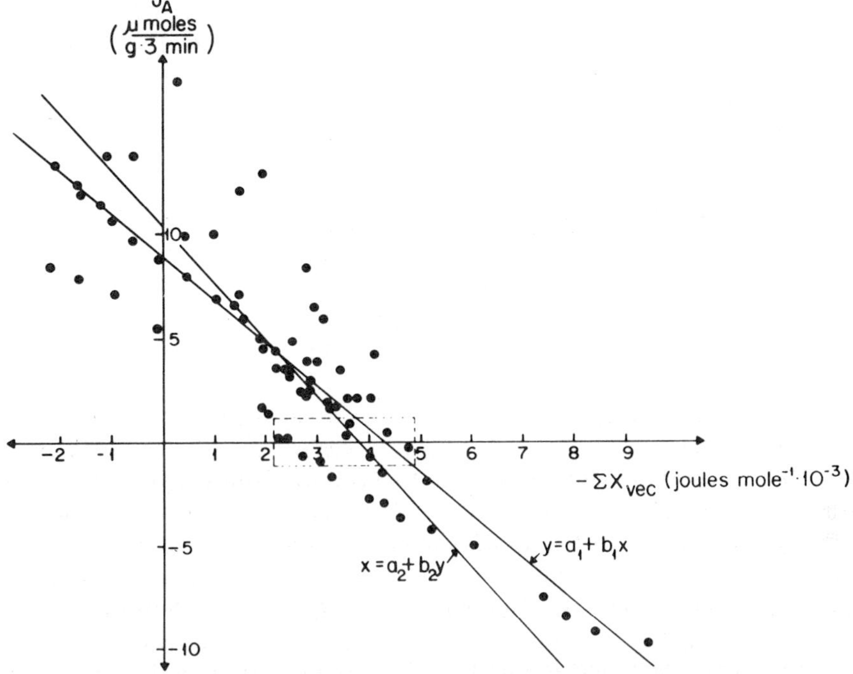

Fig. 1. The net influx of AIB as a function of the sum of the vectorial driving forces. J_A is the net flux of AIB, ΣX_{vec} is the sum of the vectorial driving forces, i.e. of the individual (negative) ECP gradients of Na^+, K^+, and AIB, respectively. The two lines represent the regression lines obtained by the method of least squares. The dotted lines enclose points for which J_a is approximately zero. (From: J.A. Schafer and E. Heinz, Biochim. Biophys. Acta 249 (1971) 15.)

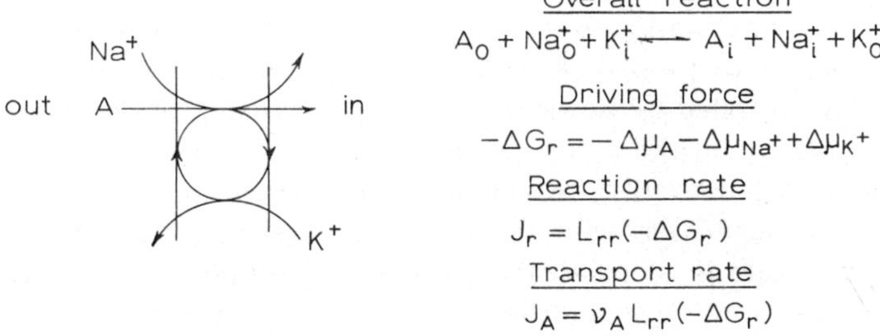

Fig. 2. Coupling between transport of amino acids (A) and the flow of Na^+ by co-transport, and the flow of K^+ by counter-transport. Assuming complete coupling $\Delta\mu$ represents the ECP difference for A, Na^+ K^+, respectively. L_{rr} is the rate coefficient for the reaction rate (J_r) which depends on the overall driving force ($-\Delta G_r$). ν_A is the stoichiometric coefficient, indicating the number of amino acid molecules transported with each reaction cycle.

ACTIVE AND PASSIVE SUGAR TRANSPORT IN YEAST

J. van STEVENINCK
Laboratory for Medical Chemistry, Rijksuniversiteit
Leiden, The Netherlands

The transport of sugars in yeast cells has been characterized as a carrier-mediated process, according to the general outlines of this model. Whereas the carrier principle itself provides the mechanism for an equilibrating, facilitated diffusion [1], coupling of metabolic energy to the carrier system may induce an active, "up-hill" transport, as e.g. in the permease-carrier model [2].

For many years there has been a tendency to describe sugar transport in yeast as a simple carrier-mediated facilitated diffusion. Although this is undoubtedly true in many cases, several examples of "up-hill" sugar transport in different yeast strains have been demonstrated [3–6].

Several years ago we postulated transport-associated phosphorylation as the mechanism of energy coupling in the active transport of glucose in yeast, with anorganic polyphosphate as high-energy phosphate donor [7–8]. This model was based on many lines of indirect experimental evidence and resembles the phosphotransferase system in bacteria, described and characterized shortly afterwards [9–10]. Kinetically transport-associated phosphorylation can be demonstrated by isotope pulsing experiments, provided that the substrate is present in the cytoplasm partly as the free sugar and partly as a phosphorylated derivative, without other metabolic conversions [6]. After adequate preincubation with the non-labeled substrate, a pulsing dose of labeled substrate is added to the medium. If transport of free sugar takes place, followed by subsequent intracellular phosphorylation, the specific activity of the intracellular free sugar will increase faster than the specific acitivity of the phosphorylated fraction. In the case of transport-associated phosphorylation, followed by partial intracellular de-phosphorylation however, the specific activity in the phosphorylated fraction should increase

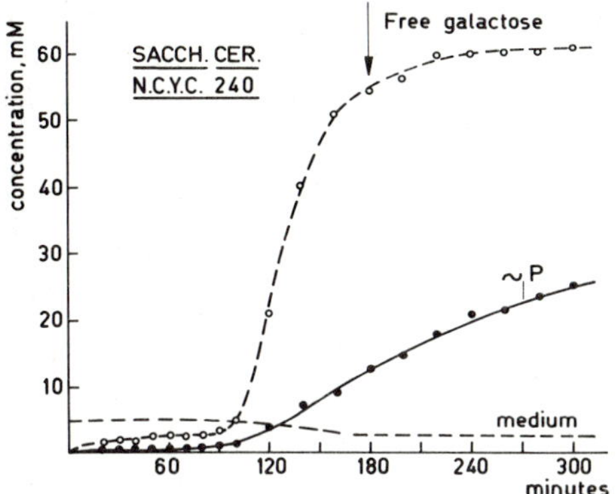

Fig. 1. Medium concentration and intracellular concentration of free and phosphorylated galactose during incubation at 25°. Yeast strain: NCYC 240.

faster than the specific activity of the free sugar. With this technique transport-associated phosphorylation could be demonstrated for 2-deoxy-d-glucose in saccharomyces cerevisiae, strain CBS 1172 [6], and for α-methyl-glucoside in strain NCYC 240 [11].

Recently Kuo, Christensen and Cirillo concluded that galactose transport in saccharomyces cerevisiae can be characterized as a facilitated diffusion [12, 13]. Such a generalization, based on observations with a particular yeast

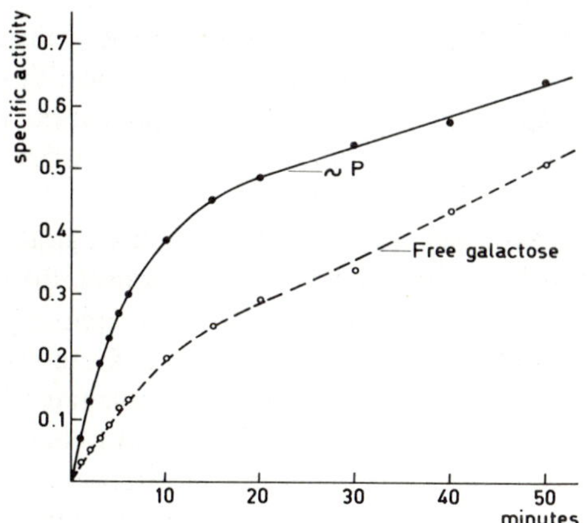

Fig. 2. Specific activity of intracellular galactose phosphate and free galactose after pulsing with ^{14}C-labeled galactose. Yeast strain: NCYC 240. Preincubation: 180 min at 25° in 5 mM unlabeled galactose.

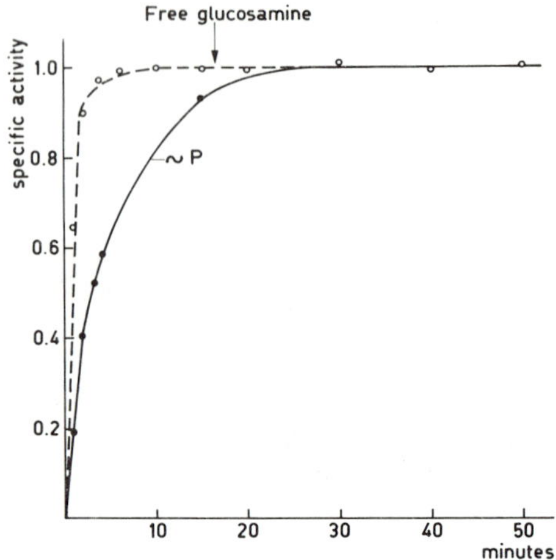

Fig. 3. Specific activity of intracellular glucosamine phosphate and free glucosamine after pulsing with ^{14}C-labeled glucosamine. Yeast strain: CBS 1172. Preincubation: 60 min at 25° in 2 mM unlabeled glucosamine.

strain, is not justified however, as shown by the following experiment. Uptake of galactose in saccharomyces cerevisiae, strain NCYC 240, proceeds as depicted in fig. 1. After an initial period of, presumably, facilitated diffusion, an active transport system has been induced, demonstrated by up-hill transport of free galactose. In addition to galactose a phosphorylated derivative is present in the cell. The results of an isotope pulsing experiment conducted after a preincubation period of about 3 hr are shown in fig. 2. The specific activity of the phosphorylated fraction increased considerably faster than the specific activity of the free fraction, indicating transport-associated phosphorylation of galactose in this yeast strain.

Incubation of yeast cells, strain CBS 1172, with glucosamine leads to an intracellular accumulation of the free sugar against the concentration gradient, demonstrating active transport. Further a phosphorylated derivative can be detected inside the cells. An isotope pulsing experiment gives the results shown in fig. 3. In this case the radioactive label first appears in the free fraction, indicating transport of free sugar with subsequent intracellular phosphorylation. Apparently in this example of active transport the coupling of metabolic energy to the uptake system is not effectuated via a transport-associated phosphorylation, but in a different way.

In conclusion, at least three different mechanisms of sugar transport in yeast cells can be distinguished:

1. Carrier-mediated facilitated diffusion;
2. Active transport with transport-associated phosphorylation;
3. Active transport with a different, as yet unknown, mechanism of energy coupling.

The actual transport mechanism that will be operative in a particular case depends on the kind of sugar, the yeast strain and the experimental conditions.

References

[1] T. Rosenberg and W. Wilbrandt, J. Theoret. Biol. 5 (1963) 288.
[2] A. Kepes, in N.E. Gibbons, 8th Intern. Congr. Microbiol., Montreal, 1962 (Univ. of Toronto Press, 1963), p. 38.
[3] H. Okada and H.O. Halvorson, Biochim. Biophys. Acta 82 (1964) 547.
[4] H. de Robichon-Szulmajster, Ann. Technol. Agric. 10 (1961) 81.
[5] G. Harris and C.C. Thompson, Biochim. Biophys. Acta 52 (1961) 176.
[6] J. van Steveninck, Biochim. Biophys. Acta 163 (1968) 386.
[7] J. van Steveninck, Thesis Leiden, 1962.
[8] J. van Steveninck, Biochem. J. 88 (1963) 25p.
[9] W. Kundig, S. Ghosh and S. Roseman, Proc. Nat. Acad. Sci. (U.S.A.) 52 (1964) 1067.
[10] W. Kundig, F.D. Kundig, B. Anderson and S. Roseman, Fed. Proc. Fed. Am. Socs. exp. Biol. 24 (1965) 658.
[11] J. van Steveninck, Biochim. Biophys. Acta 203 (1970) 376.
[12] S.C. Kuo, M.S. Christensen and V.P. Cirillo, J. Bacteriol. 103 (1970) 671.
[13] S.C. Kuo and V.P. Cirillo, J. Bacteriol. 103 (1970) 679.

THE ENERGY COUPLING TO BACTERIAL SUGAR TRANSPORT

A. KEPES
Institut de Biologie Moléculaire, Laboratoire des Biomembranes, Paris, France

1. Introduction

The molecular mechanism of membrane transport in general and of sugar transport in bacteria in particular is largely unknown. Bacteria were the first to reveal the presence of specific macromolecules involved in transport [1] first identified by their genetic deficiencies mapped in well defined structural genes. Later they became accessible to *in vitro* biochemical or biophysical characterization in phosphorylating transport systems by their enzymic activities [2] and in a large number of different transport systems by their affinity to specific substrates [3–5]. It can be generalized that a transport system must include a protein molecule capable to associate to the transported substrate, but it might include one or several other protein species indispensable for its function. The role played by the identified proteins in transport is variable in different transport systems, and it is understood to different degrees as outlined below. The way through which metabolic energy is geared to active unidirectional transport is generally even less understood.

The present article is not intended to review all relevant facts and to list more or less well established models of active transport and of energy coupling. Nevertheless it will deal with three different transport systems thought to be typical of three widely different classes and will try after a very short description of well established facts to ask some critical questions and to suggest some new avenues of approach on the basis of recent and sometimes very preliminary observations.

2. Phosphotransferase systems

The chemical basis of these transport systems was established by Roseman and his coworkers, in 1964 [2] as being due to a membrane bound enzyme E II catalyzing the following type of reaction:

$$\text{Sugar} + \text{Phosphoprotein (P}\sim\text{Hpr)} \rightarrow \text{sugar phosphate} + \text{H pr}$$

The phosphate carrier protein Hpr M. W. 9 000 is phosphorylated on a histidine residue by an auxiliary enzyme E I according to the reaction:

$$\text{Phosphoenol pyruvate} + \text{Hpr} \rightarrow \text{P}\sim\text{Hpr} + \text{pyruvate.}$$

The role of the phosphotranferase in sugar transport was proposed by the original authors [6] and it was later established that components of the system were indispensable in *E. coli* for the uptake of glucose fructose mannose and mannitol [7] and in other bacteria for the uptake of an even larger number of sugars. This mechanism is perfectly satisfactory from the thermodynamic point of view, the splitting of phosphoenol pyruvate provides the energy necessary for transport. From the mechanistic point of view, the membrane bound enzyme can offer its active site to the exogenous sugar and reject its product, the phsophorylated sugar, into the cytoplasm. The phosphorylated product is prevented from leaving the cell by the impermeability of the membrane to phosphate esters. Such a mechanism should not be described as an active transport but rather as a vectorial metabolism [8] since the transport step is unseparable from a chemical step. In favor of this vectorial metabolism is the impossibility to demonstrate any countertransport in these systems (unpublished results from several laboratories) in sharp contrast to the system of lactose transport (see below).

Strong support to this vectorial metabolism was brought by experiments of Kaback [9] with membrane vesicles derived from *E. coli* protoplasts. These vesicles are capable to phosphorylate α-methyl glucoside (αmglc) in the presence of added phosphoenolpyruvate, and the product αmglc 6P is found inside the vesicles. No dephosphorylating enzyme is present in these preparations and the sugar phosphate cannot leak out from the vesicles except at relatively high temperatures [10]. When visicles are preloaded in the cold and without added PEP with tritiated glucose and then transferred into a medium containing PEP, and ^{14}C glucose at the incubation temperature, exogenous ^{14}C glucose is phosphorylated by priority and ^{14}C glc 6P appears in the vesicles [9].

The reason why the problem of energy coupling in this system cannot be considered as entirely solved lays in our ignorance of the basis of anisotropy (or polarity) of the system. Is E II built into the membrane in such a way that it presents its substrate binding site toward the medium and its product releasing site towards the cytoplasm, or is the polarity to be attributed merely to the fact that the sugar comes from the medium and P Hpr is confined in the cytoplasm. The answer to this question is essential to understand what a vectorial enzyme might be and consequently how it is capable to channel the non-vectorial chemical energy into vectorial transport work.

First some uncertainty subsists as to the location of Hpr. It was claimed to be a periplasmic protein [11], thus approaching E II from the same side as free sugar. This hypothesis would attribute the important role in unidirectional action to the structural features of E II but would raise a number of difficult questions if, as reasonably expected, the phosphorylation of Hpr by E I and PEP occurs, in the cytoplasm, about the mechanism by which P~Hpr reaches the periplasmic space and by which Hpr is recycled in the cytoplasmic compartment (fig. 1, left).

If, on the other hand, Hpr and PHpr are both confined in the cytoplasm, as generally believed, there is sufficient anisotropy in the system without postulating a built-in anisotropy of E II.

The experiments with the membrane vesicles do not answer these questions unambiguously. Since the total phosphorylating activity found in the vesicles is less than one hundredth of the activity found in whole bacteria from which the vesicles were derived, it is reasonable to suppose that the rate limiting factor is the soluble part of the system, namely E I and Hpr, present in trace amounts presumably inside the closed vesicles, since the outside contaminants are removed by washing. Thus anisotropy is still due to the dissimilarity of the two aqueous compartments. When purified E I and Hpr are added to the vesicles the phosphorylating activity increases but the extra sugar phosphate produced is now found in the medium. However this last experiment cannot rule out the hyporhesis of anisotropy built into E II since it is possible that the unidirectional basal activity be due to only a part of the membrane particles which are tightly closed and therefore contain basal levels of E I and Hpr, while another part is formed by vesicles which are leaky and therefore do not contain E I and Hpr but are accessible to these components from both sides when they are added to the incubation medium. The only active configuration still might be the access of PHpr to E II on the inner face of the membrane. The two models on the right of fig. 1 depict two possibilities in the framework of an anisotropic E II. Since the phosphate residue comes at some phase of the cycle in close contact with the sugar, some

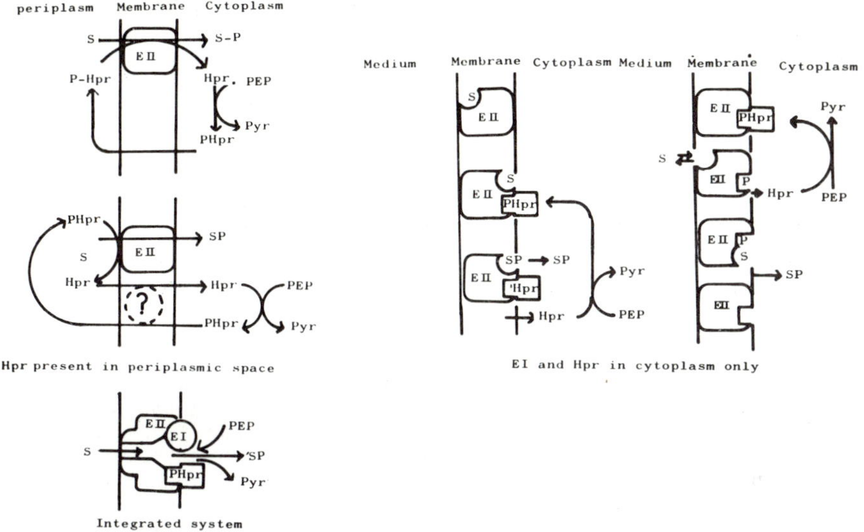

Fig. 1. Possible models of phosphotransferase function. *Top left*: Hpr phosphorylated in cytoplasm, PHpr is transported to periplasm by unknown trnasport system. E II reacts with sugar and PHpr on its outer side and releases its products sugar P and Hpr in the cytoplasm. *Middle left*: Hpr is phosphorylated in cytoplasm and PHpr reaches the periplasm by exchange diffusion against Hpr. E II reacts with donor and acceptor substrates on the outer side and releases sugar phsophate in cytoplasm. *Bottom left*: Integrated system a unique complexe of E II, E I and Hpr. *Center*: E II reacts with sugar on outer face first, then with PHpr on inner face, with concomitent translocation of sugar. Pi transfer and release occur on the inner face. *Right*: E II reacts with PHpr on the inner face first, then with sugar on the outer face. Translocation of sugar is concomitant to phosphorylation of E II and release of Hpr. Pi transfer and release of sugar phosphate are the last steps.

physical translocation step of one or the other of both is unavoidable. In the first of the two models the translocation of the sugar is triggered by the association of E II to PHpr, in the second the appearance of the sugar binding site and the translocation of the sugar are triggered by the hypothetic phospholation of E II by PHpr. This last hypothesis is so more attractive, that it clearly links the translocation to the release of a part of the chemical bond energy, it decomposes the overall reaction to reasonable elementary steps accessible to experimental approach, and experimental evidence for such an intermediate phosphoenzyme has been found in a similar system in *Micrococcus pyogenes* [12]. Last but not least this type of model makes the

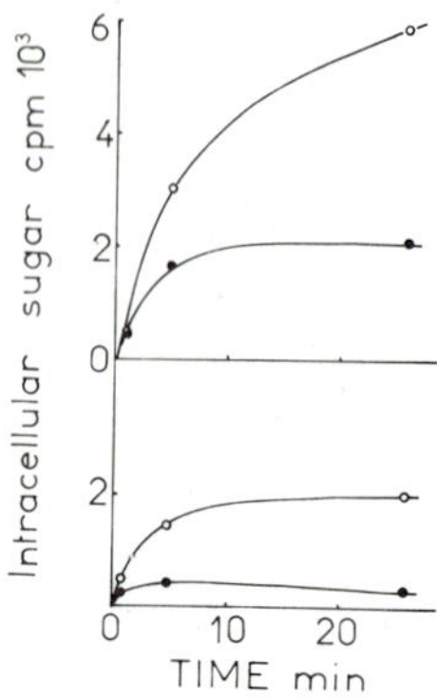

Fig. 2. Uptake and phosphorylation of α-methyl glucose by *E. coli* K_{12} 3300. Upper part: control. Lower part: bacteria treated two minutes with 1 mM *N*-ethyl maleimide. Inactivation was stopped by addition of excess β-mercapto-ethanol. ○ – ○ total radioactivity in cells; ● – ● Baryum precipitable radioactivity.

phosphorylation of the sugar an accessory step which might not be indispensable and would permit the generalization of the system to other transports where the substrate is not found phosphorylated.

Preliminary evidence from our laboratory suggests that transport by the phosphotransferase system might not be linked to obligate phosphorylation of the sugar. When *E. coli* is submitted to treatment with a sulfhydryl reagent like *N*-ethyl maleimide, the phosphotransferase system is inactivated. Following up the kinetics of this inactivation, it was found that after a limited treatment, overall uptake of radioactivity from ^{14}C αmeglc was inactivated about 50 per cent, while the appearance of phosphorylated sugar was diminished by 90–95 per cent (R. Haguenauer, unpublished) (fig. 2). The difficulty is to measure accurately initial velocities of phosphorylation at such a low level, in order to demonstrate that it does not match the initial rate of uptake. If this is demonstrated, the right hand model of fig. 1 could account for the facts by assuming after the third step, that due to the modification of the system by NEM, orthophosphate and sugar are released separately in the cytoplasm realizing an authentic active transport.

2.1. *The lactose permease system*

This system differs sharply from the one described above. Only one specific protein or M-protein, the product of the *Y* gene, is involved in its operation according to present genetic evidence. If accessory enzymes exist, they must be non-specific, non-inducible and presumably ubiquitous. The

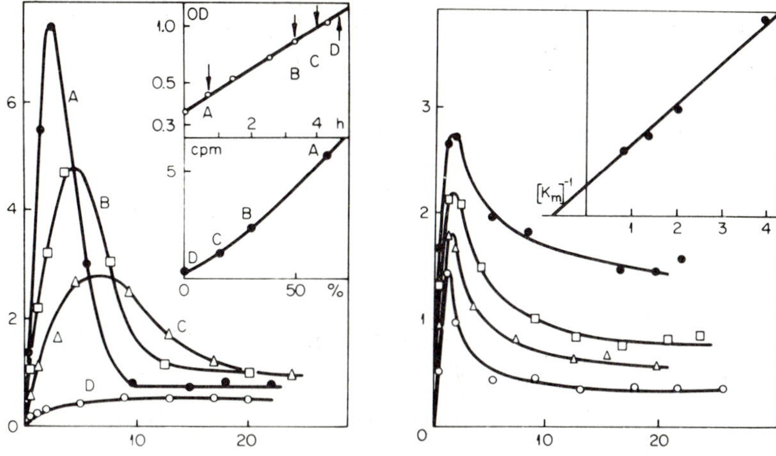

Fig. 3a. Overshoot experiments with *E. coli* 3 000 at various stages of induction of the Lac operon. Samples of the culture were arrested with 50 μg/ml chloramphenicol, centrifuged and resuspended in 1/200 vol of mineral medium containing 4 × 10^{-2} M Na azide and 10^{-1} M radioactive TMG. After 40 min at 25°C samples were diluted in 200 vol. of mineral medium containing Na azide and a small amount of ^{14}C TMG of high specific activity. Ordinates: counts per min × 10^3. *Insert top.* Growth curve showing the times of addition of inducer: IPTG 2 × 10^{-4} M to samples A, B, and C. At the arrow D all four samples were supplemented with chloramphenicol. *Insert below.* Initial velocity of overshoot measured by the uptake in 1 min 15 sec, as a function of percent induction calculated from the ratio of the increase of OD in the presence of inducer to the total OD.

Fig. 3b. Overshoot experiments with *E. coli* 3300. Load with 10^{-1} M TMG as in fig. 3a. Dilution 400 times in solutions containing radioactive TMG to make up final concentrations of 12.5, 7.5, 5.0 and 2.5 10^{-4} M, respectively, from top to bottom, all to the same final specific activity. Ordinates: counts per min × 10^3. *Insert*: Lineweaver Burk plot of initial velocity measured by the uptake in 15 sec. Apparent K_m is 1.4 mM to be compared to the K_m of uptake for TMG, which is 0.5 to 0.6 mM.

second difference with the phosphotransferase systems is the fact that the membrane bound permease protein behaves as a mobile carrier of β-galactosides [13, 14]. The best demonstration of this is the counterflux experiment made in the presence of energy inhibitors, during which the passive efflux of a substrate previously accumulated causes an uphill flux of another substrate from the medium to the cytoplasm (fig. 3). The final result of transport via a mobile carrier is the equilibration of intra- and extracellular substrates. In the presence of a normal energy metabolism the equilibrating transport is transformed into unidirectional active transport. In order to achieve this result starting from an equilibrating transport cycle

$$\begin{array}{ccc} C + S & \rightleftarrows & C\,S_{ex} \\ \downarrow\uparrow & & \downarrow\uparrow \\ C + S_{in} & \rightleftarrows & C\,S_{in} \end{array}$$

the energy coupling must dissociate the carrier substrate complex in the inner side [15], prevent the intracellular substrate from reassociating to the carrier [16] inactivate the carrier [4] or lower its affinity for the substrate [13], unless it contributes to one of the translocation steps (e.g. the back diffusion of free carrier). In recent publications [17, 18] I advocated the hypothesis that this occurs via a covalent change in the carrier protein as exemplified by the phosphorylated intermediate of the Na K ATPase. In the model proposed, this covalent change has two effects, it abolishes the binding site for the substrate and it facilitates the back diffusion of the permease protein to the outer face. This step with the non-energized permease appears to be rate limiting being two to four times slower than either the maximal flux of exchange in the energy inhibited conditions or the maximal rate of uptake in the normal energized cycle [18].

The only biological property of a carrier which can presently be measured *in vitro* in the absence of a closed cytoplasmic compartment is its substrate binding property. This has been first explored in the case of Lac permease by Kennedy [19], who incubated a crude extract with radioactive TDG and showed after centrifugation an enrichment in radioactivity in the membrane pellet. The water content of the pellet was measured with ^{32}P orthophosphate, and the enrichment in isotope ratio was currently of the order of 40 per cent using bacterial strains diploid for the Lac region and having therefore twice the amount of permease protein of a wild type haploid strain. In our experiment with the wild type strain and using tritiated water to evaluate the volume of the pellet we currently found 18 per cent enrichment in ^{35}S TDG.

For a detailed study of the binding site a method giving a signal of 18 over a backgroun of 100 was notoriously insufficient. We tried therefore a method which succeded in the study of inducer-repressor binding [20] namely the ammonium sulfate precipitation of the complex followed by washing with ammonium sulfate. Using 70–80 per cent saturated ammonium sulfate on membrane preparations derived from mechanically disrupted *E. coli*, we found radioactive TDG bound to the pellet in amounts consistent with the expected number of binding sites. Bound TDG could be displaced from the

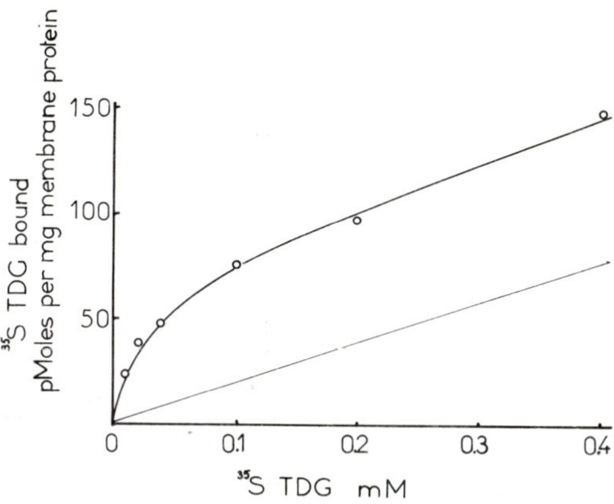

Fig. 4. TDG binding to membrane preparation of *E. coli* K_{12} 3300. Membranes were incubated in 10^{-2} M phosphate buffer *p*H 6.8 containing 1 mM Mg^{2+} and the indicated concentrations of ^{35}S TDG. Samples containing approximately 10 mg membrane protein were diluted 40 fold in 80% saturated ammonium sulfate, centrifuged and washed with 4 ml 80% saturated ammonium sulfate. The final pellet was resuspended in 1% triton X100 and counted. The thin straight line is the estimated contamination.

complex only by sugars known to be substrates of the Lac-permease (table). The binding constant of TDG as determined by this method was close to the K_m of transport measured *in vivo* (fig. 4). The background noise as measured with this method after displacement of radioactive TDG with a high excess of

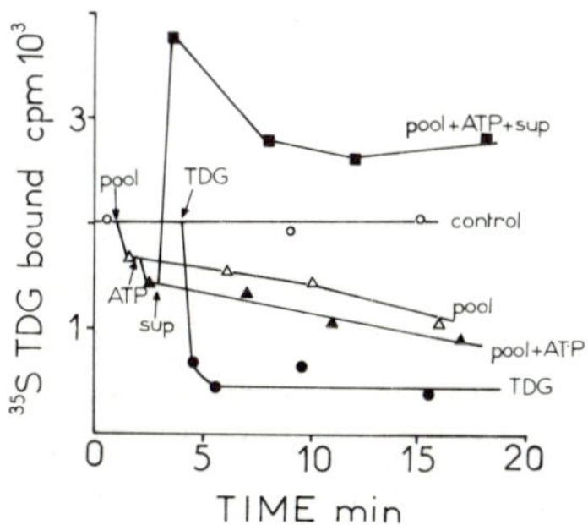

Fig. 5. TDG binding to a membrane preparation of *E. coli* K_{12} 3300. Incubation was 2 $\times$ 10^{-5} M ^{35}S TDG as described in legend to fig. 4 and aliquots were supplemented with a concentrated acid soluble pool from *E. coli* (final OD_{260} = 50), ATP 2 mM dialysed soluble extract of *E. coli* (sup.) or a combination. Non-specific contamination was measured after addition of 1 mM non radioactive TDG.

non radioactive substrate is 10 to 30% of the specifically bound radioactivity, varying slightly from one membrane preparation to the other.

With such a method it becomes possible to check the validity of the hypothesis described above about the influence of energy coupling on the number of the quality of substrate binding sites. As suggested by the report of T.H. Wilson in an energy coupling deficient permease mutant, [21] the permease protein is responsible of the energy coupling. It is possible to speculate, that only an energy donating small molecule and the permease protein are necessary to bring about the postulated modification of the substrate binding site.

In preliminary experiments, we found that the addition of an acid soluble pool extracted from *E. coli* diminished the number of substrate binding sites as measured by the method described. ATP enhanced the effect, but added alone had no significant influence on the number of binding sites (fig. 5).

The addition of a number of high energy metabolites individually at a concentration of 2 mM revealed that CTP had the most significant effect in decreasing the number of TDG binding sites.

The purity and reproducibility of membrane preparations has to be improved considerably before the final proof is made that the observed changes are connected with the energy coupling cycle of Lac-permease, but I think it of interest to report these preliminary results, which open up a new avenue in the exploration of the energy coupling in one of the typical carrier-mediated transport systems.

3. Transport systems involving a periplasmic binding protein

Treatment of *E. coli* with EDTA in the presence of a hyperosmotic sucrose solution followed by an osmotic shock in distilled water containing magnesium releases in the suspension fluid a special class of proteins [22] thought to be located originally between the cell membrane and the outer cell wall. Among the proteins liberated in these conditions and increasing number is found to posses a high affinity for small molecules or ions: sulfate [3], phosphate [23], galactose [5, 24], leucine [5, 25], arabinose [26]. Several lines of evidence show the involvement of these proteins in the transport of the respective small molecules: simultaneous loss by mutation of the binding protein and the transport system, similarity of specificity and affinity, partial impairment of the transport by osmotic shock and partial restoration by adding back the released protein.

Evidence also suggests that the binding protein alone is not the whole transport system. Many mutants which lost the transport capacity retain the binding protein. In the sulfate transport system three cistrons are essential for the function, thus presumably three different proteins only one of which is the sulfate binding protein. A non-binding fraction of the shock fluid contributes significantly to the restoration of the galactose transport system inactivated by shock, etc.

These binding proteins are generally freely soluble in water and their specific tendency to associate to the cell membrane was not shown. In these conditions it is particularly uneasy to integrate these proteins in a model of membrane transport. W. Boos [27] has recently shown that the galactose binding protein can reversibly assume two different configurations which differ in their respective K_m by a factor of 10 or more, and the author suggests that this transition might be the basis of their role in active transport. We found [28] that the relevant galactose transport system, the methyl galactose permease, is capable to concentrate galactose in the cytoplasm to levels 10^5 times higher than in the surrounding medium. A change in K_m of 10^1 to 10^2 is inappropriate to account for this phenomenon. No experiments of flux coupling (countertransport) have been reported in any of the permease systems involving a periplasmic binding protein, therefore the role of the binding protein as a transmembrane mobile carrier is lacking experimental support.

In their article on the arabinose binding protein [26], Hogg and Englesberg report a paradoxical experiment. When the protein is dialysed against increasing concentrations of radioactive arabinose, a rather typical binding curve is obtained from which a K_m can be defined. If a second dialysis is performed against non radioactive arabinose, the radioactive molecule is displaced from the complex. But if the second dialysis is made against buffer, the bound radioactive arabinose does not dissociate from the complex after fourty hours, while dialysis equilibrium was established in less that twelve hours in the first experiment. The observation looks as if the association-dissociation cycle had a high hysteresis as represented on fig. 6. If the saturation curve is taken as a proof of reversible association-dissociation, each point representing an equilibrium, the non-dissociation during the second step is a paradox. If instead the first experiment reflects an exchange between the radioactive arabinose added and the unsuspected cold arabinose bound to the protein before the experiment (the cells have been grown in the presence of arabinose): Ara P +*Ara $\rightleftharpoons$ Ara*P + Ara, then the "saturation curve" actually reflects an isotopic dilution, which under constant geometry is expected to give a hyperbola similar to that predicted by the Michaelis relation. The apparent "K_m" of course should then depend on the protein concentration.

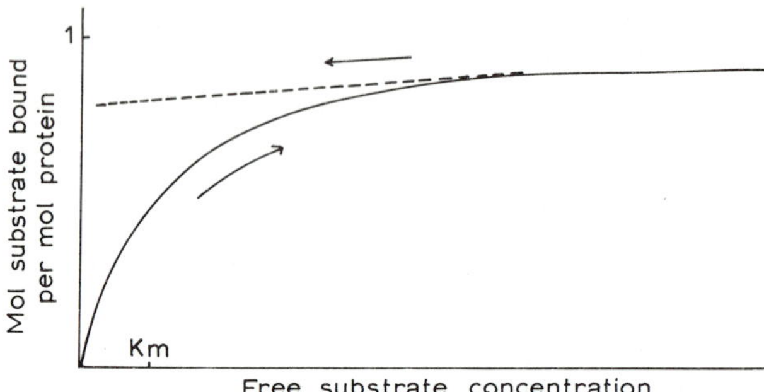

Fig. 6. Diagrammatic representation of a binding curve of galactose binding protein as a function of the concentration of free galactose. Increasing or decreasing concentrations give different results for the same final concentration.

Describing the galactose binding protein, Boos [27] mentions that the K_m here depends also on the protein concentration. In the case of the galactose binding protein (GBP) there is also indication of a slow dissociation. When GBP-gal is chromatographed on a sephadex column, radioactivity comes out in two distinctly separate peaks, the first with the protein in the void volume is followed by a valley before galactose is eluted. The dissociation of the complex was not achieved at the trailing edge of the protein peak. It is difficult from this observation to determine the time constant for dissociation but it is certainly way above 10 min (a very fast column procedure). In contrast, transport proteins usually exhibit a turnover of the order of 10^4 per minute (NaK ATPase, Skou personal communication, Lac-permease as calculated from the number of protein copies and the maximal flux of TMG uptake); and therefore the association-dissociation cycle as observed *in vitro* is certainly inadequate for a hypothetic carrier mechanism. If it is confirmed that exchange unlike dissociation is a rapid process, then an intervention of the binding protein in obligate exchange diffusion becomes possible, but no net transport.

If the binding protein plays a role in net transport, there must be a physiological circumstance where it dissociates from its substrate. This might occur with the help of another identified or suspected component of the transport system, presumably a membrane bound component, and with the help of an energy releasing metabolic reaction. The experimental approach to the *in vitro* energy coupling using purified well characterized binding proteins is particularly attractive, provided that a method faster than classical equilibrium dialysis can be found to measure the binding to a substrate.

The mechanisms involving soluble binding proteins seem to be particularly widespread. Only those having very high affinity for the substrate have been hitherto identified for the reason, that equilibrium dialysis becomes unpractical when the binding constants are low. The phenomenon of slow dissociation was only reported for the arabinose binding protein but according to preliminary experiments it is also true for galactose binding protein and rumours tend to indicate that with other binding proteins difficulties are experienced to obtain dissociation.

4. Conclusions

It appears from the description above, that in all three families of transport systems, the energy coupling step becomes accessible to *in vitro* experimental approach. The case of the phosphotransferase systems illutrates nevertheless, that the biochemical approach may not be sufficient for a complete understanding of the energy coupling. It can identify the fuel and demonstrate its combustion, but this is not a complete description of an engine. The carrier mechanism illustrated in connection with the Lac-permease is a four stroke engine and the proposed model of covalent modification with the help of an energy donating metabolite makes of it an internal combustion engine. The chemiosmotic model of transport, would in contrast be an external combustion engine, where the electrochemical gradient is built up by a separate system. This type of model has been purposely left aside, because its accessibility to *in vitro* experiments is far behind that of the covalent mechanisms. The labeling of protons and the identification of a specific proton binding site among hundreds of "trivial" proton binding sites is a formidable task. The chemiosmotic model might be operative in some transport systems, but it certainly is not universal. In our laboratory we could not detect proton fluxes during the transport of α-methyl glucoside by the phosphotransferase system or during transport of thiogalactosides by Lac-permease, whereas the ion movements linked to the uptake of an anionic substrate such as gluconate are easily accessible to a *p*H meter. The overall fluxes in these three transport systems are comparable.

Another noteworthy omission from the description is the role of membrane lipids and phospholipids. The role of phospholipids in the energy coupling cycle, as visualized in a former model by Hokin, might be revived in the future for one or another transport system. In the framework of the model of covalent change in transport proteins advocated above, the energy donating metabolite might well be among others a phospholipid derivative. The structural role of phospholipids in the membrane on the other side is obvious. Since transport proteins are built into lipid containing membranes, the interaction between the two must insure the unique active configuration of the protein by satisfying the hydrophobic bond "valencies" of the protein without undue constraint. In this conception, the phospholipid would play the role of a mortar with more or less strict specification. Actually the number of strict phospholipid specificities in the activation of membrane enzymes on lipoprotein enzymes is rather limited, in many cases the tolerance is large. In the case of transport proteins the role of the phospholipids can go farther than just the saturation of the hydrophobic areas. In most transport systems an actual motion of the substrate binding site must be postulated, and for this motion, a proper lubricant is necessary. This concept can explain the change in transition temperatures of membrane protein functions, according to the melting points of the fatty acids incorporated into the membrane [29, 30]. Again, no strict fatty acid specificities have ever been claimed, but oviously different mixtures give different temperatures profiles.

In conclusion, I believe that progress in the knowledge of membrane architecture and of membrane functions must help each other to reach a description in terms of molecules and of molecular interactions particularly in the fascinating field of active transport which features an energy transducing machine of molecular dimensions.

References

[1] G.N. Cohen and J. Monod, Bacteriol. Rev. 21 (1957) 169.
[2] W. Kundig, S. Ghosh and S. Roseman, Proc. Natl. Acad. Sci. (U.S.) 52 (1964) 1067.
[3] A.B. Pardee and L.S. Prestidge, Proc. Natl. Acad. Sci. (U.S.) 55 (1966) 189.
[4] C.F. Fox and E.P. Kennedy, Proc. Natl. Acad. Sci. (U.S.) 54 (1965) 891.
[5] Y. Anraku, J. Biol. Chem. 242 (1967) 793.
[6] W. Kundig and S. Roseman, Methods in Enzymology, 9 (1966) 396.
[7] S. Tanaka and E.C.C. Lin, Proc. Natl. Acad. Sci. (U.S.) 52 (1967) 913.
[8] P. Mitchell, in: Membrane Transport and Metabolism, eds. A. Kleinzeller and A. Kotyk (Publ. Czechoslovak Academy of Sciences, 1960, p. 101).
[9] H.R. Kaback, J. Biol. Chem. 243 (1968) 3711.
[10] H.R. Kaback, in: The Molecular Basis of Membrane Function, ed. D. Tosteson (Prentice Hall, 1970, p. 436).
[11] W. Kundig, F. Dodyk-Kundig, B. Anderson and S. Roseman, J. Biol. Chem. 241 (1966) 3243.
[12] T. Nakazawa, R.D. Simoni, J.B. Hays and S. Roseman, Biochem. Biophys. Res. Comm. 42 (1971) 836.
[13] H.H. Winkler and T.H. Wilson, J. Biol. Chem. 241 (1966) 2200.

[14] A. Kepes, in: Molecular Basis of Membrane Function, ed. D.C. Tosteson (Prentice Hall, Engleswood Cliffs, 1969 p. 353).
[15] D. Schachter and A.J. Mindlin, J. Biol. Chem. 244 (1969) 1808.
[16] A.L. Koch, Biochim. Biophys. Acta, 79 (1964) 177.
[17] A. Kepes, in: Current Topics in Membranes and Transport, Vol. 1 (Academic Press, New York and London, 1970 p. 101).
[18] A. Kepes, J. Membrane Biol. 4 (1971) 87.
[19] E.P. Kennedy, in: The Lactose Operon, eds. J. Beckwith and D. Ziprer (Cold Spring Harbor, 1970 p. 49).
[20] S. Bourgeois and A. Jobe, in: The Lactose Operon, eds. J. Beckwith and D. Ziprer (Cold Spring Harbor, 1970, p. 325).
[21] T.H. Wilson, in: This book, p. 151.
[22] H.C. Neu and L.A. Heppel, J. Biol. Chem. 240 (1965) 3685.
[23] N. Medveczky and H. Rosenberg, Biochim. Biophys. Acta, 192 (1969) 369.
[24] W. Boos, Eur. J. Biochem. 10 (1969) 66.
[25] J.R. Piperno and D.L. Oxender, J. Biol. Chem. 241 (1966) 5732.
[26] R.W. Hogg and E. Englesberg, J. Bacterial. 100 (1969) 423.
[27] W. Boos and A.S. Gordon, J. Biol. Chem. 246 (1971) 621.
[28] J. Vorisek and A. Kepes, Eur. J. Biochem. 1972 (in press).
[29] P. Overath, H.V. Schairer and W. Stoffel, Proc. Natl. Acad. Sci. (U.S.) 670 (1970) 606.
[30] G. Wilson, S.P. Rose and C.F. Fox, Biochem. Biophys. Res. Comm. 38 (1970) 623.

COUPLING AND MEMBRANE INSTABILITY

W. DORST, C. BOTRE, L. BOLIS and P. LULY
Farmacologisch Laboratorium, Vrije Universiteit, Amsterdam, The Netherlands
Istituto Chimico Farmaceutico, Universita di Roma, Italy
Istituto di Fisiologia Generale, Universita di Roma, Italy

1. Introduction

Coupling phenomena are known to constitute an important feature of biological transport processes. If the metabolic system can be left out of consideration, coupling is considered to occur between two or more species permeating a cell membrane.

In the case of synthetic membranes, coupling between ionic flows arises in the same way as in free solution; direct influence of ion exchange would lead to negative interference coefficients which have not been observed. In biomembrane transport exchange between particles on a membrane site may lead to a change of sign of the interference coefficient [1].

In the following, we briefly describe some complications arising from the simultaneous occurrence of exchange coupling and conformational changes in a biological membrane.

2. Theoretical discussion

We follow a kinetic theory, originally developed by Hill, Kedem [2] and Heckmann [3] and extended to membrane excitation by Blumenthal, Changeux and Lefever [4].

Let us consider a membrane, built up by protomeric units that can be in the conformation S or R (fig. 1). In fig. 1, x_i is the fraction of protomers in state i; ligands of the first species are represented by circles and the other

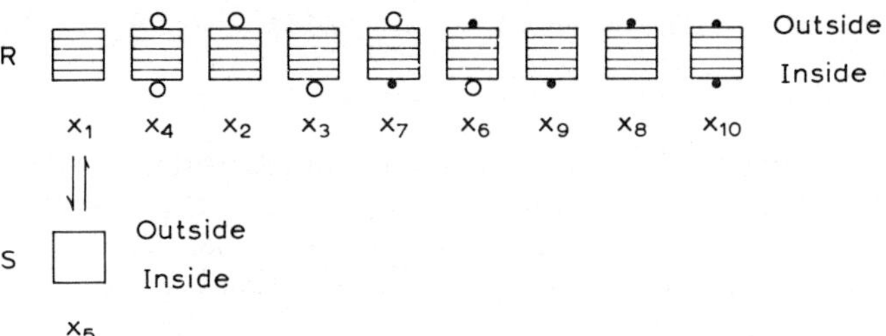

Fig. 1

species by dots. It is assumed that ligands are "bound" exclusively by conformation R; so we have for the metal-protomer complex dissociation constants

$$K_R/K_S = 0\ . \qquad (1)$$

For the x_i

$$\sum_{i=1}^{10} x_i = 1\ . \qquad (2)$$

The model presented here could be formally identified as a carrier system in which the loaded state moves much slower than the free species (ref. [2], models 5 and 12), although its identification as such is not essential. Active transport, for instance of K^+ may occur on the protomers; indeed a leakage of this ion usually accompanies "instability" of the cell's dynamic transport balance. We will not extend this consideration to active transport processes, for the sake of simplicity.

If A and A' are the ligand concentrations inside the membrane and B and B' the outside concentrations, respectively, the kinetic equations become

$$\begin{aligned}
\mathrm{d}x_2/\mathrm{d}t &= k_\mathrm{a}Bx_1 - k_\mathrm{d}x_2 - k_\mathrm{a}Ax_2 + k_\mathrm{d}x_4 - k_\mathrm{m}(x_2 - x_3) + k''_\mathrm{d}x_7 - k'_\mathrm{a}A'x_2\\
\mathrm{d}x_3/\mathrm{d}t &= -k_\mathrm{a}Bx_3 + k_\mathrm{d}x_4 + k_\mathrm{a}Ax_1 - k_\mathrm{d}x_3 + k_\mathrm{m}(x_2 - x_3) + k''_\mathrm{d}x_6 - k'_\mathrm{a}B'x_3\\
\mathrm{d}x_4/\mathrm{d}t &= k_\mathrm{a}Ax_2 + k_\mathrm{a}Bx_3 - 2k_\mathrm{d}x_4\\
\mathrm{d}x_5/\mathrm{d}t &= k'_\mathrm{c}x_1 - k_\mathrm{c}x_5\\
\mathrm{d}x_6/\mathrm{d}t &= k_\mathrm{a}Ax_8 + k'_\mathrm{a}B'x_3 - k_\mathrm{d}x_6 - k''_\mathrm{d}x_6 - k_\mathrm{e}x_6 + k_\mathrm{e}x_7 \qquad (3)\\
\mathrm{d}x_7/\mathrm{d}t &= k_\mathrm{a}Bx_9 + k'_\mathrm{a}A'x_2 - k_\mathrm{d}x_7 - k''_\mathrm{d}x_7 - k_\mathrm{e}x_7 + k_\mathrm{e}x_6\\
\mathrm{d}x_8/\mathrm{d}t &= k'_\mathrm{a}B'x_1 - k''_\mathrm{d}x_8 - k'_\mathrm{a}A'x_8 + k''_\mathrm{d}x_{10} - k'_\mathrm{m}(x_8 - x_9) + k_\mathrm{d}x_6 - k_\mathrm{a}Ax_8
\end{aligned}$$

$$\mathrm{d}x_9/\mathrm{d}t = k'_a A' x_1 - k''_d x_9 - k'_a B' x_9 + k''_d x_{10} + k'_m(x_8 - x_9) + k_d x_7 - k_a B x_9$$

$$\mathrm{d}x_{10}/\mathrm{d}t = k'_a A' x_8 + k'_a B' x_9 - 2k''_d x_{10}$$

where k_a, k'_a and k_d, k''_d are the constants of adsorption and desorption of ligand, respectively, on the protomers ($K_R = k_d/k_a$; $K'_R = k''_d/k''_a$) and k_m, k'_m the diffusion "rate constants" across the membrane. The "switch" of the two species, occurring by means of the transition between states 6 and 7 (and reverse) is represented by the exchange constant k_e. Making the molecular field approximation [5], we have for the isomerization constant of the S-R transition

$$l' = k'_c/k_c = \exp(\epsilon - \eta r)/k_B T = l\Lambda^r, \tag{4}$$

Where r (= $1 - x_5$) represents the fraction of R-protomers. In eq. (4) k_B is Boltzmann's constant, ϵ the energy required to transform a protomer from S to R when all other protomers are in state S and η the energy of interaction between protomers.

To solve eqs. (3), we consider for simplicity only the extreme case*

$$k_d \ll k''_d\,; \qquad k_m, k'_m \ll k_e\,; \qquad A \approx \text{constant}\,; \qquad B \to 0\,.$$

We also introduce the quantities

$$\alpha = A/K_R\,, \qquad a = A'/K'_R\,, \qquad b = B'/K'_R\,, \qquad \lambda = k_e/k_d\,. \tag{5}$$

The kinetic eqs. (3) are then easily solved yielding the transport equations and the state function r (for the latter see "Appendix").

The flux of A through the cell membrane is given by

$$\frac{\mathrm{d}\alpha}{\mathrm{d}t} = \frac{1-r}{l'}\lambda b(ak_e - k_m)\frac{\frac{1}{4}\alpha ab + (1+\frac{1}{2}b)[a(1+\frac{1}{2}a) + ab]}{[1+\frac{1}{2}(a+b)]\,[1+\frac{1}{2}\alpha(1+\lambda b)]}$$

$$+\frac{1-r}{l'}\alpha(1+\tfrac{1}{2}\alpha)\frac{[1+\frac{1}{2}(a+b)](1+b)}{[1+\frac{1}{2}(a+b)]\,[1+\frac{1}{2}\alpha(1+\lambda b)]}(k_m - bk_e)\,. \tag{6}$$

The existence of parameter λ includes the possibility for a change of sign

* The inequality concerning k_m, k'_m resulted from computer simulation of the given system.

in $d\alpha/dt$ as coupled to db/dt. This combined with the dependence of α not only on A but also on A' and B' may give rise to a mutual stimulation of the R and S protomers, since a decrease in α followed by an increase in b first decreases, then increases and finally again decreases r by transition of various "instability regions" (see Appendix). This may be further explained by eq. (6), if we assume an extremum in the function describing the dependence of α on b (ref. [1]) and also the possibility of hysteresis (ref. [6]).

3. Appendix

The equation of conformation is given by

$$r = \frac{1}{1 + \dfrac{l'}{1+\phi}}$$

where

$$\begin{aligned}\phi = {} & \{[1+\tfrac{1}{2}\alpha(1+\lambda b)]\,[1+\tfrac{1}{2}(a+b)]\}^{-1} \\ & \times \{\lambda b(1+\tfrac{1}{2}\alpha + a)[\tfrac{1}{4}\alpha ab + (1+\tfrac{1}{2}b)[a(1+\tfrac{1}{2}b)+ab] \\ & + \alpha(1+\tfrac{1}{2}\alpha)[(1+b)^2[1+\tfrac{1}{2}(a+b)] \\ & + b(1+\tfrac{1}{2}a)\{(1+\tfrac{1}{2}a)(1+\tfrac{1}{2}b) + (\tfrac{1}{2}\alpha+\lambda a)[1+\tfrac{1}{2}(a+b)]\} \\ & + a(1+\tfrac{1}{2}b)[1+\tfrac{1}{2}(a+b)]\,[1+\tfrac{1}{2}\alpha - \lambda b(\tfrac{1}{2}\alpha+\lambda a)]\}\ .\end{aligned}$$

Acknowledgments

The authors are indebted to Dr. R. Lefever and Dr. G.E. Fuhrmann for helpful discussion. They wish to thank Mr. P.W. Hemker for his assistance with the numerical calculations.

References

[1] J.A. Jacqez, Biochim. Biophys. Acta 71 (1963) 15.
[2] T.L. Hill and O. Kedem, J. Theor. Biol. 10 (1966) 390.
[3] K. Heckman and H. Passow, Ber. Bunsen Ges. Physik Chemie 71 (1967) 839.
[4] R. Blumenthal, J.P. Changeux and R. Lefever, J. Membrane Biol. 2 (1970) 351.
[5] S. Straessler and C. Kittel, Phys. Rev. 139A (1965) 758.
[6] C. Botre et al., Il Farmaco, 26 (1971) 607.

SUBJECT INDEX